# Fundamental Theories of Physics

## Volume 170



For further volumes:
http://www.springer.com/series/6001

Salvatore Capozziello • Valerio Faraoni

# Beyond Einstein Gravity

## A Survey of Gravitational Theories for Cosmology and Astrophysics

Springer

Prof. Salvatore Capozziello
Università di Napoli Federico II
Dipartimento di Scienze Fisiche
Via Cinthia
80126 Napoli
Italy
capozziello@na.infn.it

Prof. Valerio Faraoni
Physics Department
Bishop's University
2600 College Street
Sherbrooke, QC Canada
J1M 1Z7
vfaraoni@ubishops.ca

ISBN 978-94-007-0164-9 e-ISBN 978-94-007-0165-6
DOI 10.1007/978-94-007-0165-6
Springer Dordrecht Heidelberg London New York

Library of Congress Control Number: 2010938912



*Cover design*: SPi Publisher Services

Printed on acid-free paper

Springer is part of Springer Science+Business Media (www.springer.com)

*To Donovan, Emanuele, and Francesca Sofia*

# Preface

There are good theoretical reasons to consider seriously the possibility that gravity is not described precisely by Einstein's General Relativity but rather by some alternative theory. First, attempts to renormalize General Relativity in the 1960s and 1970s showed clearly that counterterms must be introduced which alter the theory significantly and make its field equations of fourth order instead of second. From the physical point of view, this fact implies that extra degrees of freedom, in addition to the usual spin two graviton, need to be introduced. Unfortunately, the corrected theory is not free of ghosts, which makes it non-unitary. The corrections introduced by renormalization are quadratic in the algebraic invariants of the curvature tensor and were successfully employed in $R^2$-inflation in the early universe. By retaining only corrections quadratic in the Ricci scalar $R$ or, by extension, corrections which are general non-linear functions of $R$ (and no longer motivated by renormalization), one obtains the so-called class of $f(R)$ theories of gravity.

Second, when one tries to approach gravity (and the other interactions) from the high energy side and then obtain low-energy physics, one does not recover Einstein's theory. Adopting, at least as a temporary model, string theory as a full theory of quantum gravity which also unifies the known interactions, one can take a low-energy limit which, again, does not reproduce General Relativity but gives instead a Brans-Dicke theory. Scalar-tensor theories of gravity have been known for a long time and were developed following initial suggestions by Dirac, Jordan, Fierz, and Thiery, culminating in the 1961 paper by Brans and Dicke introducing what is now known as Brans-Dicke theory. The original motivations for Brans-Dicke theory were rooted in the need to implement Mach's principle, which is not fully incorporated in General Relativity, in a relativistic theory of gravity. After Brans-Dicke theory (the prototype of scalar-tensor theories of gravity) was born, research on Mach's principle went its own way and, without doubt, the interest of modern physicists in scalar-tensor gravity arises more from string theories than from Mach's principle. Dilaton fields and their non-minimal couplings to the spacetime curvature are unavoidable features of string theories, shared with scalar-tensor gravity.

It seems, therefore, that first loop corrections or attempts to fully quantize gravity necessarily introduce significant deviations from General Relativity and extra degrees of freedom. The recent thermodynamics of spacetime approach to emergent gravity pictures General Relativity as a thermodynamical state of equilibrium among a wider spectrum of gravity theories and it only makes sense that, if extra degrees of freedom are allowed in addition to the standard spin two graviton of General Relativity, deviations from this equilibrium state will correspond to the excitation of these extra degrees of freedom and to deviations from Einstein's theory. From the theoretical point of view, going beyond General Relativity is a necessity and exploring the wider landscape of theories becomes a cultural need.

From the experimental point of view, General Relativity has been tested directly in the Solar System in its weak-field, slow motion approximation. Binary pulsars, most notably the Hulse-Taylor system PSR 1913 + 16, allow for indirect tests outside the Solar System, in the same regime. However, strong gravity tests are still missing and gravity is tested very poorly at the scale of galaxies and clusters, witnessing the fact that even Newtonian gravity is doubted at galactic scales, which has led to the introduction of *MOND* and *TeVeS* theories to replace galactic dark matter. Cosmology cannot be advocated as a precise test of General Relativity at large scales: in fact, almost all theories of gravity admit the Friedmann-Lemaitre-Robertson-Walker line element as a solution of their field equations, with perfect fluids or other reasonable matter sources. Indeed, it is from cosmology that comes the indication that gravity may not be described exactly by General Relativity. The 1998 discovery that the present expansion of the universe appears to be accelerated, made using the luminosity distance versus redshift relation of type Ia supernovae, has left cosmologists scrambling for an explanation. In order to explain the cosmic acceleration within the context of General Relativity, one needs to introduce the mysterious dark energy, which is very exotic (its pressure $P$ and energy density $\rho$ must satisfy $P \simeq -\rho$), diluted, comprises approximately 75% of the energy content of the universe, and is not detected in the laboratory. Dark energy seems very much an *ad hoc* solution of the problem of the present acceleration of the universe and, understandably, alternatives have been looked for. Attempts to explain away dark energy using the backreaction of inhomogeneities on the dynamics of the background universe have been, so far, unconvincing. In 2002, the idea was advanced by one of us, soon followed by other authors, that perhaps we are observing the first deviations from General Relativity on the largest scales. $f(R)$ theories of gravity (although not of the quadratic form obtained by renormalization) were resurrected in an attempt to explain this phenomenon. Curiously, $f(R)$ gravity admits a scalar-tensor representation. Since these first attempts, the literature on $f(R)$ and scalar-tensor gravity and their applications to cosmology has flourished, and scalar fields or $f(R)$ modifications of gravity are now even proposed as alternatives to dark matter. This book attempts to organize the available knowledge about these classes of theories and the vast literature into a coherent view. The book is not meant to be a comprehensive review of half a century of literature, including its recent explosion: it is conceived more as an advanced introduction to this expanding area of research.

It would be premature and unjustified to claim that gravity is described by any of the theories described in this book. However, even if none of these extended theories of gravity ultimately proves to be correct, they are simple enough to solve many problems while still allowing us to peek into the vast landscape of theories beyond Einstein gravity and to understand many ways in which gravity can be enlarged with respect to Einstein's conception.

**Notations and conventions:** the following notations and conventions are used in this book. The metric signature is $- + ++$ in four spacetime dimensions. Units are used in which the speed of light $c$ and the reduced Planck constant $\hbar$ assume the value unity. $G$ is Newton's constant, the Planck mass is $m_{pl} = G^{-1/2}$, while $\kappa \equiv 8\pi G$. Greek indices assume the values 0, 1, 2, 3 and Latin indices assume the spatial values 1, 2, 3. A comma denotes ordinary differentiation, $\nabla_a$ is the covariant derivative operator, and $g$ denotes the determinant of the metric tensor $g_{\mu\nu}$, while $\eta_{\mu\nu}$ is the Minkowski metric, which takes the form diag$(-1, 1, 1, 1)$ in Cartesian coordinates in four dimensions. $\varepsilon^{\alpha\beta\mu\nu}$ is the totally antisymmetric Levi-Civita tensor. Round and square brackets around indices denote symmetrization and antisymmetrization, respectively, which include division by the number of permutations of the indices, for example:

$$A_{(\alpha\beta)} \equiv \frac{A_{\alpha\beta} + A_{\beta\alpha}}{2}, \qquad A_{[\alpha\beta]} \equiv \frac{A_{\alpha\beta} - A_{\beta\alpha}}{2}.$$

The Riemann and Ricci tensors are given in terms of the Christoffel symbols $\Gamma^{\delta}_{\alpha\beta}$ by

$$R_{\alpha\beta\gamma}{}^{\delta} = \Gamma^{\delta}_{\alpha\gamma,\beta} - \Gamma^{\delta}_{\beta\gamma,\alpha} + \Gamma^{\mu}_{\alpha\gamma}\Gamma^{\delta}_{\mu\beta} - \Gamma^{\mu}_{\beta\gamma}\Gamma^{\delta}_{\mu\alpha},$$

$$R_{\alpha\gamma} \equiv R_{\alpha\beta\gamma}{}^{\beta} = \Gamma^{\beta}_{\alpha\gamma,\beta} - \Gamma^{\beta}_{\beta\gamma,\alpha} + \Gamma^{\delta}_{\alpha\gamma}\Gamma^{\mu}_{\delta\mu} - \Gamma^{\delta}_{\beta\gamma}\Gamma^{\beta}_{\delta\alpha},$$

and $R \equiv g^{\alpha\beta} R_{\alpha\beta}$ is the Ricci curvature. $\Box \equiv g^{\alpha\beta}\nabla_\alpha\nabla_\beta$ is d'Alembert's operator. A tilde usually denotes quantities defined in the Einstein frame and the subscript 0 identifies quantities evaluated at the present instant of time in the history of the universe.

Naples and Sherbrooke
November 2010

*Salvatore Capozziello*
*Valerio Faraoni*

# Acknowledgements

We are indebted to many collaborators and colleagues for discussions or remarks over the years on the works that came together in this book, or for bringing to our attention references which we would otherwise have missed in the large literature surveyed. In particular we thank J.D. Barrow, V.F. Cardone, S. Carloni, R. Cianci, A. De Felice, J.C.C. de Souza, M. De Laurentis, P.K.S. Dunsby, E. Elizalde, M. Francaviglia, R. Garattini, R. Goswami, W. Israel, S. Joras, N. Lanahan-Tremblay, S. Liberati, F.S.N. Lobo, S. Nojiri, S.D. Odintsov, C. Rubano, V. Salzano, P. Scudellaro, S. Sonego, T.P. Sotiriou, A. Stabile, C. Stornaiolo, H. Stefančić, N. Straumann, A. Troisi, S. Tsujikawa, S. Vignolo, M. Visser, V. Vitagliano, D.N. Vollick, and S. Zerbini.

Finally, we wish to thank the editor of the Springer series *Fundamental Theories of Physics*, Professor Alwyn van der Merwe, for his invitation to write this work, the Springer Publishing Editor for Mathematical and Theoretical Physics, Dr. Tobias Schwaibold, and Kirsten Theunissen from Springer for their friendly assistance and support during the completion of this book.

# Contents

# Acronyms

| | |
|---|---|
| ADM | Arnowitt-Deser-Misner |
| BAO | Baryon Acoustic Oscillation |
| *BBO* | Big Bang Observer |
| CL | Confidence Level |
| *COBE* | COsmic Background Explorer |
| CMB | Cosmic Microwave Background |
| DGP | Dvali-Gabadadze-Porrati |
| EEP | Einstein Equivalence Principle |
| EoS | Equation of State |
| ETG | Extended Theory of Gravity |
| FLRW | Friedmann-Lemaitre-Robertson-Walker |
| GR | General Relativity |
| GUT | Grand-Unified Theory |
| *LISA* | Laser Interferometer Space Antenna |
| LLR | Lunar Laser Ranging |
| LRG | Luminous Red Galaxies |
| LSB | Low Surface Brightness |
| QED | Quantum ElectroDynamics |
| SDSS | Sloan Digital Sky Survey |
| SEP | Strong Equivalence Principle |
| SNeIa | Type Ia supernovae |
| SUGRA | Supergravity |
| TOV | Tolman-Oppenheimer-Volkoff |
| UDE | Unified Dark Energy |
| UDM | Unified Dark Matter |
| WDW | Wheeler-DeWitt |
| WEP | Weak Equivalence Principle |
| *WMAP* | Wilkinson Microwave Anisotropy Probe |
| $\Lambda$CDM | $\Lambda$ Cold Dark Matter |

# Chapter 1
# Extended gravity: a primer

*Ubi materia, ibi geometria.*
*– Johannes Kepler*

## 1.1 Why extending gravity?

Einstein's theory of General Relativity (GR) provides a comprehensive and coherent description of space, time, gravity, and matter at the macroscopic level. It is formulated in such a way that space and time are not the absolute entities of classical mechanics, but, rather, dynamical quantities determined together with the distribution and motion of matter and energy. As a consequence, Einstein's approach gave rise to a new conception of the universe which, for the first time in the history of physics, came to be considered as a dynamical system susceptible of precise mathematical modeling and physical measurement. Cosmology thus left the realm of philosophy where it had been relegated until Einstein's work and was legitimately incorporated into that of science. Over the years, the possibility of investigating the universe scientifically has led to the formulation of the Standard Big Bang model of the universe [1153] which matched the available cosmological observations until recently. However, in the last thirty years several shortcomings of Einstein's theory were found and scientists began wondering whether GR is the only fundamental theory capable of successfully explaining the gravitational interaction. This new point of view comes mainly from the study of cosmology and of quantum field theory. In cosmology, the presence of the Big Bang singularity, together with the flatness, horizon, and monopole problems [564] led to the realization that the standard cosmological model based on GR and on the Standard Model of particle physics is inadequate to describe the universe at extreme regimes. On the other hand, GR is a classical theory which cannot work as a fundamental theory when a full quantum description of spacetime and gravity is sought for. For these reasons, and especially because of the lack of a definitive quantum theory of gravity, various alternative gravitational theories were proposed which attempt to formulate at least a semiclassical scheme in which GR and its successes could be replicated. One of the most fruitful approaches resulted in the so-called *Extended Theories of Gravity* (ETGs) which have, in some sense, become a paradigm in the study of the gravitational interaction. ETGs are based on corrections and enlargements of Einstein's theory.

S. Capozziello and V. Faraoni, *Beyond Einstein Gravity: A Survey of Gravitational Theories for Cosmology and Astrophysics*, Fundamental Theories of Physics 170,
DOI 10.1007/978-94-007-0165-6_1, 

The paradigm consists, essentially, of adding higher order curvature invariants and minimally or non-minimally coupled scalar fields into the dynamics emerging from some effective quantum gravity action [190].

Other reasons to modify GR are provided by the attempt to fully incorporate Mach's principle into the theory. GR contains only some of Mach's ideas and admits solutions that are explicitly anti-Machian, such as the Gödel universe [544] and exact *pp*-waves [877]. According to Mach's principle the local inertial frame is determined by the average motion of distant astronomical objects [159]. This feature implies that the gravitational coupling at a spacetime point is not absolute but is determined by surrounding matter and, therefore, becomes a function of the spacetime location, a scalar field. As a consequence, the concept of inertia and the Equivalence Principle have to be revised. Brans-Dicke theory [165] was the first fully fleshed out alternative to Einstein's GR, and the prototype of alternative theories of gravity. The variable gravitational "constant" corresponding to a scalar field coupled non-minimally to the geometry constitutes a more satisfactory implementation of Mach's principle than GR [165, 234, 999].

Furthermore, every scheme unifying the fundamental interactions, such as superstring, supergravity, or Grand-Unified Theories (GUTs) produces effective actions in which non-minimal couplings to the geometry or higher order terms in the curvature invariants are necessarily present. Such contributions are due to first or higher loop corrections in the high curvature regime approaching the full (and still speculative) quantum gravity regime [190]. This scheme was adopted in the quantization of matter fields on curved spacetimes and the result was that the interactions between quantum scalar fields and the background geometry, or the gravitational self-interactions, yield corrections to the Hilbert-Einstein Lagrangian [144]. Moreover, it has been realized that these corrective terms are unavoidable in the effective quantum gravity actions [1122]. All these approaches certainly do not constitute a full quantum gravity theory, but are needed as temporary working schemes toward it.

To summarize, higher order invariants of the curvature tensor such as $R^2$, $R^{\mu\nu}R_{\mu\nu}$, $R^{\mu\nu\alpha\beta}R_{\mu\nu\alpha\beta}$, $R\,\Box R$, $R\,\Box^k R$, or non-minimally coupled terms between matter (especially scalar) fields and geometry such as $\phi R$ have to be added to the effective gravitational Lagrangian as soon as quantum corrections are introduced. For instance, such terms occur in the low-energy limit of the Lagrangian of string theories or in Kaluza-Klein theories when extra spatial dimensions are compactified [530].

On the other hand, from the conceptual point of view, there is no *a priori* reason to restrict the gravitational Lagrangian to be a linear function of the Ricci scalar $R$, minimally coupled with matter [768]. Furthermore, the idea that there are no "exact" laws of physics has been contemplated seriously: in such a case, the effective Lagrangians of physical interactions would be average quantities arising from stochastic behavior at a more microscopic level. This feature would mean that local gauge invariances and the related conservation laws are well approximated only in the low-energy limit and the fundamental constants of physics can vary [103].

In addition to being motivated by fundamental physics, ETGs have been the subject of great interest in cosmology because they naturally exhibit an inflationary behavior capable of overcoming the shortcomings of the Standard Big Bang model based on GR. The related inflationary scenarios seem realistic and capable of matching the current observations of the cosmic microwave background (CMB) [402, 701, 1044]. It has been shown that, by means of conformal transformations, the higher order and non-minimally coupled terms always correspond to Einstein gravity plus one (or more) scalar field(s) minimally coupled to the curvature [233,550,764,1072,1144]. More precisely, higher order terms always appear as second order contributions to the field equations when they are replaced by equivalent scalar fields. For example, the term $R^2$ in the Lagrangian yields fourth order equations of motion [951], $R\,\Box R$ gives sixth order equations [34, 550], $R\,\Box^2 R$ yields eighth order equations [113], and so on. By means of a conformal transformation, any second order derivative term corresponds to a scalar field:[1] for example, fourth order gravity is equivalent to Einstein gravity plus a single scalar field; sixth order gravity to GR plus two scalar fields; and so on [550, 985]. For example, it is possible to show that $f(R)$ gravity is equivalent not only to a scalar-tensor theory but also to Einstein theory coupled to an ideal fluid [244]. This property is useful if multiple inflationary events are desired because an early stage could produce large-scale structures with very long wavelengths which later grow into the clusters of galaxies observed today, while a later stage could select smaller scale structures observed as galaxies today [34]. The philosophy is that each inflationary era is related to the dynamics of a scalar field. Finally, these extended schemes could naturally solve the graceful exit problem, avoiding the shortcomings of previous inflationary models [35,701].

## 1.2 Cosmological and astrophysical motivation

The revision of standard early cosmological scenarios leading to inflation implies that a new approach is necessary also at later epochs: ETGs could play a fundamental role also in this context. In fact, the increasing bulk of data accumulated over the past few years has paved the way for a new cosmological model usually referred to as the *Concordance Model* or $\Lambda$ Cold Dark Matter ($\Lambda$CDM) model.

The Hubble diagram of type Ia supernovae (hereafter SNeIa) measured by both the Supernova Cosmology Project [676,901] and the High-$z$ Team [936,1078] up to redshift $z \sim 1$, was the first piece of evidence that the universe is currently undergoing a phase of accelerated expansion. Later on, balloon-born experiments such as *BOOMERANG* [377] and *MAXIMA* [1055] determined the location of the first two Doppler peaks in the spectrum of CMB anisotropies, strongly suggesting a universe with flat spatial sections. When combined with the constraints on the dimensionless matter density parameter $\Omega_M$ from galaxy clusters, these data indicate that

[1] The dynamics of all these scalars are usually determined by second order Klein-Gordon-like equations.

the universe is dominated by an unclustered fluid with negative pressure commonly referred to as *dark energy*, which drives the accelerated expansion. This picture has been further strengthened by the recent precise measurements of the CMB spectrum by the *WMAP* satellite experiment [592, 1038, 1039], and by the extension of the SNeIa Hubble diagram to redshifts larger than one [940].

An overwhelming number of papers appeared following these observational pieces of evidence, which present a large variety of models attempting to explain the cosmic acceleration. The simplest explanation would be the well known cosmological constant $\Lambda$ [956]. Although the latter provides the best-fit to most of the available astrophysical data [1038], the $\Lambda$CDM model fails egregiously in explaining why the inferred value of $\Lambda$ is so tiny (120 orders of magnitude lower) in comparison with the typical value of the vacuum energy density predicted by particle physics, and why its present value is comparable to the matter density – this is the so-called *coincidence problem.*

As a tentative solution, many authors have replaced the cosmological constant with a scalar field $\phi$ rolling slowly down a flat section of a potential $V(\phi)$ and giving rise to the models known as *quintessence* [340, 882]. Albeit successful in fitting the data with many models, the quintessence approach to dark energy is still plagued by the coincidence problem since the dark energy and matter densities evolve differently and reach comparable values only during a very short time of the history of the universe, coinciding right at the present era. In other words, the quintessence dark energy is tracking matter and evolves in the same way for a long time; at late times, somehow it changes its behavior and no longer tracks the dark matter but begins to dominate in the fashion of a (dynamical) cosmological constant. This is the coincidence problem of quintessence.

To add to this puzzle, the origin of this quintessence scalar field is mysterious, although various (usually rather exotic) models have been proposed, leaving a great deal of uncertainty on the choice of the scalar field potential $V(\phi)$ necessary to achieve the late-time acceleration of the universe. The subtle and elusive nature of dark energy has led many authors to look for a completely different explanation of the acceleration behavior of the cosmos without introducing exotic components. To this end, it is worth stressing that the present-day cosmic acceleration only requires a negative pressure component that comes to dominate the dynamics late in the matter era, but does not tell us anything about the nature and the number of the cosmic fluids advocated to fill the universe. This consideration suggests that it could be possible to explain the accelerated expansion with a single cosmic fluid characterized by an equation of state causing it to act like dark matter at high densities, while behaving as dark energy at low densities. An attractive feature of these models, usually referred to as *Unified Dark Energy* (UDE) or *Unified Dark Matter* (UDM) models, is that the coincidence problem is solved naturally, at least at the phenomenological level, with such an approach. Examples are the generalized Chaplygin gas [661], the tachyon field [881], and condensate cosmology [110]. A different class of UDE models with a single fluid has been proposed [214, 259]: its energy density scales with the redshift $z$ in such a way that a radiation-dominated era, followed by a matter era and then by an accelerating phase can be naturally achieved. These models are extremely versatile since they can be interpreted both in the framework of UDE or as two-fluid scenarios containing dark matter and scalar field dark energy. A characteristic feature of this approach is that a generalized equation of state can always be obtained and the fit to the observational data can be attempted.

There is another, different, way to approach the problem of the cosmic acceleration. As stressed in [756], it is possible that the observed acceleration is not the manifestation of yet another ingredient of the cosmic pie, but rather the first signal of a breakdown, in the infrared limit, of the laws of gravitation as we understand them. From this point of view, it is tempting to modify the Einstein-Friedmann equations to see whether it is still possible to fit the astrophysical data with models containing only standard matter and without exotic fluids. Examples are the Cardassian expansion [507] and Dvali-Gabadadze-Porrati (DGP) gravity [405]. Within the same conceptual framework, it is possible to find alternative schemes in which a quintessential behavior is obtained by incorporating effective models coming from fundamental physics and giving rise to generalized or higher order gravity actions [211] (see [843] for a comprehensive review). For instance, a cosmological constant may be recovered as a consequence of a non-vanishing torsion field, leading to a model consistent with both the SNeIa Hubble diagram and observations of the Sunyaev–Zel'dovich effect in galaxy clusters [215]. SNeIa data could also be efficiently fitted by including in the gravitational sector higher order curvature invariants [212, 719, 720, 843]. These alternative models provide naturally a cosmological component with negative pressure originating in the geometry of the universe, thus overcoming the problematic nature of quintessence scalar fields.

The variety of cosmological models which are phenomenologically viable candidates to explain the observed accelerated expansion is clear from this short review. This overabundance signals that only a limited number of cosmological tests are available to discriminate between competing theories, and it is clear that there is a high degeneracy of models. Let us remark that both the SNeIa Hubble diagram and the angular size-redshift relation of compact radio sources [294, 908] are distance-based probes of the cosmological model and, therefore, systematic errors and biases could be iterated. With this point in mind, it is interesting to search for tests based on time-dependent observables. For example, one can take into account the *look-back time* to distant objects since this quantity can discriminate between different cosmological models. The lookback time is observationally estimated as the difference between the present-day age of the universe and the age of a given object at redshift $z$. This estimate is possible if the object is a galaxy observed in more than one photometric band since its color is determined by its age as a consequence of stellar evolution. Hence, it is possible to obtain an estimate of the galaxy's age by measuring its magnitude in different bands and then using stellar evolutionary codes to choose the model that best reproduces the observed colors.

Coming to the weak-field limit, which essentially means considering Solar System scales, ETGs are expected to reproduce GR which, in any case, is firmly tested only in this limit and at these scales [1167]. Even this limit is a matter of debate since several relativistic theories do not reproduce exactly the Einsteinian results in their Newtonian limit but, in some sense, generalize them. As was first noticed by Stelle

[1052], $R^2$-gravity gives rise to Yukawa-like corrections to the Newtonian potential with potentially interesting physical consequences. For example, it is claimed by certain authors that the flat rotation curves of galaxies can be explained by such terms [966]. Others [772] have shown that a conformal theory of gravity is nothing but a fourth order theory containing such terms in the Newtonian limit. Reports of an apparent anomalous long-range acceleration in the data of the Pioneer 10/11, Galileo, and Ulysses spacecrafts could be accommodated in a general theoretical scheme incorporating Yukawa corrections to the Newtonian potential [44, 131].

In general, any relativistic theory of gravitation yields corrections to the weak-field gravitational potentials (*e.g.*, [920]) which, at the post-Newtonian level and in the Parametrized Post-Newtonian (PPN) formalism, could constitute a test of these theories [1167]. Furthermore, the newborn *gravitational lensing astronomy* [991] is providing additional tests of gravity over small, large, and very large scales which will soon provide direct measurements of the variation of the Newton coupling [697], the potential of galaxies, clusters of galaxies, and several other features of self-gravitating systems. Likely, such data will be capable of confirming or ruling out as physically inconsistent GR or ETGs.

## 1.3 Mathematical motivation

In ETGs, the Einstein field equations are modified in two ways: *(i)* the geometry can be coupled non-minimally to some scalar field, and/or *(ii)* derivatives of the metric higher than second order appear. In the former case, we generically deal with scalar-tensor theories of gravity; in the latter, with higher order theories. Combinations of non-minimally coupled and higher order terms can also emerge as contributions to effective Lagrangians (higher order-scalar-tensor theories of gravity).

From the mathematical point of view, the problem of formally reducing more general theories to the Einstein form has been discussed extensively. Through a Legendre transformation on the metric, higher order theories with Lagrangians satisfying minimal regularity conditions assume the form of GR with (possibly multiple) scalar field(s) sourcing the gravitational field (*e.g.*, [484,768,769,1023]). The formal equivalence between models with variable gravitational coupling and Einstein gravity via conformal transformations has also been known for a long time [360, 392]. This has given rise to a debate on whether the mathematical equivalence between different conformal representations of the theory (called Jordan and Einstein conformal frames) is also a physical equivalence – this debate is still continuing ([473] and references therein) and we will discuss it below.

Another issue is the Palatini approach to gravity: this was first analyzed by Einstein himself [415] but was called the Palatini approach as a consequence of an historical misunderstanding [191, 485]. The fundamental idea of the Palatini formalism is to regard the (usually torsion-less) connection $\Gamma^{\mu}_{\alpha\beta}$ defining the Ricci tensor $R_{\mu\nu}$ as a quantity independent of the spacetime metric $g_{\mu\nu}$. It is well known from standard textbooks (*e.g.*, [1139, 1153]) that the Palatini formulation of GR is equivalent to the purely metric theory. This follows from the fact that the field equations for the connection $\Gamma^{\mu}_{\alpha\beta}$, regarded as a quantity independent of the metric, yield the Levi-Civita connection of the metric $g_{\mu\nu}$. Therefore, there is no particular reason to impose the Palatini variational principle in the standard Hilbert-Einstein theory instead of the metric variational principle. However, the situation changes completely if we consider the ETGs formulated in terms of functions of curvature invariants, such as $f(R)$, or coupled non-minimally to some scalar field. Then, the Palatini and the metric variational principles generate different field equations and the theories obtained describe different physics [486, 769]. The relevance of the Palatini approach in this framework has been highlighted recently in relation to its cosmological applications [211, 721, 722, 843, 1129].

The crucial problem of the Newtonian potential in alternative gravity and its relations with the conformal factor [794] have also been studied. Physically, considering the metric $g_{\mu\nu}$ and the connection $\Gamma^{\mu}_{\alpha\beta}$ as independent fields amounts to decoupling the metric structure of spacetime from its geodesic structure (with the connection $\Gamma^{\mu}_{\alpha\beta}$ being, in general, different from the Levi-Civita connection $\left\{ {\mu \atop \alpha\beta} \right\}$ of $g_{\mu\nu}$). The causal structure of spacetime is governed by $g_{\mu\nu}$, while the spacetime trajectories of particles are governed by $\Gamma^{\mu}_{\alpha\beta}$. This decoupling of causal and geodesic structures enriches the spacetime geometry and generalizes the purely metric formalism. This metric-affine structure of spacetime is naturally translated, by means of the Palatini field equations, into a bimetric structure of spacetime. In addition to the physical metric $g_{\mu\nu}$, a second metric $h_{\mu\nu}$ is present which is related, in the case of $f(R)$ gravity, to the connection. The connection $\Gamma^{\mu}_{\alpha\beta}$ turns out to be the Levi-Civita connection of this second metric $h_{\mu\nu}$ and provides the geodesic structure of spacetime.

If we consider non-minimal coupling interactions in the gravitational Lagrangian of scalar-tensor theories, the new metric $h_{\mu\nu}$ is related to this non-minimal coupling and $h_{\mu\nu}$ can be related to a different geometric and physical aspect of the gravitational theory. Through the Palatini formalism, the non-minimal coupling and the scalar field entering the evolution of the gravitational fields are separated from the metric structure of spacetime. The situation mixes when we consider the case of higher order-scalar-tensor theories.

## 1.4 Quantum gravity motivation

One of the main challenges of modern theoretical physics is to construct a theory able to describe the fundamental interactions of nature as different aspects of the same theoretical construct. This goal has led, in the past decades, to the formulation of several unification schemes which, *inter alia*, attempt to describe gravity by putting it on the same footing as the other interactions. All these schemes try to describe the fundamental fields in terms of the conceptual apparatus of quantum mechanics. This is based on the fact that the states of a physical system are described by vectors in a Hilbert space $\mathscr{H}$ and the physical fields are represented by linear operators defined on domains of $\mathscr{H}$. Until now, any attempt to incorporate gravity in this scheme has either failed or been unsatisfactory. The main conceptual problem is that the gravitational field describes simultaneously the gravitational degrees of freedom and the background spacetime in which these degrees of freedom live.

Owing to the difficulties of building a complete theory unifying interactions and particles, during the last thirty years the two fundamental theories of modern physics, GR and quantum mechanics, have been critically re-analyzed. On the one hand, one assumes that the matter fields (bosons and fermions) come out from superstructures (*e.g.*, Higgs bosons or superstrings) that, undergoing certain phase transitions, have generated the known particles. On the other hand, it is assumed that the geometry (*e.g.*, the Ricci tensor or the Ricci scalar) interacts directly with quantum matter fields which back-react on it. This interaction necessarily modifies the standard gravitational theory, that is, the Lagrangian of gravity plus the effective fields is modified with respect to the Hilbert-Einstein one, and this fact leads to the ETGs.

From the point of view of cosmology, the modifications of standard gravity provide inflationary scenarios of remarkable interest. In any case, a condition that must be satisfied in order for such theories to be physically acceptable is that GR is recovered in the low-energy limit.

Although remarkable conceptual progress has been made following the introduction of generalized gravitational theories, at the same time the mathematical difficulties have increased. The corrections introduced into the Lagrangian augment the (intrinsic) non-linearity of the Einstein equations, making them more difficult to study because differential equations of higher order than second are often obtained and because it is impossible to separate the geometric from the matter degrees of freedom. In order to overcome these difficulties and simplify the equations of motion, one often looks for symmetries of the Lagrangian and identifies conserved quantities which allow exact solutions of the dynamics to be discovered. The key step in the implementation of this program consists of passing from the Lagrangian of the relevant fields to a point-like Lagrangian or, in other words, in going from a system with an infinite number of degrees of freedom to one with a finite number of degrees of freedom. Fortunately, this is feasible in cosmology because most models of physical interest are spatially homogeneous Bianchi models and the observed universe is spatially homogeneous and isotropic to a high degree (Friedmann-Lemaitre-Robertson-Walker or FLRW model).

The need for a quantum theory of gravity was recognized at the end of the 1950s, when physicist tried for the first time to treat all interactions at a fundamental level and describe them in terms of quantum field theory. Naturally, the first attempt to quantize gravity used the canonical approach and the covariant approach, which had been applied with remarkable success to electromagnetism. In the first approach applied to electromagnetism, one considers the electric and magnetic fields satisfying the Heisenberg uncertainty principle and the quantum states are gauge-invariant functionals generated by the vector potential defined on three-surfaces of constant time. In the second approach applied to electromagnetism, one quantizes the two degrees of freedom of the Maxwell field without any $3 + 1$ decomposition of the metric, while the quantum states are elements of a Fock space [628]. These procedures are equivalent. The former allows for a well-defined measure, whereas the latter is more convenient for perturbative calculations such as, for example, the computation of the $S$-matrix in QED.

These methods have been applied also to GR, but many difficulties arise in this case due to the fact that Einstein's theory cannot be formulated in terms of a quantum field theory on a fixed Minkowski background. To be more specific, in GR the geometry of the background spacetime cannot be given *a priori*: spacetime is the dynamical variable itself. In order to introduce the notions of causality, time, and evolution, one must first solve the equations of motion and build the spacetime. For example, in order to know if a particular initial condition will give rise to a black hole, it is necessary to fully evolve it by solving the Einstein equations. Then, taking into account the causal structure of the solution obtained, one has to study the asymptotic metric at future null infinity in order to assess whether it is related, in the causal past, with that initial condition. This problem becomes intractable at the quantum level. Due to the uncertainty principle, in non-relativistic quantum mechanics particles do not move along well-defined trajectories and one can only calculate the probability amplitude $\psi(t,x)$ that a measurement at time $t$ detects a particle around the spatial point $x$. Similarly, in quantum gravity, the evolution of an initial state does not provide a specific spacetime. In the absence of a spacetime, how is it possible to introduce basic concepts such as causality, time, elements of the scattering matrix, or black holes?

The canonical and covariant approaches provide different answers to these questions. The canonical approach is based on the Hamiltonian formulation of GR and aims at using the canonical quantization procedure. The canonical commutation relations are the same that lead to the uncertainty principle; in fact, the commutation of certain operators on a spatial three-manifold of constant time is imposed, and this three-manifold is fixed in order to preserve the notion of causality. In the limit of asymptotically flat spacetime, the motion generated by the Hamiltonian must be interpreted as temporal evolution (in other words, when the background becomes the Minkowski spacetime, the Hamiltonian operator assumes again its role as the generator of translations). The canonical approach preserves the geometric features of GR without the need to introduce perturbative methods [54, 387, 802, 803, 1157].

The covariant approach, instead, employs quantum field theory concepts and methods. The basic idea is that the problems mentioned above can be easily circumvented by splitting the metric $g_{\mu\nu}$ into a kinematical part $\eta_{\mu\nu}$ (usually flat) and a dynamical part $h_{\mu\nu}$, as in

$$g_{\mu\nu} = \eta_{\mu\nu} + h_{\mu\nu} \,. \tag{1.1}$$

The geometry of the background is given by the flat metric tensor and is the same as in Special Relativity and ordinary quantum field theory, which allows one to define the concepts of causality, time, and scattering. The quantization procedure is then applied to the dynamical field, considered as a (small) deviation of the metric from the Minkowski background metric. Quanta are discovered to be particles with spin two, called *gravitons*, which propagate in flat spacetime and are defined by $h_{\mu\nu}$. Substituting the metric $g_{\mu\nu}$ into the Hilbert-Einstein action, it follows that the Lagrangian of the gravitational sector contains a sum whose terms represent, at different orders of approximation, the interaction of gravitons and, eventually, terms describing matter-graviton interaction (if matter is present). Such terms are analyzed by using the familiar techniques of perturbative quantum field theory.

These quantization programs were both pursued during the 1960s and 1970s. In the canonical approach, Arnowitt, Deser, and Misner [54] provided a Hamiltonian formulation of GR using methods proposed earlier by Dirac and Bergmann. In this Hamiltonian formalism, the canonical variables are the three-metric on the spatial submanifolds obtained by foliating the four-dimensional manifold (note that this foliation is arbitrary). The Einstein equations give constraints between the three-metrics and their conjugate momenta and the evolution equation for these fields, known as the *Wheeler-DeWitt* (WDW) *equation*. In this way, GR is interpreted as the dynamical theory of the three-geometries (*geometrodynamics*). The main difficulties arising from this approach are that the quantum equations involve products of operators defined at the same spacetime point and, in addition, they entail the construction of distributions whose physical meaning is unclear. In any case, the main problem is the absence of a Hilbert space of states and, as consequence, a probabilistic interpretation of the quantities calculated is missing.

The covariant quantization approach is closer to the known physics of particles and fields in the sense that it has been possible to extend the perturbative methods of QED to gravitation. This has allowed the analysis of the mutual interaction between gravitons and of the matter-graviton interactions. The formulation of Feynman rules for gravitons and the demonstration that the theory might be unitary at every order of the expansion was achieved by DeWitt.

Further progress was achieved with Yang-Mills theories, which describe the strong, weak, and electromagnetic interactions of quarks and leptons by means of symmetries. Such theories are renormalizable because it is possible to give the fermions a mass through the mechanism of spontaneous symmetry breaking. Then, it is natural to attempt to consider gravitation as a Yang-Mills theory in the covariant perturbation approach and check whether it is renormalizable. However, gravity does not fit into this scheme; it turns out to be non-renormalizable when one considers the graviton-graviton interactions (two-loops diagrams) and graviton-matter interactions (one-loop diagrams).[2] The covariant method allows one to construct a theory of gravity which is renormalizable at one-loop in the perturbative series [144]. Due to the non-renormalizability of gravity at different orders, its validity is restricted only to the low-energy domain, *i.e.*, to large scales, while it fails at high energy and small scales. This implies that the full unknown theory of gravity has to be invoked near or at the Planck era and that, sufficiently far from the Planck scale, GR and its first loop corrections describe the gravitational interactions. In this context it

[2] Higher order terms in the perturbative series imply an infinite number of free parameters. At the one-loop level it is sufficient to renormalize only the effective constants $G_{eff}$ and $\Lambda_{eff}$ which, at low energy, reduce to Newton's constant $G_N$ and the cosmological constant $\Lambda$.

makes sense to add higher order terms to the Hilbert-Einstein action. Besides, if the free parameters are chosen appropriately, the theory has a better ultraviolet behavior and is asymptotically free. Nevertheless, the Hamiltonian of these theories is not bounded from below and they are unstable. In particular, unitarity is violated and probability is not conserved.

An alternative approach to the search for a theory of quantum gravity comes from the study of the electroweak interaction. In this approach, gravity is treated neglecting the other fundamental interactions. The unification of the electromagnetic and the weak interactions suggests that it might be possible to obtain a consistent theory when gravitation is coupled to some kind of matter. This is the basic idea of *supergravity*. In this class of theories, the divergences due to the bosons (gravitons in this case) are cancelled exactly by those due to the fermions, leading to a renormalized theory of gravity. Unfortunately, this scheme works only at the two-loop level and for matter-gravity couplings. The Hamiltonian is positive-definite and the theory turns out to be unitary. But, including higher order loops, the infinities re-appear, destroying the renormalizability of the theory.

Perturbative methods are also used in string theories. In this case, the approach is completely different from the previous one because the concept of particle is replaced by that of an extended object, the fundamental string. The usual physical particles, including the spin two graviton, correspond to excitations of the string. The theory has only one free parameter (the string tension) and the interaction couplings are determined uniquely. It follows that string theory contains all fundamental physics and it is therefore considered as a candidate for the *theory of everything*. String theory seems to be unitary and the perturbative series converges implying finite terms. This property follows from the fact that strings are intrinsically extended objects, so that ultraviolet divergencies coming from small scales or from large transfer impulses, are naturally cured. In other words, the natural cutoff is given by the string length, which is of Planck size $l_{Pl}$. At scales larger than $l_{Pl}$, the effective string action can be rewritten in terms of non-massive vibrational modes, *i.e.*, in terms of scalar and tensor fields (*tree-level effective action*). This eventually leads to an effective theory of gravity non-minimally coupled with scalar fields, the so-called *dilaton fields*.

To conclude, let us summarize the previous considerations:

- A consistent (*i.e.*, unitary and renormalizable) theory of gravity does not yet exists.
- In the quantization program for gravity, two approaches have been used: the *covariant approach* and the *perturbative approach*. They do not lead to a definitive theory of quantum gravity.
- In the low-energy regime (with respect to the Planck energy) at large scales, GR can be generalized by introducing into the Hilbert-Einstein action terms of higher order in the curvature invariants and non-minimal couplings between matter and gravity. These lead, at the one-loop level, to a consistent and renormalizable theory.

### 1.4.1 Emergent gravity and thermodynamics of spacetime

In recent years we have witnessed the growth of theoretical efforts on *emergent gravity* theories. The central idea is that, given the complete lack of experimental data about high-energy quantum gravity, it is worth approaching gravity from the low-energy side with some kind of effective theory. In this approach, gravity "emerges" from fundamental constituents, sort of "atoms of spacetime", with the metric and the affine connection being collective variables in a way similar to hydrodynamics, in which a fluid description emerges from an aggregate of microscopic particles. The emergent gravity approach attempts the reconstruction of the microscopic system underlying classical gravity. It may be possible to constrain the microscopic properties of the fundamental constituents by requiring that the emergent theory of gravity be not too far from GR today. This approach naturally questions the principles which constitute the foundations of gravitational theories.

A related area is that of *analogue models*: if gravity emerges as a collective system made of microscopic constituents of quantum nature, it may be possible to model it with the help of physical systems in which an effective metric and a connection rule the dynamics, for example one can attempt to study Hawking radiation from black holes using acoustic analogues ("dumb holes") [858, 1099, 1125], or Bose-Einstein condensates (see [82] for a review of analogue models). While an effective metric is generated, it usually has a kinematic nature, in the sense that field equations are not generated with it. However, a recent work was able to generate a complete toy model theory of scalar gravity [541] and it is hoped that progress will be made in this direction. A common feature to all known emergent spacetimes is that they exhibit Lorentz invariance in the low-energy limit. The Lorentz symmetry is expected to be broken in the ultraviolet limit in which the fundamental quantum constituents of gravity make their effects being felt. We mention these modern approaches here because they question the foundations of gravitational theory and do not insist that GR is the theory to be reproduced with large scale coarse-graining: different theories with similar features are possible as well.

Another line of research related to the previous one is based on the idea that gravity can be reproduced through a sort of *thermodynamics of spacetime*. It was shown in [631] that the Einstein equations could be derived through local considerations of equilibrium thermodynamics. Using thermodynamical considerations on local Rindler horizons associated to the worldlines of physical observers and assuming the GR relation $S_{BH} = A/4$ between entropy and horizon area (which is believed to be more fundamental than the Einstein equations, or the field equations of any gravitational theory), Jacobson was able to derive the Einstein equations more or less in the same manner that an equation of state is derived for an ideal gas [631]. The implication of this result would be that it does not make much sense to quantize directly the Einstein equations in order to learn about the fundamental ingredients of quantum gravity, the same way that by quantizing the equation of state of an atomic hydrogen gas one would not learn anything about the hydrogen atom and its energy levels. From our perspective it is interesting that, if a similar thermodynamics of spacetime approach is applied to $f(R)$ gravity, it is then necessary to resort to near-equilibrium thermodynamics in order to derive the correct field equations [419]. This fact seems to demonstrate that GR is just one state of gravity corresponding to thermodynamic equilibrium and, when this equilibrium is disturbed at higher energies, near-equilibrium configurations correspond to alternative theories of gravity, and further justify the study of ETGs.

A result with a conceptually similar flavour is found in the cosmology of scalar-tensor theories: scalar-tensor cosmologies appear to relax to GR during the evolution of the universe [361–363]. This is another hint that GR may be only a state of equilibrium, while an entire spectrum of theories could be found as higher energy excitations.

These conclusions are still very speculative and based on results that require further study; however, they underline the need to think about gravity outside of the GR box and hint to the fact that much more work needs to be done before one can claim a full understanding of gravity even at lower energy scales.

## 1.5 What a good theory of gravity should do: General Relativity and its extensions

A relativistic theory of gravity must satisfy certain minimal requirements from the phenomenological point of view. First, it must explain astronomical observations mapping the orbits of planets and the potential well of self-gravitating structures such as galaxies and clusters. This means that the theory must reproduce the Newtonian dynamics in its weak-field, slow-motion limit. Then, at the Post-Newtonian level, the theory must pass the classical Solar System tests, which have by now become very precise [1167]. Second, the theory should reproduce correctly the observed galactic dynamics accounting for the known baryonic constituents including luminous (stars) and sub-luminous (planets, dust, and gas) components, and radiation, and reproduce the Newtonian potential which is, by assumption, extrapolated to galactic scales. Third, the theory must address the problem of the generation of large scale structures (galaxy clusters, superclusters, voids, and filaments). Finally, the cosmological dynamics must be reproduced: this means predicting in a self-consistent way the Hubble parameter $H_0$, the deceleration parameter $q_0$, the density parameters, *etc.* Astronomical observations and experiments probe directly standard baryonic matter, radiation, and indirectly the overall attraction of gravity acting at all scales and depending on distance.

The simplest theory satisfying the above requirements at least to a certain degree was Einstein's GR [414]. It is based on the assumption that space and time are entangled into a single spacetime structure which, in the limit of zero gravity, reduces to the flat Minkowski spacetime. Einstein did benefit of earlier ideas of Riemann, who had stated that the universe should be a curved manifold and that its curvature must be measured by means of astronomical observations [416]. The matter distribution determines, point by point, the local curvature of this spacetime manifold. The theory, eventually formulated by Einstein in 1915, relies on three basic assumptions on gravity:

1. The *Principle of Relativity* is the requirement that all observers be equally valid for describing physics. In particular, inertial frames (which do not exist globally) are not *a priori* preferred. This postulate addresses the main shortcoming of Special Relativity, being based on preferred inertial frames and Lorentz boosts between them.
2. The *Equivalence Principle* (EP) requires acceleration effects to be locally indistinguishable from gravitational effects (roughly speaking, the equivalence between inertial and gravitational mass).[3]
3. The *Principle of General Covariance* requires that the field equations be "generally covariant" tensor equations whose form is the same in all coordinate systems, and states that all coordinate systems are in principle equivalent in the description of physics [994]. In modern language, the field equations are required to be invariant under the action of the group of spacetime diffeomorphisms.

In addition to these three principles one imposes that causality is preserved (*Principle of Causality*, *i.e.*, that each spacetime point should admit a notion of past, present, and future which is the same for all physical observers. It is generally felt that the notion of causality forbids the presence of closed timelike curves and time travel, although this belief is rather superficial (see [748] for a discussion and references). In any case, to enforce the absence of closed timelike curves it is necessary to impose restrictions on the matter distribution (*energy conditions*) [83,273,1139]).

The old Newtonian theory of gravitation that Einstein needed to recover in the limit of weak gravitational forces and slow motions regarded space and time as absolute entities and required particles to move, in a preferred inertial frame, along curved trajectories, the curvature of which (*i.e.*, the acceleration) was a function of the strength of the sources through the "forces". With this premise, Einstein was led to postulate that gravitational forces should be described by the curvature of a metric tensor field $g_{\mu\nu}$ related to the line element $ds^2 = g_{\mu\nu}dx^\mu dx^\nu$ of a four-dimensional spacetime manifold, and having the same signature of the Minkowski metric (the *Lorentz signature* herewith assumed to be $(-,+,+,+)$). He also postulated that spacetime curves onto itself and that its curvature is locally determined by the distribution of the sources, *i.e.*, – spacetime being a continuum – by the four-dimensional generalization of the matter stress-energy tensor $T^{(m)}_{\mu\nu}$ (a rank-two symmetric tensor) of continuum mechanics.

Once a metric $g_{\mu\nu}$ is given, its curvature is expressed by the Riemann (or curvature) tensor

$$R_{\alpha\beta\mu}{}^{\nu} = \Gamma^{\nu}_{\alpha\mu,\beta} - \Gamma^{\nu}_{\beta\mu,\alpha} + \Gamma^{\sigma}_{\alpha\beta}\Gamma^{\nu}_{\sigma\mu} - \Gamma^{\sigma}_{\mu\beta}\Gamma^{\nu}_{\sigma\alpha} \tag{1.2}$$

[3] The Equivalence Principle admits (at least) three distinct formulations which are discussed later in this chapter [1167].

where the commas denote partial differentiation. Its contraction

$$R_{\alpha\mu} = R_{\alpha\beta\mu}{}^{\beta}\,, \tag{1.3}$$

is the *Ricci tensor*, while

$$R \equiv R^{\mu}{}_{\mu} = g^{\mu\nu} R_{\mu\nu} \tag{1.4}$$

is the *scalar* (or *Ricci*) *curvature* of $g_{\mu\nu}$. Einstein initially contemplated the equations for the dynamics of gravity

$$R_{\mu\nu} = \frac{\kappa}{2}\, T^{(m)}_{\mu\nu}\,, \tag{1.5}$$

where $\kappa = 8\pi G$ (in units in which $c = 1$) contains the gravitational coupling constant $G$. These equations turned out to be physically and mathematically inconsistent. As pointed out by Hilbert [994], they do not derive from an action principle; there is no action which reproduces them exactly through a variation.[4] Einstein's reply was that he knew that the equations were physically unsatisfactory, since they were incompatible with the continuity equation deemed to be satisfied by any reasonable form of matter. Assuming that the latter consists of a perfect fluid with stress-energy tensor

$$T^{(m)}_{\mu\nu} = (P + \rho)\, u_\mu u_\nu + P\, g_{\mu\nu}\,, \tag{1.6}$$

where $u^\mu$ is the four-velocity of the fluid particles and $P$ and $\rho$ are the pressure and energy density of the fluid, respectively, the continuity equation requires $T^{(m)}_{\mu\nu}$ to be covariantly constant, *i.e.*, to satisfy the conservation law

$$\nabla^\mu T^{(m)}_{\mu\nu} = 0\,, \tag{1.7}$$

where $\nabla_\alpha$ denotes the covariant derivative operator of the metric $g_{\mu\nu}$. In fact, $\nabla^\mu R_{\mu\nu}$ does not vanish, except in the special case $R \equiv 0$. Einstein and Hilbert independently concluded that the incorrect field equations (1.5) had to be replaced by the correct ones

$$G_{\mu\nu} = \kappa\, T^{(m)}_{\mu\nu} \tag{1.8}$$

where

$$G_{\mu\nu} \equiv R_{\mu\nu} - \frac{1}{2}\, g_{\mu\nu} R \tag{1.9}$$

is now called the *Einstein tensor* of $g_{\mu\nu}$. These equations can be derived by minimizing an action and satisfy the conservation law (1.7) since the relation

$$\nabla^\mu G_{\mu\nu} = 0\,, \tag{1.10}$$

holds as a contraction of the Bianchi identities that the curvature tensor of $g_{\mu\nu}$ has to satisfy [1153].

[4] This is not entirely correct but this point is not essential.

The Lagrangian that, when varied, produces the field equations (1.8) is the sum of a "matter" Lagrangian density $\mathscr{L}^{(m)}$, the variational derivative of which is

$$T^{(m)}_{\mu\nu} = -\frac{2}{\sqrt{-g}}\frac{\delta\left(\sqrt{-g}\,\mathscr{L}^{(m)}\right)}{\delta g^{\mu\nu}}, \tag{1.11}$$

and of the gravitational (*Hilbert-Einstein*) *Lagrangian*

$$\sqrt{-g}\,\mathscr{L}_{HE} = \sqrt{-g}\,R\,, \tag{1.12}$$

where $g$ is the determinant of the metric $g_{\mu\nu}$.

Hilbert's and Einstein's choice was rather arbitrary, as it became clear a few years later, but it was certainly the simplest from both the mathematical and the physical points of view. As clarified by Levi-Civita in 1919, curvature is not a purely metric notion but is also related to the linear connection defining parallel transport and covariant differentiation [717]. In a sense, this idea is the precursor of what would later be called a "gauge-theoretical framework" [673], following the pioneering work of Cartan in 1925 [279]. But in Einstein's times only metric concepts were available to mathematicians and physicists alike and his solution was the only viable one.

It was later clarified that the three principles of relativity, equivalence, and covariance, together with causality, require only that the spacetime structure be determined by either one or both of two fields, a Lorentzian metric $g_{\mu\nu}$ and a linear connection $\Gamma^{\alpha}_{\mu\nu}$, assumed to be torsionless for simplicity. The metric $g_{\mu\nu}$ fixes the causal structure of spacetime (the light cones) as well as its metric relations measured by clocks and rods and the lengths of four-vectors. The connection $\Gamma^{\alpha}_{\mu\nu}$ determines the laws of free fall, the four-dimensional spacetime trajectories followed by locally inertial observers. These, of course, must satisfy a number of compatibility relations including the requirement that photons follow null geodesics of $\Gamma^{\alpha}_{\mu\nu}$, so that $\Gamma^{\alpha}_{\mu\nu}$ and $g_{\mu\nu}$ can *a priori* be independent, but constrained *a posteriori* by the physics. These physical constraints, however, do not necessarily impose that $\Gamma^{\alpha}_{\mu\nu}$ is the Levi-Civita connection of $g_{\mu\nu}$ [887].

The previous considerations illustrate the fact that one can envisage alternative gravitational theories, which we prefer to call "extended gravitational theories" (ETGs) because their basic assumptions are exactly the same as those used by Einstein and Hilbert in the construction of GR. These are theories in which gravitation is described by either a metric (*purely metric theories*), or by a linear connection (*purely affine theories*), or by both fields (metric-affine theories, also known as *first order formalism* theories). In these theories, the Lagrangian is a scalar density of the curvature invariants constructed out of both $g_{\mu\nu}$ and $\Gamma^{\alpha}_{\mu\nu}$. The choice (1.12) is by no means unique and it turns out that the Hilbert-Einstein Lagrangian is in fact the only choice that produces an invariant linear in the second derivatives of the metric (or the first derivatives of the connection). Unfortunately, this Lagrangian is rather singular from the Hamiltonian point of view in the same way of Lagrangians linear in the canonical momenta in classical mechanics (*e.g.*, [53]).

A number of attempts to generalize GR and unify it with electromagnetism along these lines were followed by Einstein and many others, including Eddington, Weyl, and Schrödinger to mention a few [51]. These attempts were eventually abandoned in the 1950s, mainly because of a number of difficulties related to the definitely more complicated structure of a non-linear theory (where by "non-linear" we mean one based on non-linear invariants of the curvature tensor), and also because of the discovery of two new physical interactions, the strong and the weak nuclear forces that required the more general framework of gauge theory [656]. Still, sporadic investigations of alternative theories continued after 1960 (see [1167] and the references therein for a short history). The search for a coherent quantum theory of gravitation, or the belief that gravity has to be considered as a sort of low-energy limit of string or other quantum theories [553] – something that we will not discuss here – has more recently revived the idea that it is not mandatory to adhere to the simple prescription of Einstein and Hilbert and to assume that classical gravity is governed by a Lagrangian linear in the curvature. Further curvature invariants or non-linear functions of them can also be contemplated, especially in view of the fact that their inclusion is required in both the semiclassical expansion of a quantum Lagrangian and in the low-energy limit of stringy actions. Moreover, it is clear from recent astrophysical observations and from the current cosmological investigations that it is legitimate to doubt the paradigmatic role played by the Einstein equations at Solar System, galactic, extragalactic, and cosmological scales, unless one is willing to admit that the right hand side of Eq. (1.8) contains some types of exotic energy, the *dark matter* and *dark energy* components of our universe.

The idea discussed in this section is, in principle, much simpler. Instead of changing the matter side of the Einstein equations (1.8) and introducing the missing matter-energy content of the observed universe (up to 95% of its total energy content), while adding mysterious and odd-behaving states of the matter fields, we contemplate the fact that it is *a priori* simpler and more convenient to change the geometric/gravitational sector of these equations by inserting non-linear corrections to the Lagrangian. This procedure could be regarded as a mere matter of taste; however, there is no reason to discard this approach *a priori*, and this possibility is intriguing and worth exploring. In principle, the action belongs to a vast family of permissible actions and, from the purely phenomenological point of view, this freedom allows it to be chosen on the basis of its best-fit with the available observational data at all scales (solar, galactic, extragalactic, and cosmological). The down side of this approach is that too many models fit well the observations because of the relatively large number of free functions and parameters that they contain, and predictive power may be lost. However, it is hoped that theoretical work will provide guidelines pointing to a preferred action and will discriminate between huge classes of models which fit the data, of which already too many are known. From the theoretical point of view, it makes perfect sense to give serious consideration to rather well-motivated non-linear theories of gravity based on non-singular Lagrangians. Instead, the $\Lambda$CDM model is accompanied by exotic matter completely different from the known baryons, never detected in our laboratories, and segregated at astrophysical scales.

## 1.6 Quantum field theory in curved space

At small scales and high energies, an hydrodynamic description of matter as a perfect fluid is inadequate: a more accurate description requires quantum field theory formulated on a curved space, in the framework of either GR or another relativistic theory of gravity. Since, at scales comparable to the Compton wavelength of the relevant particles, matter must be quantized, one can employ a semiclassical description of gravitation in which the Einstein equations assume the form

$$G_{\mu\nu} \equiv R_{\mu\nu} - \frac{1}{2} g_{\mu\nu} R = < T_{\mu\nu} > , \tag{1.13}$$

where the usual Einstein tensor $G_{\mu\nu}$ appears on the left hand side whereas the right hand side contains the expectation value of a quantum stress-energy tensor sourcing the gravitational field. More precisely, if $|\psi >$ is a quantum state describing the early universe, then

$$< T_{\mu\nu} > \equiv < \psi | \hat{T}_{\mu\nu} | \psi > , \tag{1.14}$$

where $\hat{T}_{\mu\nu}$ is the quantum operator associated with the classical energy-momentum tensor of the matter field and the right hand side is an appropriately regularized expectation value. In general, a quantized matter field $\hat{\phi}$ is subject to self-interactions and it interacts also with other fields and with the gravitational background. Such interaction terms may be included in the definition of an effective potential[5]

$$V_{eff}(\phi) = < a | \hat{\mathscr{H}} | a > \tag{1.15}$$

with

$$\phi = < a | \hat{\phi} | a > , \tag{1.16}$$

where $|a >$ represents a normalized state of the theory under consideration (*i.e.*, $< a|a >= 1$) and $\hat{\mathscr{H}}$ is the Hamiltonian operator satisfying $\delta < a|\hat{\mathscr{H}}|a >= 0$, where $\delta$ is the variation on the average of $\mathscr{H}$-eigenstates. This condition corresponds to energy conservation.

In a curved spacetime, even in the absence of classical matter and radiation, quantum fluctuations of matter fields give non-vanishing contributions to $< T_{\mu\nu} >$, an effect similar to the vacuum of QED [144, 705]. When matter fields are free, massless and conformally invariant, these corrections assume the form

$$< T_{\mu\nu} > = k_1 \, {}^{(1)}H_{\mu\nu} + k_3 \, {}^{(3)}H_{\mu\nu} . \tag{1.17}$$

[5] Hereafter, scalar fields and potentials are understood as their effective values, obtained averaging over quantum states. In this sense, classical fields and potentials are the expectation values of quantum fields and potentials.

Here $k_1$ and $k_3$ are numerical coefficients, while

$$^{(1)}H_{\mu\nu} = 2R_{;\mu\nu} - 2g_{\mu\nu}\Box R + 2RR_{\mu\nu} - \frac{1}{2}g_{\mu\nu}R^2\,, \tag{1.18}$$

$$^{(3)}H_{\mu\nu} = R^{\sigma}{}_{\mu}R_{\nu\sigma} - \frac{2}{3}RR_{\mu\nu} - \frac{1}{2}g_{\mu\nu}R^{\sigma\tau}R_{\sigma\tau} + \frac{1}{4}g_{\mu\nu}R^2\,. \tag{1.19}$$

The divergence of the tensor $^{(1)}H_{\mu\nu}$ vanishes,

$$^{(1)}H^{\nu}_{\mu;\nu} = 0\,. \tag{1.20}$$

This tensor can be obtained by varying a quadratic contribution to the local action,

$$^{(1)}H_{\mu\nu} = \frac{2}{\sqrt{-g}}\frac{\delta}{\delta g^{\mu\nu}}\left(\sqrt{-g}\,R^2\right). \tag{1.21}$$

In order to remove the infinities coming from $< T_{\mu\nu} >$ and obtain a renormalizable theory, one has to introduce infinitely many counterterms in the Lagrangian density of gravity. One of these terms is $CR^2\sqrt{-g}$, where $C$ is a parameter that diverges logarithmically. Equation (1.13) cannot be generated by a finite action because then the gravitational field would be completely renormalizable, *i.e.*, it would suffice to eliminate a finite number of divergences to make gravity similar to QED. Instead, one can only construct a truncated quantum theory of gravity. The expansion in loops is done in terms of $\hbar$, so the truncated theory at the one-loop level contains all terms of order $\hbar$. In this sense, this is the first quantum correction to GR. It assumes that matter fields are *free* and, due to the Equivalence Principle, all forms of matter couple in the same way to gravity. It also implies an *intrinsic* non-linearity of gravity, so that a number of loops are needed in order to take into account self-interactions or mutual interactions between matter and gravitational fields. At the one-loop level, divergences can be removed by renormalizing the cosmological constant $\Lambda_{eff}$ and the gravitational constant $G_{eff}$. The one-loop contributions to $< T_{\mu\nu} >$ are the quantities $^{(1)}H_{\mu\nu}$ and $^{(3)}H_{\mu\nu}$ above. In addition, one has to consider

$$^{(2)}H_{\mu\nu} = 2R^{\sigma}{}_{\mu;\nu\sigma} - \Box R_{\mu\nu} - \frac{1}{2}g_{\mu\nu}\Box R + R^{\sigma}{}_{\mu}R_{\sigma\nu} - \frac{1}{2}R^{\sigma\tau}R_{\sigma\tau}g_{\mu\nu}\,. \tag{1.22}$$

It is shown in [144] that the relation

$$^{(2)}H_{\mu\nu} = \frac{1}{3}\,{}^{(1)}H_{\mu\nu} \tag{1.23}$$

holds in conformally flat spacetimes. In this case, only the first and third terms of $H_{\mu\nu}$ are present in Eq. (1.17).

Since one can add to the parameter $C$ in the Lagrangian term $C\sqrt{-q}\,R^2$ an arbitrary constant, the coefficient $k_1$ can assume any value – the latter should be determined experimentally [144].

The tensor ${}^{(3)}H_{\mu\nu}$ is conserved only in conformally flat spacetimes (for example, FLRW spaces) and it cannot be obtained by varying a local action. Finally, one has

$$k_3 = \frac{1}{1440\pi^2}\left(N_0 + \frac{11}{2}N_{1/2} + 31N_1\right), \tag{1.24}$$

where the $N_i$'s ($i = 0, 1/2, 1$) are determined by the number of quantum fields with spin $0, 1/2$, and 1. Vector fields contribute more to $k_3$ due to the larger coefficient 31 of $N_1$. These massless fields, as well as the spinorial ones, are described by conformally invariant equations and appear in $< T_{\mu\nu} >$ in the form (1.17).

The trace of the energy-momentum tensor vanishes for conformally invariant classical fields while, owing to the term weighted by $k_3$, one finds that the expectation value of the tensor (1.17) has non-vanishing trace. This fact is at the origin of the so-called *trace anomaly*.

Let us discuss briefly how the conformal anomalies are generated when the origin of the tensor $T_{\mu\nu}$ is not classical, *i.e.*, when quantum field theories are formulated in curved spacetime. As we will see in more detail later, if a theory is conformally invariant, under the conformal transformation

$$g_{\mu\nu}(x) \to \tilde{g}_{\mu\nu}(x) \equiv \Omega^2(x) g_{\mu\nu}(x) . \tag{1.25}$$

the action in $(n+1)$ spacetime dimensions satisfies the functional equation

$$S[\tilde{g}_{\mu\nu}] = S[g_{\mu\nu}] + \int d^{n+1}x\, \frac{\delta S[\tilde{g}_{\mu\nu}]}{\delta \tilde{g}^{\rho\sigma}}\delta\tilde{g}^{\rho\sigma}, \tag{1.26}$$

where the use of

$$\delta\tilde{g}^{\mu\nu}(x) = -2\Omega^{-1}(x)\,\tilde{g}^{\mu\nu}(x)\delta\Omega(x), \tag{1.27}$$

and of the classical variational principle

$$T^{(m)}_{\mu\nu} = -\frac{2}{\sqrt{-g}}\frac{\delta S^{(m)}}{\delta g^{\mu\nu}}, \tag{1.28}$$

yields

$$S[\tilde{g}_{\mu\nu}] = S[g_{\mu\nu}] - \int d^{n+1}x\,\sqrt{-\tilde{g}}\,T^{\rho}{}_{\rho}(\tilde{g}_{\mu\nu})\Omega^{-1}\delta\Omega . \tag{1.29}$$

From this, it follows that

$$T^{\rho}{}_{\rho}[g_{\mu\nu}(x)] = -\left.\frac{\Omega(x)}{\sqrt{-g}}\frac{\delta S[\tilde{g}_{\mu\nu}]}{\delta\Omega(x)}\right|_{\Omega=1} . \tag{1.30}$$

Hence, if the classical action is invariant under conformal transformations, the trace of the energy-momentum tensor vanishes. At the quantum level this situation could not occur for the following reason. A conformal transformation is, essentially, a rescaling of lengths with a different rescaling factor at each spacetime point $x$; the presence of a mass, and hence of a length scale, in the theory breaks conformal invariance and generates the trace anomaly. To preserve conformal invariance one has to consider massless fields, as done in (1.17). In this case one obtains the condition

$$< T_\rho^\rho >= 0\,, \tag{1.31}$$

which allows one to consider a conformally invariant theory. Note that gravity is not renormalizable in the usual way; because of this, divergences appear as soon as quantum effects are considered. A loop expansion yields

$$< T_\rho^\rho >=< T_\rho^\rho >_{div} + < T_\rho^\rho >_{ren} = 0\,, \tag{1.32}$$

confirming the validity of Eq. (1.31). In this case conformal invariance is preserved only if the divergent part is equal (up to the sign) to the renormalized tensor. An anomalous trace term will appear on the right hand side of the field equations (1.13) which, at one-loop and in the zero mass limit of the fields, is given by

$$< T_\rho^\rho >_{div} = \left[\tilde{k}_1\left(M - \frac{2}{3}\Box R\right) + \tilde{k}_3 \mathcal{G}\right] = - < T_\rho^\rho >_{ren}\,, \tag{1.33}$$

for a four-dimensional theory. Here $\tilde{k}_1$ and $\tilde{k}_3$ are proportional to $k_1$ and $k_3$, while $M$ and $\mathcal{G}$ are obtained from ${}^{(1)}H_{\mu\nu}$ and ${}^{(3)}H_{\mu\nu}$ as

$$M = R^{\alpha\beta\gamma\delta} R_{\alpha\beta\gamma\delta} - 2R^{\alpha\beta} R_{\alpha\beta} + \frac{1}{3} R^2\,, \tag{1.34}$$

$$\mathcal{G} = R^2 - 4R^{\alpha\beta} R_{\alpha\beta} + R^{\alpha\beta\gamma\delta} R_{\alpha\beta\gamma\delta}\,. \tag{1.35}$$

$\mathcal{G}$ is the *Gauss-Bonnet* term. In four dimensions, the integral

$$\int d^4x \sqrt{-g}\, \mathcal{G} \tag{1.36}$$

is an invariant (*Euler characteristic*) which provides information about the topology of the spacetime manifold on which the theory is formulated (*Gauss-Bonnet theorem*). In a FLRW background $M$ vanishes identically but $\mathcal{G}$ gives non-vanishing contributions to (1.17) even if the variation of (1.36) is zero (in four dimensions).

In general, by summing all the geometric terms deduced from the Riemann tensor and of the same order in $< T_\rho^\rho >_{ren}$, one derives the right hand side of (1.17). If the background metric is conformally flat, this can be expressed by means of

Eqs. (1.18) and (1.19). Then, one can conclude that the trace anomaly due to the geometric terms arises because the one-loop approach is an attempt to formulate quantum field theories on curved spacetime.[6] Cosmological models arising from (1.17) are studied in [137].

The masses of the matter fields and their mutual interactions can be neglected in the high curvature limit because $R \gg m^2$. The matter-graviton interactions generate non-minimal coupling terms in the effective Lagrangian. The one-loop contributions of such terms are comparable to the ones due to the trace anomaly and generate, from the conformal point of view, the same effects on gravity. The simplest effective Lagrangian that takes into account these corrections is

$$\mathscr{L}_{NMC} = -\frac{1}{2}\nabla^{\alpha}\phi\nabla_{\alpha}\phi - V(\phi) - \frac{\xi}{2}R\phi^2\,, \tag{1.37}$$

where $\xi$ is a dimensionless coupling constant between the scalar and the gravitational fields. The scalar field stress-energy tensor will be modified accordingly (see Sect. 1.7.2 below) but a conformal transformation can be found such that the modifications due to curvature terms can, at least formally, be cast in the form of a matter-curvature interaction. The same argument holds for the trace anomaly.

Certain Grand-Unified theories lead to a polynomial coupling of the form $1 + \xi\phi^2 + \zeta\phi^4$ generalizing the one of (1.37), while an exponential coupling $e^{-\alpha\varphi}R$ between a scalar field (dilaton) $\varphi$ and the Ricci scalar appears instead in the effective Lagrangian of string theories.

The field equations obtained by varying the Lagrangian density $\sqrt{-g}\,\mathscr{L}_{NMC}$ are

$$\begin{aligned}\left(1-\kappa\xi\phi^2\right)G_{\mu\nu} &= \kappa\left\{\nabla_\mu\phi\nabla_\nu\phi - \frac{1}{2}g_{\mu\nu}\nabla^\alpha\phi\,\nabla_\alpha\phi - V\,g_{\mu\nu}\right.\\ &\left.\quad + \xi\left[g_{\mu\nu}\Box\left(\phi^2\right) - \nabla_\mu\nabla_\nu\left(\phi^2\right)\right]\right\}\,,\end{aligned} \tag{1.38}$$

$$\Box\phi - \frac{dV}{d\phi} - \xi R\phi = 0. \tag{1.39}$$

The non-minimal coupling of the scalar field is reminiscent of that exhibited by the four-vector potential of curved space Maxwell theory in Eq. (1.56) below.

Motivation for the non-minimal coupling in the Lagrangian $\mathscr{L}_{NMC}$ comes from many directions. A nonzero $\xi$ is generated by first loop corrections even if it is absent in the classical action [143, 144, 499, 500, 833, 892]. Renormalization of a classical theory with $\xi = 0$ shifts this coupling constant to a value which is typically small [27, 626] but can, however, affect drastically an inflationary

[6] Equations (1.18) and (1.19) can include terms containing derivatives of the metric of order higher than fourth (fourth order corresponding to the $R^2$ term) if all possible Feynman diagrams are included. For example, corrections such as $R\Box R$ or $R^2\Box R$ can be present in ${}^{(3)}H_{\mu\nu}$ implying equations of motion that contain sixth order derivatives of the metric. Also these terms can be treated by making use of conformal transformations [34].

cosmological scenario and determine its success or failure [3, 442, 446, 453, 515, 518]. A non-minimal coupling term is expected at high curvatures [499, 500], and it has been argued that classicalization of the universe in quantum cosmology indeed requires $\xi \neq 0$ [866]. Moreover, non-minimal coupling can solve potential problems of primordial nucleosynthesis [295, 297] and the absence of pathologies in the propagation of $\phi$-waves seems to require conformal coupling for all non-gravitational scalar fields ([474,556,557,1028], see also [376,466]).[7] The conformal value $\xi = 1/6$ is also an infrared fixed point of the renormalization group in finite GUTs [156, 157, 185–190, 423, 820, 864, 933]. Non-minimally coupled scalar fields have been widely used in relation with specific inflationary scenarios [89, 111, 437–440, 519, 522, 620, 691, 692, 713, 964]. The approach adopted was largely one in which $\xi$ is regarded as a free parameter to be used at will in order to fix possible problems of specific inflationary scenarios; see [446, 449] for more general treatments. Geometric reheating of the universe with strong coupling $\xi \gg 1$ has also been studied [111, 1090, 1091] and non-minimally coupled scalar fields have been considered in relation with wormholes [347, 349, 569], black holes [593, 1110], and boson stars [639, 731, 1109]. The value of the coupling $\xi$ is not, in general, a free parameter but it depends on the physical nature of the particular scalar field $\phi$ [499, 500, 589, 600, 893, 1131] (see [446, 449, 453] for reviews of the available theoretical prescriptions for the value of $\xi$).

To conclude, any attempt to formulate quantum field theory on a curved space-time necessarily leads to modifying the Hilbert-Einstein action. This means adding terms containing non-linear invariants of the curvature tensor or non-minimal couplings between matter and the curvature originating in the perturbative expansion. In cosmology, all these modifications may affect deeply inflationary scenarios originally proposed using minimally coupled scalars [449,453]. Although rare and very speculative alternatives have been proposed to the inflationary paradigm, the latter is currently accepted by most authors as the "canonical" cure to the shortcomings of the Standard Big Bang model, with the added bonus of providing a viable mechanism for the generation of density perturbations to seed the structures observed today in the universe. However, the effects of non-minimal coupling on the inflationary paradigm need to be assessed carefully [446, 449].

## 1.7 Mach's principle and other fundamental issues

We now comment on the role played by ETGs in connection with fundamental problems which are only partially addressed by GR. We will consider first the concept of inertia and the problem of how to incorporate Mach's principle in the gravitational theory. We focus on the time period between the Lense-Thirring 1918

---

[7] Note, however, that the distinction between gravitational and non-gravitational fields becomes representation-dependent in ETGs, together with the various formulations of the EP [1035].

*gedanken experiment* and the formulation of Brans-Dicke theory in 1961. This scalar-tensor theory contained a new important feature, the variability of the gravitational "constant".

It is interesting that theories originally developed to better understand the concept of inertia, the most peculiar property of mass, were recovered in the low-energy limit of string theory in the 1980s and, independent of this fact, are used nowadays to cure the shortcomings of standard cosmology and explain the large-scale distribution of matter observed in the universe. This curious fact becomes even more intriguing in view of the dark matter problem. This is why we begin by discussing some of the older features of ETGs: in the following chapters we also consider features of these theories which were discovered more recently and play a role in our current understanding of cosmology.

Although two of the three classical tests of GR were performed only a few years after Einstein's 1916 paper, during the next forty-five years the progress in the area of experimental gravity was very slow, essentially because the technology was inadequate for the detection and the study of the extremely small effects predicted. From the 1960s onward the technological and theoretical efforts in the study of GR were renewed. On the experimental side, the new interest was spurred by astronomical discoveries indicating the role of relativistic gravity in astrophysics and cosmology, and by the technological advances in laboratory and space experiments which made high precision tests possible. Dirac was explicit about the relevance of this technological progress in a 1973 article about his Large Number Hypothesis, in which he stated: "the gravitational constant will be inversely proportional to the epoch, ... moreover the variability of $G$ is now ... not too small to be beyond the capability of present day technology" [395].

On the theoretical side, we also find a similar atmosphere. Theoretical research in GR did not languish as much as its experimental counterpart in the same years, and important results were obtained gaining a better understanding of crucial aspects of GR (*e.g.*, the discovery of the singularity theorems, the Kerr-Newman metric, and black hole thermodynamics). New theories of gravity were also proposed, most notably the Brans-Dicke theory which appeared in 1961 [165]. We do not consider here the progress of standard GR but we focus on those aspects of ETGs which were proposed and studied in the period under consideration, such as:

- Multi-dimensional Kaluza-Klein-type theories proposed to link gravity and microphysical models;
- Attempts to quantize the classical world described by GR;
- The introduction and development of inflationary scenarios.[8]

[8] We do not discuss here the extension to the gravitational interaction of the gauge theory approach developed in relation with the other fundamental interactions. Such an approach is designed to address different problems, for example how to embody particle spin in GR, and the role of torsion (in this case the theory of the gravitational interaction is formulated in a spacetime which differs from the standard one because the connection is not symmetric), also in connection with early universe cosmology, or the relevance of such geometrical ingredients for elementary particles physics (see [583] for an exhaustive exposition of the relevant Einstein-Cartan-Sciama-Kibble theory).

### 1.7.1 *Higher order corrections to Einstein's theory*

Two particular features recur in all the theoretical constructions mentioned above: the first is the role of higher order theories of gravity, while the second is the high relevance of scalar fields in gravity, which is exemplified by (multi-)scalar-tensor theories of gravity. Both types of theories were present in the gravitational arena before 1960 but after that time their relevance grew larger and larger (see [978] for a history of these theories and [128] for a short presentation of Jordan-Fierz-Thiery-Brans-Dicke theories). Concerning the first aspect, we stress that higher order gravity theories necessarily appear when quantization is performed on a curved spacetime and the renormalization problem is addressed [144]. This class of theories also appears in studies of inflation in the early universe (*e.g.*, [550]). The increasing interest in (multi-)scalar-tensor theories is intimately connected with the success that the inflationary paradigm had in cosmology since 1980 and due to the fact that inflation provides very reasonable answers to the puzzles of Standard Big Bang cosmology, as well as providing a physical mechanism for the generation of large scale structures in the universe. Finally, the presence of the scalar field is connected with the relevance of multi-dimensional gravity as an essential ingredient of string and superstring theories, in which dilaton and moduli scalar fields similar to that of scalar-tensor gravity appear in the low-energy limit.

We will discuss higher order gravity as well as string cosmology in the context of the Noether symmetry approach in the following chapters: for the moment, we focus our attention on a bridge between the, apparently different, higher order and scalar-tensor gravities. We begin with the particular case of fourth order theories described by the Lagrangian density

$$\sqrt{-g}\,\mathscr{L} = \sqrt{-g}\, f(R)\,. \tag{1.40}$$

The variation of this Lagrangian with respect to $g^{\mu\nu}$ yields the fourth order field equations

$$f'(R)\, R_{\mu\nu} - \frac{1}{2}\, f(R)\, g_{\mu\nu} - \nabla_\mu \nabla_\nu f'(R) + g_{\mu\nu} \Box f'(R) = 0, \tag{1.41}$$

with $f' \equiv df(R)/dR$. The new set of variables

$$p = f'(R) = f'\left(g_{\mu\nu}, \partial_\sigma g_{\mu\nu}, \partial_\sigma \partial_\rho g_{\mu\nu}\right), \qquad \tilde{g}_{\mu\nu} = p\, g_{\mu\nu}\,, \tag{1.42}$$

links the *Jordan frame* variable $g_{\mu\nu}$ to the *Einstein frame* variables $\left(p, \tilde{g}_{\mu\nu}\right)$ ($p$ must be positive-definite). The widely used word "frame" is rather misleading: it denotes the use of one variable or the others and has nothing to do with choosing some spacetime reference frames (such as, for example, the inertial frames of Special Relativity). The terminology "Einstein frame" comes from the fact that, using the transformation $g \to (p, \tilde{g})$, Eqs. (1.41) are transformed into equations very similar to the Einstein equations of GR. The Einstein frame equations in the absence of ordinary matter ($T^{(m)}_{\mu\nu} = 0$) are

$$\tilde{G}_{\mu\nu} = \frac{1}{p^2}\left[\frac{3}{2}\,p_{,\mu}p_{,\nu} - \frac{3}{4}\,\tilde{g}_{\mu\nu}\tilde{g}^{\alpha\beta}\,p_{,\alpha}p_{,\beta} + \frac{1}{2}\tilde{g}_{\mu\nu}\,(f(R) - Rp)\right]. \tag{1.43}$$

Defining $\varphi = \sqrt{\frac{3}{2}}\ln p$, these turn into

$$\tilde{G}_{\mu\nu} = \nabla_\mu\varphi\nabla_\nu\varphi - \frac{1}{2}\,\tilde{g}_{\mu\nu}\nabla_\alpha\varphi\nabla^\alpha\varphi - \tilde{g}_{\mu\nu}V(\varphi)\,, \tag{1.44}$$

where

$$V(\varphi) = \frac{Rf'(R) - f(R)}{2f'^2(R)}\Big|_{R=R(p(\varphi))}\,. \tag{1.45}$$

In these expressions the function $R = R(p(\varphi))$ is obtained by inverting the relation $p = f'(R)$, which is always possible if $f''(R) \neq 0$. Equation (1.44) can be obtained from the Lagrangian rewritten in terms of $\varphi$ and the tilded geometrical quantities

$$\sqrt{-g}\,\mathcal{L} = \sqrt{-\tilde{g}}\left[\frac{\tilde{R}}{2} - \frac{1}{2}\,\tilde{g}^{\mu\nu}\nabla_\mu\varphi\nabla_\nu\varphi - V(\varphi)\right], \tag{1.46}$$

which has the same form as that of Einstein gravity coupled to a self-interacting minimally coupled scalar field. Looking at Eq. (1.46), it is clear why the set of variables $(\tilde{g}_{\mu\nu}, p)$ is called Einstein frame. The examples below further clarify these considerations.

*Example 1:* $f(R) = R^2\,, \qquad p \equiv f' = 2R.$

In this case $R(p) = p/2$ and the potential

$$V(\varphi) \equiv V(p(\varphi)) = \frac{R(p(\varphi))\,p(\varphi) - f(R(p(\varphi))}{2p^2(\varphi)} = \frac{1}{8} \tag{1.47}$$

is constant.

*Example 2:* $f(R) = R + \alpha R^2\,, \qquad p \equiv f'(R) = 1 + 2\alpha R.$

It is $R(p) = (p-1)/2\alpha$. The potential expressed in terms of $p$ is

$$V(p) = \frac{(p-1)^2}{8\alpha p^2}\,. \tag{1.48}$$

The sign of $V$ depends on $\alpha$; the case $\alpha > 0$ is studied in [1057] while $\alpha < 0$ is discussed in [769] and is subject to a violent local instability [460, 851]. From the cosmological point of view, both cases and the more general situation $f(R) = R + \alpha R^2 + \beta R^N$ (with $N$ a generic integer), are discussed in [246]. This $R^2$ model is considered again later in connection with inflation. As shown by Whitt [1160], the equations of motion arising from the Lagrangian (1.40) with $f(R) = R + \alpha R^2$ coincide with those for a system with the conformally transformed metric

$$\tilde{g}_{\mu\nu} = (1 + 2\alpha R)\, g_{\mu\nu} \,. \tag{1.49}$$

So far we have seen how to go from the Jordan to the Einstein frame: is the inverse procedure always possible? This is a relevant question that will be encountered also in the discussion of scalar-tensor theories. Two points need to be emphasized in this regard: first, beginning with the Einstein frame, it is in principle possible to go to a Jordan-type Lagrangian. Second, when standard matter is present in these models, it is important to look at its dynamics. For example, the photon worldlines are geodesics in the Jordan frame as well as in the Einstein frame, but the case of massive particles is different: their Jordan frame geodesics are no longer transformed into Einstein frame geodesics, and *vice-versa*. In this regard, the two frames are not equivalent (see [769] and the discussion of Chap. 3). The above considerations can easily be extended to theories of order higher than fourth [191, 550].

### 1.7.2 *Minimal and non-minimal coupling and the Equivalence Principle*

Gell-Mann seems to have been the first to introduce the expression "minimal coupling" in connection with the electromagnetic interaction: "We shall postulate a principle that is given wide, though usually tacit acceptance, that of minimal electromagnetic interaction" [532]. What Gell-Mann says concerning this principle is that, given a Lagrangian with all electric charges switched off (all the other effects being included in that Lagrangian), the coupling with the electromagnetic interaction is obtained via the substitution

$$\partial_\mu \rightarrow \partial_\mu + ieA_\mu, \tag{1.50}$$

that is, the electromagnetic interaction is introduced by replacing every partial derivative with the covariant derivative in (1.50) (see also [628]). A similar scheme is used to introduce the gravitational interaction: beginning from the special-relativistic description of the physical interaction (which is equivalent to switching off the gravitational charge), the gravitational interaction is switched on by the substitution

$$\eta_{\mu\nu} \rightarrow g_{\mu\nu}\,, \qquad \partial_\mu \rightarrow \nabla_\mu\,, \qquad \sqrt{-\eta}\, d^4x \rightarrow \sqrt{-g}\, d^4x\,, \tag{1.51}$$

in the Lagrangian ($\eta_{\mu\nu}$ is the flat Minkowski metric and $g_{\mu\nu}$ is the Lorentzian one, while $\eta$ and $g$ are their determinants). An important aspect of these "comma goes to semicolon" and "$\eta_{\mu\nu}$ goes to $g_{\mu\nu}$" rules is the following (*e.g.*, [804, 1139]). Let us consider the curved space Maxwell equations

$$F^{\alpha\beta}{}_{;\beta} = -4\pi J^{\alpha}\,, \qquad F_{\alpha\beta;\gamma} + F_{\beta\gamma;\alpha} + F_{\gamma\alpha;\beta} = 0\,, \tag{1.52}$$

and the four-vector potential $A^{\mu}$ related to the Maxwell field by

$$F_{\alpha\beta} = \nabla_{\alpha} A_{\beta} - \nabla_{\beta} A_{\alpha}\,. \tag{1.53}$$

We have the following problem: using the above-mentioned rule one obtains two possible equations from the first of Eqs. (1.52):

$$A^{\beta;\alpha}{}_{;\beta} - A^{\alpha;\beta}{}_{;\beta} = 4\pi J^{\alpha}\,, \tag{1.54}$$

or

$$A^{\beta;\alpha}{}_{;\beta} - A^{\alpha;\beta}{}_{;\beta} + R^{\alpha}{}_{\beta} A^{\beta} = 4\pi J^{\alpha}\,; \tag{1.55}$$

while the second of Eqs. (1.52) yields, using the Lorentz gauge $\nabla_{\mu} A^{\mu} = 0$,

$$(\Delta_{dR} A)^{\alpha} = 4\pi J^{\alpha}\,, \tag{1.56}$$

where

$$(\Delta_{dR} A)^{\alpha} = -\Box A^{\alpha} + R^{\alpha}{}_{\beta} A^{\beta} \tag{1.57}$$

and $\Delta_{dR}$ is the de Rham vector wave operator. Now the question is: both Maxwell equations for the four-potential $A^{\mu}$ are obtained using the "comma goes to semicolon" rule, but which is the correct one? The answer is: the one obtained using the de Rham operator. As consequence, we see that "correspondence rules" are not sufficient to write down equations in curved space from known physics in flat space when second derivatives are involved (that is, in most situations of physical interest). In such cases, extra caution is needed.

As stressed, for example, in [999] such a prescription is also insufficient in the presence of interactions which do not have a "Minkowskian" counterpart at all, that is, when typical general-relativistic interactions are relevant. These are interactions expressed via the Riemann tensor or some function of it and occur, for example, in the study of the free fall of a particle with spin: the corresponding equations of motion (Papapetrou-Mathisson equations) involve a contribution containing a coupling between the spin tensor and the Riemann tensor [999]. Such a contribution cannot be obtained from the prescriptions given above. This motion is described by the corrected geodesic equation [1153]

$$\frac{d^2 x^{\lambda}}{d\tau^2} + \Gamma^{\lambda}_{\mu\nu} \frac{dx^{\mu}}{d\tau}\frac{dx^{\nu}}{d\tau} + a R^{\lambda}{}_{\mu\nu\alpha} \frac{d\dot{x}_{\mu}}{d\tau}\frac{d\dot{x}_{\nu}}{d\tau} S^{\alpha} = 0 \tag{1.58}$$

($a$ is a parameter measuring the strength of the spin contribution $S^{\alpha}$ coupled with the curvature tensor $R^{\lambda}{}_{\alpha\beta\gamma}$). This replaces the usual geodesic equation

$$\frac{d^2 x^{\lambda}}{d\tau^2} + \Gamma^{\lambda}_{\mu\nu} \frac{dx^{\mu}}{d\tau}\frac{dx^{\nu}}{d\tau} = 0 \tag{1.59}$$

obtained from the special-relativistic equation of motion with gravity switched off

$$\frac{d^2x^\lambda}{d\tau^2} = 0\,, \tag{1.60}$$

by using the previous minimal coupling procedure.

In order to give a satisfactory formulation of the notion of non-minimal coupling, some preliminary steps are needed, namely we have to discuss the Equivalence Principle. An exhaustive treatment can be found in [1167]. The first step in the formulation of this principle is the equivalence between inertial and gravitational mass found already in Galilei' s experiments and in Newton's work,

$$m_I = m_G\,, \tag{1.61}$$

which implies that all bodies fall with the same acceleration independent of their mass and internal structure, in a given gravitational field. This statement is called "universality of free fall" or *Weak Equivalence Principle* (WEP). A more precise statement of the WEP is:

> "*If an uncharged body is placed at an initial event in spacetime and given an initial velocity there, then its subsequent trajectory will be independent of its internal structure and composition*" [1167].

Einstein added to this a new fundamental part: in his formulation, using as an example the famous freely falling elevator, not only the laws of mechanics behave in it as if gravity were absent, but *all* physical laws (except those of gravitational physics) have the same behavior. Following current terminology, we refer to this principle as the *Einstein Equivalence Principle* (EEP). A more precise statement is:

> "*the outcome of any local non-gravitational test experiment is independent of the velocity of the (free falling) apparatus and the outcome of any local non-gravitational test experiment is independent of where and when in the universe it is performed*" [1167].

A "local non-gravitational experiment" is defined as an experiment performed in a small size freely falling laboratory, in order to avoid the inhomogeneities of the external gravitational field, and in which any gravitational self-interaction can be ignored. For example, the measurement of the fine structure constant is a local non-gravitational experiment, while the Cavendish experiment is not.

From the EEP it follows that the gravitational interaction must be described in terms of a curved spacetime, that is, the postulates of the so-called metric theories of gravity have to be satisfied [1167]:

1. Spacetime is endowed with a metric $g_{\mu\nu}$;
2. The world lines of test bodies are geodesics of that metric;
3. In local freely falling frames (called *local Lorentz frames*), the non-gravitational laws of physics are those of Special Relativity.

Both GR and Brans-Dicke-like theories are metric theories of gravity. However, in the context of ETGs, these definitions meant to characterize the most fundamental feature of GR, the Equivalence Principle, and the physical properties discriminating between GR and other metric theories of gravity turn out to ultimately depend on the conformal representation of the theory adopted. More precisely, in scalar-tensor gravity, massive test particles follow geodesics in the Jordan frame, satisfying the WEP, but the same particles deviate from geodesic motion in the Einstein frame (a property referred to as non-metricity of the theory). This difference shows that the EP is formulated in a representation-dependent way [1035]. This serious shortcoming has not yet been addressed properly; for the moment we set aside this problem and proceed. Three comments are necessary at this point:

1. Let us assume that the WEP is violated; of course this implies that also the EEP is violated. Let us assume, for example, that the inertial masses ($m_{Ii}$) in a system differ from the passive ones according to

$$m_{Pi} = m_{Ii}\left(1 + \Sigma_A \eta^A \frac{E_i^A}{m_{Ii}c^2}\right), \tag{1.62}$$

where $E^A$ is the internal energy of the body connected with the A-interaction and $\eta^A$ is a dimensionless parameter quantifying the violation of the WEP. It is then convenient to introduce a new dimensionless parameter (the *Eötvös ratio*) considering, for example, two bodies moving with accelerations

$$a_i = \left(1 + \Sigma_A \eta^A \frac{E_i^A}{m_{Ii}c^2}\right) g \quad (i = 1, 2)\,, \tag{1.63}$$

where $g$ is now the acceleration of gravity. Then we define the Eötvös ratio as

$$\eta = 2\frac{|a_1 - a_2|}{|a_1 + a_2|} = \Sigma_A \eta^A \left|\frac{E_1^A}{m_{I1}c^2} - \frac{E_2^A}{m_{I2}c^2}\right|. \tag{1.64}$$

The measured value of $\eta$ provides information on the WEP-violation parameters $\eta^A$. Experimentally, the equivalence between inertial and gravitational masses is strongly confirmed [1167].
2. The minimal coupling prescriptions given in our previous discussion are connected precisely with the mathematical formulation of the EEP (actually, in order to implement the EEP we have to put in special-relativistic form the laws under consideration and then proceed to find the general-relativistic formulation, switching on gravity. In other words, we have to apply minimal coupling prescriptions with the *caveat* already discussed).
3. The final comment is strictly related with the theories that we will deal with: since we are interested in ETGs, do these theories satisfy the EEP?

In order to address this question we must introduce new concepts and generalize the two principles reported above. First, still following Will [1167], we introduce the notion of "purely dynamical metric theory"; this is a theory such that "the behaviour of each field is influenced to some extent by a coupling to at least one of the other fields in the theory" [1167]. It is obvious that GR is a purely dynamical theory; Brans-Dicke theory also belongs to this class since the equations for the metric involve the scalar field, and *vice-versa*.

Let us consider then an experimental situation such as the Einstein freely falling elevator. We require that the frame used be sufficiently large to encompass a gravitational system (for example, the Cavendish apparatus). The first step is to compute the metric, and this is done in two stages. First we have to assign boundary conditions "far" from the local system, then we have to solve the equations for the fields generated by the local system. Since the metric couples to the other fields of the theory, it will be influenced somehow by these fields, in particular the metric will be related to the boundary values assumed "far away" by these fields. The world surrounding the local gravitating system can influence the metric generated by the local system via the values taken by those other fields on the boundary. Then, local gravitational experiments can depend on where the laboratory is located in the universe, as well as on its velocity relative to the external world. Local non-gravitational experiments are unaffected by such a behavior because they couple only with the metric which is locally Minkowskian. Of course, in a purely metric theory the only field coupling the local system with the environment is $g_{\mu\nu}$ and it is always possible to find a Minkowskian coordinate system at the boundary between the local system and the external world. In this way the asymptotic behavior of the metric is Lorentz-invariant, *i.e.*, independent of the velocity and flat, *i.e.*, independent of the location. The status of Brans-Dicke-like theories is different: in this case it is still possible to choose an asymptotically Minkowskian (Lorentz-invariant) metric which is independent of the velocity and of the scalar field(s), but now the asymptotic value of these scalar(s) can give rise to a dependence on the location of the laboratory. An example of this situation is given by Brans-Dicke-like theories in which the gravitational coupling "constant" actually depends on the asymptotic value assumed by the scalar field.

All these considerations can be summarized in the *Strong Equivalence Principle* (SEP), which states:

*(i)* *"WEP is valid for self-gravitating bodies as well for test bodies;*
*(ii)* *The outcome of any local test experiment is independent of the velocity of the (freely falling) apparatus;*
*(iii)* *The outcome of any local test experiment is independent on where and when in the universe it is performed" [1167].*

The SEP differs from the EEP because of the inclusion of bodies with self-gravitating interactions, such as planets or stars, and because of experiments involving gravitational forces (*e.g.*, the Cavendish experiment). If gravitational forces are ignored, the SEP reduces to the EEP.

Finally, many authors have conjectured that the only theory compatible with the Strong Equivalence Principle is GR, that is,

$$SEP \longrightarrow GR \text{ only}. \tag{1.65}$$

Before concluding our considerations on minimal and non-minimal couplings, we return to what is probably the most important conceptual ingredient of this subject, the Mach principle.

### 1.7.3 *Mach's principle and the variation of G*

Following Bondi [159] there are, at least in principle, two entirely different ways of measuring the rotational velocity of Earth. The first is a purely terrestrial experiment (*e.g.*, a Foucault pendulum), while the second is an astronomical observation consisting of measuring the terrestrial rotation with respect to the fixed stars. In the first type of experiment the motion of the Earth is referred to an idealized inertial frame in which Newton's laws are verified. In the second kind of experiment the frame of reference is connected with a matter distribution surrounding the Earth and the motion of the latter is referred to this matter distribution. In this way we face the problem of Mach's principle, which essentially states that the local inertial frame is determined by some average motion of distant astronomical objects [159, 999].[9] Trying to incorporate Mach's principle into metric gravity, Brans and Dicke constructed a theory alternative to GR [165]. Taking into account the influence that the total matter has at each point (constructing the "inertia"), these two authors introduced, together with the standard metric tensor, a new scalar field of gravitational origin as the effective gravitational coupling. This is why the theory is referred to as a "scalar-tensor" theory; actually, theories in this spirit had already been proposed years earlier by Jordan, Fierz, and Thiery [128]. An important ingredient of this approach is that the gravitational "constant" is actually a function of the total mass distribution and of the scalar field, and is actually variable. In this picture, gravity is described by the Lagrangian density

$$\sqrt{-g}\,\mathscr{L}_{BD} = \sqrt{-g}\left[\phi R - \frac{\omega}{\phi}\,\nabla^{\mu}\phi\nabla_{\mu}\phi + \mathscr{L}^{(m)}\right], \tag{1.66}$$

where $\omega$ is the dimensionless Brans-Dicke parameter and $\mathscr{L}^{(m)}$ is the matter Lagrangian including all the non-gravitational fields. As stressed by Dicke [392], the Lagrangian (1.66) has a property similar to one already discussed in the context of higher order gravity. Under the conformal transformation $g_{\mu\nu} \rightarrow \tilde{g}_{\mu\nu} = \Omega^2 g_{\mu\nu}$ with $\Omega = \sqrt{G_0\varphi}$, the Lagrangian density (1.66) is mapped into

$$\sqrt{-g}\,\mathscr{L} = \sqrt{-\tilde{g}}\left(\tilde{R} + G_0\tilde{\mathscr{L}}^{(m)} + G_0\tilde{\mathscr{L}}^{(\Omega)}\right), \tag{1.67}$$

[9] An interesting discussion on this topic, also connected with different theories of space, both in philosophy and in physics, is found in Dicke's contribution "The Many Faces of Mach" in *Gravitation and Relativity* [391]. This discussion presents also the problematic position that Einstein had on Mach's principle.

where

$$\tilde{\mathscr{L}}^{(\Omega)} = -\frac{(2\omega+3)}{4\pi G_0 \Omega} \nabla^\alpha \left(\sqrt{\Omega}\right) \nabla_\alpha \left(\sqrt{\Omega}\right), \tag{1.68}$$

and $\tilde{\mathscr{L}}^{(m)}$ is the conformally transformed Lagrangian density of matter. In this way the total matter Lagrangian density $\tilde{\mathscr{L}}_{tot} = \tilde{\mathscr{L}}^{(m)} + \tilde{\mathscr{L}}^{(\Omega)}$ has been introduced. The field equations are now written in the form of Einstein-like equations as

$$\tilde{R}_{\mu\nu} - \frac{1}{2} \tilde{g}_{\mu\nu} \tilde{R} = G_0 \tilde{\tau}_{\mu\nu}, \tag{1.69}$$

where the stress-energy tensor is now the sum of two contributions,

$$\tilde{\tau}_{\mu\nu} = T^{(m)}_{\mu\nu} + \Lambda_{\mu\nu}(\Omega). \tag{1.70}$$

Dicke noted that this new (tilded, or Einstein frame) form of the scalar-tensor theory has certain advantages over the theory expressed in the previous (non-tilded, or Jordan frame) form; the Einstein frame representation, being similar to the Einstein standard description is familiar and easier to handle in some respects. But, in this new form, Brans-Dicke theory also exhibits unpleasant features. If we consider the motion of a spinless, electrically neutral, massive particle, we find that in the conformally rescaled world its trajectory is no longer a geodesic. Only null rays are left unchanged by the conformal rescaling. This is a manifestation of the fact that the rest mass is not constant in the conformally transformed world and the equation of motion of massive particles is modified by the addition of an extra force proportional to $\nabla^\mu \Omega$ [392]. Photon trajectories, on the other hand, are not modified because the vanishing of the photon mass implies the absence of a preferred physical scale and photons stay massless under the conformal rescaling, therefore their trajectories are unaffected.

This new approach to gravitation has increased the relevance of theories with varying gravitational coupling. They are of particular interest in cosmology since, as we discuss in detail in the following chapters, they have the potential to circumvent many shortcomings of the standard cosmological model. We list here the Lagrangians of this type which are most relevant for this book.

- The low-energy limit of the bosonic string theory [553,977,1086,1087] produces the Lagrangian density

$$\mathscr{L} = \mathrm{e}^{-2\phi} \left(R + 4g^{\mu\nu} \nabla_\mu \phi \nabla_\nu \phi - \Lambda\right). \tag{1.71}$$

- The general scalar-tensor Lagrangian density [128, 1133] is

$$\mathscr{L}_{ST} = \left[ f(\phi) R - \frac{\omega(\phi)}{2} g^{\alpha\beta} \nabla_\alpha \phi \nabla_\beta \phi - V(\phi) \right], \tag{1.72}$$

where $f(\phi)$ and $\omega(\phi)$ are arbitrary coupling functions and $V(\phi)$ is a scalar field potential. The original Brans-Dicke Lagrangian is contained as the special case $f(\phi) = \phi, \omega(\phi) = \omega_0/\phi$ (with $\omega_0$ a constant), and $V(\phi) \equiv 0$.

- A special case of the previous general theory is that of a scalar field non-minimally coupled to the Ricci curvature, which has received so much attention in the literature to deserve a separate mention,

$$\mathscr{L}_{NMC} = \left( \frac{1}{16\pi G} - \frac{\xi\phi^2}{2} \right) R - \frac{1}{2} g^{\mu\nu} \nabla_\mu \phi \nabla_\nu \phi - V(\phi) \,, \tag{1.73}$$

where $\xi$ is a dimensionless constant. This explicit non-minimal coupling was originally introduced in the context of classical radiation problems [298] and, later, conformal coupling with $\xi = 1/6$ was discovered to be necessary for the renormalizability of the $\lambda\phi^4$ theory on a curved spacetime [144, 205]. The corresponding stress-energy tensor ("improved energy-momentum tensor") and the relevant equations will be discussed later. In particular, the theory is conformally invariant in four dimensions when $\xi = 1/6$ and either $V \equiv 0$ or $V = \lambda\phi^4$ [144, 205, 897, 1139].

All these theories exhibit a non-constant gravitational coupling. The Newton constant $G_N$ is replaced by the effective gravitational coupling

$$G_{eff} = \frac{1}{f(\phi)} \,, \tag{1.74}$$

in Eq. (1.72) which, in general, is different from $G_N$ (we use $\phi$ as the generic function describing the effective gravitational coupling). In string theory or with non-minimally coupled scalars, such functions are specified in (1.71) and (1.73). In particular, in spatially homogeneous and isotropic cosmology, the coupling $G_{eff}$ can only be a function of the epoch, *i.e.*, of the cosmological time.

We stress that all these scalar-tensor theories of gravity do not satisfy the SEP because of the above-mentioned feature: the variation of $G_{eff}$ implies that local gravitational physics depends on the scalar field via $\phi$. We have then motivated the introduction of a stronger version of the Equivalence Principle, the SEP. General theories with such a peculiar aspect are called *non-minimally coupled theories*. This generalizes older terminology in which the expression "non-minimally coupled scalar" referred specifically to the field described by the Lagrangian $\mathscr{L}_{NMC}$ of Eq. (1.73), which is a special case of (1.72).

Let us consider, as in Eq. (1.72), a general scalar-tensor theory in the presence of "standard" matter with total Lagrangian density $\phi R + \mathscr{L}^{(\phi)} + \mathscr{L}^{(m)}$, where $\mathscr{L}^{(m)}$ describes ordinary matter. The dynamical equations for this matter are contained in the covariant conservation equation $\nabla^\nu T^{(m)}_{\mu\nu} = 0$ for the matter stress-energy tensor $T^{(m)}_{\mu\nu}$, which is derived from the variation of the total Lagrangian with respect to $g^{\mu\nu}$. In other words: concerning standard matter, everything goes as in GR (*i.e.*, $\eta_{\mu\nu} \to g_{\mu\nu}$, $\partial_\mu \to \nabla_\mu$) following the minimal coupling prescription. What is new in these theories is the way in which the scalar and the metric degrees of freedom appear: now there is a direct coupling between the scalar degree of freedom and a function of the tensor degree of freedom (the metric) and its derivatives (specifically, with the Ricci scalar of the metric $R\left(g, \partial g, \partial^2 g\right)$). Then, confining our analysis to the cosmological arena, we are presented with two alternatives. The first is

$$\lim_{t \to \infty} G_{eff}\left(\phi(t)\right) = G_N \,, \tag{1.75}$$

in which standard GR cosmology is recovered at the present time in the history of the universe. The second possibility occurs if the gravitational coupling is not constant today, *i.e.*, $G_{eff}$ is still varying with the epoch and $\dot{G}_{eff}/G_{eff}|\text{now}$ (in brief $\dot{G}/G$) is non-vanishing.

In many theories of gravity, then, it is perfectly conceivable that $G_{eff}$ varies with time: in some solutions $G_{eff}$ does not even converge to the value observed today. What do we know, from the observational point of view, about this variability? There are three main avenues to analyze the variability of $G_{eff}$: the first is *lunar laser ranging* (LLR) monitoring the Earth-Moon distance; the second is information from solar astronomy; the third consists of data from binary pulsars. The LLR consists of measuring the round trip travel time and thus the distances between a transmitter and a reflector, and monitoring them over an extended period of time. The change of round trip time contains information about the Earth-Moon system. This round trip travel time has been measured for more than twenty-five years in connection with the Apollo 11, 14, 15, and the Lunakhod 2 lunar missions. Combining these data with those coming from the evolution of the Sun (the luminosity of main sequence stars is quite sensitive to the value of $G$) and the Earth-Mars radar ranging, the current bounds on $\left|\dot{G}/G\right|$ allow at most $0.4 \times 10^{-11}$ to $1.0 \times 10^{-11}$ per year [394]. The third source of information on $G$-variability is given by binary pulsar systems. In order to extract data from this type of system (the prototype is the famous binary pulsar PSR 1913+16 of Hulse and Taylor [1069]), it has been necessary to extend the post-Newtonian approximation, which can be applied only to a (gravitationally) weakly interacting $n$-body system, to (gravitationally) strongly interacting systems. The order of magnitude of $\left|\dot{G}/G\right|$ allowed by these strongly interacting systems is $2 \times 10^{-11}\ \text{yr}^{-1}$ [394].

## 1.8 Extended gravity from higher dimensions and area metric approach

In addition to the reasons that we have already discussed for extending gravity, we mention here additional motivation for Brans-Dicke and scalar-tensor gravity which, although not as compelling, is nevertheless at least of some interest.

It is sometimes remarked that the gravitational scalar field of Brans-Dicke theory has no geometric origin, while in GR the only gravitational field, the metric $g_{\mu\nu}$, has purely geometric character, an aesthetically appealing feature. In actual fact, also the Brans-Dicke scalar can be derived, at least in special cases, from the geometry in Kaluza-Klein and in Lyra's theories and in the area metric approach. These theories, however, exhibit features that are not desirable in the spirit of Mach's principle.

We begin with the well known fact that Brans-Dicke gravity with the special value of the Brans-Dicke parameter $\omega = -\frac{(d-1)}{d}$ can be derived from Kaluza-Klein theory with $d$ extra spatial dimensions, in which the Brans-Dicke scalar originates from the determinant of the metric defined on the submanifold of the extra dimensions [51, 69, 335, 657, 674, 876]. In the simplest version of this theory, the total spacetime $(M \otimes K, \hat{g}_{AB})$ contains a four-dimensional submanifold $M$ with one timelike and three spacelike dimensions and a $d$-dimensional spatial submanifold $K$ $(d \geq 1)$. Denoting $(4+d)$-dimensional quantities with a caret, the $(4+d)$-dimensional metric

$$(\hat{g}_{AB}) = \begin{pmatrix} \hat{g}_{\mu\nu} & 0 \\ 0 & \hat{\phi}_{ab} \end{pmatrix} \tag{1.76}$$

(with $A, B, \ldots = 0, 1, \ldots, (3+d)$, $\mu, \nu, \ldots = 0, 1, 2, 3$, and $a, b, \ldots = 4, 5, \ldots, (3+d)$) is assumed to be diagonal (off-diagonal components appearing in the original theories of Kaluza and Klein [657,674], designed to unify gravity and electromagnetism, generated a gauge field), with $g_{\mu\nu}$ a FLRW metric on $M$ and $\hat{\phi}_{ab}$ a diagonal Riemannian metric on $K$.

By assuming that the extra spatial dimensions curl up on a microscopic scale $l$, the Hilbert-Einstein action integral in $(4+d)$ dimensions *in vacuo*[10]

$$S = \int d^{(4+d)}x\, \sqrt{-\hat{g}}\, \mathscr{L}^{(4+d)} = \frac{1}{16\pi \hat{G}} \int d^{(4+d)}x\, \sqrt{-\hat{g}} \left( \hat{R} + \hat{\Lambda} \right) \tag{1.77}$$

is split into the product of an integral over the four spacetime dimensions and one over the remaining $d$ dimensions. If

$$\varphi \equiv \left| \det(\hat{\phi}_{ab}) \right| \tag{1.78}$$

is the determinant of the metric of the extra dimensions and

$$\rho_{ab} \equiv \varphi^{-1/d} \hat{\phi}_{ab} \tag{1.79}$$

(with $|\det(\rho_{ab})| = 1$), then

$$S = \frac{V^{(l)}}{16\pi \hat{G}} \int d^4x\, \sqrt{-g}\, \sqrt{\varphi} \left[ \left( R + R_K + \hat{\Lambda} \right) + \frac{(d-1)}{4d} \frac{g^{\mu\nu} \nabla_\mu \varphi \nabla_\nu \varphi}{\varphi^2} \right], \tag{1.80}$$

---

[10] Here $\hat{g} = \det(\hat{g}_{AB})$, $\hat{R}$ is the Ricci curvature of $\hat{g}_{AB}$, $\hat{\Lambda}$ and $\hat{G}$ are the $(4+d)$-dimensional cosmological constant and gravitational constant, and $g_{AB} = g_{AB}(x^\mu)$ only.

where $V^{(l)}$ is the volume of the compact manifold $K$ and $R_K$ is the Ricci curvature of $K$. Using $G = \hat{G}/V^{(l)}$ and $\phi \equiv \sqrt{\varphi}$, the action becomes

$$S = \frac{1}{16\pi G} \int d^4x \sqrt{-g} \left[ \phi \left( R + R_K + \hat{\Lambda} \right) + \frac{(d-1)}{d} \frac{g^{\mu\nu} \nabla_\mu \phi \nabla_\nu \phi}{\phi} \right], \quad (1.81)$$

which describes a Brans-Dicke theory with parameter $\omega = -\frac{(d-1)}{d}$ (the scalar $\phi$ does not carry dimensions).[11]

Another possible way of rooting the Brans-Dicke scalar field in the geometry makes use of the derivation of this theory from a Lyra manifold [135, 637, 759, 914, 1003–1006, 1015, 1026]. An $n$-dimensional Lyra manifold $\left(M, \psi, g_{\mu\nu}\right)$ consists of a smooth manifold $M$, a smooth scalar field $\psi$ (the *gauge function* with dimensions of an inverse length), and a *Lyra connection*

$$\Gamma^\gamma_{\alpha\beta} = \frac{1}{\psi} \left\{ {\gamma \atop \alpha\beta} \right\} + \frac{s+1}{\psi^2} g^{\gamma\rho} \left( g_{\beta\rho}\, \partial_\alpha \psi - g_{\alpha\beta}\, \partial_\rho \psi \right), \quad (1.82)$$

where $s$ is a constant and $\left\{ {\gamma \atop \alpha\beta} \right\}$ are the Christoffel symbols of $g_{\mu\nu}$, and where $\nabla_\alpha g_{\mu\nu} = 0$. Torsion

$$T_{\alpha\beta}{}^\gamma = \frac{s}{\psi^2} \left( g^\gamma_\beta \nabla_\alpha \psi - g^\gamma_\alpha \nabla_\beta \psi \right) \quad (1.83)$$

is generated by the torsion potential $\psi$. The dimensionless *Lyra curvature*

$$K_{\rho\alpha\beta}{}^\gamma \equiv \frac{1}{\psi^2} \left[ \partial_\alpha \left( \psi \Gamma^\gamma_{\rho\beta} \right) - \partial_\rho \left( \psi \Gamma^\gamma_{\alpha\beta} \right) + \Gamma^\sigma_{\rho\beta} \Gamma^\gamma_{\sigma\alpha} - \Gamma^\sigma_{\alpha\beta} \Gamma^\gamma_{\sigma\rho} \right] \quad (1.84)$$

and its contractions $K_{\alpha\gamma} \equiv K_{\alpha\beta\gamma}{}^\beta$ and $K \equiv g^{\mu\nu} K_{\mu\nu}$ are analogous to, but distinct from, the Riemann tensor and its contractions and are used to define the gravitational Lyra action

$$S = \int d^4x\, \psi^4 \sqrt{-g}\, K. \quad (1.85)$$

By using the expression [759, 1026]

$$K = \frac{R}{\psi^2} + \frac{2(s+1)}{\psi^3} (1-n) \Box \psi$$
$$+ \frac{1}{\psi^4} \left[ (s+1)^2 \left(3n - n^2 - 2\right) - 2(s+1)(2-n) \right] \nabla^\alpha \psi\, \nabla_\alpha \psi \quad (1.86)$$

[11] This derivation of Brans-Dicke theory can be used as a solution-generating technique [138].

with $n = 4$, discarding a total divergence, and setting $\phi \equiv \psi^2$ one obtains [1026] the Brans-Dicke action

$$S = \int d^4x \sqrt{-g} \left( \phi R - \frac{\omega}{\phi} g^{\mu\nu} \nabla_\mu \phi \nabla_\nu \phi \right) . \tag{1.87}$$

The Lyra action contains no dimensional coupling and agrees with the idea that a fundamental theory could be scale-invariant whereas dimensional effective couplings arise as vacuum expectation values of certain fields, as exemplified by Sakharov's induced gravity [960]. In Lyra's theory, it is matter that breaks scale invariance. A mass or potential for the Brans-Dicke field, however, does not arise naturally in the Lyra geometry (it would be akin to a graviton mass in GR), while an exponential potential arises naturally in Kaluza-Klein compactifications described in the Einstein frame.

Another way of giving a geometric meaning to the scalar field of Brans-Dicke theory is through the area metric approach in which both $g_{\mu\nu}$ and $\phi$ derive from a higher rank geometric structure [916–919, 996]. A Lorentzian *area metric* on a smooth manifold is a non-degenerate rank four tensor field $G^{\alpha\beta\gamma\delta}$ with the symmetries

$$G^{\alpha\beta\gamma\delta} = G^{\gamma\delta\alpha\beta} , \tag{1.88}$$

$$G^{\beta\alpha\gamma\delta} = -G^{\alpha\beta\gamma\delta} , \tag{1.89}$$

$$G^{\alpha\beta\delta\gamma} = -G^{\alpha\beta\gamma\delta} . \tag{1.90}$$

The inverse area metric $G_{\alpha\beta\gamma\delta}$ is such that $G^{\alpha\beta\mu\nu} G_{\mu\nu\gamma\delta} = 4\delta_\gamma^{[\alpha} \delta_\delta^{\beta]}$. If $X^\alpha$ and $Y^\beta$ are two vectors at the same spacetime point spanning a parallelogram, its area is given by $\sqrt{G_{\alpha\beta\delta\gamma} X^\alpha Y^\beta X^\gamma Y^\delta}$. An *almost metric* area metric has the structure

$$G^{\alpha\beta\delta\gamma} = g^{\alpha\gamma} g^{\beta\delta} - g^{\alpha\delta} g^{\beta\gamma} + \frac{h(\phi)}{\sqrt{-g}} \varepsilon^{\alpha\beta\gamma\delta} , \tag{1.91}$$

where $\varepsilon^{\alpha\beta\gamma\delta}$ is the Levi-Civita symbol and $\phi$ a scalar field. If $h(\phi) = 0$ at a point, then the area $\sqrt{G_{\alpha\beta\delta\gamma} X^\alpha Y^\beta X^\gamma Y^\delta}$ coincides with the one derived from the metric $g_{\mu\nu}$. An area metric volume form $\chi_G$ and an area metric curvature scalar $R_G$ can be constructed so that the analogue of the Hilbert-Einstein action

$$S_G = \int \chi_G R_G \tag{1.92}$$

can be varied with respect to $G^{\alpha\beta\gamma\delta}$ yielding the field equation of vacuum $\omega = 0$ Brans-Dicke theory with $h^2 = (2\kappa\phi)^{-1} - 1$ [917, 918, 996]. This technique singles out $\omega = 0$ Brans-Dicke theory *in vacuo*. However, when matter is included in the picture by adding the corresponding action, as in

$$S = \int \chi_G R_G + S^{(m)} \equiv S_G + S^{(m)}, \tag{1.93}$$

self-consistency of the theory imposes restrictions on the coupling of matter to the area metric. Specifically, if $K_{\alpha\beta\gamma\delta} = (\det(G))^{-1/6} \frac{\delta(\kappa S_G)}{\delta G^{\alpha\beta\gamma\delta}}$, the field equations assume the form

$$K_{\alpha\beta\gamma\delta} = T_{\alpha\beta\gamma\delta}, \tag{1.94}$$

where $T_{\alpha\beta\gamma\delta}$ is the generalized energy-momentum tensor of matter on an area metric manifold

$$\begin{aligned} T_{\alpha\beta\gamma\delta} &= -(\det(G))^{-1/6} \frac{\delta S^{(m)}}{\delta G^{\alpha\beta\gamma\delta}} \\ &= 2\, T^{(m)}_{[\alpha[\gamma} g_{\delta]\beta]} - \frac{1}{3} T^{(m)} g_{\alpha[\gamma} g_{\delta]\beta} - \frac{1}{24} \alpha \sqrt{-g}\, \varepsilon_{\alpha\beta\gamma\delta}, \end{aligned} \tag{1.95}$$

where $T^{(m)} = g^{\mu\nu} T^{(m)}_{\mu\nu}$ is the trace of the usual symmetric energy-momentum tensor $T^{(m)}_{\mu\nu}$ and $\alpha$ is a scalar weighting the antisymmetric part of $T_{\alpha\beta\gamma\delta}$. In general, only restricted forms of matter possess a $T_{\alpha\beta\gamma\delta}$ with such properties (perfect fluids do, and have been used in area metric cosmology and in the weak-field limit [996]). The modified $\omega = 0$ Brans-Dicke field equations are

$$G_{\mu\nu} = \frac{1}{\phi}\left(\nabla_\mu \nabla_\nu \phi - \frac{1}{2} g_{\mu\nu} \Box\phi\right) + \kappa\left(4T^{(m)}_{\mu\nu} - \frac{h(\phi)\alpha}{2} g_{\mu\nu}\right), \tag{1.96}$$

$$\Box\phi = \frac{4\kappa}{3}\phi T^{(m)} + \frac{1 - 8\kappa^2\phi^2}{6\sqrt{1 - 4\kappa^2\phi^2}}\alpha, \tag{1.97}$$

which reduce to the usual $\omega = 0$ Brans-Dicke field equations *in vacuo* ($T^{(m)}_{\mu\nu}=0, \alpha = 0$) and clearly exhibit a non-standard coupling between matter and gravity. It is claimed that this non-standard coupling makes the theory compatible with the experimental limits on the first order PPN parameters [918] (remember that in pure Brans-Dicke theory without a mass or a potential for the scalar field, $\omega$ cannot be zero but is forced to have values $|\omega| > 40{,}000$ [133]).

Finally, we mention that scalar-tensor gravity occurs also in brane-world scenarios in which, contrary to string theories, extra spatial dimensions are allowed to be much larger than the Planck scale (see, *e.g.*, [168, 526]).

## 1.9 Conclusions

We have not yet presented the specific contents of the models that we are going to study: this will be done in the forthcoming chapters. Before going into details, we want the reader to have a general idea of the conceptual novelty and the qualitative features that ETGs have in comparison with GR, and to understand the motivation and the historical developments of theoretical physics in the recent past that stimulate research beyond Einstein gravity.

It seems reasonable, to say the least, to enlarge GR to more general schemes because in this way it is possible to explain several theoretical and observational facts which otherwise call for rather *ad hoc* explanations. As we shall see in the following, cosmology is a field which has seen many fruitful applications of these generalizations of Einstein gravity. Following sheer curiosity early on, high energy physics and attempts to renormalize gravity have provided much of the original motivation to extend gravity. We have not yet reached any definitive conclusion on what is the "correct" theory of gravity. It is quite possible that all the theories formulated so far eventually prove to be wrong, and they are known to be left wanting in many regards. Here we identify the search for "the theory of gravity" as a pressing problem of theoretical and experimental physics. The theories developed thus far, and those currently under development, should probably not be taken too seriously, but they are useful at least as toy models to learn how gravity could be different from Einstein's theory and to get a glimpse of the difficulties and phenomena one could expect in a more advanced theory. Moreover, the reader should not forget that much of the recent interest has been motivated by new experimental data such as the observations of type Ia supernovae and that much speculation is simply ruled out by the observations. It is with this understanding that we proceed to look more closely at the terrain not covered by Einstein in his times.

# Chapter 2
# Mathematical tools

*With my full philosophical rucksack I can only climb slowly up the mountain of mathematics.*
*– Ludwig Wittgenstein*

In this chapter we discuss certain mathematical tools which are used extensively in the following chapters. Some of these concepts and methods are part of the standard baggage taught in undergraduate and graduate courses, while others enter the toolbox of more advanced researchers. These mathematical methods are very useful in formulating ETGs and in finding analytical solutions. We begin by studying conformal transformations, which allow for different representations of scalar-tensor and $f(R)$ theories of gravity, in addition to being useful in GR. We continue by discussing variational principles in GR, which are the basis for presenting ETGs in the following chapters. We close the chapter with a discussion of Noether symmetries, which are used elsewhere in this book to obtain analytical solutions.

## 2.1 Conformal transformations

A mathematical tool that has proved very useful in alternative gravitational theories as well as in GR is that of conformal transformations (see [467, 472, 769] for reviews). The idea is to perform a conformal rescaling of the spacetime metric $g_{\mu\nu} \to \tilde{g}_{\mu\nu}$. Often a scalar field is present in the theory and the metric rescaling is accompanied by a (nonlinear) redefinition of this field $\phi \to \tilde{\phi}$. New dynamical variables $(\tilde{g}_{\mu\nu}, \tilde{\phi})$ are thus obtained. The scalar field redefinition serves the purpose of casting the kinetic energy density of this field in canonical form. The new set of variables $(\tilde{g}_{\mu\nu}, \tilde{\phi})$ is called the *Einstein conformal frame*, while $(g_{\mu\nu}, \phi)$ constitute the *Jordan frame*. When a scalar degree of freedom $\phi$ is present in the theory, as in scalar-tensor or $f(R)$ gravity, it generates the transformation to the Einstein frame in the sense that the rescaling is completely determined by a function of $\phi$. In principle, infinitely many conformal frames could be introduced, giving rise to as many representations of the theory. From the physical point of view, these different representations have been the subject of many debates and misinterpretations, which will be discussed later. For the moment we expose the mathematical technique.

S. Capozziello and V. Faraoni, *Beyond Einstein Gravity: A Survey of Gravitational Theories for Cosmology and Astrophysics*, Fundamental Theories of Physics 170,
DOI 10.1007/978-94-007-0165-6_2, © Springer Science+Business Media B.V. 2011

Let the pair $(M, g_{\mu\nu})$ be a spacetime, with $M$ a smooth manifold of dimension $n \geq 2$ and $g_{\mu\nu}$ a Lorentzian or Riemannian metric on $M$. The point-dependent rescaling of the metric tensor

$$g_{\mu\nu} \longrightarrow \tilde{g}_{\mu\nu} = \Omega^2 g_{\mu\nu} \,, \tag{2.1}$$

where the *conformal factor* $\Omega(x)$ is a nowhere vanishing, regular[1] function, is called a *Weyl* or *conformal* transformation. Due to this metric rescaling, the lengths of spacelike and timelike intervals and the norms of spacelike and timelike vectors are changed, while null vectors and null intervals of the metric $g_{\mu\nu}$ remain null in the rescaled metric $\tilde{g}_{\mu\nu}$. The light cones are left unchanged by the transformation (2.1) and the spacetimes $(M, g_{\mu\nu})$ and $(M, \tilde{g}_{\mu\nu})$ exhibit the same causal structure; the converse is also true [1139]. A vector that is timelike, spacelike, or null with respect to the metric $g_{\mu\nu}$ has the same character with respect to $\tilde{g}_{\mu\nu}$, and *vice-versa*.

In the Arnowitt-Deser-Misner (ADM) [54] decomposition of the metric

$$g_{\mu\nu}\, dx^\mu\, dx^\nu = -\left(N^2 - N_i N^i\right) dt^2 + 2N_j dt\, dx^j + h_{ij} dx^i\, dx^j \tag{2.2}$$

using the lapse function $N$ and the shift vector $N^i$, the transformation properties of these quantities follow from Eq. (2.1):

$$\tilde{N} = \Omega\, N \,, \qquad \tilde{N}^i = N^i \,, \qquad \tilde{h}_{ij} = \Omega^2 h_{ij} \,. \tag{2.3}$$

The ADM mass of an asymptotically flat spacetime [54] does not change under the conformal transformation and scalar field redefinition [282].

The transformation properties of various geometrical quantities are useful [1065, 1139]. We list them here, leaving their proof to the reader as an exercise:

$$\tilde{g}^{\mu\nu} = \Omega^{-2}\, g^{\mu\nu} \,, \qquad \tilde{g} = \Omega^{2n}\, g \tag{2.4}$$

for the inverse metric and the metric determinant,

$$\tilde{\Gamma}^\alpha_{\beta\gamma} = \Gamma^\alpha_{\beta\gamma} + \Omega^{-1}\left(\delta^\alpha_\beta \nabla_\gamma \Omega + \delta^\alpha_\gamma \nabla_\beta \Omega - g_{\beta\gamma} \nabla^\alpha \Omega\right) \tag{2.5}$$

for the Christoffel symbols,

$$\begin{aligned} \widetilde{R_{\alpha\beta\gamma}{}^\delta} = R_{\alpha\beta\gamma}{}^\delta &+ 2\,\delta^\delta_{[\alpha} \nabla_{\beta]} \nabla_\gamma (\ln \Omega) - 2g^{\delta\sigma} g_{\gamma[\alpha} \nabla_{\beta]} \nabla_\sigma (\ln \Omega) \\ &+ 2\nabla_{[\alpha} (\ln \Omega)\, \delta^\delta_{\beta]} \nabla_\gamma (\ln \Omega) - 2\nabla_{[\alpha} (\ln \Omega)\, g_{\beta]\gamma}\, g^{\delta\sigma}\, \nabla_\sigma (\ln \Omega) \\ &- 2g_{\gamma[\alpha} \delta^\delta_{\beta]}\, g^{\sigma\rho} \nabla_\sigma (\ln \Omega)\, \nabla_\rho (\ln \Omega) \end{aligned} \tag{2.6}$$

[1] See [171, 172, 180] for the possibility of continuation beyond singular points of the conformal factor.

for the Riemann tensor,

$$\tilde{R}_{\alpha\beta} = R_{\alpha\beta} - (n-2)\nabla_\alpha \nabla_\beta (\ln \Omega) - g_{\alpha\beta}\, g^{\rho\sigma}\, \nabla_\sigma \nabla_\rho (\ln \Omega) + (n-2)\, \nabla_\alpha (\ln \Omega) \nabla_\beta (\ln \Omega) - (n-2)\, g_{\alpha\beta}\, g^{\rho\sigma}\, \nabla_\rho (\ln \Omega) \nabla_\sigma (\ln \Omega) \tag{2.7}$$

for the Ricci tensor, and

$$\tilde{R} \equiv \tilde{g}^{\alpha\beta}\, \tilde{R}_{\alpha\beta} = \Omega^{-2} \left[ R - 2\,(n-1)\, \Box\, (\ln \Omega) - (n-1)\,(n-2)\, \frac{g^{\alpha\beta}\, \nabla_\alpha \Omega\, \nabla_\beta \Omega}{\Omega^2} \right] \tag{2.8}$$

for the Ricci scalar. In the case of $n = 4$ spacetime dimensions, the transformation property of the Ricci scalar can be written as

$$\tilde{R} = \Omega^{-2} \left[ R - \frac{6 \Box \Omega}{\Omega} \right] = \Omega^{-2} \left[ R - \frac{12 \Box (\sqrt{\Omega})}{\sqrt{\Omega}} + \frac{3 g^{\alpha\beta} \nabla_\alpha \Omega \nabla_\beta \Omega}{\Omega^2} \right] . \tag{2.9}$$

The Weyl tensor ${C_{\alpha\beta\gamma}}^{\delta}$ with the last index contravariant is conformally invariant,

$$\widetilde{{C_{\alpha\beta\gamma}}^{\delta}} = {C_{\alpha\beta\gamma}}^{\delta} , \tag{2.10}$$

but the same tensor with indices raised or lowered with respect to ${C_{\alpha\beta\gamma}}^{\delta}$ is not. This property explains the name *conformal tensor* sometimes used for ${C_{\alpha\beta\gamma}}^{\delta}$ [749]. If the original metric $g_{\alpha\beta}$ is Ricci-flat (*i.e.*, $R_{\alpha\beta} = 0$), the conformally transformed metric $\tilde{g}_{\alpha\beta}$ is not (cf. Eq. (2.7)). In the conformally transformed world the conformal factor $\Omega$ plays the role of an effective form of matter and this fact has consequences for the physical interpretation of the theory. A vacuum metric in the Jordan frame is not such in the Einstein frame, and the interpretation of what is matter and what is gravity becomes frame-dependent [1035]. However, if the Weyl tensor vanishes in one frame, it also vanishes in the conformally related frame. Conformally flat metrics are mapped into conformally flat metrics, a property used in cosmology when mapping FLRW universes (which are conformally flat) into each other. In particular, de Sitter spaces with scale factor $a(t) = a_0 \exp{(H_0 t)}$ and a constant scalar field as the material source are mapped into similar de Sitter spaces.

Since, in general, tensorial quantities are not invariant under conformal transformations, neither are the tensorial equations describing geometry and physics. An equation involving a tensor field $\psi$ is said to be *conformally invariant* if there exists a number $w$ (the *conformal weight* of $\psi$) such that, if $\psi$ is a solution of a tensor equation with the metric $g_{\mu\nu}$ and the associated geometrical quantities, $\tilde{\psi} \equiv \Omega^w \psi$ is a solution of the corresponding equation with the metric $\tilde{g}_{\mu\nu}$ and the associated geometry.

In addition to geometric quantities, one needs to consider the behavior of common forms of matter under conformal transformations. It goes without saying that most forms of matter or fields are not conformally invariant: invariance under conformal transformations is a very special property. In general, the covariant conservation equation for a (symmetric) stress-energy tensor $T_{\alpha\beta}^{(m)}$ representing ordinary matter,

$$\nabla^{\beta} T_{\alpha\beta}^{(m)} = 0 \tag{2.11}$$

is not conformally invariant [1139]. The conformally transformed $\tilde{T}_{\alpha\beta}^{(m)}$ satisfies the equation

$$\tilde{\nabla}^{\beta} \tilde{T}_{\alpha\beta}^{(m)} = -\tilde{T}^{(m)} \tilde{\nabla}_{\alpha} (\ln \Omega) \, . \tag{2.12}$$

Clearly, the conservation equation (2.11) is conformally invariant only for a matter component that has vanishing trace $T^{(m)}$ of the energy-momentum tensor. This feature is associated with light-like behavior; examples are the electromagnetic field and a radiative fluid with equation of state $P^{(m)} = \rho^{(m)}/3$. Unless $T^{(m)} = 0$, Eq. (2.12) describes an exchange of energy and momentum between matter and the scalar field $\Omega$, reflecting the fact that matter and the geometric factor $\Omega$ are directly coupled in the Einstein frame description.

Since the geodesic equation ruling the motion of free particles in GR can be derived from the conservation equation (2.11) (*geodesic hypothesis*), it follows that timelike geodesics of the original metric $g_{\alpha\beta}$ are not geodesics of the rescaled metric $\tilde{g}_{\alpha\beta}$ and *vice-versa*. Particles in free fall in the world $(M, g_{\alpha\beta})$ are subject to a force proportional to the gradient $\tilde{\nabla}^{\alpha}\Omega$ in the rescaled world $(M, \tilde{g}_{\alpha\beta})$ – this is often identified as a fifth force acting on all massive particles and, therefore, it can be said that no massive test particles exist in the Einstein frame. The stress-energy tensor definition in terms of the matter action $S^{(m)} = \int d^4x \sqrt{-g}\, \mathscr{L}^{(m)}$,

$$\tilde{T}_{\alpha\beta}^{(m)} = \frac{-2}{\sqrt{-\tilde{g}}} \frac{\delta\left(\sqrt{-\tilde{g}}\, \mathscr{L}^{(m)}\right)}{\delta \tilde{g}^{\alpha\beta}} \, , \tag{2.13}$$

together with the rescaling (2.1) of the metric, yields

$$\tilde{T}_{\alpha}\beta^{(m)} = \Omega^{-2}\, T_{\alpha\beta}^{(m)} \, , \;\; \widetilde{T_{\alpha}{}^{\beta\,(m)}} = \Omega^{-4}\, T_{\alpha}{}^{\beta\,(m)} \, , \;\; \tilde{T}^{\alpha\beta} = \Omega^{-6}\, T^{\alpha\beta\,(m)} \, , \tag{2.14}$$

and

$$\tilde{T}^{(m)} = \Omega^{-4}\, T^{(m)} \, . \tag{2.15}$$

The last equation makes it clear that the trace vanishes in the Einstein frame if and only if it vanishes in the Jordan frame.

**Perfect fluids.** Now let us consider the stress-energy tensor of a perfect fluid,[2]

$$T^{(m)}_{\alpha\beta} = \left(P^{(m)} + \rho^{(m)}\right) u_\alpha u_\beta + P^{(m)} g_{\alpha\beta} : \tag{2.16}$$

the corresponding tensor in the rescaled world is

$$\tilde{T}^{(m)}_{\alpha\beta} = \left(\tilde{P}^{(m)} + \tilde{\rho}^{(m)}\right) \tilde{u}_\alpha \tilde{u}_\beta + \tilde{P}^{(m)} \tilde{g}_{\alpha\beta} , \tag{2.17}$$

where the four-velocity $\tilde{u}^\mu$ of the fluid satisfies

$$\tilde{g}_{\alpha\beta}\, \tilde{u}^\alpha \tilde{u}^\beta = -1. \tag{2.18}$$

Together with the metric rescaling (2.1), this normalization gives the transformation properties of the fluid four-velocity and of its inverse, which are widely used in the literature,

$$\tilde{u}^\mu = \Omega^{-1}\, u^\mu , \qquad \tilde{u}_\mu = \Omega\, u_\mu . \tag{2.19}$$

By comparing Eqs. (2.14) and (2.17) and using Eq. (2.19), one obtains

$$\left(\tilde{P}^{(m)} + \tilde{\rho}^{(m)}\right) \tilde{u}_\alpha \tilde{u}_\beta + \tilde{P}^{(m)} \tilde{g}_{\alpha\beta} = \Omega^{-2} \left[\left(P^{(m)} + \rho^{(m)}\right) u_\alpha u_\beta \right. \\ \left. + P^{(m)} g_{\alpha\beta}\right] , \tag{2.20}$$

and the transformation properties of the energy density and pressure of the fluid under the conformal transformation (2.1) are

$$\tilde{\rho}^{(m)} = \Omega^{-4}\, \rho^{(m)} , \qquad \tilde{P}^{(m)} = \Omega^{-4}\, P^{(m)} . \tag{2.21}$$

If, in the Jordan frame, the fluid has a barotropic equation of state of the form

$$P^{(m)} = (\gamma - 1)\, \rho^{(m)} \tag{2.22}$$

with $\gamma$ =constant, then the same equation of state is valid in the Einstein frame thanks to the relations (2.21) between $\rho^{(m)}$, $P^{(m)}$ and their conformal cousins $\tilde{\rho}^{(m)}$ and $\tilde{P}^{(m)}$. However, this property does not hold true for a more general barotropic equation of state $P = P(\rho)$ which is not of the form (2.22).

---

[2] See [317] for the transformation properties of an imperfect fluid under a conformal transformation.

In the case of FLRW metrics the usual Jordan frame conservation equation for a fluid

$$\frac{d\rho^{(m)}}{dt} + 3H\left(P^{(m)} + \rho^{(m)}\right) = 0 \tag{2.23}$$

is modified in the Einstein frame to

$$\frac{d\tilde{\rho}^{(m)}}{dt} + 3\tilde{H}\left(\tilde{P}^{(m)} + \tilde{\rho}^{(m)}\right) = \left(3\tilde{P}^{(m)} - \tilde{\rho}^{(m)}\right)\frac{d\left(\ln \Omega\right)}{dt}, \tag{2.24}$$

as follows from Eq. (2.12).

Let us now review some fundamental fields:

**The Klein-Gordon field.** The source-free Klein-Gordon equation $\Box\phi = 0$ in the absence of self-interactions is not conformally invariant. However, its generalization

$$\Box\phi - \frac{n-2}{4(n-1)} R\phi = 0 \tag{2.25}$$

for $n \geq 2$ is conformally invariant [298, 898, 1139]. It is reasonable to allow for the possibility that the scalar $\phi$ acquires a mass or other potential at high energies and, accordingly, in particle physics and in cosmology it is customary to introduce a potential energy density $V(\phi)$ for the Klein-Gordon scalar. We have already discussed how a non-minimal coupling between $\phi$ and the Ricci curvature arises. Taking both of these into account, the relevant equation for $\phi$ becomes

$$\Box\phi - \xi\, R\phi - \frac{dV}{d\phi} = 0\,, \tag{2.26}$$

where $\xi$ is the dimensionless coupling constant. The introduction of non-minimal coupling with $\xi \neq 0$ makes the theory a scalar-tensor one.

Equation (2.26) is conformally invariant in four spacetime dimensions if $\xi = 1/6$ and $V = 0$ or $V = \lambda\phi^4$ [205, 898, 1139]. Even a constant potential $V$, equivalent to a cosmological constant, corresponds to an effective mass for the scalar (not to be identified with a real mass [464]) which breaks conformal invariance [762].

Although unintuitive, it is not difficult to understand why a quartic potential preserves conformal invariance on the basis of dimensional considerations. Conformal invariance corresponds to the absence of a characteristic length (or mass) scale in the physics. In general, the potential $V(\phi)$ contains dimensional parameters (such as the mass $m$ in $V = m^2\phi^2/2$) but, when $V = \lambda\phi^4$, the dimension of $V$ (a mass to the fourth power) is carried by $\phi^4$ and the self-coupling constant $\lambda$ is dimensionless, *i.e.*, there is no scale associated to $V$ in this case.

**The Maxwell field.** In four spacetime dimensions the Maxwell equations are conformally invariant, while the equation satisfied by the electromagnetic four-potential $A^{\mu}$,

$$\Box A_{\mu} - R^{\nu}{}_{\mu} A_{\nu} = -4\pi j_{\mu} \tag{2.27}$$

(where $j^\mu$ is the four-current) is not [112, 353]. However, this quantity is gauge-dependent and is not an observable. As already discussed, quantum corrections to classical electrodynamics, including the generation of mass terms and the conformal anomaly, break the conformal invariance.

**Higher spin fields.** The conditions for conformal invariance of fields of arbitrary spin in general spacetime dimensions are varied and, generally, complicated; we refer the reader to [623].

## 2.2 Variational principles in General Relativity

Variational principles are used to formulate the equations of motion of particles and fields in theoretical physics, and GR is no exception. We first discuss the variational principle for test particles and then the one leading to the Einstein equations.

### 2.2.1 *Geodesics*

In GR the spacetime metric is related to geodesic motion because the Equivalence Principle requires that the motion of a point-like body in free fall be described by the geodesic equation. The latter can be derived from the variational principle

$$\delta S = \delta \int_A^B ds = 0\,, \tag{2.28}$$

where $ds$ is the line element and $A$ and $B$ are the initial and final points along the spacetime trajectory, respectively. The line element is written as

$$ds = \left| g_{\alpha\beta}\, dx^\alpha dx^\beta \right|^{1/2} = \left| g_{\alpha\beta} \frac{dx^\alpha}{ds} \frac{dx^\beta}{ds} \right|^{1/2} ds\,, \tag{2.29}$$

with $s$ playing the role of an affine parameter, and from which it follows that

$$g_{\alpha\beta} u^\alpha u^\beta = -1\,, \tag{2.30}$$

where $u^\alpha = \dfrac{dx^\alpha}{ds}$ is the four-velocity of the particle. Substitution into Eq. (2.28) yields

$$\delta S = \delta \int_A^B \left| g_{\alpha\beta} \frac{dx^\alpha}{ds} \frac{dx^\beta}{ds} \right|^{1/2} ds = 0\,. \tag{2.31}$$

By performing this variation, one obtains

$$\delta S = \int_A^B \frac{1}{2\sqrt{|g_{\alpha\beta}u^\alpha u^\beta|}} \left[ g_{\alpha\beta,\lambda}\delta x^\lambda \frac{dx^\alpha}{ds}\frac{dx^\beta}{ds} + 2g_{\alpha\beta}\frac{d}{ds}(\delta x^\alpha)\frac{dx^\beta}{ds} \right] ds = 0 \,. \tag{2.32}$$

The second term in square brackets is $g_{\alpha\beta}\delta\left(\frac{dx^\alpha}{ds}\frac{dx^\beta}{ds}\right)$ as a consequence of the fact that $\delta(ds) = d(\delta s)$, hence

$$g_{\alpha\beta}\delta\left(\frac{dx^\alpha}{ds}\frac{dx^\beta}{ds}\right) = g_{\alpha\beta}\frac{dx^\alpha}{ds}\delta\left(\frac{dx^\beta}{ds}\right) + g_{\alpha\beta}\frac{dx^\beta}{ds}\delta\left(\frac{dx^\alpha}{ds}\right) = 2g_{\alpha\beta}\frac{dx^\beta}{ds}\frac{d}{ds}(\delta x^\alpha) \,. \tag{2.33}$$

Using $g_{\alpha\beta}u^\alpha u^\beta = -1$, it is

$$\delta S = \int_A^B \frac{1}{2}\left[ g_{\alpha\beta,\lambda}\delta x^\lambda \frac{dx^\alpha}{ds}\frac{dx^\beta}{ds} + 2g_{\alpha\beta}\frac{dx^\beta}{ds}\frac{d}{ds}(\delta x^\alpha) \right] ds = 0 \tag{2.34}$$

and integration by parts of the second term yields

$$\delta S = \int_A^B \frac{1}{2}\left( g_{\alpha\beta,\lambda}\delta x^\lambda \frac{dx^\alpha}{ds}\frac{dx^\beta}{ds} \right) ds + \left[ g_{\alpha\beta}\frac{dx^\beta}{ds}\delta x^\alpha \right]_A^B - \int_A^B \frac{d}{ds}\left( g_{\alpha\beta}\frac{dx^\beta}{ds} \right)\delta x^\alpha ds = 0 \,. \tag{2.35}$$

By imposing that, at the endpoints, it is $\delta x^\alpha(A) = \delta x^\alpha(B) = 0$, the second term vanishes and

$$\delta S = \int_A^B \frac{1}{2}\left( g_{\alpha\beta,\lambda}\frac{dx^\alpha}{ds}\frac{dx^\beta}{ds}\delta x^\lambda \right) ds - \int_A^B \left( g_{\alpha\beta}\frac{d^2x^\beta}{ds^2} + g_{\alpha\beta,\lambda}\frac{dx^\lambda}{ds}\frac{dx^\beta}{ds} \right)\delta x^\alpha ds = 0 \,. \tag{2.36}$$

This equation can be written as

$$\delta S = \int_A^B \left[ \left( \frac{1}{2}g_{\alpha\beta,\lambda} - g_{\lambda\beta,\alpha} \right)\frac{dx^\alpha}{ds}\frac{dx^\beta}{ds} - g_{\lambda\beta}\frac{d^2x^\beta}{ds^2} \right]\delta x^\lambda ds = 0 \,. \tag{2.37}$$

This integral vanishes for all variations $\delta x^\lambda$ with fixed endpoints if

$$g_{\lambda\beta}\frac{d^2x^\beta}{ds^2} = \left( \frac{1}{2}g_{\alpha\beta,\lambda} - g_{\lambda\beta,\alpha} \right)u^\alpha u^\beta \,. \tag{2.38}$$

Since

$$g_{\lambda\beta,\alpha}u^{\alpha}u^{\beta} = g_{\lambda\alpha,\beta}u^{\beta}u^{\alpha} = \frac{1}{2}\left(g_{\lambda\beta,\alpha} + g_{\lambda\alpha,\beta}\right)u^{\alpha}u^{\beta}\,, \tag{2.39}$$

whereas

$$g_{\lambda\beta}\frac{d^2x^{\beta}}{ds^2} = \frac{1}{2}\left(g_{\alpha\beta,\lambda} - g_{\lambda\beta,\alpha} - g_{\lambda\alpha,\beta}\right)u^{\alpha}u^{\beta} \tag{2.40}$$

we have

$$\{\lambda,\alpha\beta\} = \frac{1}{2}\left(g_{\lambda\alpha,\beta} + g_{\lambda\beta,\alpha} - g_{\alpha\beta,\lambda}\right) \tag{2.41}$$

and

$$g_{\lambda\beta}\frac{d^2x^{\beta}}{ds^2} + \{\lambda,\alpha\beta\}\,u^{\alpha}u^{\beta} = 0\,. \tag{2.42}$$

Multiplying by $g^{\lambda\tau}$ and remembering that

$$g^{\lambda\tau}g_{\lambda\beta} = \delta^{\tau}_{\beta}\,, \qquad g^{\lambda\tau}\{\lambda,\alpha\beta\} = \Gamma^{\tau}_{\alpha\beta}\,, \tag{2.43}$$

one has

$$\frac{d^2x^{\tau}}{ds^2} + \Gamma^{\tau}_{\alpha\beta}u^{\alpha}u^{\beta} = 0\,, \tag{2.44}$$

which is the geodesic equation describing the free fall motion of a point-like body in the gravitational field $\Gamma^{\tau}_{\alpha\beta}$.

### 2.2.2 *Field equations*

The Einstein equations or the gravitational field equations of any ETG can be derived from a variational principle. Of course, the description is more involved than for point particles because we are discussing a field theory, *i.e.*, a distributed physical system with an infinite number of degrees of freedom. We illustrate the derivation of the Einstein field equations *in vacuo* as the starting point.

Let us consider

$$\delta\int d\Omega\sqrt{-g}\,\mathscr{L} = 0\,, \tag{2.45}$$

where $\sqrt{-g}\,d\Omega$ is the invariant volume element and $\mathscr{L}$ is the desired Lagrangian density. In fact, under the coordinate transformation $\overline{x}^{\alpha} \rightarrow x^{\alpha} = x^{\alpha}\left(\overline{x}^{\mu}\right)$, where $\overline{x}^{\mu}$ are the "initial" local coordinates, we have

$$d\Omega = J\,d\overline{\Omega}\,, \qquad J = \det\left(\frac{\partial x^{\alpha}}{\partial \overline{x}^{\mu}}\right), \tag{2.46}$$

with $J$ the Jacobian determinant of the transformation. Moreover, we have

$$\overline{g}_{\alpha\beta} = \text{diag}\ (-1, 1, 1, 1)\,, \tag{2.47}$$

$$\overline{g}_{\alpha\beta} = \frac{\partial x^\mu}{\partial \overline{x}^\alpha}\frac{\partial x^\nu}{\partial \overline{x}^\beta}\, g_{\mu\nu}\,, \tag{2.48}$$

$\overline{g} = -1 = J^2 g$ and, therefore,

$$d\overline{\Omega} = \frac{d\Omega}{J} = \sqrt{-g}\, d\Omega\,. \tag{2.49}$$

Since we want the Euler-Lagrange equations deriving from the variational principle to be of second order, the Lagrangian must be quadratic in the first order derivatives of $g_{\mu\nu}$. These first order derivatives contain the Christoffel symbols, which are not coordinate-invariant. Then we have to choose for the Lagrangian density $\mathscr{L}$ expressions containing higher order derivatives and, *a priori*, this brings the danger that the field equations could become of order higher than second (we will discuss in detail this point for ETGs). The obvious choice of Hilbert and Einstein for the Lagrangian density $\mathscr{L}$ was the Ricci scalar curvature $R$. The variational principle is then

$$\delta \int d\Omega\, \sqrt{-g}\, R = 0\,. \tag{2.50}$$

The relations

$$\delta g = g\, g^{\mu\nu}\delta g_{\mu\nu} = -g\, g_{\mu\nu}\delta g^{\mu\nu} \tag{2.51}$$

yield

$$\delta\left(\sqrt{-g}\right) = -\frac{\delta g}{2\sqrt{-g}} = -\frac{1}{2}\sqrt{-g}\, g_{\alpha\beta}\delta g^{\alpha\beta}\,, \tag{2.52}$$

from which it follows that

$$\begin{aligned}&\int \left[\left(\delta\sqrt{-g}\right) R + \sqrt{-g}\, R_{\mu\nu}\delta g^{\mu\nu} + \sqrt{-g}\, g^{\mu\nu}\delta R_{\mu\nu}\right] d\Omega \\ &\quad = \int \sqrt{-g}\, \delta g^{\mu\nu}\left[R_{\mu\nu} - \tfrac{1}{2} R g_{\mu\nu}\right] d\Omega + \int \sqrt{-g}\, g^{\mu\nu}\delta R_{\mu\nu} d\Omega = 0\,.\end{aligned} \tag{2.53}$$

The second integral can be evaluated in the local inertial frame, obtaining

$$R_{\mu\nu}(0) = \Gamma^\alpha_{\mu\nu,\alpha} - \Gamma^\alpha_{\mu\alpha,\nu}\,, \tag{2.54}$$

$$\delta R_{\mu\nu}(0) = \frac{\partial}{\partial x^\alpha}\left(\delta\Gamma^\alpha_{\mu\nu}\right) - \frac{\partial}{\partial x^\nu}\left(\delta\Gamma^\alpha_{\mu\alpha}\right)\,, \tag{2.55}$$

$$g^{\mu\nu}(0)\delta R_{\mu\nu}(0) = g^{\mu\nu}(0)\,\frac{\partial}{\partial x^{\alpha}}\left(\delta\Gamma^{\alpha}_{\mu\nu}\right) - g^{\mu\nu}(0)\,\frac{\partial}{\partial x^{\nu}}\left(\delta\Gamma^{\alpha}_{\mu\alpha}\right)$$
$$= g^{\mu\nu}(0)\,\frac{\partial}{\partial x^{\rho}}\left(\delta\Gamma^{\rho}_{\mu\nu}\right) - g^{\mu\rho}(0)\,\frac{\partial}{\partial x^{\rho}}\left(\delta\Gamma^{\alpha}_{\mu\alpha}\right)$$
$$= \frac{\partial}{\partial x^{\rho}}\left[g^{\mu\nu}(0)\delta\Gamma^{\rho}_{\mu\nu} - g^{\mu\rho}(0)\delta\Gamma^{\alpha}_{\mu\alpha}\right]. \tag{2.56}$$

Then, we can write

$$g^{\mu\nu}(0)\delta R_{\mu\nu}(0) = \frac{\partial W^{\rho}}{\partial x^{\rho}}\,, \qquad W^{\rho} = g^{\mu\nu}(0)\delta\Gamma^{\rho}_{\mu\nu} - g^{\mu\rho}(0)\delta\Gamma^{\alpha}_{\mu\alpha}\,. \tag{2.57}$$

The second integral in Eq. (2.53) can be discarded since its argument is a pure divergence; in fact, in general coordinates it is

$$\int\sqrt{-g}\,g^{\mu\nu}\,\delta R_{\mu\nu}d\Omega = \int\sqrt{-g}\,\frac{\partial W^{\rho}}{\partial x^{\rho}}\,d\Omega$$
$$= \int\sqrt{-g}\,W^{\rho}{}_{;\rho}d\Omega = \int\frac{\partial}{\partial x^{\rho}}\left(\sqrt{-g}\,W^{\rho}\right)d\Omega = 0, \tag{2.58}$$

and then

$$\int\sqrt{-g}\,\delta g^{\mu\nu}\left(R_{\mu\nu} - \frac{1}{2}Rg_{\mu\nu}\right)d\Omega = 0\,, \tag{2.59}$$

from which we obtain the vacuum field equations of GR

$$G_{\mu\nu} \equiv R_{\mu\nu} - \frac{1}{2}\,g_{\mu\nu}R = 0 \tag{2.60}$$

as Euler-Lagrange equations of the Hilbert-Einstein action. *Vice-versa*, starting from Eq. (2.60) and retracing the previous steps in inverse order (*i.e.*, integrating the Einstein equations), one can re-obtain the Hilbert-Einstein action (2.50), thus demonstrating the equivalence between this action and the field equations (2.60). Introducing matter fields as sources is straightforward, producing Eqs. (1.8) as a result.

## 2.3 Adding torsion

Several questions of interest in modern physics could depend on the fact that GR is a classical theory that does not include ultraviolet quantum effects. Quantum effects should be considered in any theory dealing with gravity at a fundamental level and even in effective theories. Assuming a $\mathbf{U}_4$ manifold instead of the usual $\mathbf{V}_4$ (see below) is a straightforward generalization of GR which attempts to include fields with non-zero spin in the geometrical framework of GR. The Einstein-Cartan-Sciama-Kibble (ECSK) theory is one of the most serious attempts in this direction [584]. However, the mere inclusion of spin matter fields does not exhaust the role of torsion, which can give important contributions in any fundamental theory. For example, a torsion field appears in (super)string theories if we consider the fundamental string modes. One needs at least a scalar and two tensor modes, a symmetric and an antisymmetric one. In the low-energy limit, the latter is a torsion field [553].

Several attempts to unify gravity with electromagnetism have taken into account torsion in four- and higher-dimensional theories such as Kaluza-Klein models [700]. Any theory of gravity incorporating twistors needs to include torsion [601], while supergravity is the natural arena in which torsion, curvature, and matter fields enter on the same footing [888].

Several authors agree that curvature and torsion could play various roles in the cosmological dynamics at both early and late epochs. In fact, the interplay of curvature and torsion produces naturally repulsive contributions to the energy-momentum tensor, hence cosmological models become singularity-free and accelerating [385].

All these reasons suggest considering torsion in any comprehensive theory of gravity which takes into account non-gravitational fundamental interactions. However, in most papers in the literature, a clear distinction between the different kinds of torsion is not made. Usually torsion is simply related to the spin density of matter but, very often, it assumes more general meanings. There are more than one independent torsion tensors with different properties [240]. The problem of extending GR to actions more general than the Hilbert-Einstein one is naturally related to the consideration of torsion. In this section we illustrate the general features of torsion and the associated quantities defined in $\mathbf{U}_4$ spacetimes [584]. This formalism can be applied, in general, to any alternative theory of gravity.

The torsion tensor $S_{\mu\nu}{}^{\rho}$ is the antisymmetric part of the affine connection coefficients $\Gamma^{\rho}_{\mu\nu}$,

$$S_{\mu\nu}{}^{\rho} = \frac{1}{2}\left(\Gamma^{\rho}_{\mu\nu} - \Gamma^{\rho}_{\nu\mu}\right) \equiv \Gamma^{\rho}_{[\mu\nu]} \,. \tag{2.61}$$

In GR it is postulated that $S_{\mu\nu}{}^{\rho} = 0$. It is a general convention to call $\mathbf{U}_4$ a four-dimensional spacetime manifold endowed with metric and torsion, while four-dimensional manifolds with metric and without torsion are labelled $\mathbf{V}_4$. In general, torsion occurs in linear combinations as the *contortion tensor*

$$K_{\mu\nu}{}^{\rho} = -S_{\mu\nu}{}^{\rho} - S^{\rho}{}_{\mu\nu} + S_{\nu}{}^{\rho}{}_{\mu} = -K_{\mu}{}^{\rho}{}_{\nu} \,, \tag{2.62}$$

and the *modified torsion tensor*

$$T_{\mu\nu}{}^{\rho} = S_{\mu\nu}{}^{\rho} + 2\delta_{[\mu}{}^{\rho} S_{\nu]} \,, \tag{2.63}$$

where $S_{\mu} \equiv S_{\mu\nu}{}^{\nu}$. According to these definitions, the affine connection can be written as

$$\Gamma^{\rho}_{\mu\nu} = \left\{{}^{\rho}_{\mu\nu}\right\} - K_{\mu\nu}{}^{\rho} \,, \tag{2.64}$$

where $\left\{ {\rho \atop \mu\nu} \right\}$ are the usual Christoffel symbols of the symmetric connection. The presence of torsion in the affine connection implies that the covariant derivatives of a scalar field $\phi$ do not commute, *i.e.*,

$$\tilde{\nabla}_{[\mu}\tilde{\nabla}_{\nu]}\phi = -S_{\mu\nu}{}^{\rho}\tilde{\nabla}_{\rho}\phi\,. \tag{2.65}$$

For a vector $v^a$ and a covector $w_a$, the relations

$$(\tilde{\nabla}_{\mu}\tilde{\nabla}_{\nu} - \tilde{\nabla}_{\nu}\tilde{\nabla}_{\mu})v^{\rho} = R_{\mu\nu\alpha}{}^{\rho}v^{\alpha} - 2S_{\mu\nu}{}^{\alpha}\tilde{\nabla}_{\alpha}v^{\rho} \tag{2.66}$$

and

$$(\tilde{\nabla}_{\mu}\tilde{\nabla}_{\nu} - \tilde{\nabla}_{\nu}\tilde{\nabla}_{\mu})w_{\rho} = R_{\mu\nu\rho}{}^{\alpha}w_{\alpha} - 2S_{\mu\nu}{}^{\alpha}\tilde{\nabla}_{\alpha}w_{\rho} \tag{2.67}$$

hold. The torsion contribution to the Riemann tensor $R_{\mu\nu\rho}{}^{\sigma}$ is given explicitly by

$$R_{\mu\nu\rho}{}^{\sigma} = R_{\mu\nu\rho}{}^{\sigma}(\{\}) - \nabla_{\mu}K_{\nu\rho}{}^{\sigma} + \nabla_{\nu}K_{\mu\rho}{}^{\sigma} + K_{\mu\beta}{}^{\sigma}K_{nu\rho}{}^{\beta} - K_{\nu\beta}{}^{\sigma}K_{\mu\rho}{}^{\beta}\,, \tag{2.68}$$

where $R_{\mu\nu\rho}{}^{\sigma}(\{\})$ is the tensor generated by the Christoffel symbols. The symbols $\tilde{\nabla}_{\mu}$ and $\nabla_{\mu}$ denote the covariant derivative operators with and without torsion, respectively. Using Eq. (2.68), the expressions for the Ricci tensor and the curvature scalar are

$$R_{\mu\rho} = R_{\mu\rho}(\{\}) - 2\nabla_{\mu}S_{\rho} + \nabla_{\nu}K_{\mu\rho}{}^{\nu} + K_{\mu\beta}{}^{\nu}K_{\nu\rho}{}^{\beta} - 2S_{\beta}K_{\mu\rho}{}^{\beta} \tag{2.69}$$

and

$$R = R(\{\}) - 4\nabla_{\mu}S^{\mu} + K_{\rho\beta\nu}K^{\nu\rho\beta} - 4S_{\mu}S^{\mu}\,. \tag{2.70}$$

Torsion can be decomposed with respect to the Lorentz group into three irreducible tensors

$$S_{\mu\nu}{}^{\rho} = {}^{T}S_{\mu\nu}{}^{\rho} + {}^{A}S_{\mu\nu}{}^{\rho} + {}^{V}S_{\mu\nu}{}^{\rho}\,, \tag{2.71}$$

where

$${}^{A}S_{\mu\nu}{}^{\rho} = g^{\rho\sigma}S_{[\mu\nu\sigma]} \tag{2.72}$$

is called the axial (or totally antisymmetric) torsion and

$${}^{T}S_{\mu\nu}{}^{\rho} = S_{\mu\nu}{}^{\rho} - {}^{A}S_{\mu\nu}{}^{\rho} - {}^{V}S_{\mu\nu}{}^{\rho} \tag{2.73}$$

is the traceless non-totally antisymmetric part of torsion. Torsion has 24 components, of which ${}^{T}S_{\mu\nu}$ has 16 components, ${}^{A}S_{\mu\nu}$ has 4, and ${}^{V}S_{\mu\nu}$ has the remaining 4. It is also

$${}^{V}S_{\mu\nu}{}^{\rho} = \frac{1}{3}\left(S_{\mu}\delta^{\rho}_{\nu} - S_{\nu}\delta^{\rho}_{\mu}\right). \tag{2.74}$$

It is clear that relating torsion to the spin density of matter is only one of its possible applications [240].

## 2.4 Noether symmetries

The celebrated theorem of Emmy Noether states that conserved quantities in the dynamics of a physical system are related to the existence of symmetries and cyclic variables in its Lagrangian [53, 774, 808]. Here we review the Noether symmetry approach for dynamical systems with a finite number of degrees of freedom. We will use it later to obtain exact solutions of ETGs.

Let $L\left(q^i, \dot{q}^i\right)$ be a canonical, non-degenerate, point-like Lagrangian satisfying

$$\frac{\partial L}{\partial \lambda} = 0, \qquad \det\left(H_{ij}\right) \equiv \det\left\|\frac{\partial^2 L}{\partial \dot{q}^i \partial \dot{q}^j}\right\| \neq 0, \tag{2.75}$$

where $H_{ij}$ is the Hessian matrix of $L$ and an overdot denotes differentiation with respect to the affine parameter $\lambda$ (which usually corresponds to the time $t$). In the Lagrangian formalism for point particles and rigid bodies, the Lagrangian $L$ assumes the form

$$L = T(\mathbf{q}, \dot{\mathbf{q}}) - V(\mathbf{q}), \tag{2.76}$$

where $T$ and $V$ are the kinetic and potential energies, respectively. $T$ is a quadratic form of the $\dot{q}^i$. The Hamiltonian associated with $L$ is

$$E_L \equiv \frac{\partial L}{\partial \dot{q}^i}\dot{q}^i - L, \tag{2.77}$$

it coincides with the total energy $T + V$, and is a constant of motion. Any smooth and invertible transformation[3] of the generalized coordinates $q^i \rightarrow Q^i(\mathbf{q})$ induces a transformation of the generalized velocities

$$\dot{q}^i \rightarrow \dot{Q}^i(\mathbf{q}) = \frac{\partial Q^i}{\partial q^j}\dot{q}^j. \tag{2.78}$$

We assume that the Jacobian matrix $\mathcal{J} = ||\partial Q^i / \partial q^j||$ of the coordinate transformation does not vanish. The Jacobian $\tilde{\mathcal{J}}$ of the "induced" transformation is easily derived and $\mathcal{J} \neq 0$ implies that $\tilde{\mathcal{J}} \neq 0$. In general, this transformation is local because the condition $\mathcal{J} \neq 0$ cannot be satisfied on the entire space but only in the neighbourhood of a given point. If the transformation is extended to the maximal submanifold on which $\mathcal{J} \neq 0$, problems can arise for the whole manifold due to the possibility of different topologies [808].

A point transformation $Q^i = Q^i(\mathbf{q})$ can depend on one or more parameters. Let us assume that a point transformation depends on a parameter $\varepsilon$, $Q^i = Q^i(\mathbf{q}, \varepsilon)$, and that it defines a one-parameter Lie group. For infinitesimal values of $\varepsilon$, the transformation is then generated by a vector field. Examples are the vector field $\partial/\partial x$

[3] Here we consider only point transformations.

associated with a translation along the $x$-axis, and the field $x(\partial/\partial y) - y(\partial/\partial x)$ associated with a rotation about the $z$-axis. In general, an infinitesimal point transformation is represented by a generic vector field on $Q$

$$\mathbf{X} = \alpha^i(\mathbf{q}) \frac{\partial}{\partial q^i} \,. \tag{2.79}$$

The induced transformation (2.78) is then represented by

$$\mathbf{X}^{(c)} = \alpha^i(\mathbf{q}) \frac{\partial}{\partial q^i} + \left(\frac{d}{d\lambda}\alpha^i(\mathbf{q})\right) \frac{\partial}{\partial \dot{q}^j} \,. \tag{2.80}$$

The vector field $\mathbf{X}^{(c)}$ is called the *complete lift* of $\mathbf{X}$ [808]. A function $f(\mathbf{q}, \dot{\mathbf{q}})$ is invariant under the transformation $\mathbf{X}^{(c)}$ if

$$\pounds_{\mathbf{X}^{(c)}} f \equiv \alpha^i(\mathbf{q}) \frac{\partial f}{\partial q^i} + \left(\frac{d}{d\lambda}\alpha^i(\mathbf{q})\right) \frac{\partial f}{\partial \dot{q}^j} = 0 \,, \tag{2.81}$$

where $\pounds_{\mathbf{X}^c} f$ is the Lie derivative of $f$ along $\mathbf{X}^{(c)}$. If, in particular, $\pounds_{\mathbf{X}^{(c)}} L = 0$, then $\mathbf{X}^{(c)}$ is said to be a *symmetry* for the dynamics described by $L$.

In order to fully flesh out the relation between Noether's theorem and cyclic variables, let us consider a Lagrangian $L$ yielding the Euler-Lagrange equations

$$\frac{d}{d\lambda}\left(\frac{\partial L}{\partial \dot{q}^j}\right) - \frac{\partial L}{\partial q^j} = 0 \tag{2.82}$$

and the vector field (2.80). By contracting Eq. (2.82) with the $\alpha^i$'s, one obtains

$$\alpha^j \left[\frac{d}{d\lambda}\left(\frac{\partial L}{\partial \dot{q}^j}\right) - \frac{\partial L}{\partial q^j}\right] = 0 \,. \tag{2.83}$$

By using the fact that (as follows from Eq. (2.83))

$$\alpha^j \frac{d}{d\lambda}\left(\frac{\partial L}{\partial \dot{q}^j}\right) = \frac{d}{d\lambda}\left(\alpha^j \frac{\partial L}{\partial \dot{q}^j}\right) - \left(\frac{d\alpha^j}{d\lambda}\right) \frac{\partial L}{\partial \dot{q}^j} \,, \tag{2.84}$$

one obtains that

$$\frac{d}{d\lambda}\left(\alpha^i \frac{\partial L}{\partial \dot{q}^i}\right) = \pounds_{\mathbf{X}} L \,. \tag{2.85}$$

For brevity, from now on we abuse notations when there is no possibility of confusion and we write $\mathbf{X}$ instead of $\mathbf{X}^{(c)}$. A straightforward consequence of Eq. (2.85) is the

**Noether theorem:**

*If* $\pounds_{\mathbf{X}} L = 0$, *then the function*

$$\Sigma_0 = \alpha^i \frac{\partial L}{\partial \dot{q}^i} \tag{2.86}$$

*is a constant of motion.*

A few remarks are in order. First, Eq. (2.86) can be expressed in a coordinate-independent way as the contraction of **X** with the Cartan one-form

$$\theta_L \equiv \frac{\partial L}{\partial \dot{q}^i}\, dq^i \,. \tag{2.87}$$

Given a generic vector field $\mathbf{Y} = y^i \partial/\partial x^i$ and a one-form $\beta = \beta_i dx^i$ it is, by definition,

$$i_{\mathbf{Y}} \beta = y^i \beta_i \tag{2.88}$$

and Eq. (2.86) can then be written as

$$i_{\mathbf{X}} \theta_L = \Sigma_0 \,. \tag{2.89}$$

Using a point transformation, the vector field **X** is rewritten as

$$\widetilde{\mathbf{X}} = \left( i_{\mathbf{X}} dQ^k \right) \frac{\partial}{\partial Q^k} + \left[ \frac{d}{d\lambda} \left( i_x dQ^k \right) \right] \frac{\partial}{\partial \dot{Q}^k} \,, \tag{2.90}$$

hence $\widetilde{\mathbf{X}}'$ is still the lift of a vector field defined on the configuration space. If **X** is a symmetry and a point transformation is chosen such that

$$i_{\mathbf{X}} dQ^1 = 1 \,, \quad i_{\mathbf{X}} dQ^i = 0 \quad (i \neq 1) \,, \tag{2.91}$$

it follows that

$$\widetilde{\mathbf{X}} = \frac{\partial}{\partial Q^1} \,, \qquad \frac{\partial L}{\partial Q^1} = 0 \,. \tag{2.92}$$

Therefore, $Q^1$ is a cyclic coordinate and the dynamics can be reduced [53, 774].

The coordinate transformation (2.91) is not unique and a clever choice can be very advantageous. Moreover, the solution of Eq. (2.91) is, in general, not defined on the entire space but only locally, as noted above. It is possible that multiple vector fields **X** are found, say $\mathbf{X}_1$ and $\mathbf{X}_2$. If these commute, $[\mathbf{X}_1, \mathbf{X}_2] = 0$, then two cyclic coordinates can be found by solving the system

$$i_{\mathbf{X}_1} dQ^1 = 1 \,, \quad i_{\mathbf{X}_2} dQ^2 = 1 \,, \quad i_{\mathbf{X}_1} dQ^i = 0 \;\; (i \neq 1) \,, \quad i_{\mathbf{X}_2} dQ^i = 0 \quad (i \neq 2) \,. \tag{2.93}$$

The transformed fields are then $\partial/\partial Q^1$ and $\partial/\partial Q^2$. If $\mathbf{X}_1$ and $\mathbf{X}_2$ do not commute, this procedure cannot be applied, as is clear from the fact that diffeomorphisms preserve the commutation relations. To proceed, let us note that the commutator $\mathbf{X}_3 = [\mathbf{X}_1, \mathbf{X}_2]$ is also a symmetry because

$$\pounds_{\mathbf{X}_3} L = \pounds_{\mathbf{X}_1} \pounds_{\mathbf{X}_2} L - \pounds_{\mathbf{X}_2} \pounds_{\mathbf{X}_1} L = 0 . \tag{2.94}$$

If $\mathbf{X}_3$ does not depend on $\mathbf{X}_1$ and $\mathbf{X}_2$, the procedure is repeated until the vector fields close the Lie algebra. The usual way to treat this situation consists of performing a Legendre transformation to switch to the Hamiltonian formalism and to a Lie algebra of Poisson brackets. If a reduction to cyclic coordinates is sought for, this procedure can be achieved by:

1. choosing arbitrarily one of the symmetries or a linear combination of them and obtaining new coordinates as above. After the reduction, the new Lagrangian $\widetilde{L}(\mathbf{Q})$ is obtained.
2. Repeating the search for symmetries in this new space, performing a new reduction, and repeating this procedure until possible.
3. If the search for symmetries fails, another attempt is made with a different existing symmetry.

Let us now assume that $L$ is of the form (2.76). Since $\mathbf{X}$ is of the form (2.80), $\pounds_{\mathbf{X}} L$ will consists of the sum of a second degree homogeneous polynomial in the velocities and of an inhomogeneous term in $q^i$. Since such a polynomial must vanish identically, all its coefficients vanish. If the configuration space has dimension $n$, one obtains $1 + \dfrac{n(n+1)}{2}$ partial differential equations; the system is then overdetermined and, if any solution exists, it must be expressed in terms of integration constants instead of boundary conditions. Clearly, an overall constant factor in the Lie vector $\mathbf{X}$ is irrelevant.

The Noether approach will be used in Chaps. 4 and 8 to obtain exact solutions with symmetries of ETGs.

## 2.5 Conclusions

Armed with the mathematical tools described in this chapter, we are now ready to explore in more detail the landscape of gravitational theories that lie beyond Einstein's GR. These theories are conveniently described in terms of their actions satisfying the variational principle, and the search for analytical solutions can be performed using Noether symmetries. In addition, general solutions in cosmology can be discussed using qualitative analysis, which is presented in Chap. 6.

# Chapter 3
# The landscape beyond Einstein gravity

*What is a scientist after all? It is a curious man looking through a keyhole, the keyhole of nature, trying to know what's going on.*
*– Jacques Cousteau*

The two main classes of ETGs considered in this book, scalar-tensor and $f(R)$ gravity, are the subject of much of this chapter. After exposing the metric formalism, due consideration is given to the Palatini version of $f(R)$ theories, emphasizing its bimetric nature. Specifically, we present the actions describing ETGs, derive the field equations from a variational principle, and then discuss their different conformal representations. In this chapter the emphasis is on the general structure of these theories, while their application to astrophysics and cosmology is studied in later chapters.

As is the case for GR, alternative theories of gravity are best expressed using actions and variational principles for the degrees of freedom that they contain. In this chapter we discuss the action and field equations of Brans-Dicke theory first: this is the prototype of scalar-tensor theories and, historically, was the first complete and successful alternative to GR. Then, metric $f(R)$ gravity is presented, beginning with the case of quadratic corrections to the Hilbert-Einstein Lagrangian which was employed in the first scenario of inflation in the early universe [1044]. The discussion of more general ETGs follows. We then examine the different conformal representations of ETGs and discuss the effective equation of state appearing in these theories and their initial value problem.

## 3.1 The variational principle and the field equations of Brans-Dicke gravity

The Brans-Dicke theory of gravity [165, 490, 648, 649] is the prototype of gravitational theories alternative to GR. The action in the Jordan frame (the set of variables $(g_{\mu\nu}, \phi)$) is

$$S_{(BD)} = \frac{1}{16\pi} \int d^4x \sqrt{-g} \left[ \phi R - \frac{\omega}{\phi} g^{\mu\nu} \nabla_\mu \phi \nabla_\nu \phi - V(\phi) \right] + S^{(m)} , \quad (3.1)$$

S. Capozziello and V. Faraoni, *Beyond Einstein Gravity: A Survey of Gravitational Theories for Cosmology and Astrophysics*, Fundamental Theories of Physics 170,
DOI 10.1007/978-94-007-0165-6_3, 

where

$$S^{(m)} = \int d^4x \sqrt{-g}\, \mathcal{L}^{(m)} \tag{3.2}$$

is the action of ordinary matter and $\omega$ is the dimensionless Brans-Dicke parameter. The factor $\phi$ in the denominator of the kinetic term of $\phi$ in the action (3.1) is purely conventional and has the only purpose of making $\omega$ dimensionless. Matter does not couple directly to $\phi$, *i.e.*, the Lagrangian density $\mathcal{L}^{(m)}$ is independent of $\phi$ ("minimal coupling" of matter). However, $\phi$ couples directly to the Ricci scalar. The gravitational field is described by both the metric tensor $g_{\mu\nu}$ and the Brans-Dicke scalar $\phi$ which, together with the matter variables, constitute the degrees of freedom of the theory. As usual for scalar fields, the potential $V(\phi)$ generalizes the cosmological constant and may reduce to a constant, or to a mass term.[1]

As discussed in Chap. 1, the original motivation for introducing Brans-Dicke theory was the implementation of Mach's principle. This is achieved in Brans-Dicke theory by making the effective gravitational coupling strength $G_{\mathit{eff}} \sim \phi^{-1}$ depend on the spacetime position and being governed by distant matter sources, as in Eq. (3.9) below. As already remarked, modern interest in Brans-Dicke and scalar-tensor theories is motivated by the fact that they are obtained as low-energy limits of string theories.

The variation of the action (3.1) with respect to $g^{\mu\nu}$ and the well known properties [705]

$$\delta\left(\sqrt{-g}\right) = -\frac{1}{2}\sqrt{-g}\, g_{\mu\nu}\, \delta g^{\mu\nu}\,, \tag{3.3}$$

$$\delta\left(\sqrt{-g}\, R\right) = \sqrt{-g}\left(R_{\mu\nu} - \frac{1}{2}\, g_{\mu\nu} R\right)\delta g^{\mu\nu} \equiv \sqrt{-g}\, G_{\mu\nu}\, \delta g^{\mu\nu}\,, \tag{3.4}$$

yield the field equation

$$G_{\mu\nu} = \frac{8\pi}{\phi}\, T^{(m)}_{\mu\nu} + \frac{\omega}{\phi^2}\left(\nabla_\mu\phi\,\nabla_\nu\phi - \frac{1}{2} g_{\mu\nu}\nabla^\alpha\phi\nabla_\alpha\phi\right)$$
$$+\frac{1}{\phi}\left(\nabla_\mu\nabla_\nu\phi - g_{\mu\nu}\,\Box\phi\right) - \frac{V}{2\phi}\, g_{\mu\nu}\,, \tag{3.5}$$

where

$$T^{(m)}_{\mu\nu} \equiv \frac{-2}{\sqrt{-g}}\,\frac{\delta}{\delta g^{\mu\nu}}\left(\sqrt{-g}\,\mathcal{L}^{(m)}\right) \tag{3.6}$$

---

[1] Due to the particular equation (3.9) satisfied by the Brans-Dicke field $\phi$, its mass is not the coefficient of the quadratic term in the expansion of $V(\phi)$, as for minimally coupled scalar fields, but rather the quantity $m$ defined by $m^2 = \frac{1}{2\omega+3}\left(\phi\,\frac{d^2V}{d\phi^2} - \frac{dV}{d\phi}\right)$ [443].

is the energy-momentum tensor of ordinary matter. By varying the action with respect to $\phi$, one obtains

$$\frac{2\omega}{\phi}\Box\phi + R - \frac{\omega}{\phi^2}\nabla^\alpha\phi\nabla_\alpha\phi - \frac{dV}{d\phi} = 0\,. \tag{3.7}$$

Taking now the trace of Eq. (3.5),

$$R = \frac{-8\pi\, T^{(m)}}{\phi} + \frac{\omega}{\phi^2}\nabla^\alpha\phi\nabla_\alpha\phi + \frac{3\Box\phi}{\phi} + \frac{2V}{\phi}\,, \tag{3.8}$$

and using the resulting Eq. (3.8) to eliminate $R$ from Eq. (3.7) leads to

$$\Box\phi = \frac{1}{2\omega+3}\left(8\pi\, T^{(m)} + \phi\frac{dV}{d\phi} - 2V\right)\,. \tag{3.9}$$

According to this equation, the scalar $\phi$ is sourced by non-conformal matter (*i.e.*, by matter with trace $T^{(m)} \neq 0$), however the scalar does not couple directly to $\mathscr{L}^{(m)}$: the Brans-Dicke scalar $\phi$ reacts on ordinary matter only indirectly through the metric tensor $g_{\mu\nu}$, as dictated by Eq. (3.5). The term proportional to $\phi\, dV/d\phi - 2V$ on the right hand side of Eq. (3.9) vanishes if the potential has the form $V(\phi) = m^2\phi^2/2$ familiar from the Klein-Gordon equation and from particle physics.

The action (3.1) and the field equation (3.5) suggest that the field $\phi$ be identified with the inverse of the effective gravitational coupling

$$G_{\mathit{eff}}(\phi) = \frac{1}{\phi}\,, \tag{3.10}$$

a function of the spacetime location. In order to guarantee a positive gravitational coupling, only the range of values $\phi > 0$ corresponding to attractive gravity is considered. The dimensionless Brans-Dicke parameter $\omega$ is a free parameter of the theory: a value of $\omega$ of order unity would be natural in principle (and it does appear in the low-energy limit of the bosonic string theory). However, values of $\omega$ of this order of magnitude are excluded by Solar System experiments, for a massless or light field $\phi$ (*i.e.*, one that has a range larger than the size of the Solar System).

The larger the value of $\omega$, the closer Brans-Dicke gravity is to GR [1153]; there are, however, exceptions such as vacuum Brans-Dicke solutions,[2] and solutions sourced by conformal matter [43, 75, 446, 451, 454, 780, 944, 946, 948, 975]. The most stringent experimental limit, $\omega > 40{,}000$, was set by the Cassini probe in 2003 [133].

Brans-Dicke theory with a free or light scalar field is viable in the limit of large $\omega$, but the large value of this parameter required to satisfy the experimental bounds is

[2] One should keep in mind, however, that the limit of particular spacetime solutions of the field equations of a gravitational theory should be taken in a coordinate-independent way [535,885,886].

certainly fine-tuned and makes Brans-Dicke theory unappealing. However, this fine-tuning becomes unnecessary if the scalar field is given a sufficiently large mass and, therefore, a short range. This means that a self-interaction potential $V(\phi)$ has to be considered in discussing the limits on $\omega$ and this fact is an adjustment of the original Brans-Dicke theory [165].

## 3.2 The variational principle and the field equations of metric $f(R)$ gravity

We now examine the variational principle and the field equations of another class of ETGs, $f(R)$ gravity in the metric formalism. The salient feature of these ETGs is that the field equations are of fourth order and, therefore, more complicated than those of GR (which is recovered as the special case $f(R) = R$). Due to their higher order, these field equations admit a much richer variety of solutions than the Einstein equations. For simplicity, we begin by discussing quadratic corrections to the Hilbert-Einstein theory, which provide interesting cosmology.

### 3.2.1 $f(R) = R + \alpha R^2$ theory

Quadratic corrections in the Ricci scalar motivated by attempts to renormalize GR, as discussed in Chap. 1, constitute a straightforward extension of GR and have been particularly relevant in cosmology since they allow a self-consistent inflationary model to be constructed [1044]. We will use this model as an example before discussing general metric $f(R)$ gravity.

Let us begin by deriving the field equations for the Lagrangian density

$$\mathscr{L} = R + \alpha R^2 + 2\kappa \mathscr{L}^{(m)} \tag{3.11}$$

from the variational principle $\delta \int d^4x \sqrt{-g}\,\mathscr{L} = 0$. We consider vacuum first. The variation gives

$$\int d^4x \sqrt{-g}\, G_{\alpha\beta} \delta g^{\alpha\beta} + \alpha\, \delta \int d^4x \sqrt{-g}\, R^2 = 0\,, \tag{3.12}$$

in which the variation of $R\sqrt{-g}$ produces the Einstein tensor. We now compute the second term on the right hand side of Eq. (3.12). We have

$$\delta \int d^4x \sqrt{-g}\, R^2 = -\frac{1}{2} \int d^4x \sqrt{-g}\; g_{\alpha\beta} \delta g^{\alpha\beta} R^2 + 2 \int d^4x \sqrt{-g}\; R \delta R \tag{3.13}$$

and

$$\int d^4x \sqrt{-g}\, R\delta R = \int d^4x \sqrt{-g}\, R \left(\delta g^{\alpha\beta} R_{\alpha\beta} + g^{\alpha\beta}\delta R_{\alpha\beta}\right) . \tag{3.14}$$

By using the fact that

$$g^{\alpha\beta}\delta R_{\alpha\beta} = \nabla_\alpha \nabla_\beta h^{\alpha\beta} - \Box h , \tag{3.15}$$

where

$$h^{\alpha\beta} \equiv -\delta g^{\alpha\beta}, \qquad h \equiv -g_{\alpha\beta}\delta g^{\alpha\beta} , \tag{3.16}$$

one has

$$\int d^4x \sqrt{-g}\, R\, g^{\alpha\beta}\delta R_{\alpha\beta} = \int d^4x \sqrt{-g}\, R \left(\nabla_\alpha \nabla_\beta h^{\alpha\beta} - \Box h\right) . \tag{3.17}$$

Integrating by parts twice, the operators $\nabla_\alpha \nabla_\beta$ and $\Box$ acting on $h^{\alpha\beta}$ and $h$, respectively, transfer their action onto $R$ and

$$\int d^4x \sqrt{-g}\, R\, g^{\alpha\beta}\delta R_{\alpha\beta} = \int d^4x \sqrt{-g} \left(h^{\alpha\beta}\nabla_\alpha \nabla_\beta R - h\Box R\right) . \tag{3.18}$$

Using Eq. (3.16), Eq. (3.18) becomes

$$\int d^4x \sqrt{-g}\, R\, g^{\alpha\beta}\delta R_{\alpha\beta} = \int d^4x \sqrt{-g} \left(-\delta g^{\alpha\beta}\nabla_\alpha \nabla_\beta R + g_{\alpha\beta}\Box R\delta g^{\alpha\beta}\right) . \tag{3.19}$$

Upon substitution of Eq. (3.19) into Eq. (3.14), one obtains

$$\int d^4x \sqrt{-g}\, R\delta R = \int d^4x \sqrt{-g} \left(R\delta g^{\alpha\beta} R_{\alpha\beta} - \delta g^{\alpha\beta}\nabla_\alpha \nabla_\beta R + g_{\alpha\beta}\Box R\delta g^{\alpha\beta}\right) \tag{3.20}$$

and Eq. (3.13) takes the form

$$\begin{aligned} \delta\int d^4x \sqrt{-g}\, R^2 &= -\frac{1}{2}\int d^4x \sqrt{-g}\, g_{\alpha\beta}\delta g^{\alpha\beta} R^2 \\ &\quad +2\int d^4x \sqrt{-g} \left(R\delta g^{\alpha\beta} R_{\alpha\beta} - \delta g^{\alpha\beta}\nabla_\alpha \nabla_\beta R + g_{\alpha\beta}\Box R\delta g^{\alpha\beta}\right) \\ &= \int d^4x \sqrt{-g}\, (2RR_{\alpha\beta} - \tfrac{1}{2}\, g_{\alpha\beta} R^2)\delta g^{\alpha\beta} \\ &\quad +2\int d^4x \sqrt{-g}\, (g_{\alpha\beta}\Box R - \nabla_\alpha \nabla_\beta R)\, \delta g^{\alpha\beta} . \end{aligned} \tag{3.21}$$

Substituting this equation into Eq. (3.12) and including the matter part of the Lagrangian $\mathscr{L}^{(m)}$ which produces the energy-momentum tensor $T_{\mu\nu}^{(m)}$, the field equations

$$G_{\alpha\beta} + \alpha \left[ 2R \left( R_{\alpha\beta} - \frac{1}{4} g_{\alpha\beta} R \right) + 2 \left( g_{\alpha\beta} \Box R - \nabla_\alpha \nabla_\beta R \right) \right] = \kappa \, T_{\alpha\beta}^{(m)} \tag{3.22}$$

are obtained; they are fourth-order PDEs for the metric components.

The trace of Eq. (3.22) is

$$\Box R - \frac{1}{6\alpha} \left( R + \kappa \, T^{(m)} \right) = 0 \, , \tag{3.23}$$

which shows that $\alpha$ must be positive. One can also define an angular frequency $\omega$ (equivalent to a mass $m$) so that

$$\frac{1}{6\alpha} = \omega^2 = m^2 \, . \tag{3.24}$$

Following this definition, Eq. (3.23) becomes

$$\Box R - m^2 \left( R + \kappa \, T^{(m)} \right) = 0 \, . \tag{3.25}$$

After the early phases of the universe, as the temperature decreases with the expansion, the term proportional to $m^2$ becomes dominant. Equation (3.25) can be seen as an effective Klein-Gordon equation for the effective scalar field degree of freedom $R$ (sometimes called *scalaron*).

### 3.2.2 Metric $f(R)$ gravity in general

Let us discuss now a generic analytical[3] function $f(R)$ in the metric formalism, beginning with the vacuum case, as described by the Lagrangian density $\sqrt{-g}\,\mathscr{L} = \sqrt{-g}\, f(R)$ obeying the variational principle $\delta \int d^4x \sqrt{-g}\, f(R) = 0$. We have

$$\begin{aligned} \delta \int d^4x \sqrt{-g}\, f(R) &= \int d^4x \left[ \delta \left( \sqrt{-g}\, f(R) \right) + \sqrt{-g}\, \delta \left( f(R) \right) \right] \\ &= \int d^4x \sqrt{-g} \left[ f'(R) R_{\mu\nu} - \tfrac{1}{2} g_{\mu\nu} f(R) \right] \delta g^{\mu\nu} \\ &\quad + \int d^4x \sqrt{-g}\, f'(R) g^{\mu\nu} \delta R_{\mu\nu} \, , \end{aligned} \tag{3.26}$$

[3] This assumption is not, strictly speaking, necessary and is sometimes relaxed in the literature.

where the prime denotes differentiation with respect to $R$. We now compute these integrals in the local inertial frame. By using

$$g^{\mu\nu}\delta R_{\mu\nu} = g^{\mu\nu}\partial_\sigma\left(\delta\Gamma^\sigma_{\mu\nu}\right) - g^{\mu\sigma}\partial_\sigma\left(\delta\Gamma^\nu_{\mu\nu}\right) \equiv \partial_\sigma W^\sigma \tag{3.27}$$

where

$$W^\sigma \equiv g^{\mu\nu}\delta\Gamma^\sigma_{\mu\nu} - g^{\mu\sigma}\delta\Gamma^\nu_{\mu\nu}\,, \tag{3.28}$$

the second integral in Eq. (3.26) can be written as

$$\int d^4x\sqrt{-g}\, f'(R)g^{\mu\nu}\delta R_{\mu\nu} = \int d^4x\sqrt{-g}\, f'(R)\partial_\sigma W^\sigma\,. \tag{3.29}$$

Integration by parts yields

$$\begin{aligned}\int d^4x\sqrt{-g}\, f'(R)g^{\mu\nu}\delta R_{\mu\nu} = &\int d^4x\frac{\partial}{\partial x^\sigma}\left[\sqrt{-g}\, f'(R)W^\sigma\right]\\ &-\int d^4x\partial_\sigma\left[\sqrt{-g}f'(R)\right]W^\sigma\,.\end{aligned} \tag{3.30}$$

The first integrand is a total divergence and can be discarded by assuming that the fields vanish at infinity, obtaining

$$\int d^4x\sqrt{-g}\, f'(R)g^{\mu\nu}\delta R_{\mu\nu} = -\int d^4x\partial_\sigma\left[\sqrt{-g}\, f'(R)\right]W^\sigma\,. \tag{3.31}$$

Let us calculate now the term $W^\sigma$ appearing in Eq. (3.31). We have

$$\begin{aligned}\delta\Gamma^\sigma_{\mu\nu} &= \delta\left[\frac{1}{2}g^{\sigma\alpha}\left(\partial_\mu g_{\alpha\nu} + \partial_\nu g_{\mu\alpha} - \partial_\alpha g_{\mu\nu}\right)\right]\\ &= \frac{1}{2}g^{\sigma\alpha}\left[\partial_\mu\left(\delta g_{\alpha\nu}\right) + \partial_\nu\left(\delta g_{\mu\alpha}\right) - \partial_\alpha\left(\delta g_{\mu\nu}\right)\right],\end{aligned} \tag{3.32}$$

since in the locally inertial frame considered here it is

$$\partial_\alpha g_{\mu\nu} = \nabla_\alpha g_{\mu\nu} = 0\,. \tag{3.33}$$

Similarly, it is

$$\delta\Gamma^\nu_{\mu\nu} = \frac{1}{2}\, g^{\nu\alpha}\partial_\mu\left(\delta g_{\nu\alpha}\right)\,. \tag{3.34}$$

By combining Eqs. (3.33) and (3.34), one obtains

$$g^{\mu\nu}\delta\Gamma^\sigma_{\mu\nu} = \frac{1}{2}\, g^{\mu\nu}\left[-\partial_\mu\left(g_{\alpha\nu}\delta g^{\alpha\sigma}\right) - \partial_\nu\left(g_{\mu\alpha}\delta g^{\sigma\alpha}\right) - g^{\sigma\alpha}\partial_\alpha\left(\delta g_{\mu\nu}\right)\right]$$

$$= \frac{1}{2}\,\partial^\sigma\left(g_{\mu\nu}\delta g^{\mu\nu}\right) - \partial^\mu\left(g_{\alpha\mu}\delta g^{\nu\alpha}\right), \tag{3.35}$$

$$g^{\mu\sigma}\delta\Gamma^\nu_{\mu\nu} = -\frac{1}{2}\,\partial^\sigma\left(g_{\nu\alpha}\delta g^{\nu\alpha}\right), \tag{3.36}$$

from which it follows immediately that

$$W^\sigma = \partial^\sigma\left(g_{\mu\nu}\delta g^{\mu\nu}\right) - \partial^\mu\left(g_{\mu\nu}\delta g^{\sigma\nu}\right). \tag{3.37}$$

Using this equation one can write

$$\begin{aligned}&\int d^4x\sqrt{-g}\,f'(R)g^{\mu\nu}\delta R_{\mu\nu}\\ &\quad= \int d^4x\partial_\sigma\left[\sqrt{-g}\,f'(R)\right]\left[\partial^\mu\left(g_{\mu\nu}\delta g^{\sigma\nu}\right) - \partial^\sigma\left(g_{\mu\nu}\delta g^{\mu\nu}\right)\right].\end{aligned} \tag{3.38}$$

Integrating by parts and discarding total divergences, one obtains

$$\begin{aligned}\int d^4x\,\sqrt{-g}\,f'(R)g^{\mu\nu}\delta R_{\mu\nu} &= \int d^4x\,g_{\mu\nu}\partial^\sigma\partial_\sigma\left[\sqrt{-g}f'(R)\right]\delta g^{\mu\nu}\\ &\quad-\int d^4x\,g_{\mu\nu}\partial^\mu\partial_\sigma\left[\sqrt{-g}\,f'(R)\right]\delta g^{\sigma\nu}.\end{aligned} \tag{3.39}$$

The variation of the action is then

$$\begin{aligned}\delta\int d^4x\sqrt{-g}\,f(R) &= \int d^4x\sqrt{-g}\left[f'(R)R_{\mu\nu} - \tfrac{1}{2}f(R)g_{\mu\nu}\right]\delta g^{\mu\nu}\\ &\quad+\int d^4x\left[g_{\mu\nu}\partial^\sigma\partial_\sigma\left(\sqrt{-g}\,f'(R)\right) - g_{\sigma\nu}\partial^\mu\partial_\sigma\left(\sqrt{-g}f'(R)\right)\right]\delta g^{\mu\nu}.\end{aligned} \tag{3.40}$$

The vanishing of the variation implies the fourth order vacuum field equations

$$f'(R)R_{\mu\nu} - \frac{f(R)}{2}\,g_{\mu\nu} = \nabla_\mu\nabla_\nu f'(R) - g_{\mu\nu}\Box f'(R). \tag{3.41}$$

These equations can be re-arranged in the Einstein-like form

$$f'(R)R_{\mu\nu}-\frac{f'(R)}{2}\,g_{\mu\nu}R+\frac{f'(R)}{2}\,g_{\mu\nu}R-\frac{f(R)}{2}\,g_{\mu\nu} = \nabla_\mu\nabla_\nu f'(R)-g_{\mu\nu}\Box f'(R), \tag{3.42}$$

and then

$$G_{\mu\nu} = \frac{1}{f'(R)}\left\{\nabla_\mu\nabla_\nu f'(R) - g_{\mu\nu}\Box f'(R) + g_{\mu\nu}\frac{\left[f(R) - f'(R)R\right]}{2}\right\}. \tag{3.43}$$

The right hand side of Eq. (3.43) is then regarded as an effective stress-energy tensor, which we call *curvature fluid* energy-momentum tensor $T^{(curv)}_{\mu\nu}$ sourcing the effective Einstein equations. Although this interpretation is questionable in principle because the field equations describe a theory different from GR, and one is forcing upon them the interpretation as effective Einstein equations, this approach is quite useful in practice.

## 3.3 A more general class of ETGs

ETGs exhibit two main features: first, the geometry can couple non-minimally to some scalar field; second, derivatives of the metric components of order higher than second may appear. In the first case, we say that we have scalar-tensor theories of gravity, and in the second case we have higher order theories. Combinations of non-minimally coupled and higher order terms can also emerge in effective Lagrangians, producing mixed higher order/scalar-tensor gravity.

From the mathematical point of view, the reduction of more general theories to the Einstein-like form is common practice. Through a Legendre transformation of the metric, higher-order theories take the form of effective Einstein gravity under suitable regularity conditions of the Lagrangian, with (possibly multiple) scalar field(s) as the source of the gravity, but with important differences if matter is present [484, 768, 769, 1023]. The mathematical equivalence between models with variable gravitational coupling and standard Einstein gravity through conformal transformations has also been known for a long time [360,392]. This mathematical equivalence gives rise to different conformal representations of scalar-tensor theories, the most well known being the Jordan and the Einstein conformal frames. The issue of the physical equivalence of these conformal frames has been debated at length, and probably blown out of proportion. However, there are still open questions to this regard ([473] and references therein). Several authors claim a true physical difference between these two conformal frames on the basis that experimental and observational data support the Jordan frame as a better fit with theoretical solutions. Other authors support the viewpoint that only the Einstein frame is physical on the basis of energy considerations [769]. These points of view are discussed later in this chapter.

## 3.4 The Palatini formalism

The Palatini approach to gravitational theories was introduced by Einstein himself [415] but received its name because of an historical misunderstanding [191, 485]. The fundamental idea of the Palatini formalism is to regard the (usually torsion-free) connection $\Gamma^{\mu}_{\nu\alpha}$ entering the definition of the Ricci tensor as a variable independent of the spacetime metric $g_{\mu\nu}$. The Palatini formulation of GR is equivalent to the metric version of this theory as a consequence of the fact that the field equations for the connection $\Gamma^{\alpha}_{\mu\nu}$ give the Levi-Civita connection of the metric $g_{\mu\nu}$ [1139]. As a consequence, there is no particular reason to impose the Palatini variational principle in GR instead of the metric variational principle.

The situation is different in ETGs depending on functions of curvature invariants, such as $f(R)$, or for gravity non-minimally coupled to a scalar field. In these cases, the Palatini and the metric variational principle yield different field equations and different physics [486, 769]. The Palatini approach in the context of ETGs has been the subject of much (and, as we will see, ill-fated) interest in cosmological applications [211, 721, 722, 843, 1129].

The Newtonian potential obtained in the weak-field limit of alternative theories of gravity and its relations with a conformal factor have also been studied [794]. From the physical point of view, considering the metric $g_{\mu\nu}$ and the connection $\Gamma^{\alpha}_{\mu\nu}$ as independent fields amounts to decoupling the metric structure of spacetime and its geodesic structure with the connection $\Gamma^{\alpha}_{\mu\nu}$ being distinct from the Levi-Civita connection of $g_{\alpha\beta}$. The causal structure of spacetime is defined by $g_{\mu\nu}$, while the spacetime trajectories of particles are governed by $\Gamma^{\alpha}_{\mu\nu}$. In principle, this decoupling enriches the geometric structure of spacetime and generalizes the purely metric formalism. By means of the Palatini field equations, this dual structure of spacetime is naturally translated into a bimetric structure of the theory: instead of a metric and an independent connection, the Palatini formalism can be seen as containing two independent metrics $g_{\mu\nu}$ and $h_{\mu\nu} = f'(R)\, g_{\mu\nu}$. In Palatini $f(R)$ gravity the new metric $h_{\mu\nu}$ determining the geodesics is related to the connection $\Gamma^{\alpha}_{\mu\nu}$ by the fact that the latter turns out to be the Levi-Civita connection of $h_{\mu\nu}$.

In scalar-tensor gravity, the second metric $h_{\mu\nu}$ is related to the non-minimal coupling of the Brans-Dicke-like scalar. In the Palatini formalism the non-minimal coupling and the scalar field are separated from the metric structure of spacetime. The situation mixes when we consider higher order/scalar-tensor theories.

### *3.4.1 The Palatini approach and the conformal structure of the theory*

Let us work out examples showing the role of conformal transformations in the Palatini approach to ETGs [26], beginning with fourth order gravity in which the difference between metric and Palatini variational principles is evident. The Ricci scalar in $f(\mathscr{R})$ is $\mathscr{R} \equiv \mathscr{R}(g,\Gamma) \equiv g^{\alpha\beta}\mathscr{R}_{\alpha\beta}(\Gamma)$ and is a generalized Ricci scalar, whereas $\mathscr{R}_{\mu\nu}(\Gamma)$ is the Ricci tensor of a torsion-free connection $\Gamma^{\alpha}_{\mu\nu}$ which, *a priori*, has no relations with the spacetime metric $g_{\mu\nu}$. The gravitational sector of the theory is described by the analytical function $f(\mathscr{R})$, while $\sqrt{-g}$ denotes the usual scalar density of weight 1. The field equations derived with the Palatini variational principle are

$$f'(\mathscr{R})\mathscr{R}_{(\mu\nu)}(\Gamma) - \frac{f(\mathscr{R})}{2}\, g_{\mu\nu} = T^{(m)}_{\mu\nu}\,, \tag{3.44}$$

$$\nabla^{\Gamma}_{\alpha}\left[\sqrt{-g}\, f'(\mathscr{R}) g^{\mu\nu}\right] = 0\,, \tag{3.45}$$

where $\nabla^{\Gamma}_{\mu}$ is the covariant derivative of the non-metric connection $\Gamma^{\alpha}_{\mu\nu}$, and we use units in which $8\pi G = 1$.

It is important to stress that Eq. (3.45) is obtained under the assumption that the matter sector described by $\mathscr{L}^{(m)}$ is functionally independent of the (non-metric) connection $\Gamma^{\alpha}_{\mu\nu}$; however it may contain metric covariant derivatives $\overset{g}{\nabla}$ of the matter fields. This means that the matter stress-energy tensor $T^{(m)}_{\mu\nu}[g, \Psi]$ depends on the metric $g_{\mu\nu}$ and on the matter fields collectively denoted by $\Psi$, together with their covariant derivatives with respect to the Levi-Civita connection of $g_{\mu\nu}$. It is easy to see from Eq. (3.45) that $\sqrt{-g}\, f'(\mathscr{R}) g^{\mu\nu}$ is a symmetric tensor density of weight 1, which naturally leads to the introduction of a new metric $h_{\mu\nu}$ conformally related to $g_{\mu\nu}$ by [26,486]

$$\sqrt{-g}\, f'(\mathscr{R})\, g^{\mu\nu} = \sqrt{-h}\, h^{\mu\nu}\,. \tag{3.46}$$

With this definition $\Gamma^{\alpha}_{\mu\nu}$ is the Levi-Civita connection of the metric $h_{\mu\nu}$, with the only restriction that the conformal factor $\sqrt{-g}\, f'(\mathscr{R}) g^{\mu\nu}$ relating $g_{\mu\nu}$ and $h_{\mu\nu}$ be non-degenerate. In the case of the Hilbert-Einstein Lagrangian it is $f'(\mathscr{R}) = 1$ and the statement is trivial.

The conformal transformation

$$g_{\mu\nu} \longrightarrow h_{\mu\nu} = f'(\mathscr{R})\, g_{\mu\nu} \tag{3.47}$$

implies that $\mathscr{R}_{(\mu\nu)}(\Gamma) = \mathscr{R}_{\mu\nu}(h)$. It is useful to consider the trace of the field equations (3.44)

$$f'(\mathscr{R})\mathscr{R} - 2f(\mathscr{R}) = g^{\alpha\beta} T^{(m)}_{\alpha\beta} \equiv T^{(m)}\,, \tag{3.48}$$

which controls the solutions of Eq. (3.45). We refer to this scalar equation as the *structural equation* of spacetime. *In vacuo* and in the presence of conformally invariant matter with $T^{(m)} = 0$, this scalar equation admits constant solutions. In these cases, Palatini $f(\mathscr{R})$ gravity reduces to GR with a cosmological constant [486,956].

In the case of interaction with matter fields, the structural equation (3.47), if explicitly solvable, provides in principle an expression $\mathscr{R} = F(T^{(m)})$ and, as a result, both $f(\mathscr{R})$ and $f'(\mathscr{R})$ can be expressed in terms of $T^{(m)}$. This fact allows one to express, at least formally, $\mathscr{R}$ in terms of $T^{(m)}$, which has deep consequences for the description of physical systems, as we will see later. Matter rules the bimetric structure of spacetime and, consequently, both the geodesic and metric structures which are intrinsically different. This behavior generalizes the vacuum case.

Let us now extend the Palatini formalism to non-minimally coupled scalar-tensor theories, with the goal of understanding the bimetric structure of spacetime in these theories and its possible geometric and physical interpretation. We denote by $S_1$ the action functional of Palatini scalar-tensor theories, while non-minimal interaction between scalar-tensor and $f(R)$ gravities will be considered later, calling $S_2$ the respective action. Then, we will finally consider the case of scalar fields $\phi$ non-minimally coupled to the gravitational fields $(g_{\mu\nu}, \Gamma^{\alpha}_{\mu\nu})$, denoting by $S_3$ the corresponding action. In this case, the low curvature limit $\mathcal{R} \to 0$, which is relevant for the present epoch of the history of the universe, is particularly significant.

The scalar-tensor action can be generalized, in order to better develop the Palatini approach, as

$$S_1 = \int d^4x\sqrt{-g}\left[F(\phi)\mathcal{R} - \frac{\varepsilon}{2}\overset{g}{\nabla}_\mu \phi \overset{g}{\nabla}{}^{\mu} \phi - V(\phi) + \mathcal{L}^{(m)}\left(\Psi, \overset{g}{\nabla}\Psi\right)\right], \tag{3.49}$$

with $\varepsilon = \pm 1$ corresponding to an ordinary scalar or a phantom field, respectively. The field equations for the metric $g_{\mu\nu}$ and the connection $\Gamma^\alpha_{\mu\nu}$ are

$$F(\phi)\left(\mathcal{R}_{(\mu\nu)} - \frac{1}{2}g_{\mu\nu}\mathcal{R}\right) = T^{(\phi)}_{\mu\nu} + T^{(m)}_{\mu\nu}, \tag{3.50}$$

$$\nabla^\Gamma_\alpha\left[\sqrt{-g}\,F(\phi)g^{\mu\nu}\right] = 0, \tag{3.51}$$

where $\mathcal{R}_{(\mu\nu)}$ is defined by Eq. (3.44). The equation of motion of the matter fields is

$$\varepsilon\Box\phi = V_\phi(\phi) + F_\phi(\phi)\mathcal{R}, \tag{3.52}$$

$$\frac{\delta\mathcal{L}^{(m)}}{\delta\Psi} = 0. \tag{3.53}$$

In this case, the structural equation of spacetime implies that

$$\mathcal{R} = -\frac{\left(T^{(\phi)} + T^{(m)}\right)}{F(\phi)}, \tag{3.54}$$

where we must require that $F(\phi) > 0$. The bimetric structure of spacetime is thus defined by the *ansatz*

$$\sqrt{-g}\;F(\phi)g^{\mu\nu} = \sqrt{-h}\;h^{\mu\nu} \tag{3.55}$$

so that $h_{\mu\nu}$ is conformal to $g_{\mu\nu}$,

$$h_{\mu\nu} = F(\phi)\,g_{\mu\nu}. \tag{3.56}$$

It follows from Eq. (3.54) that *in vacuo* $T^{(\phi)} = 0$ and $T^{(m)} = 0$ this theory is equivalent to vacuum GR. If $F(\phi) = F_0 =$ const. we recover GR with a minimally coupled scalar field, which means that the Palatini approach intrinsically gives rise to the conformal structure (3.56) of the theory which is trivial in the Einsteinean, minimally coupled, case.

As a further step, let us generalize the previous results to the case of non-minimal coupling in the framework of $f(R)$ theories. The action functional can be written as

$$S_2 = \int d^4x\sqrt{-g}\left[F(\phi)f(\mathcal{R}) - \frac{\varepsilon}{2}\overset{g}{\nabla}_\mu \phi \overset{g}{\nabla}{}^{\mu} \phi - V(\phi) + \mathcal{L}^{(m)}(\Psi, \overset{g}{\nabla}\Psi)\right] \tag{3.57}$$

where $f(\mathscr{R})$ is, as usual, an analytical function of $\mathscr{R}$. The Palatini field equations for the gravitational sector are

$$F(\phi)\left[f'(\mathscr{R})\mathscr{R}_{(\mu\nu)} - \frac{f(\mathscr{R})}{2} g_{\mu\nu}\right] = T^{(\phi)}_{\mu\nu} + T^{(m)}_{\mu\nu}\,, \tag{3.58}$$

$$\nabla^{\Gamma}_{\alpha}\left[\sqrt{-g}F(\phi)f'(\mathscr{R})g^{\mu\nu}\right] = 0\,. \tag{3.59}$$

The equations of motion for the scalar and matter fields are

$$\varepsilon\Box\phi = V_{\phi}(\phi) + F_{\phi}(\phi)f(\mathscr{R})\,, \tag{3.60}$$

$$\frac{\delta\mathscr{L}^{(m)}}{\delta\Psi} = 0\,, \tag{3.61}$$

in which the non-minimal interaction term enters the modified Klein-Gordon equations. In this case, the structural equation of spacetime implies that

$$f'(\mathscr{R})\mathscr{R} - 2f(\mathscr{R}) = \frac{T^{(\phi)} + T^{(m)}}{F(\phi)}\,. \tag{3.62}$$

The bimetric structure of spacetime is given by

$$\sqrt{-g}\,F(\phi)f'(\mathscr{R})g^{\mu\nu} = \sqrt{-h}\,h^{\mu\nu} \tag{3.63}$$

with $g_{\mu\nu}$ and $h_{\mu\nu}$ again conformally related,

$$h_{\mu\nu} = F(\phi)f'(\mathscr{R})\,g_{\mu\nu}\,. \tag{3.64}$$

Once the structural equation is solved, the conformal factor depends on the values of the matter fields $(\phi, \Psi)$ or, more precisely, on the traces of their stress-energy tensors and the value of $\phi$. *In vacuo*, Eq. (3.62) implies that the theory reduces again to Einstein gravity as for minimally interacting $f(R)$ theories [486]. The validity of this property is related to the decoupling of the scalar field from the metric.

Finally, let us discuss the situation in which the gravitational Lagrangian is a general function of $\phi$ and $\mathscr{R}$, as in

$$S_3 = \int d^4x\sqrt{-g}\left[K(\phi,\mathscr{R}) - \frac{\varepsilon}{2}\overset{g}{\nabla}_{\mu}\phi\overset{g}{\nabla}{}^{\mu}\phi - V(\phi) + \mathscr{L}^{(m)}\left(\Psi, \overset{g}{\nabla}\Psi\right)\right], \tag{3.65}$$

which yields the gravitational field equations

$$\frac{\partial K(\phi,\mathscr{R})}{\partial\mathscr{R}}\mathscr{R}_{(\mu\nu)} - \frac{K(\phi,\mathscr{R})}{2}g_{\mu\nu} = T^{(\phi)}_{\mu\nu} + T^{(m)}_{\mu\nu}\,, \tag{3.66}$$

$$\nabla^{\Gamma}_{\alpha}\left[\sqrt{-g}\,\frac{\partial K(\phi,\mathscr{R})}{\partial\mathscr{R}}\,g^{\mu\nu}\right] = 0\,, \tag{3.67}$$

while the scalar and matter fields obey

$$\varepsilon\Box\phi = V_\phi(\phi) + \frac{\partial\, K(\phi,\mathscr{R})}{\partial\phi}\,, \tag{3.68}$$

$$\frac{\delta\mathscr{L}^{(m)}}{\delta\Psi} = 0\,. \tag{3.69}$$

The structural equation of spacetime can be expressed as

$$\frac{\partial K(\phi,\mathscr{R})}{\partial\mathscr{R}}\mathscr{R} - 2K(\phi,\mathscr{R}) = T^{(\phi)} + T^{(m)}\,. \tag{3.70}$$

When solved, Eq. (3.70) provides again the form of the Ricci scalar in terms of the traces of the stress-energy tensors of matter and of the scalar field (with $K(\phi,\mathscr{R}) > 0$). The bimetric structure of spacetime is defined by

$$\sqrt{-g}\,\frac{\partial K(\phi,\mathscr{R})}{\partial\mathscr{R}}\,g^{\mu\nu} = \sqrt{-h}\,h^{\mu\nu} \tag{3.71}$$

with

$$h_{\mu\nu} = \frac{\partial K(\phi,\mathscr{R})}{\partial\mathscr{R}}\,g_{\mu\nu}\,. \tag{3.72}$$

The conformal factor depends on the matter fields only through the traces of their stress-energy tensors. The conformal factor and the bimetric structure are ruled by these traces and by the value of the scalar field $\phi$. In this case, in general, one does not recover GR, as is evident from Eq. (3.70) in which the strong coupling between $\mathscr{R}$ and $\phi$ prevents, even *in vacuo*, the possibility of obtaining constant solutions.

Let us discuss the $R \to 0$ regime, a good approximation to the present epoch of the observed universe. The linear expansion of the analytical function $K(\phi,\mathscr{R})$

$$K(\phi,\mathscr{R}) = K_0(\phi) + K_1(\phi)\mathscr{R} + \mathrm{O}(\mathscr{R}^2) \tag{3.73}$$

with

$$K_0(\phi) = K(\phi,\mathscr{R})\,|_{\mathscr{R}=0}\,, \qquad K_1(\phi) = \left(\frac{\partial K(\phi,\mathscr{R})}{\partial\mathscr{R}}\right)|_{\mathscr{R}=0}\,, \tag{3.74}$$

can be substituted into Eqs. (3.70) and (3.72) obtaining, to first order, the structural equation and the bimetric structure. The structural equation yields

$$\mathscr{R} = \frac{-1}{K_1(\phi)}\left[T^{(\phi)} + T^{(m)}) + 2K_0(\phi)\right] \tag{3.75}$$

and the value of the Ricci scalar is always determined, in the linear approximation, in terms of $T^{(\phi)}$, $T^{(m)}$, and $\phi$. The bimetric structure is, otherwise, simply defined by the first term of the Taylor expansion, which is

$$h_{\mu\nu} = K_1(\phi)\, g_{\mu\nu} \tag{3.76}$$

reproducing, as expected, the scalar-tensor case (3.56). Scalar-tensor theories can then be recovered as the linear approximation of a general theory in which gravity and the non-minimal couplings are arbitrary (cf. Eqs. (3.75) and (3.62)). This fact agrees with the above considerations when the Lagrangians of physical interactions can be considered as locally gauge-invariant stochastic functions [103].

Finally, there exist also bimetric theories which cannot be conformally related [1167] and torsion will also appear in the most general framework [240,583]. These more general theories will not be discussed here.

### 3.4.2 *Problems with the Palatini formalism*

Palatini $f(R)$ gravity suffers from two serious problems: *(i)* the presence of curvature singularities at the surface of stars [78–80], and *(ii)* incompatibilities with the Standard Model of particle physics [497, 622, 870] (see also the discussion of the initial value problem of Palatini modified gravity later in this chapter).

Let us begin by discussing the first problem, which occurs when attempting to build static spherically symmetric interior solutions and matching them with the spherically symmetric exterior metric while satisfying the Darmois-Israel junction conditions [627]. Since Palatini $f(R)$ gravity *in vacuo* reduces to GR with a cosmological constant, the unique exterior solution is the Schwarzschild-de Sitter metric. We then search for an interior solution with matter described by a perfect fluid with stress-energy tensor $T_{\mu\nu} = (P+\rho)\, u_\mu u_\nu + P g_{\mu\nu}$. Following [78–80], the most general static and spherically symmetric line element is

$$ds^2 \equiv -e^{2\mu(r)}\mathrm{d}t^2 + e^{2\nu(r)}\mathrm{d}r^2 + r^2\, d\Omega_2^2 \tag{3.77}$$

(where $d\Omega_2^2 = d\theta^2 + \sin^2\theta d\varphi^2$ is the line element on the unit two-sphere) for which the Palatini field equations yield

$$\mu' = \frac{-1}{2(1+\gamma)}\left(\frac{1-\mathrm{e}^{2\nu}}{r} - \frac{\mathrm{e}^{2\nu}}{F}8\pi G r P + \frac{\alpha}{r}\right), \tag{3.78}$$

$$\nu' = \frac{1}{2(1+\gamma)}\left(\frac{1-\mathrm{e}^{2\nu}}{r} + \frac{\mathrm{e}^{2\nu}}{F}8\pi G r \rho + \frac{\alpha+\beta}{r}\right), \tag{3.79}$$

$$\alpha \equiv r^2\left[\frac{3}{4}\left(\frac{F'}{F}\right)^2 + \frac{2F'}{rF} + \frac{\mathrm{e}^{2\nu}}{2}\left(\mathscr{R} - \frac{f}{F}\right)\right], \tag{3.80}$$

$$\beta \equiv r^2\left[\frac{F''}{F} - \frac{3}{2}\left(\frac{F'}{F}\right)^2\right], \tag{3.81}$$

$$\gamma \equiv \frac{r\,F'}{2F}, \tag{3.82}$$

where, in this section, a prime denotes differentiation with respect to the radial coordinate $r$ and $F \equiv \partial f/\partial \mathscr{R}$. A generalization to Palatini $f(R)$ gravity of the Tolman-Oppenheimer-Volkoff equation of hydrostatic equilibrium was found in [86, 197, 651]. By introducing

$$m_{\text{tot}}(r) \equiv \frac{r}{2}\left(1 - e^{-2\nu}\right) \tag{3.83}$$

and using the Euler equation

$$P' = -\nu'\,(P + \rho)\,, \tag{3.84}$$

Equations (3.78) and (3.79) yield the generalized Tolman-Oppenheimer-Volkoff equation

$$P' = -\frac{1}{1+\gamma}\,\frac{(\rho + P)}{r\,(r - 2m_{\text{tot}})}\left[m_{\text{tot}} + \frac{4\pi r^3 P}{F} - \frac{\alpha}{2}\,(r - 2m_{\text{tot}})\right], \tag{3.85}$$

$$m'_{\text{tot}} = (1+\gamma)^{-1}\left[\frac{4\pi r^2 \rho}{F} + \frac{\alpha + \beta}{2} - \frac{m_{\text{tot}}}{r}\,(\alpha + \beta - \gamma)\right]. \tag{3.86}$$

The system (3.84), (3.85), and (3.86) for the four functions $\mu$, $\nu$ (or $m_{\text{tot}}$), $P$, and $\rho$ is closed by prescribing an equation of state, for example of the barotropic form $P = P(\rho)$. The authors of [78–80] choose the simple polytropic equation of state $P = k\rho_0^{\Gamma}$ (where $\rho_0$, $k$, and the polytropic index $\Gamma$ are constants).[4] Thermodynamical considerations allow the rest mass density $\rho_0$ to be eliminated, casting the polytropic equation of state in the form [78–80]

$$\rho = \left(\frac{P}{k}\right)^{1/\Gamma} + \frac{P}{\Gamma - 1}. \tag{3.87}$$

The Darmois-Israel junction conditions at the surface of the star $r = r_{\text{out}}$ (defined as the radius where $P = \rho = 0$) require the continuity of the metric and its normal derivative, hence of $\mu'$ there. The unique exterior solution is the Schwarzschild-de Sitter metric with a cosmological constant $\Lambda = \mathscr{R}_0/4$, where $\mathscr{R}_0$ is the constant value of $\mathscr{R}$, therefore in the exterior it must be

$$\exp\left(-2\nu(r)\right) = b\exp\left(2\mu(r)\right) = 1 - \frac{2m}{r} - \frac{\mathscr{R}_0 r^2}{12}\,, \tag{3.88}$$

where $b$ and $m$ are integration constants to be determined. Equation (3.86) yields

$$m_{\text{tot}}(r) = m + \frac{r^3}{24}\,\mathscr{R}_0 \tag{3.89}$$

[4] A polytropic equation of state is inappropriate to model main sequence stars but is used for white dwarfs and neutron stars.

approaching the star surface from the interior, while approaching from the exterior it must be

$$\mu'(r_{out}) = \frac{1}{r_{out}} \frac{r_{out}^3 \mathscr{R}_0 - 12m}{\mathscr{R}_0 r_{out}^3 - 12 r_{out} + 24m} . \tag{3.90}$$

The continuity of $\mu'(r)$ across the star surface requires that $F'(r_{out}) = 0$ for $r \to r_{out}^-$ [78]; but then

$$m'_{tot}(r_{out}) = \frac{2F_0 \mathscr{R}_0 r_{out}^2 + \left(r_{out}^3 \mathscr{R}_0 - 8m_{tot}\right)\mathscr{C}'}{16F_0}, \tag{3.91}$$

where

$$\mathscr{C} = \frac{dF}{dP}(P + \rho) . \tag{3.92}$$

At the star surface the derivative $m'_{tot}$ is ill-behaved for a wide range of values of the polytropic index $\Gamma$. If $1 < \Gamma < 3/2$, then $\mathscr{C}' = d\mathscr{C}/dP\, P' \propto d\mathscr{C}/dP\, (P + \rho)$ tends to zero at the surface. As a consequence, $m'_{tot}(r_{out})$ given by Eq. (3.91) is finite and $m'_{tot}$ is continuous across the star surface. In the range of values $3/2 < \Gamma < 2$ of the polytropic index, $\mathscr{C}'$ diverges as this surface is approached when $dF/d\mathscr{R}(\mathscr{R}_0), d\mathscr{R}/dT(T_0) \neq 0$ (these conditions are satisfied by generic forms of $f(\mathscr{R})$ [78–80]). In spite of the fact that $m_{tot}$ stays finite, the divergence of $m'_{tot}$ implies the divergence of the Riemann tensor $\mathscr{R}_{\mu\nu\rho}{}^{\sigma}$, the Ricci curvature $\mathscr{R}$, and the Kretschmann scalar $\mathscr{R}^{\mu\nu\alpha\beta}\mathscr{R}_{\mu\nu\alpha\beta}$. The divergence occurs in all regimes ranging from Newtonian to strong gravity. Now, it is expected that theoretical models of simple polytropic stars (with $\frac{3}{2} < \Gamma < 2$) can be built in any acceptable theory of gravity, as in the Newtonian case.[5]

The reason for the occurrence of these singularities in Palatini $f(R)$ gravity has been identified clearly in [78–80] and could be related to the non-dynamical nature of the effective scalar field $f'(\mathscr{R})$. Consider the second order field equations (3.44): since $\mathscr{R}$ can be expressed in terms of the trace of the energy-momentum tensor $T^{(m)}$ using the trace equation (3.48), the right hand side of Eq. (3.44) contains second derivatives of $T^{(m)}$ which, in turn, contain first and zero order derivatives of the matter fields. Therefore, Eq. (3.44) can contain derivatives of the matter fields up to third order. This situation is very different from that occurring in GR and in most ETGs, in which the field equations contain at most first order derivatives of the matter fields. As a consequence, the metric in GR is generated by an integral over the matter distribution and discontinuities or singularities in the matter fields and their derivatives do not translate into discontinuities or singularities of the metric [79]. In Palatini $f(R)$ gravity we find instead an algebraic dependence of the metric on the matter fields and a discontinuity in the matter fields or their derivatives translates in a discontinuity of the metric and a curvature singularity.

[5] A few unphysical exceptions are presented in [78] but, otherwise, the singular behavior is generic in Palatini $f(R)$ gravity.

It is not the polytropic description that causes the divergence: in fact, a perfect fluid energy-momentum tensor arising from a kinetic theory includes the matter fields (energy density and pressure) but not their derivatives, making the problem less severe because the microscopic matter distribution is already smoothed out. The problem will be more severe in matter distributions such as the scalar field whose stress-energy tensor contains first order derivatives of the scalar (or possibly second derivatives, if this field couples non-minimally to the curvature).

The root of the problem for Palatini $f(R)$ gravity is due to the fact that the independent connection is actually an auxiliary field that appears to be without dynamics, as is clear from the fact that the trace equation (3.48) is not an evolution equation for the effective scalar field $\phi = f'(\mathcal{R})$ (it contains no derivatives) but, rather, is an algebraic or transcendental equation. This effective scalar field is related algebraically to derivatives of the matter fields and of the metric and, as a result, the theory has a higher differential order in the matter than the metric.

Let us discuss now the difficulties that Palatini modified gravity encounters with respect to the Standard Model of particle physics. A theory satisfying the Equivalence Principle,[6] but with a higher differential structure than its matter fields is likely to exhibit unexpected phenomenology in local non-gravitational experiments, which in turn causes the difficulties with the Standard Model. The algebraic dependence of the connection on the derivatives of matter fields introduces strong couplings between matter and gravity and, therefore, self-interactions of the matter fields. If one attempts to eliminate the connection completely in the action, then this action necessarily includes higher order derivatives of the matter field (self-interactions) [1035].

The problems were first pointed out in [497] using Dirac particles as matter, and re-discussed in [622] and [870] using a Higgs field. Both calculations were performed in the Einstein frame of the scalar-tensor equivalent of Palatini $f(R)$ gravity, but were also repeated in the Jordan frame ([79], see also [1033] – we follow these two references here). Assuming a scalar field $H$ as a form of matter described by the action

$$S^{(m)} = \frac{1}{2\hbar} \int d^4x \sqrt{-g} \left( g^{\mu\nu} \partial_\mu H \partial_\nu H - \frac{m_{\mathrm{H}}^2}{\hbar^2} H^2 \right) \tag{3.93}$$

in units $G = c = 1$ and taking, as an example, $f(\mathcal{R}) = \mathcal{R} - \mu^4/\mathcal{R}$ [275, 1129] the potential for the effective scalar field $\phi = f'(\mathcal{R})$ is $V(\phi) = 2\mu^2\sqrt{\phi - 1}$. To describe the physics in the local frame we must expand the action to second order around vacuum. The vacuum of the Palatini action with (3.93) as the matter action is given by $H = 0$, $\phi = 4/3$, and $g_{\mu\nu} \simeq \eta_{\mu\nu}$ (with $\mu^2 \sim \Lambda$ acting as an effective cosmological constant which can be ignored in the local frame). The point is that the field $\phi$ cannot be expanded perturbatively because it is related algebraically to the matter fields. As a consequence, $\delta\phi \simeq T^{(m)}/\mu^2 \simeq m_{\mathrm{H}}^2 \delta H^2/(\hbar^3 \mu^2)$ at energies

[6] The matter energy-momentum tensor is covariantly conserved with respect to the metric covariant derivative, which implies the geodesic equation for freely falling particles [1167].

lower than the mass $m_{\rm H}$ (if $H$ is a Higgs boson, it is $m_{\rm H} \sim 10^2 - 10^3$ GeV). The second order expansion of the action for the $H$-field for energies much lower than $m_{\rm H}$ is

$$S^{(m)} \simeq \int d^4x \, \frac{\sqrt{-g}}{2\hbar} \left[ g^{\mu\nu} \partial_\mu(\delta H) \partial_\nu(\delta H) - \frac{m_{\rm H}^2}{\hbar^2} \delta H^2 \right] \left[ 1 + \frac{m_{\rm H}^2 \delta H^2}{\mu^2 \hbar^3} + \frac{m_{\rm H}^2 (\partial \delta H)^2}{\mu^4 \hbar^3} \right] . \tag{3.94}$$

Using the estimate $\mu^2 \simeq \Lambda \simeq H_0^2$ with the Hubble radius $H_0^{-1} = 3000$ Mpc and $\delta H \sim m_{\rm H}$, the order of magnitude of the corrections can be estimated. At energies $\sim 10^{-3}$ eV the first correction is of order

$$\frac{m_{\rm H}^2 \delta H^2}{\mu^2 \hbar^3} \sim \left( \frac{H_0^{-1}}{\lambda_{\rm H}} \right)^2 \left( \frac{m_{\rm H}}{m_{Pl}} \right)^2 \gg 1 , \tag{3.95}$$

where $\lambda_{\rm H} = \hbar / m_{\rm H} \sim 2 \times 10^{-19} - 2 \times 10^{-16}$ cm is the Compton wavelength of the Higgs boson and $m_{Pl} = 1.2 \times 10^{19}$ GeV is the Planck mass. The second correction is of order

$$\frac{m_{\rm H}^2 (\partial \delta H)^2}{\mu^4 \hbar^3} \sim \left( \frac{H_0^{-1}}{\lambda_{\rm XH}} \right)^2 \left( \frac{m_{\rm H}}{m_{Pl}} \right)^2 \left( \frac{H_0^{-1}}{L} \right)^2 \gg 1 . \tag{3.96}$$

Such large, non-perturbative, corrections in the physics of $H$-matter in the local frame are bound to violate the constraints deriving from the high precision experiments testing the Standard Model.

## 3.5 Equivalence between $f(R)$ and scalar-tensor gravity

Metric and Palatini $f(R)$ gravities are equivalent to scalar-tensor theories with the derivative of the function $f(R)$ playing the role of the Brans-Dicke scalar, as has been re-discovered several times [301, 587, 1072, 1144, 1160]. We illustrate this equivalence beginning with the metric formalism.

### *3.5.1 Equivalence between scalar-tensor and metric $f(R)$ gravity*

In metric $f(R)$ gravity, we introduce the scalar $\phi \equiv R$; then the action

$$S = \frac{1}{2\kappa} \int d^4x \sqrt{-g} \, f(R) + S^{(m)} \tag{3.97}$$

is rewritten in the form [301, 587, 1072, 1144, 1160]

$$S = \frac{1}{2\kappa} \int d^4x \sqrt{-g} \left[ \psi(\phi) R - V(\phi) \right] + S^{(m)} \tag{3.98}$$

when $f''(R) \neq 0$, where

$$\psi = f'(\phi) \,, \qquad V(\phi) = \phi f'(\phi) - f(\phi) \,. \tag{3.99}$$

It is trivial to see that the action (3.98) coincides with (3.97) if $\phi = R$. *Vice-versa*, let us vary the action (3.98) with respect to $\phi$, which leads to

$$R \frac{d\psi}{d\phi} - \frac{dV}{d\phi} = (R - \phi)\, f''(R) = 0 \,. \tag{3.100}$$

Equation (3.100) implies that $\phi = R$ when $f''(R) \neq 0$. The action (3.98) has the Brans-Dicke form

$$S = \frac{1}{2\kappa} \int d^4x \sqrt{-g} \left[ \psi R - \frac{\omega}{2} \nabla^\mu \psi \nabla_\mu \psi - U(\psi) \right] + S^{(m)} \tag{3.101}$$

with Brans-Dicke field $\psi$, Brans-Dicke parameter $\omega = 0$, and potential $U(\psi) = V[\phi(\psi)]$. An $\omega = 0$ Brans-Dicke theory was originally studied for the purpose of obtaining a Yukawa correction to the Newtonian potential in the weak-field limit [878] and called "O'Hanlon theory" or "massive dilaton gravity". The variation of the action (3.98) yields the field equations

$$G_{\mu\nu} = \frac{\kappa}{\psi} T^{(m)}_{\mu\nu} - \frac{1}{2\psi} U(\psi) g_{\mu\nu} + \frac{1}{\psi} \left( \nabla_\mu \nabla_\nu \psi - g_{\mu\nu} \Box \psi \right) , \tag{3.102}$$

$$3\Box\psi + 2U(\psi) - \psi \frac{dU}{d\psi} = \kappa T^{(m)} \,. \tag{3.103}$$

### 3.5.2 *Equivalence between scalar-tensor and Palatini $f(R)$ gravity*

Palatini $f(R)$ gravity is also equivalent to a special Brans-Dicke theory with a scalar field potential. The Palatini action

$$S = \frac{1}{2\kappa} \int d^4x \sqrt{-g}\, f(\mathscr{R}) + S^{(m)} \tag{3.104}$$

is equivalent to

$$S = \frac{1}{2\kappa} \int d^4x \sqrt{-g} \left[ f(\chi) + f'(\chi)(\mathscr{R} - \chi) \right] + S^{(m)} \,. \tag{3.105}$$

It is straightforward to see that the variation of this action with respect to $\chi$ yields $\chi = \mathscr{R}$. We can now use the field $\phi \equiv f'(\chi)$ and the fact that the curvature $\mathscr{R}$ is the (metric) Ricci curvature of the new metric $h_{\mu\nu} = f'(\mathscr{R})\, g_{\mu\nu}$ conformally related to $g_{\mu\nu}$, as already explained. Using now the well known transformation property of the Ricci scalar under conformal rescalings [1065, 1139]

$$\mathscr{R} = R + \frac{3}{2\phi}\nabla^{\alpha}\phi\nabla_{\alpha}\phi - \frac{3}{2}\Box\phi \tag{3.106}$$

and discarding a boundary term, the action (3.105) can be presented in the form

$$S = \frac{1}{2\kappa}\int d^4x\sqrt{-g}\left[\phi R + \frac{3}{2\phi}\nabla^{\alpha}\phi\nabla_{\alpha}\phi - V(\phi)\right] + S^{(m)}, \tag{3.107}$$

where

$$V(\phi) = \phi\chi(\phi) - f\left[\chi(\phi)\right]. \tag{3.108}$$

This action is clearly that of a Brans-Dicke theory with Brans-Dicke parameter $\omega = -3/2$ and a potential. This theory has been studied occasionally in the literature [46, 81, 354, 366, 373, 878, 879], but it turns out to be a pathological case [78–80, 452, 497, 622, 870].

## 3.6 Conformal transformations applied to extended gravity

We have already mentioned the Jordan and the Einstein frame on several occasions: it is now time to look in detail at the conformal transformations providing different representations of ETGs and a solution-generating technique. In the following chapters we will apply the tool of conformal transformations to ETGs. The basic properties of conformal transformations were introduced in Sect. 2.1; in this section we present their application to Brans-Dicke gravity first, and then to more general scalar-tensor and $f(R)$ theories.

### 3.6.1 *Brans-Dicke gravity*

In Brans-Dicke theory the choice of conformal factor [392]

$$\Omega = \sqrt{G\phi} \tag{3.109}$$

in the conformal transformation $g_{\mu\nu} \rightarrow \tilde{g}_{\mu\nu} = \Omega^2 g_{\mu\nu}$ brings the gravitational sector of the Brans-Dicke action

$$S_{BD} = \int d^4x\,\sqrt{-g}\left[\phi R - \frac{\omega}{\phi}\,g^{\alpha\beta}\nabla_{\alpha}\phi\nabla_{\beta}\phi - V(\phi)\right] + S^{(m)} \tag{3.110}$$

into the Einstein frame form. Then the scalar field redefinition

$$\tilde{\phi}(\phi) = \sqrt{\frac{2\omega+3}{16\pi G}} \ln\left(\frac{\phi}{\phi_0}\right), \tag{3.111}$$

with $\phi > 0$ and $\omega > -3/2$ transforms the scalar field kinetic energy density into canonical form. In terms of the variables $(\tilde{g}_{\mu\nu}, \tilde{\phi})$, the Brans-Dicke action assumes its *Einstein frame* form

$$S_{BD} = \int d^4x \left\{ \sqrt{-\tilde{g}} \left[ \frac{\tilde{R}}{16\pi G} - \frac{1}{2} \tilde{g}^{\alpha\beta} \tilde{\nabla}_\alpha \tilde{\phi} \tilde{\nabla}_\beta \tilde{\phi} - U\left(\tilde{\phi}\right) \right] \right.$$
$$\left. + \exp\left(-8\sqrt{\frac{\pi G}{2\omega+3}}\,\tilde{\phi}\right) \mathscr{L}^{(m)}\left[\tilde{g}\right] \right\}, \tag{3.112}$$

where $\tilde{\nabla}_\alpha$ is the covariant derivative operator of the rescaled metric $\tilde{g}_{\alpha\beta}$ and

$$U\left(\tilde{\phi}\right) = V\left[\phi\left(\tilde{\phi}\right)\right] \exp\left(-8\sqrt{\frac{\pi G}{2\omega+3}}\,\tilde{\phi}\right) = \frac{V(\phi)}{(G\phi)^2} \tag{3.113}$$

is the Einstein frame potential. The restriction of the parameter range to $\omega > -3/2$ is sometimes attributed to the need of guaranteeing that it is possible to perform the conformal transformation. However, one could take the absolute value $|2\omega+3|$ there, but in actual fact $\omega$ cannot cross the barrier $-3/2$: the $\omega = -3/2$ Brans-Dicke theory is pathological. With a special potential, $\omega = -3/2$ Brans-Dicke theory is equivalent to Palatini $f(R)$ gravity.

The Jordan frame scalar has the dimensions of $G^{-1}$, while the Einstein frame scalar $\tilde{\phi}$ has the dimensions of $G^{-1/2}$ and is usually measured in Planck masses. In the GR limit $\phi \rightarrow$const., the Jordan and the Einstein frames coincide.

The inspection of the action (3.112) often leads people to state that, in the Einstein frame, gravity is described by GR, but there are two important differences between Einstein frame Brans-Dicke gravity and Einstein's theory. First, the free scalar $\tilde{\phi}$ acting as a source of gravity on the right hand side of the field equations is always present, *i.e.*, in the Einstein frame solutions of the vacuum field equations $\tilde{R}_{\mu\nu} = 0$ cannot be obtained as in vacuum GR because the scalar $\tilde{\phi}$ pervades the spacetime manifold and cannot be removed. This persistence is a reminder of the cosmological origin of $\phi \sim G_{eff}^{-1}$ in the original (Jordan frame) Brans-Dicke theory [165]. The scalar $\tilde{\phi}$ is always present even if, formally, the gravitational field is only described by the metric tensor $\tilde{g}_{\alpha\beta}$ in the Einstein frame. The conformal transformation shifts the Jordan frame gravitational variable $\phi$ into Einstein frame matter[7] $\tilde{\phi}$.

[7] This property makes it clear that the distinction between gravitational and non-gravitational degrees of freedom depends on the conformal representation of a gravitational theory.

The second difference between GR and Einstein frame Brans-Dicke theory consists of the fact that the matter Lagrangian $\mathscr{L}^{(m)}$ is now multiplied by the exponential factor in Eq. (3.112). This factor is described as an anomalous coupling of matter to the scalar $\tilde{\phi}$ which has no counterpart in GR. It is because of this coupling that the matter energy-momentum tensor $\tilde{T}^{(m)}_{\alpha\beta}$ in the Einstein frame obeys Eq. (2.12) instead of the GR conservation equation $\nabla^\beta\, T^{(m)}_{\alpha\beta} = 0$. The modified conservation equation implies changes to the geodesic equation and to the equation of geodesic deviation, and the violation of the Equivalence Principle in the Einstein frame.

Under the conformal transformation (2.1), the matter energy-momentum tensor $T^{(m)}_{\mu\nu}$ scales as

$$\tilde{T}^{\alpha\beta}_{(m)} = \Omega^s\, T^{\alpha\beta}_{(m)}\,, \qquad \tilde{T}^{(m)}_{\alpha\beta} = \Omega^{s+4}\, T^{(m)}_{\alpha\beta}\,, \tag{3.114}$$

where $s$ is an appropriate conformal weight. The conservation equation $\nabla^\beta\, T^{(m)}_{\alpha\beta} = 0$ transforms (in four spacetime dimensions) as [1139]

$$\tilde{\nabla}_\alpha \left( \Omega^s\, T^{\alpha\beta}_{(m)} \right) = \Omega^s\, \nabla_\alpha T^{\alpha\beta}_{(m)} + (s+6)\, \Omega^{s-1}\, T^{\alpha\beta}_{(m)} \nabla_a \Omega - \Omega^{s-1} g^{\alpha\beta}\, T^{(m)} \nabla_\alpha \Omega\,. \tag{3.115}$$

It is convenient to choose the conformal weight $s = -6$ which yields, consistently with Eq. (2.15),

$$\tilde{T}^{(m)} \equiv \tilde{g}^{\alpha\beta}\, \tilde{T}^{(m)}_{\alpha\beta} = \Omega^{-4}\, T^{(m)}\,, \tag{3.116}$$

and $\tilde{T}^{(m)}$ vanishes if and only if $T^{(m)} = 0$. Equation (3.115) assumes the form

$$\tilde{\nabla}_\alpha \tilde{T}^{\alpha\beta}_{(m)} = -\tilde{T}^{(m)}\, \tilde{g}^{\alpha\beta}\, \tilde{\nabla}_\alpha \left( \ln \Omega \right)\,. \tag{3.117}$$

Since $\Omega = \sqrt{G\phi}$ it is

$$\tilde{\nabla}_\alpha \tilde{T}^{\alpha\beta}_{(m)} = -\frac{1}{2\phi}\, \tilde{T}^{(m)}\, \tilde{\nabla}^\beta \phi \tag{3.118}$$

or, in terms of the Einstein frame scalar [1133],

$$\tilde{\nabla}_\alpha \tilde{T}^{\alpha\beta}_{(m)} = -\sqrt{\frac{4\pi G}{2\omega+3}}\, \tilde{T}^{(m)}\, \tilde{\nabla}^\beta \tilde{\phi}\,. \tag{3.119}$$

The geodesic equation receives corrections as a consequence of eq. (3.119). Consider a dust fluid with energy-momentum tensor

$$\tilde{T}^{(m)}_{\alpha\beta} = \tilde{\rho}^{(m)}\, \tilde{u}_\alpha\, \tilde{u}_\beta\,; \tag{3.120}$$

Equation (3.119) yields

$$\tilde{u}_\alpha \, \tilde{u}_\beta \, \tilde{\nabla}^\beta \tilde{\rho}^{(m)} + \tilde{\rho}^{(m)} \, \tilde{u}_\alpha \, \tilde{\nabla}^\beta \tilde{u}_\beta + \tilde{\rho}^{(m)} \, \tilde{u}_\gamma \, \tilde{\nabla}^\gamma \, \tilde{u}_\alpha = \sqrt{\frac{4\pi G}{2\omega + 3}} \, \tilde{\rho}^{(m)} \, \tilde{\nabla}_\alpha \tilde{\phi} \, . \quad (3.121)$$

Using an affine parameter $\lambda$ along the fluid worldlines with tangent $\tilde{u}^\mu$, Eq. (3.121) becomes

$$\tilde{u}_\alpha \left( \frac{d\tilde{\rho}^{(m)}}{d\lambda} + \tilde{\rho}^{(m)} \, \tilde{\nabla}^\gamma \tilde{u}_\gamma \right) + \tilde{\rho}^{(m)} \left( \frac{d\tilde{u}_a}{d\lambda} - \sqrt{\frac{4\pi G}{2\omega + 3}} \, \tilde{\nabla}_\alpha \phi \right) = 0 \, . \quad (3.122)$$

This equations splits into the two equations

$$\frac{d\tilde{\rho}^{(m)}}{d\lambda} + \tilde{\rho}^{(m)} \, \tilde{\nabla}^\gamma \tilde{u}_\gamma = 0 \quad (3.123)$$

and

$$\frac{d\tilde{u}^\alpha}{d\lambda} = \sqrt{\frac{4\pi G}{2\omega + 3}} \, \tilde{\nabla}^\alpha \tilde{\phi} \, . \quad (3.124)$$

The geodesic equation is then modified in the Einstein frame as [312, 313, 1133]

$$\frac{d^2 x^\mu}{d\lambda^2} + \tilde{\Gamma}^\mu_{\nu\sigma} \frac{dx^\nu}{d\lambda} \frac{dx^\sigma}{d\lambda} = \sqrt{\frac{4\pi G}{2\omega + 3}} \, \tilde{\nabla}^\mu \tilde{\phi} \, . \quad (3.125)$$

The correction on the right hand side is often described as a fifth force proportional to the gradient $\tilde{\nabla}^\mu \tilde{\phi}$ that couples universally to all massive test particles. The Weak Equivalence Principle (universality of free fall) is violated by this fifth force because of the spacetime dependence of $\tilde{\nabla}^\mu \tilde{\phi}$. Due to this coupling, scalar-tensor theories in the Einstein frame appear to be non-metric theories. On the other hand, it is well known that all metric theories of gravity satisfy the Weak Equivalence Principle [1167], and the (non-)metricity becomes a statement on whether a theory satisfies or not the WEP. Therefore, the metric character of ETGs, and whether they satisfy or not the Equivalence Principle, become properties dependent on the conformal frame representation. This fact leaves the foundation of relativistic gravity on a rather shaky ground, which is a problem especially when trying to isolate the fundamental properties of classical gravity which should be preserved in approaches to quantum or emergent gravity. We do not discuss this issue further and we refer the reader to [1035] and the references therein.

As expected, the equation of null geodesics is left unaffected by the conformal transformation: null geodesics receive no fifth force correction in the Einstein frame. This invariance is consistent with the fact that the equation of null geodesics can be derived from the Maxwell equations in the high frequency limit of the geometric optics approximation, in conjunction with the fact that Maxwell's equations are conformally invariant in a four-dimensional manifold. A more direct way of looking at conformal invariance for null geodesics is by noting that the electromagnetic field stress-energy tensor has vanishing trace $T = 0$ and the corresponding conservation equation $\nabla^\beta T_{\alpha\beta} = 0$ is unaffected by the conformal rescaling $g_{\mu\nu} \to \tilde{g}_{\mu\nu} = \Omega^2 g_{\mu\nu}$, together with the geodesic equation for a null dust described by Eq. (3.120) when $\tilde{u}_\mu \tilde{u}^\mu = 0$.

A correction to the timelike geodesic equation similar to the one discovered in Brans-Dicke theory appears in the low-energy limit of string theory [527, 1068], in which the dilaton replaces the Brans-Dicke field and a similar coupling violates the Equivalence Principle [312, 313, 364, 1068]. The violation is kept small in order not to violate the Solar System bounds [1167]. In the low-energy limit of string theory the dilaton couples with different strengths to bodies of different nuclear composition which carry a dilatonic charge $q$, contrary to the Brans-Dicke field which couples universally to all forms of non-conformal matter. The formal substitution of the dilatonic charge $q$ with the factor $2\sqrt{\pi G/(2\omega+3)}$ allows a parallel between the two theories, but in string theory it may be possible to eliminate the coupling by setting the dilatonic charge $q$ to zero in certain cases, whereas the coupling of the Einstein frame Brans-Dicke scalar cannot be eliminated.

### 3.6.2 *Scalar-tensor theories*

More general scalar-tensor theories are described by the Jordan frame action

$$S_{ST} = \int d^4x \sqrt{-g} \left\{ \frac{1}{16\pi G} \left[ \phi R - \frac{\omega(\phi)}{\phi} g^{\alpha\beta} \nabla_\alpha \phi \nabla_\beta \phi \right] - V(\phi) + \alpha_m \mathcal{L}^{(m)} \right\}, \tag{3.126}$$

where $\alpha_m$ is the coupling constant of ordinary matter. The conformal factor is still given by Eq. (3.109) while the Einstein frame scalar field is defined by the differential relation

$$d\tilde{\phi} = \sqrt{\frac{2\omega(\phi)+3}{16\pi G}} \frac{d\phi}{\phi}. \tag{3.127}$$

The Einstein frame scalar-tensor action is

$$S_{ST} = \int d^4x \sqrt{-\tilde{g}} \left[ \frac{\tilde{R}}{16\pi G} - \frac{1}{2} \tilde{g}^{\alpha\beta} \tilde{\nabla}_\alpha \tilde{\phi} \tilde{\nabla}_\beta \tilde{\phi} - U(\tilde{\phi}) + \tilde{\alpha}_m(\phi) \mathcal{L}^{(m)} \right] \tag{3.128}$$

with scalar field potential

$$U(\tilde{\phi}) = \frac{V[\phi(\tilde{\phi})]}{(G\phi)^2} \tag{3.129}$$

and coupling

$$\tilde{\alpha}_m\left(\tilde{\phi}\right)=\frac{\alpha_m}{(G\phi)^2}\,. \qquad (3.130)$$

Again, Eq. (3.128) can be seen as the action for GR with a canonical scalar field which has positive-definite kinetic energy density, but with the important difference that the matter Lagrangian density is multiplied by the factor $\Omega^{-4}=(G\phi)^{-2}$, which can be interpreted as a variation of the coupling constant $\alpha_m$ with space and/or time. Again, this matter-$\tilde{\phi}$ coupling is responsible for the non-conservation of $\tilde{T}^{(m)}_{\mu\nu}$ as in Eq. (3.119), and for violating the Equivalence Principle.

The conformal transformation technique has been used as a tool for generating exact solutions of a scalar-tensor theory beginning from known solutions of GR [100, 123, 574, 751, 1103–1108], and for deriving approximate solutions of the linearized theory [88]. This solution-generating technique is most convenient for solutions with vanishing potential: in fact, when a potential $V(\phi)$ is present, solutions that correspond to a physically well motivated potential in one frame generate, via the conformal mapping, solutions in the other frame which rarely correspond to a physical potential. Consider, for example, Brans-Dicke theory with a mass term $V(\phi)=m^2\phi^2/2$ in the Jordan frame. The Einstein frame potential is

$$U=\frac{1}{2}\left(\frac{m}{G}\right)^2\,, \qquad (3.131)$$

*i.e.*, a cosmological constant. A given functional form of $V(\phi)$ in the Jordan frame corresponds to a very different form of $U\left(\tilde{\phi}\right)$ in the Einstein frame. Reversing the problem, which Jordan frame potential $V(\phi)$ produces a mass term $U\left(\tilde{\phi}\right)=m^2\tilde{\phi}^2/2$ in the Einstein frame? Equations (3.111) and (3.129) yield the answer

$$V(\phi)=U\left(\tilde{\phi}\right)\phi^2=\frac{m^2G}{32\pi}\,(2\omega+3)\left[\phi\ln\left(\frac{\phi}{\phi_1}\right)\right]^2\,, \qquad (3.132)$$

with $\phi_1$ is a constant. It would be difficult to motivate this potential from a known theory of particle physics. As a conclusion, it is legitimate to use exact solutions in the Einstein frame to generate solutions in the Jordan frame, but this procedure usually produces solutions of limited physical interest.

In $D>2$ spacetime dimensions, the scalar-tensor theory described by the action

$$S_{ST}^{(D)}=\int d^Dx\,\sqrt{-g}\left[f(\phi)R-\omega(\phi)\nabla^\alpha\phi\nabla_\alpha\phi-V(\phi)+\alpha_m\,\mathcal{L}^{(m)}\right]\,, \qquad (3.133)$$

can be conformally transformed according to

$$g_{\alpha\beta}\longrightarrow\tilde{g}_{\alpha\beta}=f(\phi)^{\frac{2}{D-2}}\,g_{\alpha\beta}\,, \qquad (3.134)$$

and

$$d\tilde{\phi} = \frac{d\phi}{f(\phi)} \sqrt{f(\phi) + \frac{D-1}{D-2}\left(\frac{dF}{d\phi}\right)^2} \tag{3.135}$$

producing the new scalar field potential in the Einstein frame

$$U(\tilde{\phi}) = \frac{V[\phi(\tilde{\phi})]}{f(\phi)^{\frac{D}{D-2}}} . \tag{3.136}$$

### 3.6.3 Mixed $f(R)$/scalar-tensor gravity

The action of generalized scalar-tensor gravity

$$S = \frac{1}{16\pi} \int d^4x \sqrt{-g} \left[ f(\phi, R) - \frac{\varepsilon}{2} g^{\alpha\beta} \nabla_\alpha \phi \nabla_\beta \phi \right] \tag{3.137}$$

is mapped into its Einstein frame form by a conformal transformation which was rediscovered many times in particular realizations [97,346,550,764,984,1047,1072, 1144,1160]. The conformal factor is

$$\Omega = \left[ 16\pi G \left| \frac{\partial f}{\partial R} \right| + \text{constant} \right]^{1/2} \tag{3.138}$$

and, together with the scalar field redefinition

$$\tilde{\phi} = \frac{1}{\sqrt{8\pi G}} \sqrt{\frac{3}{2}} \ln \left[ \sqrt{32\pi}\, G \left| \frac{\partial f}{\partial R} \right| \right] , \tag{3.139}$$

allows the action (3.137) to be rewritten in the Einstein frame form

$$S = \alpha \int d^4x \sqrt{-\tilde{g}} \left\{ \frac{\tilde{R}}{16\pi G} - \frac{1}{2} \tilde{g}^{\alpha\beta} \tilde{\nabla}_\alpha \tilde{\phi} \tilde{\nabla}_\beta \tilde{\phi} \right.$$

$$\left. - \frac{\varepsilon\, \alpha}{2} \exp\left[ -\sqrt{\frac{16\pi G}{3}}\, \tilde{\phi} \right] - U(\phi, \tilde{\phi}) \right\} , \tag{3.140}$$

where [610,764]

$$\alpha = \text{sign}\left( \frac{\partial f}{\partial R} \right) , \tag{3.141}$$

and

$$U\left(\phi,\tilde{\phi}\right) = \alpha\, e^{-8\sqrt{\frac{\pi G}{3}}\,\tilde{\phi}} \left[\frac{\alpha R\left(\phi,\tilde{\phi}\right)}{16\pi G} \exp\left(\sqrt{\frac{16\pi G}{3}}\,\tilde{\phi}\right) - F\left(\phi,\tilde{\phi}\right)\right], \tag{3.142}$$

$$F\left(\phi,\tilde{\phi}\right) = f\left[\phi, R\left(\phi,\tilde{\phi}\right)\right]. \tag{3.143}$$

This action describes a non-linear $\sigma$-model with canonical gravity and two scalar fields $\phi$ and $\tilde{\phi}$ which reduce to a single one if $f\left(\phi,\tilde{\phi}\right)$ is linear in $R$. In this case the Einstein frame action is

$$S = \frac{|f|}{f} \int d^4x \sqrt{-\tilde{g}} \left[\frac{\tilde{R}}{16\pi G} - \frac{1}{2}\,\tilde{g}^{\alpha\beta}\,\tilde{\nabla}_\alpha\tilde{\phi}\,\tilde{\nabla}_\beta\tilde{\phi} - U\left(\tilde{\phi}\right)\right], \tag{3.144}$$

where

$$\tilde{\phi} = \frac{1}{\sqrt{8\pi G}} \int d\phi \left[\frac{2\,\varepsilon f(\phi) + 6\,(dF/d\phi)^2}{4f^2(\phi)}\right]^{1/2}, \tag{3.145}$$

$$U\left(\tilde{\phi}\right) = \frac{\mathrm{sign}(f)\, V(\phi)}{(16\pi G\, f)^2}, \tag{3.146}$$

in which $\phi = \phi\left(\tilde{\phi}\right)$.

### *3.6.4 The issue of the conformal frame*

Some considerations are in order at this point. The conformal transformation from the Jordan to the Einstein frame is a mathematical map which allows one to study several aspects of scalar-tensor gravity, $f(R)$ gravity and, in general, any ETG. However, having now available both the Jordan and the Einstein conformal frames (and infinitely many other conformal frames could be defined by choosing the conformal factor $\Omega$ arbitrarily), one wonders whether the two frames are also *physically* equivalent or only mathematically related. In other words, the problem is whether the physical meaning of the theory is "preserved" or not by the use of conformal transformations. One has now the metric $g_{\mu\nu}$ and its conformal cousin $\tilde{g}_{\mu\nu}$ and the question has been posed of which one is the "physical metric", *i.e.*, the metric from which curvature, geometry, and physical effects should be calculated and compared with experiment [227]. The issue of "which frame is the physical one" has been debated for a long time and it regularly resurfaces in the literature, with authors arguing in favor of one frame against the other, and others supporting the view that the two frames are physically equivalent and that the issue is a pseudo-problem. Many errors in the literature over the years, from advocates of both points of view, have contributed to confusion.

The first to approach this issue seems to have been Fierz (see [558]) but the first popular argument is due to Dicke, who presented it in the paper introducing the conformal transformation for Brans-Dicke theory [392]. Dicke's argument is that physics must be invariant under a rescaling of units and the conformal transformation is merely a local rescaling: units are not changed rigidly over the entire spacetime manifold, but by amounts which are different at different spacetime points. In Dicke's view, the two frames are equivalent provided that the units of mass, length, and time, and quantities derived from them scale with appropriate powers of the conformal factor in the Einstein frame [392].

With this view in mind, it is not difficult to see why many authors consider the issue of which conformal frame is physical a pseudo-problem. In principle, it is difficult to object to this argument, but there are two difficulties:

1. Even though Dicke's argument is clear in principle, its application to practical situations is a different matter. The view that the two conformal frames are merely different *representations* of the same theory, similar to different gauges of a gauge theory, should be checked explicitly using the equations describing the physics. "Physical equivalence" is a vague concept because one can consider many different matter (or test) fields in curved spacetime and different types of physics, or different physical aspects of a problem. When checking explicitly the physical equivalence between the two frames, one has to specify which physical field, or physical process is considered and the equations describing it. The equivalence could then be shown explicitly, but there is no proof that holds for all of physics, for example for Klein-Gordon fields, spinors, for cosmology, black holes, *etc.* While physical equivalence has been proved for various physical aspects, no proof comprehensive of all physical fields and different physical applications exists.
2. Dicke's argument is purely classical. In cosmology, black hole physics, and ETGs quantum fields in curved space play a significant role and the equivalence of conformal frames is not clear at all at the quantum level. Of course, not much is known about this equivalence in quantum gravity due to the lack of a definitive quantum gravity theory, but when the metric $g_{\mu\nu}$ is quantized in full quantum gravity approaches, inequivalent quantum theories are found [56, 421, 496, 558]. One can consider the semiclassical regime in which gravity is classical and the matter fields are quantized: again, one would expect the conformal frames to be inequivalent because the conformal transformation can be seen as a Legendre transformation [769], similar to the Legendre transformation of the classical mechanics of point particles which switches from the canonical Lagrangian coordinates $q$ to the variables $\{q, p\}$ of the Hamiltonian formalism. Now, it is well known that Hamiltonians that are classically equivalent become inequivalent when quantized, producing different energy spectra and scattering amplitudes [207,368,543]. However, the conformal equivalence between Jordan and Einstein frame seems to hold to some extent at

the semiclassical level [496].[8] Again, only a particular kind of physics has been considered and one cannot make statements about all possible physical situations.

Unfortunately, the scaling of units in the Einstein frame is often forgotten, producing results that either do not make sense or are partially or totally incorrect, or sometimes the error is inconsequential,[9] reinforcing the opposite view that the two frames are completely equivalent. While Dicke's explanation is very appealing and several claims supporting the view that the two frames are inequivalent turned out to be incorrect because they simply neglected the scaling of units in the Einstein frame, one should not forget that Dicke's argument is not inclusive of all areas of physics and it is better to check explicitly that the physics of a certain field does not depend on the conformal representation and not make sweeping statements. Certain points have been raised in the literature which either constitute a problem for Dicke's view, or, at least, indicate that this viewpoint cannot be applied blindly, including the following.

- Massive test particles follow timelike geodesics in the Jordan frame, while they deviate from geodesic motion in the Einstein frame due to a force proportional to the gradient of the scalar field (equivalently, of the conformal factor or of the varying mass unit [473]). Hence, the Weak Equivalence Principle is satisfied in the Jordan frame but not in the Einstein frame due to the coupling of the scalar field to ordinary matter, or to the variation of units. Since the Equivalence Principle is the foundation of relativistic gravity, this aspect is important and there are two ways to look at it. One can cherish the view that the two conformal frames are equivalent also with respect to the Equivalence Principle, which implies that the latter is formulated in a way that depends on the conformal frame representation. Then, a representation-independent formulation must be sought for [1035]; however, no progress has been made in this direction. Or, one could view the violation of the Weak Equivalence Principle in the Einstein frame more pragmatically by saying that "physical equivalence" of the two frames is a vague term which must be defined precisely and this concept cannot be used blindly, in fact the Equivalence Principle of standard textbooks holds only in one frame but not in the other. This fact could be used as an argument against the physical equivalence of the frames.

---

[8] A common argument among particle physicists relies on the equivalence theorem of Lagrangian field theory stating that the $S$-matrix is invariant under local (nonlinear) field redefinitions [147, 311, 332, 406, 660, 703, 961]. Since the conformal transformation is, essentially, a field redefinition, it would seem that quantum physics is invariant under change of the conformal frame. However, the field theory in which the equivalence theorem is derived applies to gravity only in the perturbative regime in which the fields deviate slightly from Minkowski space. In this regime, tree level quantities can be calculated in any conformal frame with the same result, but in the non-perturbative regime field theory and the equivalence theorem do not apply.

[9] Dicke himself applied the conformal transformation and the scaling of units incorrectly [393] in GR cosmology (see [268, 473]).

- The Brans-Dicke-like scalar field easily violates all of the energy conditions in the Jordan frame, but satisfies them in the Einstein frame. While this fact does not eliminate singularities in one frame leaving them in the other [473] (*i.e.*, the two frames are equivalent with respect to the presence of singularities), one cannot say that the two frames are "equivalent" with respect to the energy conditions. This difficulty arises because part of the matter sector of the theory, in the Einstein frame, comes from the conformal factor; in other words, the conformal transformation mixes matter and geometric degrees of freedom, which is the source of many interpretational problems [232, 1035]. Thus, even if the theory turns out to be independent of the conformal representation, its interpretation is not.
- There are studies of FLRW cosmology in which the universe accelerates in one frame but not in the other. From the pragmatic point of view of an astronomer attempting to fit observational data (for example, type Ia supernovae data to a model of the present acceleration of the universe), the two frames certainly do not appear to be "physically equivalent" [242, 245].

To approach correctly the problem of physical equivalence under conformal transformations, one can compare physics in different conformal frames at the level of the Lagrangian, of the field equations, and of their solutions. This comparison may not always be easy but, in certain cases, it is extremely useful to discriminate between frames. It has been adopted, for example, in [244], to compare cosmological models in the Einstein and the Jordan frame. Specifically, it has been shown that solutions of $f(R)$ and scalar-tensor gravity cannot be assumed to be physically equivalent to those in the Einstein frame when matter fields are given by generalized Equations of State (EoS). The situation is summarized in Table 3.1.

In these, and in other situations, one must specify precisely what "physical equivalence" means. In certain situations physical equivalence is demonstrated simply by taking into account the coupling of the Brans-Dicke-like scalar field to matter and the varying units in the Einstein frame, but in other cases the physical equivalence is not obvious and it does not seem to hold. At the very least, this equivalence, if it

**Table 3.1** Three approaches (EoS, scalar-tensor and $f(R)$)) compared at the level of Lagrangians, field equations, and their solutions. Mathematical equivalence of the three levels does not automatically imply physical equivalence of the solutions.

| | | | | |
|---|---|---|---|---|
| $EoS$ | $\longleftrightarrow$ | $\mathscr{L}_{ST}$ | $\longleftrightarrow$ | $\mathscr{L}_{f(R)}$ |
| $\updownarrow$ | | $\updownarrow$ | | $\updownarrow$ |
| Einstein eqs. | $\longleftrightarrow$ | ST eqs. | $\longleftrightarrow$ | $f(R)$ eqs. |
| $\updownarrow$ | | $\updownarrow$ | | $\updownarrow$ |
| E frame sol. | $\longleftrightarrow$ | E frame sol.+$\phi$ | $\longleftrightarrow$ | J frame sol. |

is valid at all, must be defined in precise terms and discussed in ways that are far from obvious. For this reason, it would be too simplistic to dismiss the issue of the conformal frame entirely as a pseudo-problem that has been solved for all physical situations of interest. It is fair to say that there have been surprises and non-trivial difficulties have been uncovered.

## 3.7 The initial value problem

Another important problem is the initial value formulation. There are several criteria for the viability of a theory of gravity, and one of them is certainly the property of having a well-posed initial value problem in order to guarantee that the theory has predictive power [1139]. The Cauchy problem of theories described by the action

$$S = \int d^4x\sqrt{-g}\left(\frac{R}{2\kappa} + \alpha R_{\mu\nu}R^{\mu\nu} + \beta R^2 + \gamma R_{\mu\nu\rho\sigma}R^{\mu\nu\rho\sigma}\right) + S^{(m)} \tag{3.147}$$

was studied in [840, 1072] and found to be well-posed (in four spacetime dimensions, the Gauss-Bonnet identity allows one to drop the Kretschmann scalar $R_{\mu\nu\rho\sigma}R^{\mu\nu\rho\sigma}$ from the action). Reference [704] discusses the initial value problem of metric and Palatini $f(R)$ gravity with a general (*i.e.*, not restricted to the quadratic form considered in [840,1072]) function $f(R)$ but dropping $R_{\mu\nu}R^{\mu\nu}$ and $R_{\mu\nu\rho\sigma}R^{\mu\nu\rho\sigma}$. It is useful to begin with the Cauchy problem of scalar-tensor gravity, which was studied in [962], and then to use the results and the equivalence of metric and Palatini $f(R)$ gravity with an $\omega = 0, -3/2$ Brans-Dicke theory.

Let us provide some terminology [1027]: the system of $3+1$ equations of motion of GR or ETGs is *well-formulated* if it can be recast as a system of only first order equations in time and space in the scalar field variables. The goal is to write this system in the full first order form

$$\partial_t \mathbf{u} + M^i \nabla_i \mathbf{u} = \mathbf{S}(\mathbf{u}) \,, \tag{3.148}$$

where $\mathbf{u}$ denotes the fundamental variables $h_{ij}$, $K_{ij}$, *etc.* of the usual $3+1$ Arnowitt-Deser-Misner (ADM) splitting, $M^i$ is called the *characteristic matrix* of the system, and $\mathbf{S}(\mathbf{u})$ describes source terms and contains only the fundamental variables but not their derivatives. The initial value formulation is then said to be *well-posed* if the system of partial differential equations is *symmetric hyperbolic* (*i.e.*, $M^i$ is symmetric) and *strongly hyperbolic* (*i.e.*, $s_i M^i$ has a real set of eigenvalues and a complete set of eigenvectors for any one-form $s_i$, and obeys some boundedness conditions).

For a physical theory to be viable, it must admit an appropriate initial value formulation to guarantee its predictability [350]. This means that, starting from suitably prescribed initial data, the subsequent dynamical evolution of the physical system is completely and uniquely determined. In this case, the problem is said to be *well-formulated.* For example, in classical mechanics, given the initial positions and velocities of the particles (or of the constituents) composing a physical system with a finite number of degrees of freedom and knowing the interactions between them, if the system evolves without external interferences the dynamical evolution is determined. This is true also for field theories, for example, for the Maxwell field. However, even if the initial value problem is well-formulated, the theory must possess additional properties in order to be viable. First, small changes of the initial data must produce only small perturbations in the subsequent dynamics over reasonably short time scales, in other words the evolution equations should exhibit a continuous dependence on the initial data in order to be predictive. Second, for hyperbolic equations, changes in the initial data must preserve the causal structure of the theory. If both these requirements are satisfied, the initial value problem of the theory is also *well-posed.*

GR has been shown to admit a well-formulated and well-posed initial value problem *in vacuo* and in the presence of "reasonable" forms of matter (perfect fluids, minimally coupled scalar fields, *etc.*) but, for other relativistic field theories, the initial value formulation must be studied carefully. One needs to satisfy constraints between the initial data and perform wise gauge choices in order to cast the field equations in a form suitable to correctly formulate the Cauchy problem. The consequence of well-posedness is that GR is a "stable" theory with a robust causal structure in which singularities can be classified (for a detailed discussion see [1065, 1139]).

Here we focus on whether the initial value problem of ETGs (including scalar-tensor and $f(R)$ theories in both the metric and metric-affine formulation) is well-formulated. It is not *a priori* obvious that standard GR methods are suitable for the discussion of the Cauchy problem in every ETG and it is doubtful that the full Cauchy problem can be properly addressed using only the results available in the literature for the fourth order theories described by a quadratic Lagrangian [402, 1072]. $f(R)$ gravity, like GR, is a gauge theory with constrained dynamics and establishing results on the initial value formulation relies on solving the constraints on the initial data and on finding suitable gauges, coordinate choices in which the Cauchy problem can be demonstrated to well-formulated and, possibly, well-posed. In [840, 1072] the initial value problem is studied for quadratic Lagrangians in the metric approach with the conclusion that it is well-posed. The Cauchy problem for generic $f(R)$ models is studied below in the metric and Palatini approaches with the result that the problem is well-formulated for the metric theory in the presence of "reasonable" matter and well-posed *in vacuo*.

It is shown below that the Cauchy problem of metric-affine $f(R)$ gravity is well-formulated and well-posed *in vacuo*, while it can be at least well-formulated for various forms of matter including perfect fluids, Klein-Gordon, and Yang-Mills fields. We use the $3+1$ ADM formulation and the Gaussian normal coordinates approach, both of which prove useful in the discussion of whether the Cauchy problem is well-formulated. Of course, in order to prove the complete viability of a theory, also well-posedness has to be demonstrated.

### 3.7.1 The Cauchy problem of scalar-tensor gravity

Early work on the initial value problem of scalar-tensor gravity includes [329, 840, 1072]. Noakes [840] proved well-posedness of the Cauchy problem for a non-minimally coupled scalar field $\phi$ with vacuum action

$$S_{NMC} = \int d^4x \sqrt{-g} \left[ \left( \frac{1}{2\kappa} - \xi\phi^2 \right) R - \frac{1}{2} \nabla^\mu\phi \nabla_\mu\phi - V(\phi) \right] . \tag{3.149}$$

Cocke and Cohen [329] used Gaussian normal coordinates to study the Cauchy problem of Brans-Dicke theory without potential $V(\phi)$. A systematic approach to the Cauchy problem of scalar-tensor theories of the form

$$S = \int d^4x \sqrt{-g} \left[ \frac{f(\phi)R}{2\kappa} - \frac{1}{2} \nabla^\alpha\phi \nabla_\alpha\phi - V(\phi) \right] + S^{(m)} \tag{3.150}$$

independent of particular coordinate choices was proposed by Salgado [962], obtaining the result that the Cauchy problem is well-posed *in vacuo* and well-formulated otherwise. Slightly more general scalar-tensor theories of the form

$$S = \int d^4x \sqrt{-g} \left[ \frac{f(\phi)R}{2\kappa} - \frac{\omega(\phi)}{2} \nabla^\mu\phi \nabla_\mu\phi - V(\phi) \right] + S^{(m)} , \tag{3.151}$$

containing the additional coupling function $\omega(\phi)$ were studied in [704].

In the notation of [962], and setting $\kappa = 1$ in this section, the field equations are

$$\begin{aligned} G_{\mu\nu} = & \frac{1}{f} \left[ f'' \left( \nabla_\mu\phi \nabla_\nu\phi - g_{\mu\nu} \nabla^\alpha\phi \nabla_\alpha\phi \right) + f' \left( \nabla_\mu \nabla_\nu \phi - g_{\mu\nu} \Box\phi \right) \right] \\ & + \frac{1}{f} \left[ \omega \left( \nabla_\mu\phi \nabla_\nu\phi - \frac{1}{2} g_{\mu\nu} \nabla^\alpha\phi \nabla_\alpha\phi \right) - V(\phi) g_{\mu\nu} + T^{(m)}_{\mu\nu} \right] , \end{aligned} \tag{3.152}$$

$$\omega \Box\phi + \frac{f'}{2} R - V'(\phi) + \frac{\omega'}{2} \nabla^\alpha\phi \nabla_\alpha\phi = 0 , \tag{3.153}$$

where a prime denotes differentiation with respect to $\phi$. Equation (3.152) is in the form of an effective Einstein equation [962]

$$G_{\mu\nu} = T^{(eff)}_{\mu\nu} = \frac{1}{f(\phi)} \left( T^{(f)}_{\mu\nu} + T^{(\phi)}_{\mu\nu} + T^{(m)}_{\mu\nu} \right) , \tag{3.154}$$

where

$$T^{(f)}_{\mu\nu} = f''(\phi) \left( \nabla_\mu\phi \nabla_\nu\phi - g_{\mu\nu} \nabla^\alpha\phi \nabla_\alpha\phi \right) + f'(\phi) \left( \nabla_\mu \nabla_\nu \phi - g_{\mu\nu} \Box\phi \right) , \tag{3.155}$$

and

$$T_{\mu\nu}^{(\phi)} = \omega(\phi)\left(\nabla_\mu \phi \nabla_\nu \phi - \frac{1}{2} g_{\mu\nu} \nabla^\alpha \phi \, \nabla_\alpha \phi\right) - V(\phi) g_{\mu\nu} \tag{3.156}$$

has canonical structure. The trace of Eq. (3.154) yields

$$\Box\phi = \frac{\frac{f' T^{(m)}}{2} - 2f'V(\phi) + fV'(\phi) + \left[-\frac{\omega' f}{2} - \frac{f'}{2}(\omega + 3f'')\right] \nabla^\alpha \phi \nabla_\alpha \phi}{f\left[\omega + \frac{3(f')^2}{2f}\right]} . \tag{3.157}$$

One then proceeds in the usual 3 + 1 ADM formulation of the theory in terms of lapse, shift, extrinsic curvature, and gradient of $\phi$ [932, 962, 1139]. It is assumed that a time function $t$ is defined so that the spacetime $(M, g_{\mu\nu}, \phi)$ is foliated by a family of hypersurfaces $\Sigma_t$ of constant $t$ with unit timelike normal $n^\mu$. The three-dimensional metric is $h_{\mu\nu} \equiv g_{\mu\nu} + n_\mu n_\nu$, $h^\mu{}_\nu$ is the projection operator on $\Sigma_t$, and $n^\mu$ and $h_\mu{}^\nu$ satisfy

$$n^\mu n_\mu = -1 , \qquad h_{\mu\nu} n^\mu = h_{\mu\nu} n^\nu = 0 , \qquad h_\mu{}^\nu h_{\nu\rho} = h_{\mu\rho} . \tag{3.158}$$

The metric decomposition in terms of lapse function $N$ and shift vector $N^\mu$ is

$$ds^2 = -\left(N^2 - N^i N_i\right) dt^2 - 2N_i dt dx^i + h_{ij} dx^i dx^j , \tag{3.159}$$

where $i, j, k$ are spatial indices assuming the values 1, 2, and 3, $N > 0$, $n_\mu = -N \nabla_\mu t$, and

$$N^\mu = -h^\mu{}_\nu t^\nu , \tag{3.160}$$

and where the time flow vector $t^\mu$ obeys

$$t^\mu \nabla_\mu t = 1 , \tag{3.161}$$

$$t^\mu = -N^\mu + N n^\mu . \tag{3.162}$$

As a consequence, $N = -n_\mu t^\mu$ and $N^\mu n_\mu = 0$. The extrinsic curvature of the hypersurfaces $\Sigma_t$ is

$$K_{\mu\nu} = -h_\mu{}^\rho h_\nu{}^\sigma \nabla_\rho n_\sigma \tag{3.163}$$

and the three-dimensional covariant derivative of $h_{\mu\nu}$ on $\Sigma_t$ is given by

$$D_i^{(3)} T^{\mu_1 \dots}{}_{\nu_1 \dots} = h^{\mu_1}{}_{\rho_1} \dots h^{\sigma_1}{}_{\nu_1} \dots h^f{}_i \nabla_f^{(3)} T^{\rho_1 \dots}{}_{\sigma_1 \dots} \tag{3.164}$$

for any three-tensor ${}^{(3)}T^{\mu_1 \dots}{}_{\nu_1 \dots}$, with $D_i h_{\mu\nu} = 0$. The spatial gradient of the scalar field is

$$Q_\mu \equiv D_\mu \phi , \tag{3.165}$$

its momentum is

$$\Pi = \mathscr{L}_n \phi = n^\mu \nabla_\mu \phi \tag{3.166}$$

and

$$K_{ij} = -\nabla_i n_j = -\frac{1}{2N}\left(\frac{\partial h_{ij}}{\partial t} + D_i N_j + D_j N_i\right), \tag{3.167}$$

$$\Pi = \frac{1}{N}\left(\partial_t \phi + N^\alpha Q_\alpha\right), \tag{3.168}$$

while

$$\partial_t Q_i + N^l \partial_l Q_i + Q_l \partial_i N^l = D_i\left(N\Pi\right). \tag{3.169}$$

The ADM decomposition of the effective energy-momentum tensor $T_{\mu\nu}^{(eff)}$ is then

$$T_{\mu\nu}^{(eff)} = \frac{1}{f}\left(S_{\mu\nu} + J_\mu n_\nu + J_\nu n_\mu + E n_\mu n_\nu\right), \tag{3.170}$$

where

$$S_{\mu\nu} \equiv h_\mu{}^\rho h_\nu{}^\sigma T_{\rho\sigma}^{(eff)} = \frac{1}{f}\left(S_{\mu\nu}^{(f)} + S_{\mu\nu}^{(\phi)} + S_{\mu\nu}^{(m)}\right), \tag{3.171}$$

$$J_\mu \equiv -h_\mu{}^\rho T_{\rho\sigma}^{(eff)} n^\sigma = \frac{1}{f}\left(J_\mu^{(f)} + J_\mu^{(\phi)} + J_\mu^{(m)}\right), \tag{3.172}$$

$$E \equiv n^\mu n^\nu T_{\mu\nu}^{(eff)} = \frac{1}{f}\left(E^{(f)} + E^{(\phi)} + E^{(m)}\right), \tag{3.173}$$

and $T^{(eff)} = S - E$, where $T^{(eff)} \equiv T^{(eff)\mu}{}_\mu$ and $S \equiv S^\mu{}_\mu$. Using the Gauss-Codazzi equations [1139], the effective Einstein equations projected tangentially and orthogonally to $\Sigma_t$ yield the Hamiltonian constraint [962, 1139]

$${}^{(3)}R + K^2 - K_{ij}K^{ij} = 2E, \tag{3.174}$$

the momentum constraint

$$D_l K^l{}_i - D_i K = J_i, \tag{3.175}$$

and the dynamical equations

$$\begin{aligned} &\partial_t K^i{}_j + N^l \partial_l K^i{}_j + K^i{}_l \partial_j N^l - K^l{}_j \partial_l N^i + D^i D_j N \\ &-{}^{(3)}R^i{}_j N - NKK^i{}_j = \frac{N}{2}\left[(S-E)\,\delta^i_j - 2S^i_j\right], \end{aligned} \tag{3.176}$$

where $K \equiv K^i{}_i$. The trace of Eq. (3.176) leads to

$$\partial_t K + N^l \partial_l K + {}^{(3)}\Delta N - NK_{ij}K^{ij} = \frac{N}{2}\left(S + E\right), \tag{3.177}$$

where ${}^{(3)}\Delta \equiv D^i D_i$. The second order derivatives of $\phi$ are in principle troublesome because they could make the initial value problem ill-formulated, but they can be eliminated in most cases [962]. The $f$- and $\phi$-quantities of Eqs. (3.171)–(3.173) turn out to be

$$E^{(f)} = f' \left( D^\alpha Q_\alpha + K\Pi \right) + f'' Q^2 , \tag{3.178}$$

$$J_\mu^{(f)} = -f' \left( K_\mu^\rho Q_\rho + D_\mu \Pi \right) - f'' \Pi Q_\mu , \tag{3.179}$$

$$S_{\mu\nu}^{(f)} = f' \left( D_\mu Q_\nu + \Pi K_{\mu\nu} - h_{\mu\nu} \Box\phi \right) - f'' \left[ h_{\mu\nu} \left( Q^2 - \Pi^2 \right) - Q_\mu Q_\nu \right] , \tag{3.180}$$

where $Q^2 \equiv Q^\alpha Q_\alpha$. The quantities

$$S^{(f)} = f' \left( D_\alpha Q^\alpha + K\Pi - 3\Box\phi \right) + f'' \left( 3\Pi^2 - 2Q^2 \right) , \tag{3.181}$$

$$S^{(f)} - E^{(f)} = -3f'\Box\phi - 3f'' \left( Q^2 - \Pi^2 \right) , \tag{3.182}$$

are also useful [962], and the introduction of $\omega$ and $\omega'$ yields the further quantities

$$E^{(\phi)} = \frac{\omega}{2} \left( \Pi^2 + Q^2 \right) + V(\phi) , \tag{3.183}$$

$$J_\mu^{(\phi)} = -\omega \Pi Q_\mu , \tag{3.184}$$

$$S_{\mu\nu}^{(\phi)} = \omega Q_\mu Q_\nu - h_{\mu\nu} \left[ \frac{\omega}{2} \left( Q^2 - \Pi^2 \right) + V(\phi) \right] , \tag{3.185}$$

while

$$S^{(\phi)} = \frac{\omega}{2} \left( 3\Pi^2 - Q^2 \right) - 3V(\phi) \tag{3.186}$$

and

$$S^{(\phi)} - E^{(\phi)} = \omega \left( \Pi^2 - Q^2 \right) - 4V(\phi) . \tag{3.187}$$

Finally, the quantities appearing on the right hand sides of the 3 + 1 field equations are

$$E = \frac{1}{f} \left[ f' \left( D^\alpha Q_\alpha + K\Pi \right) + \left( f'' + \frac{\omega}{2} \right) Q^2 + \frac{\omega}{2} \Pi^2 + V(\phi) + E^{(m)} \right] , \tag{3.188}$$

$$J_\mu = \frac{1}{f} \left[ -f' \left( K_\mu{}^\alpha Q_\alpha + D_\mu \Pi \right) - \left( f'' + \omega \right) \Pi Q_\mu + J_\mu^{(m)} \right] , \tag{3.189}$$

$$S_{\mu\nu} = \frac{1}{f} \left\{ f' \left( D_\mu Q_\nu + \Pi K_{\mu\nu} \right) - h_{\mu\nu} \left[ \left( f'' + \frac{\omega}{2} \right) \left( Q^2 - \Pi^2 \right) + V(\phi) + f'\Box\phi \right] \right\} + \frac{1}{f} \left[ \left( \omega + f'' \right) Q_\mu Q_\nu + S_{\mu\nu}^{(m)} \right] , \tag{3.190}$$

while

$$S = \frac{1}{f}\left[f'\left(D^{\alpha}Q_{\alpha} + \Pi K\right) - 3V(\phi) - 3f'\Box\phi\right]$$
$$-\frac{1}{f}\left[\left(2f'' + \frac{\omega}{2}\right)Q^2 - 3\left(f'' + \frac{\omega}{2}\right)\Pi^2 + S^{(m)}\right], \tag{3.191}$$

$$S - E = \frac{1}{f}\left[\left(3f'' + \omega\right)\left(\Pi^2 - Q^2\right) - 4V(\phi) - 3f'\Box\phi + S^{(m)} - E^{(m)}\right], \tag{3.192}$$

$$S + E = \frac{1}{f}\left[2f'\left(D^{\alpha}Q_{\alpha} + K\Pi\right) - f''Q^2 + \left(3f'' + 2\omega\right)\Pi^2\right]$$
$$+\frac{1}{f}\left(-2V(\phi) - 3f'\Box\phi + S^{(m)} + E^{(m)}\right). \tag{3.193}$$

The Hamiltonian and the momentum constraints assume the form

$${}^{(3)}R + K^2 - K_{ij}K^{ij} - \frac{2}{f}\left[f'\left(D_{\alpha}Q^{\alpha} + K\Pi\right) + \frac{\omega}{2}\Pi^2 + \frac{Q^2}{2}\left(\omega + 2f''\right)\right]$$
$$= \frac{2}{f}\left(E^{(m)} + V(\phi)\right), \tag{3.194}$$

$$D_l K^l{}_i - D_i K + \frac{1}{f}\left[f'\left(K_i{}^{\alpha}Q_{\alpha} + D_i\Pi\right) + \left(\omega + f''\right)\Pi Q_i\right] = \frac{J_i^{(m)}}{f}, \tag{3.195}$$

respectively, and the dynamical equation (3.176) is written as

$$\partial_t K^i{}_j + N^l\partial_l K^i{}_j + K^i{}_l\partial_j N^l - K_j{}^l\partial_l N^i + D^iD_jN - {}^{(3)}R^i{}_jN - NKK^i{}_j$$
$$+\frac{N}{2f}\left[f''\left(Q^2 - \Pi^2\right) + 2V(\phi) + f'\Box\phi\right]\delta^i_j + \frac{Nf'}{f}\left(D^iQ_j + \Pi K^i{}_j\right)$$
$$+\frac{N}{f}\left(\omega + f''\right)Q^iQ_j = \frac{N}{2f}\left[\left(S^{(m)} - E^{(m)}\right)\delta^i_j - 2S^{(m)\,i}{}_j\right]. \tag{3.196}$$

The trace of this equation is

$$\partial_t K + N^l\partial_l K + {}^{(3)}\Delta N - NK_{ij}K^{ij} - \frac{Nf'}{f}\left(D^{\alpha}Q_{\alpha} + \Pi K\right)$$
$$+\frac{N}{2f}\left[f''Q^2 - \left(2\omega + 3f''\right)\Pi^2\right] = \frac{N}{2f}\left(-2V(\phi) - 3f'\Box\phi + S^{(m)} + E^{(m)}\right) \tag{3.197}$$

where [962]

$$\mathcal{L}_n \Pi - \Pi K - Q^\alpha D_\alpha (\ln N) - D_\alpha Q^\alpha = -\Box\phi$$
$$= -\frac{1}{f\left[\omega + \frac{3(f')^2}{2f}\right]}\left(\frac{f' T^{(m)}}{2} - 2f'V(\phi) + fV'(\phi)\right)$$
$$-\frac{1}{f\left[\omega + \frac{3(f')^2}{2f}\right]}\left\{\left[\frac{-\omega' f}{2} - (\omega + 3f'')\frac{f'}{2}\right]\nabla^\alpha\phi\nabla_\alpha\phi\right\}. \qquad (3.198)$$

The initial data *in vacuo* $(h_{ij}, K_{ij}, \phi, Q_i, \Pi)$ on an initial hypersurface $\Sigma_0$ must satisfy the constraints (3.194) and (3.195), in addition to

$$Q_i - D_i\phi = 0\,, \qquad (3.199)$$

$$D_i Q_j = D_j Q_i\,. \qquad (3.200)$$

When matter is present, the additional variables $E^{(m)}$, $J_\mu^{(m)}$, $S_{\mu\nu}^{(m)}$ must be assigned on the initial hypersurface. Prescribing lapse $N$ and shift $N^\mu$ is equivalent to fixing a gauge.[10] The differential system (3.194)–(3.197) contains only first-order derivatives in both space and time once the d'Alembertian $\Box\phi$ is written in terms of $\phi, \nabla^\mu\phi\nabla_\mu\phi$, $f$, and its derivatives by means of Eq. (3.198) [704, 962]. Following [962], the reduction to a first-order system indicates that the Cauchy problem is well-posed *in vacuo* and well-formulated in the presence of matter.

### 3.7.2 *The initial value problem of $f(R)$ gravity in the ADM formulation*

In the notations of this section (cf. [962]), Brans-Dicke theory corresponds to $\omega(\phi) = \omega_0/\phi$, where $\omega_0$ is the constant Brans-Dicke parameter, $f(\phi) = \phi$, and $V \to 2V$. The Hamiltonian and momentum constraints are then

$${}^{(3)}R + K^2 - K_{ij}K^{ij} - \frac{2}{\phi}\left[D^\alpha Q_\alpha + K\Pi + \frac{\omega_0}{2\phi}\left(\Pi^2 + Q^2\right)\right]$$
$$= \frac{2}{\phi}\left[E^{(m)} + V(\phi)\right], \qquad (3.201)$$

$$D_l K^l{}_i - D_i K + \frac{1}{\phi}\left(K_i{}^l Q_l + D_i \Pi + \frac{\omega_0}{\phi}\Pi Q_i\right) = \frac{J_i^{(m)}}{\phi}. \qquad (3.202)$$

---

10 Various gauge conditions employed in the literature are surveyed in [962].

The dynamical equations read

$$\partial_t K^i{}_j + N^l \partial_l K^i{}_j + K^i{}_l \partial_j N^l - K_j{}^l \partial_l N^i + D^i D_j N$$
$$-{}^{(3)}R^i{}_j N - NKK^i_j + \frac{N}{2\phi}\delta^i_j \left(2V(\phi) + \Box\phi\right) + \frac{N}{\phi}\left(D^i Q_j + \Pi K^i{}_j\right)$$
$$+\frac{N\omega_0}{\phi^2} Q^i Q_j = \frac{N}{2\phi}\left(\left(S^{(m)} - E^{(m)}\right)\delta^i_j - 2S^{(m)\,i}{}_j\right), \tag{3.203}$$

$$\partial_t K + N^l \partial_l K + {}^{(3)}\Delta N - NK_{ij}K^{ij} - \frac{N}{\phi}\left(D^\alpha Q_\alpha + \Pi K\right) - \frac{\omega_0 N}{\phi^2}\Pi^2$$
$$= \frac{N}{2\phi}\left[-2V(\phi) - 3\Box\phi + S^{(m)} + E^{(m)}\right], \tag{3.204}$$

where

$$\left(\omega_0 + \frac{3}{2}\right)\Box\phi = \frac{T^{(m)}}{2} - 2V(\phi) + \phi V'(\phi) + \frac{\omega_0}{\phi}\left(\Pi^2 - Q^2\right). \tag{3.205}$$

Remembering the results of the previous subsection for scalar-tensor gravity, metric $f(R)$ gravity theories equivalent to $\omega_0 = 0$ Brans-Dicke gravities have a well-formulated Cauchy problem in general, which is also well-posed *in vacuo*. Palatini $f(R)$ theories, which are equivalent to Brans-Dicke gravities with $\omega_0 = -3/2$ are different. For $\omega_0 = -3/2$, the d'Alembertian $\Box\phi$ disappears from Eq. (3.205) and the field $\phi$ is not dynamical – it can be assigned arbitrarily subject only to the constraint that its gradient satisfies the degenerate equation (3.205). Unless $\Box\phi = 0$, in which case the argument does not apply, it is impossible to eliminate $\Box\phi$ from the system (3.201)–(3.204) in Palatini $f(R)$ gravity. The condition $\Box\phi = 0$ includes some important cases [255, 444]: first, the possibility that $\phi =$ constant reduces the theory to GR, which has a well-posed initial value problem. Second, Palatini $f(R)$ gravity *in vacuo* reduces to GR with a cosmological constant (which is known to have a well-posed Cauchy problem), the Ricci curvature is constant, and $\Box\phi$ vanishes identically [255–257, 444, 1033]. The argument can be extended to any situation in which the stress-energy tensor of matter has a constant trace $T^{(m)}$ [444], although this occurrence does not seem to be very physical apart from *vacuum*, the cosmological constant, and conformal matter which has $T^{(m)} = 0$. When $T^{(m)} \neq 0$, it seems virtually impossible to eliminate time derivatives and all second derivatives from the $3+1$ field equations and no theorem is known about the well-posedness of the Cauchy problem in this case.

### 3.7.3 The Gaussian normal coordinates approach

A different approach to the initial value problem uses Gaussian normal coordinates (also called *synchronous coordinates*) instead of the ADM decomposition. Before discussing ETGs, we recall the initial value formulation of GR in these coordinates, which is well-formulated and well-posed, as shown in [1139]. We adopt the formalism developed in [1065].

#### 3.7.3.1 The Cauchy problem of GR

Let us consider a system of Gaussian normal coordinates [1139], in which the metric tensor has components $g_{00} = -1$ and $g_{0i} = 0$. These coordinates serve the purpose of splitting the spacetime manifold $M$ into a spatial hypersurface $\Sigma_3$ of constant time from the orthogonal time direction.

Given a second rank symmetric tensor $W_{\mu\nu}$ on the globally hyperbolic spacetime $(M, g_{\mu\nu})$, we define its (symmetric) conjugate tensor

$$W^*_{\mu\nu} = W_{\mu\nu} - \frac{W}{2}\, g_{\mu\nu}\,, \tag{3.206}$$

where | denotes the covariant derivative with respect to the Levi-Civita connection induced by $g_{\mu\nu}$ and $W \equiv W^{\mu\nu} g_{\mu\nu}$ is the trace of $W_{\mu\nu}$. If $V_0$ is a spacetime domain in $M$ in which $g_{00} \neq 0$ and $\Sigma_3$ is the three-surface of equation $x^0 = 0$, then the following statements are equivalent:

1. $W_{\mu\nu} = 0$ in $V_0$;
2. $W^*_{ij} = 0$ and $W_{0\alpha} = 0$ in $V_0$;
3. $W^*_{ij} = 0$ and $W^{\mu}_{\nu|\mu} = 0$ in $V_0$ with $W_{o\mu} = 0$ in $\Sigma_3$.

Let us consider the Einstein equations $G_{\mu\nu} = \kappa T^{(m)}_{\mu\nu}$ and the contracted Bianchi identities $\nabla^{\nu} T^{(m)}_{\mu\nu} = 0$; introducing the tensor

$$W_{\mu\nu} \equiv G_{\mu\nu} - \kappa\, T^{(m)}_{\mu\nu}\,, \tag{3.207}$$

the conjugate tensor is

$$W^*_{\mu\nu} = R_{\mu\nu} - \kappa\, T^*_{\mu\nu}\,, \tag{3.208}$$

and the Einstein equations are

$$W_{\mu\nu} = 0\,. \tag{3.209}$$

These are 10 equations for the 20 unknown functions $g_{\mu\nu}$ and $T^{(m)}_{\mu\nu}$. We assign the 10 functions $g_{0\mu}$ and $T^{(m)}_{ij}$; the remaining 10 functions $g_{ij}$ and $T^{(m)}_{0\mu}$ are determined by Eq. (3.209). These functions can be expressed in the equivalent form

$$R_{ij} - \kappa\, T^*_{ij} = 0\,, \qquad W^{\mu}_{\nu|\mu} = T^{\mu}_{\nu|\mu} = 0\,, \tag{3.210}$$

with the condition

$$G_{0\mu} - \kappa\, T^{(m)}_{0\mu} = 0 \tag{3.211}$$

on the hypersurface $x^0 = 0$. Eqs. (3.210) can be rewritten as

$$g_{ij,00} = 2\,\bar{R}_{ij} - \frac{A}{2}\,g_{ij,0} + g^{lm}\,g_{il,0}g_{jm,0} + 2\kappa\,T^*_{ij}\,, \tag{3.212}$$

$$T^{(m)}_{0\nu,0} = -T^{(m)0}{}_{\nu,0} = T^{(m)i}{}_{\nu,i} + \Gamma_{i\mu}{}^{i}\,T^{(m)\mu}{}_{\nu} - \Gamma_{i\nu}{}^{\mu}\,T^{(m)i}{}_{\nu}\,, \tag{3.213}$$

where $\bar{R}_{ij}$ is the intrinsic Ricci tensor of the hypersurface $x^0 = 0$, $\Gamma_{\mu\nu}{}^{\rho}$ is the Levi-Civita connection of the metric $g_{\mu\nu}$, and

$$A \equiv g^{ij}\,g_{ij,0}\,. \tag{3.214}$$

In the same way, the constraint equation (3.211) becomes

$$A_{,i} - D^j\,g_{ij,0} + 2\kappa\,T^{(m)}_{0i} = 0\,, \tag{3.215}$$

$$\bar{R} - \frac{A^2}{4} + \frac{B}{4} + 2\kappa\,T_{00} = 0\,, \tag{3.216}$$

where $\bar{R}$ is the intrinsic Ricci scalar of the hypersurface $x^0 = 0$, $D_i$ denotes the covariant derivative operator on this hypersurface associated with the Levi-Civita connection of the intrinsic metric $g_{ij}\,|_{\Sigma_0}$ and

$$B = g^{ij}\,g^{lm}\,g_{il,0}g_{im,0}\,. \tag{3.217}$$

Let us assign now the Cauchy data

$$g_{ij}\,, \quad g_{ij,0}\,, \quad T^{(m)}_{\mu 0} \tag{3.218}$$

on the hypersurface $x^0 = 0$; they must satisfy the constraint equations (3.215), (3.216), (3.212), and Eq. (3.213) gives the quantities

$$g_{ij,00}\,, \qquad T^{(m)}_{0\mu,0}\,, \tag{3.219}$$

as functions of the Cauchy data. By differentiating Eqs. (3.212) and (3.213), it is straightforward to obtain time derivatives of higher order as functions of the initial data. This procedure allows one to locally reconstruct the solution of the field equations as a power series of $x^0$. The initial three-surface $\Sigma_3$ is then a Cauchy hypersurface for the globally hyperbolic spacetime $(M, g_{\mu\nu})$ and the initial value problem is well-formulated in GR. Our goal is now to extend these results to $f(R)$ gravity in the metric-affine formalism.

#### 3.7.3.2 The Cauchy problem of vacuum $f(\mathscr{R})$ gravity in the metric-affine formalism

In the metric-affine formulation of $f(R)$ gravity the independent variables are $(g_{\mu\nu}, \Gamma^{\alpha}_{\mu\nu})$, where $g_{\mu\nu}$ is the metric and $\Gamma^{\alpha}_{\mu\nu}$ is the linear connection. *In vacuo*, the field equations are obtained by varying the action

$$S[g,\Gamma] = \int d^4x \sqrt{-g}\, f(\mathscr{R}) \tag{3.220}$$

with respect to the metric and the connection, where $\mathscr{R}(g,\Gamma) = g^{\mu\nu}\mathscr{R}_{\mu\nu}$ is the scalar curvature of the connection $\Gamma^{\alpha}_{\mu\nu}$ and $\mathscr{R}_{\mu\nu}$ is the Ricci tensor constructed with this connection. The metric connection $\Gamma^{\alpha}_{\mu\nu}$ can have a non-vanishing torsion while, in the Palatini approach, $\Gamma^{\alpha}_{\mu\nu}$ is a non-metric but torsion-free connection [220].

*In vacuo*, the field equations of $f(R)$ gravity with torsion are [220–222]

$$f'(\mathscr{R})\,\mathscr{R}_{\mu\nu} - \frac{f(\mathscr{R})}{2}\, g_{\mu\nu} = 0\,, \tag{3.221}$$

$$T_{\mu\nu}{}^{\sigma} = -\frac{1}{2f'}\frac{\partial f'}{\partial x^{\rho}}\left(\delta^{\rho}_{\mu}\delta^{\sigma}_{\nu} - \delta^{\rho}_{\nu}\delta^{\sigma}_{\mu}\right), \tag{3.222}$$

while the field equations of $f(R)$ gravity *à la* Palatini are [767,768,868,1030,1034]

$$f'(\mathscr{R})\,\mathscr{R}_{\mu\nu} - \frac{f(\mathscr{R})}{2}\, g_{\mu\nu} = 0\,, \tag{3.223}$$

$$\nabla_{\mu}\left[f'(\mathscr{R})g_{\mu\nu}\right] = 0\,. \tag{3.224}$$

In both cases, the trace of the field equations (3.221) and (3.223) yields

$$f'(\mathscr{R})\,\mathscr{R} - 2f(\mathscr{R}) = 0\,. \tag{3.225}$$

When this equation admits solutions, the scalar curvature $\mathscr{R}$ is a constant; then Eqs. (3.222) and (3.224) imply that both connections coincide with the Levi-Civita connection of the metric $g_{\mu\nu}$ which solves the field equations and both theories reduce to GR with a cosmological constant, for which the Cauchy problem is well-formulated and well-posed [1139].

#### 3.7.3.3 The Cauchy problem in the metric-affine formalism with matter

Let us allow now a perfect fluid and study the Cauchy problem of $f(R)$ gravity. We discuss simultaneously the Palatini approach and a non-vanishing torsion, but we assume that the matter Lagrangian does not couple explicitly to the connection. Then the field equations are

$$f'(\mathscr{R})\,\mathscr{R}_{\mu\nu} - \frac{f(\mathscr{R})}{2}\,g_{\mu\nu} = T^{(m)}_{\mu\nu}\,, \tag{3.226}$$

$$T_{\mu\nu}{}^{\alpha} = -\frac{1}{2f'(\mathscr{R})}\frac{\partial f'(\mathscr{R})}{\partial x^{\rho}}\left(\delta^{\rho}_{\mu}\delta^{\alpha}_{\nu} - \delta^{\rho}_{\nu}\delta^{\alpha}_{\mu}\right) \tag{3.227}$$

in the case of $f(R)$ gravity with torsion, and

$$f'(\mathscr{R})\,\mathscr{R}_{\mu\nu} - \frac{f(\mathscr{R})}{2}\,g_{\mu\nu} = T^{(m)}_{\mu\nu}\,, \tag{3.228}$$

$$\nabla_{\alpha}\left[f'(\mathscr{R})g_{\mu\nu}\right] = 0\,, \tag{3.229}$$

for Palatini $f(R)$ gravity, where $T^{(m)}_{\mu\nu} \equiv -\dfrac{2}{\sqrt{-g}}\dfrac{\delta\left(\sqrt{-g}\,\mathscr{L}^{(m)}\right)}{\delta g^{\mu\nu}}$ is the matter energy-momentum tensor. The trace of Eqs. (3.226) and (3.228) yields the relation between $\mathscr{R}$ and $T^{(m)} \equiv g^{\mu\nu}T^{(m)}_{\mu\nu}$

$$f'(\mathscr{R})\,\mathscr{R} - 2f(\mathscr{R}) = T^{(m)}\,. \tag{3.230}$$

When $T^{(m)}$ = const. the theory reduces to GR with a cosmological constant and the initial value problem is identical to the vacuum case. Assuming that the relation (3.230) is invertible and $T^{(m)} \neq$ const., the Ricci scalar can be expressed as a function of $T^{(m)}$

$$\mathscr{R} = F\left(T^{(m)}\right)\,. \tag{3.231}$$

It is then easy to show that the field equations of both the Palatini and the metric-affine theory with torsion can be expressed in the form [220, 221, 868]

$$\begin{aligned}\mathscr{R}_{\mu\nu} - \frac{1}{2}\,g_{\mu\nu}\,\mathscr{R} = \frac{1}{\varphi}\,T^{(m)}_{\mu\nu} + \frac{1}{\varphi^2}\Bigg[&-\frac{3}{2}\frac{\partial\varphi}{\partial x^{\mu}}\frac{\partial\varphi}{\partial x^{\nu}} + \varphi\tilde{\nabla}_{\nu}\frac{\partial\varphi}{\partial x^{\mu}} \\ &+ \frac{3}{4}\frac{\partial\varphi}{\partial x^{\alpha}}\frac{\partial\varphi}{\partial x^{\beta}}\,g^{\alpha\beta}\,g_{\mu\nu} \\ &- \varphi\tilde{\nabla}^{\alpha}\frac{\partial\varphi}{\partial x^{\alpha}}\,g_{\mu\nu} - V(\varphi)\,g_{\mu\nu}\Bigg]\,,\end{aligned} \tag{3.232}$$

where

$$V(\varphi) \equiv \frac{1}{4}\left[\varphi F^{-1}\left(\left(f'\right)^{-1}(\varphi)\right) + \varphi^2\left(f'\right)^{-1}(\varphi)\right]\,, \tag{3.233}$$

is an effective potential for the scalar field

$$\varphi \equiv f'\left(F\left(T^{(m)}\right)\right)\,. \tag{3.234}$$

By performing the conformal transformation $g_{\mu\nu} \longrightarrow \tilde{g}_{\mu\nu} = \varphi\, g_{\mu\nu}$, Eq. (3.232) assumes the simpler form [220, 534, 868]

$$\tilde{R}_{\mu\nu} - \frac{1}{2}\,\tilde{g}_{\mu\nu}\,\tilde{R} = \frac{1}{\varphi}\,T^{(m)}_{\mu\nu} - \frac{1}{\varphi^3}\,V(\varphi)\,\tilde{g}_{\mu\nu}\,, \tag{3.235}$$

where $\tilde{R}_{\mu\nu}$ and $\tilde{R}$ are the Ricci tensor and the Ricci scalar of the conformal metric $\tilde{g}_{\mu\nu}$, respectively.

The connection $\Gamma_{\mu\nu}{}^{\alpha}$, solution of the field equations with torsion, is

$$\Gamma_{\mu\nu}{}^{\alpha} = \tilde{\Gamma}_{\mu\nu}{}^{\alpha} + \frac{1}{2\varphi}\frac{\partial\varphi}{\partial x^{\nu}}\,\delta^{\alpha}_{\mu} - \frac{1}{2\varphi}\frac{\partial\varphi}{\partial x^{\rho}}\,g^{\rho\alpha}\,g_{\mu\nu}\,, \tag{3.236}$$

where $\tilde{\Gamma}_{\mu\nu}{}^{\alpha}$ is the Levi-Civita connection of the metric $g_{\mu\nu}$ while $\tilde{\Gamma}^{\alpha}_{\mu\nu}$, solution of the Palatini field equations, coincides with the Levi-Civita connection of the conformal metric $\tilde{g}_{\mu\nu}$. $\Gamma_{\mu\nu}{}^{\alpha}$ and $\tilde{\Gamma}^{\alpha}_{\mu\nu}$ satisfy the relation

$$\tilde{\Gamma}_{\mu\nu}{}^{\alpha} = \Gamma_{\mu\nu}{}^{\alpha} + \frac{1}{2\varphi}\frac{\partial\varphi}{\partial x^{\mu}}\,\delta^{\alpha}_{\nu} \tag{3.237}$$

and the Levi-Civita connections induced by the metrics $g_{\mu\nu}$ and $\tilde{g}_{\mu\nu}$ are related by the identity

$$\tilde{\Gamma}_{\mu\nu}{}^{\alpha} = \Gamma_{\mu\nu}{}^{\alpha} + \frac{1}{2\varphi}\frac{\partial\varphi}{\partial x^{\nu}}\,\delta^{\alpha}_{\mu} - \frac{1}{2\varphi}\frac{\partial\varphi}{\partial x^{\rho}}\,g^{\rho\alpha}\,g_{\mu\nu} + \frac{1}{2\varphi}\frac{\partial\varphi}{\partial x^{\mu}}\,\delta^{\alpha}_{\nu}\,. \tag{3.238}$$

The field equations (3.232) have to be considered together with the matter field equations and it must be kept in mind that the conservation equations for both the metric-affine theories (with torsion and *à la* Palatini) coincide with the standard conservation laws of GR [257]

$$\tilde{\nabla}_{\nu}T^{\mu\nu} = 0\,. \tag{3.239}$$

It is straightforward to show that Eq. (3.239) is equivalent to the conservation law

$$\tilde{\nabla}_{\nu}T^{\mu\nu} = 0 \tag{3.240}$$

where

$$T_{\mu\nu} = \frac{1}{\varphi}\,T^{(m)}_{\mu\nu} - \frac{1}{\varphi^3}\,V(\varphi)\,\tilde{g}_{\mu\nu}\,, \tag{3.241}$$

for the conformally transformed theories (3.235). In fact, by an explicit calculation of the divergence $\tilde{\nabla}_{\nu}T^{\mu\nu}$ where the relations (3.238) have been used, we obtain

$$\tilde{\nabla}^{\nu} T_{\mu\nu} = \frac{1}{\varphi^2} \tilde{\nabla}^{\nu} T^{(m)}_{\mu\nu} + \frac{1}{\varphi^3} \frac{\partial \varphi}{\partial x^{\mu}} \left[ -\frac{T^{(m)}}{2} + \frac{3V(\varphi)}{\varphi} - V'(\varphi) \right] . \tag{3.242}$$

The constraint equations (3.239) and (3.240) are then mathematically equivalent in view of the relation

$$T^{(m)} - \frac{6V(\varphi)}{\varphi} + 2V'(\varphi) = 0 \,, \tag{3.243}$$

which is equivalent to the definition $\varphi = f'(F(T^{(m)}))$ [220].

With these results in mind, the Cauchy problem for Eq. (3.232) and the related equations of motion for matter can be approached by discussing the equivalent initial value problem of the conformally transformed theories. Using, as in GR, Gaussian normal coordinates and beginning with Eqs. (3.235) and (3.240), it is easy to conclude that the Cauchy problem is well-formulated also in this case.

In general, the equations of motion for matter imply the Levi-Civita connection of the metric $g_{\mu\nu}$ and not the connection induced from the conformal metric $\tilde{g}_{\mu\nu}$. Thanks to Eq. (3.238), this is not a problem since the connection $\tilde{\Gamma}^{\alpha}_{\mu\nu}$ can be expressed in terms of ${\Gamma_{\mu\nu}}^{\alpha}$ and the scalar field $\varphi$ which, on the other hand, is a function of the matter fields. As a result, we could obtain slightly more complicated equations implying further constraints on the initial data but, in any case, the same equations can always be rewritten in "normal form" with respect to the maximal order time derivatives of the matter fields, determining a well-formulated Cauchy problem [1139].

As an example, let us examine in detail the perfect fluid case with barotropic equation of state $P = P(\rho)$. The corresponding energy-momentum tensor is

$$T^{(m)}_{\mu\nu} = (P + \rho)\, u_{\mu} u_{\nu} + P g_{\mu\nu} \,, \tag{3.244}$$

and satisfies Eq. (3.239) with the normalization

$$g_{\mu\nu} u^{\mu} u^{\nu} = -1 \,, \tag{3.245}$$

of the fluid four-velocity. Equation (3.239) gives

$$(\rho + P\, u^{\nu})_{|\nu}\, u_{\mu} + (\rho + P)\, u_{\mu|\nu} u^{\nu} + \frac{\partial P}{\partial x^{\mu}} = 0 \,. \tag{3.246}$$

Contraction with $u^{\alpha}$ yields

$$(\rho u^{\nu})_{|\nu} = -P\, u^{\nu}_{|\nu} \tag{3.247}$$

while, substituting Eq. (3.247) into Eq. (3.246) for $\alpha = 1, 2$, and 3, we obtain

$$(\rho + P)\, u^{\nu} u^{i}_{|\nu} = -\frac{\partial P}{\partial x^{\nu}} \left( u^{i} u^{\nu} + g^{i\nu} \right) . \tag{3.248}$$

Equations (3.245), (3.247), and (3.248) involve the metric $g_{\mu\nu}$ and its first derivatives; using the relation (3.238) we can rewrite them in terms of the conformal metric $\tilde{g}_{\mu\nu}$, the scalar $\varphi = \varphi(\rho)$, and their first derivatives, obtaining

$$\frac{1}{\varphi}\,\tilde{g}_u u^\mu u^\nu = -1\,, \tag{3.249}$$

$$\frac{\partial}{\partial x^\nu}(\rho V^\nu) + \tilde{\Gamma}_{\nu\sigma}{}^{\nu}\rho u^\sigma - \frac{2}{\varphi}\frac{\partial\varphi}{\partial x^\sigma}\rho u^\sigma$$

$$= -P\left(\frac{\partial u^\nu}{\partial x^\nu} + \tilde{\Gamma}_{\nu\sigma}{}^{\nu}u^\sigma - \frac{2}{\varphi}\frac{\partial\varphi}{\partial x^\sigma}u^\sigma\right), \tag{3.250}$$

$$(\rho + P)\,u^\nu\left[\frac{\partial u^i}{\partial x^\nu} + \tilde{\Gamma}_{\nu\sigma}{}^{i}u^\sigma + \frac{1}{2\varphi}\left(-\frac{\partial\varphi}{\partial x^\sigma}\,\delta^i_\nu + \frac{\partial\varphi}{\partial x^\nu}\,\delta^i_\sigma - \frac{\partial\varphi}{\partial x^\rho}\,g^{\rho i}g_{\nu\sigma}\right)u^\sigma\right]$$

$$= -\frac{\partial P}{\partial x^\nu}\left(u^i u^\nu + g^{i\nu}\right). \tag{3.251}$$

In Gaussian normal coordinates in which $\tilde{g}_{00} = -1$ (assuming $\varphi > 0$) and $\tilde{g}_{0i} = 0$, Eq. (3.249) yields the expression of $u^0$ in terms of the remaining components $u^i$. Eqs. (3.249) and (3.250) can be regarded as linear equations for the functions $\partial u^i/\partial x^0$ and $\partial\rho/\partial x^0$. The explicit solution of these equations, in terms of the unknown functions, could originate further constraints on the initial data and on the form of the function $f(\mathscr{R})$. In Gaussian normal coordinates, Eqs. (3.249) and (3.250) allow one to obtain $\partial u^i/\partial x^0$ and $\partial\rho/\partial x^0$ as functions of the initial data $\tilde{g}_{ij}$, $\partial\tilde{g}_{ij}/\partial x^0$, $u_i$, and $\rho$ allowing the equations of motion of matter to be cast in normal form, hence the Cauchy problem is well-formulated.

Consider, as another example, the initial value formulation of $f(R)$ gravity coupled with Yang-Mills fields, in particular with the electromagnetic field. Also in this case, the problem is well-formulated. In fact, the stress-energy tensor of a Yang-Mills field has vanishing trace. Using Eq. (3.230), it is easy to prove that the Ricci scalar is constant and then, using Eqs. (3.227) and (3.229) one concludes that the connection coincides with the Levi-Civita connection of $g_{\mu\nu}$. In this situation, both theories (*à la* Palatini and with torsion) reduce to GR with a cosmological constant and the Cauchy problem is well-formulated (this conclusions was already reached for the Maxwell field). Moreover, the initial value problem is well-posed for any theory in which the trace of the matter energy-momentum tensor is constant, which is reduced to Einstein gravity with a cosmological constant. The Cauchy problem for perfect fluid and scalar field sources is discussed in [256, 257].

To conclude, we have shown that the initial value problem for ETGs can be at least well-formulated, passing another test for the viability of these theories. Well-posedness is also necessary in order to achieve a complete control of the dynamics but it depends on the specific matter fields adopted and the discussion becomes specific to them.

Since ETGs, like GR, are gauge theories, the choice of suitable coordinates may be crucial to show that the Cauchy problem is formulated correctly. We have discussed the two approaches using the 3 + 1 ADM decomposition and Gaussian normal coordinates, which can be defined when the covariant derivative operator $\nabla_\mu$ arises from a metric. These coordinates are useful for calculations on a given non-null surface $\Sigma_3$, *i.e.*, a three-dimensional embedded submanifold of the four-dimensional manifold $M$. Gaussian normal coordinates allow one to define uniquely timelike geodesics orthogonal to $\Sigma_3$ and to formulate correctly the conditions for the validity of the Cauchy-Kowalewski theorem [1139].

In the metric-affine formalism a given $f(R)$ theory *in vacuo* is equivalent to GR plus a cosmological constant, hence the initial value problem is well-formulated and well-posed. The same conclusion holds with matter sources whenever the trace of the energy-momentum tensor is constant. As shown in [220–222], by introducing matter fields in the Palatini and in the metric-affine approach with torsion, one can define $R = F\left(T^{(m)}\right)$ and then the scalar field $\varphi \equiv f'\left(F\left(T^{(m)}\right)\right)$, which allows one to reduce the theory to scalar-tensor gravity and to relate the form of $f(R)$ to the trace of the matter energy-momentum tensor. In this case, it is always possible to show that the initial value problem is well-formulated avoiding the singularities which could emerge with different gauge choices [704]. Moreover, matter fields could induce further constraints on the Cauchy hypersurface $x^0 = 0$ which, if suitably defined, lead to the normal form of the equations of motion for the matter sources. This is one of the main requirements for a well-formulated initial value problem. However different sources of the gravitational field, such as perfect fluids, Yang-Mills, and Klein-Gordon fields, could generate different constraints on the initial hypersurface $\Sigma_3$. These constraints could also imply restrictions on the possible form of $f(R)$. In conclusion, as in GR, the choice of gauge is essential for a correct formulation of the initial value problem, while the source fields have to be discussed carefully.

## 3.8 Conclusions

Through the Lagrangian formulation, we have obtained the field equations of various theories of gravity. We have seen how the metric and Palatini variations produce different field equations in ETGs, contrary to what happens in GR. Conformal transformations have been applied to ETGs, and an overview of the Cauchy problem has been given. It is beginning to be clear that several aspects of a gravitational theory need to be taken into account before the latter can be claimed to be viable. Other aspects of ETGs and other criteria for their viability are examined in the following chapters.

# Chapter 4
# Spherical symmetry

*Like a great poet, Nature knows how to produce the greatest effects with the most limited means.*
*– Heinrich Heine*

In all areas of physics and mathematics it is common to search for insight into a theory by finding exact solutions of its fundamental equations and by studying these solutions in detail. This goal is particularly difficult in non-linear theories and the usual approach consists of assuming particular symmetries and searching for solutions with these symmetries. Stripped of inessential features and simplified in this way, the search for exact solutions becomes easier. In a sense, this approach betrays a reductionist point of view but, pragmatically, it is often crucial to gain an understanding of the theory that cannot be obtained otherwise and that no physicist or mathematician would want to renounce to. In this chapter we discuss exact solutions of ETGs with spherical symmetry. In addition to gaining insight into the theory, spherically symmetric solutions are particularly important in astrophysics as models for stars and compact objects, including black holes, which are important theoretical laboratories for theories of quantum gravity. The next section discusses spherical symmetry in GR and in metric $f(R)$ gravity and presents static spherically symmetric solutions and a Noether symmetry approach. Then, the more difficult issue of non-static and non-asymptotically flat solutions is discussed. The second part of the chapter is devoted to the study of spherically symmetric solutions in general scalar-tensor theories and of the Jebsen-Birkhoff theorem. The chapter ends with a discussion of black holes in ETGs and of a map from spherical to axially symmetric solutions. An example is given. A spherically symmetric solution in Palatini $f(R)$ gravity has already been given in Sect. 3.4.2.

## 4.1 Spherically symmetric solutions of GR and metric $f(R)$ gravity

The physically relevant spherically symmetric solutions of GR include the asymptotically flat Schwarzschild solution describing an isolated body, which was derived in the very early days of GR. Other relevant solutions include the Schwarzschild-de

S. Capozziello and V. Faraoni, *Beyond Einstein Gravity: A Survey of Gravitational Theories for Cosmology and Astrophysics*, Fundamental Theories of Physics 170,
DOI 10.1007/978-94-007-0165-6_4, © Springer Science+Business Media B.V. 2011

Sitter (or Kottler) metric representing a black hole embedded in a de Sitter universe, and the Lemaitre-Tolman-Bondi class of solutions describing spherical objects embedded in a dust-dominated cosmological background [160, 716, 1053, 1076], to which one should add the McVittie metric and its generalizations [789] (see [695] for a survey of inhomogeneous solutions including spherically symmetric ones, and [1053] for exact solutions of GR). Spherical solutions representing black holes have been instrumental in the development of black hole mechanics and thermodynamics [1139, 1142].

### 4.1.1 Spherical symmetry

The classical tests of GR pertain to the realm of spherically symmetric solutions and the weak-field limit [1167]. One of the fundamental properties of a gravitational theory is the possibility of asymptotic flatness, *i.e.*, spacetime being Minkowskian far away from a localized distribution of mass-energy. Alternative gravitational theories may or may not exhibit this physical property which allows for a consistent comparison with GR. This point is sometimes forgotten in the study of the weak-field limit of alternative theories of gravity and can be discussed in general by considering the meaning of spherical solutions in ETGs when the standard results of GR are recovered in the limits $r \to \infty$ and $f(R) \to R$. Spherical solutions can be classified using the Ricci curvature $R$ as

- solutions with $R = 0$,
- solutions with constant Ricci scalar $R = R_0 \neq 0$,
- solutions with Ricci scalar $R(r)$ depending only on the radial coordinate $r$, and
- solutions with time-dependent $R(t,r)$.

In the first three cases the Jebsen-Birkhoff theorem is valid, meaning that stationary spherically symmetric solutions are necessarily static. However, as shown in the following, this theorem does not hold for every situation in $f(R)$ gravity because temporal evolution can emerge already in perturbation theory at some order of approximation.

A crucial role for the existence of exact spherical solutions is played by the relation between the metric potentials and by the relations between the latter and the Ricci scalar. The relation between the metric potentials and $R$ can be regarded as a constraint which assumes the form of a Bernoulli equation [249]. In principle, spherically symmetric solutions can be obtained for any analytic function $f(R)$ by solving this Bernoulli equation, for both the case of constant Ricci scalar and $R = R(r)$. These spherically symmetric solutions can be used as backgrounds to test how generic $f(R)$ gravity may deviate from GR. Theories that imply $f(R) \to R$ in the weak-field limit are particularly interesting. In such cases, the comparison with GR is straightforward and the experimental results evading the GR constraints can be framed in a self-consistent picture [131]. Finally, a perturbation approach can be developed to obtain spherical solutions at zero order, after which first order solutions are searched for. This scheme is iterative and can, in principle, be extended to any order in the perturbations. It is crucial to consider $f(R)$ theories which can be Taylor-expanded about a constant value $R_0$ of the curvature scalar $R$.

### 4.1.2 The Ricci scalar in spherical symmetry

By imposing that the spacetime metric is spherically symmetric,

$$ds^2 = -A(t,r)dt^2 + B(t,r)dr^2 + r^2 d\Omega_2^2\,, \tag{4.1}$$

the Ricci scalar can be expressed as

$$R\,(t,r) = \left\{ B\left(\dot{A}\dot{B} - A'^2\right) r^2 + A\left[ r\left(\dot{B}^2 - A'B'\right) + 2B\left(2A' + rA'' - r\ddot{B}\right)\right] \right.$$
$$\left. -4A^2\left[B^2 - B + rB'\right]\right\} \left(2r^2A^2B^2\right)^{-1}, \tag{4.2}$$

where a prime and an overdot denote differentiation with respect to $r$ and $t$, respectively. If the metric (4.1) is time-independent, *i.e.*, $A(t,r) = a(r)$ and $B(t,r) = b(r)$, then Eq. (4.2) assumes the simple form

$$R(r) = \left\{ a(r)\left[2b(r)\left(2a'(r) + ra''(r)\right) - ra'(r)b'(r)\right] - b(r)a'(r)^2r^2 \right.$$
$$\left. -\,4a^2(r)\left(b(r)^2 - b(r) + rb'(r)\right)\right\} \left(2r^2a^2(r)b^2(r)\right)^{-1}. \tag{4.3}$$

One can see Eq. (4.3) as a constraint on the functions $a(r)$ and $b(r)$ once a specific form of the Ricci scalar is given. Equation (4.3) reduces to the Bernoulli equation of index two [249]

$$b'(r) + h(r)b(r) + l(r)b^2(r) = 0 \tag{4.4}$$

for the metric component $b(r)$, *i.e.*,

$$b'(r) + \left\{ \frac{r^2a'(r)^2 - 4a(r)^2 - 2ra(r)[2a(r)' + ra(r)'']}{ra(r)[4a(r) + ra'(r)]} \right\} b(r)$$
$$+\left\{ \frac{2a(r)}{r}\left[\frac{2 + r^2R(r)}{4a(r) + ra'(r)}\right]\right\} b(r)^2 = 0\,. \tag{4.5}$$

The general solution of Eq. (4.5) is

$$b(r) = \frac{\exp\left[-\int dr\, h(r)\right]}{K + \int dr\, l(r)\, \exp\left[-\int dr\, h(r)\right]}\,, \tag{4.6}$$

where $K$ is an integration constant and $h(r)$ and $l(r)$ are the coefficients of the linear and quadratic terms in $b(r)$, respectively. Inspection of this Bernoulli equation reveals that solutions corresponding to $l(r) = 0$ exist, which have a Ricci curvature scaling as $R \sim -2/r^2$ as spatial infinity is approached. No real solutions exist if $h(r)$ vanishes identically. The limit $r \to +\infty$ deserves special care: in order for the gravitational potential $b(r)$ to have the correct Minkowskian limit, both functions $h(r)$ and $l(r)$ must go to zero provided that the quantity $r^2 R(r)$ is constant. This fact implies that $b'(r) = 0$, and, finally, also the metric potential $b(r)$ has the correct Minkowskian limit.

If asymptotic flatness of the metric is imposed, the Ricci curvature must scale as $r^{-n}$ when $r \to +\infty$, where $n \geqslant 2$ is an integer,

$$r^2 R(r) \simeq r^{-n} \quad \text{as } r \to +\infty\,. \tag{4.7}$$

Any other behavior of the Ricci scalar would compromise asymptotic flatness, as can be seen from Eq. (4.5). In fact, let us consider the simplest spherically symmetric case in which

$$ds^2 = -a(r)dt^2 + \frac{dr^2}{a(r)} + r^2 d\Omega_2^2\,. \tag{4.8}$$

The Bernoulli equation (4.5) is easily integrated and the most general metric potential $a(r)$ compatible with the constraint (4.3) is

$$a(r) = 1 + \frac{k_1}{r} + \frac{k_2}{r^2} + \frac{1}{r^2}\int dr \left[\int r^2 R(r) dr\right], \tag{4.9}$$

where $k_1$ and $k_2$ are integration constants. The Minkowskian limit $a(r) \to 1$ as $r \to \infty$ is obtained only if the condition (4.7) is satisfied, otherwise the gravitational potential diverges.

### 4.1.3 *Spherical symmetry in metric $f(R)$ gravity*

Let us specialize now to metric $f(R)$ theories by considering an analytic function $f(R)$, the fourth order field equations

$$f'(R)R_{\mu\nu} - \frac{1}{2}f(R)g_{\mu\nu} - f'(R)_{;\mu\nu} + g_{\mu\nu}\Box f'(R) = \kappa\, T_{\mu\nu}\,, \tag{4.10}$$

and the corresponding trace equation

$$3\Box f'(R) + f'(R)R - 2f(R) = \kappa\, T\,. \tag{4.11}$$

By rewriting Eq. (4.10) as

$$G_{\mu\nu} = T^{(curv)}_{\mu\nu} + T^{(m)}_{\mu\nu}\,, \tag{4.12}$$

$$T^{(curv)}_{\mu\nu} = \frac{1}{f'(R)}\left\{g_{\mu\nu}\left[f(R) - Rf'(R)\right] + f'(R)^{;\rho\sigma}\left(g_{\mu\rho}g_{\nu\sigma} - g_{\rho\sigma}g_{\mu\nu}\right)\right\} \tag{4.13}$$

matter enters Eq. (4.12) through the modified stress-energy tensor

$$T^{(m)}_{\mu\nu} = \frac{\kappa\, T_{\mu\nu}}{f'(R)}\,. \tag{4.14}$$

The most general spherically symmetric metric can be written as

$$ds^2 = -m_1(t',r')dt'^2 + m_2(t',r')dr'^2 + m_3(t',r')dt'dr' + m_4(t',r')d\Omega_2^2\,, \tag{4.15}$$

where $m_i$ are functions of the radius $r'$ and of the time $t'$. A coordinate transformation $t = U_1(t',r')$, $r = U_2(t',r')$ diagonalizes the metric (4.15) and introduces the areal radius $r$ such that $m_4(t',r') = r^2$, giving

$$ds^2 = -A(t,r)dt^2 + B(t,r)dr^2 + r^2 d\Omega_2^2\,, \tag{4.16}$$

hence Eq. (4.16) can be taken as the most general torsion-free Lorentzian spherically symmetric metric without loss of generality. The field equations (4.10) and (4.11) for this metric reduce to

$$f'(R)R_{\mu\nu} - \frac{1}{2}f(R)g_{\mu\nu} + \mathcal{H}_{\mu\nu} = \kappa\, T_{\mu\nu}\,, \tag{4.17}$$

$$g^{\sigma\tau}H_{\sigma\tau} = f'(R)R - 2f(R) + \mathcal{H} = \kappa\, T\,, \tag{4.18}$$

where

$$\begin{aligned}\mathcal{H}_{\mu\nu} = &-f''(R)\Bigg\{R_{,\mu\nu} - \Gamma^t_{\mu\nu}R_{,t} - \Gamma^r_{\mu\nu}R_{,r} - g_{\mu\nu}\Bigg[\left({g^{tt}}_{,t} + g^{tt}\ln\sqrt{-g}_{,t}\right)R_{,t} \\ &+\left({g^{rr}}_{,r} + g^{rr}\ln\sqrt{-g}_{,r}\right)R_{,r} + g^{tt}R_{,tt} + g^{rr}R_{,rr}\Bigg]\Bigg\} \\ &-f'''(R)\left[R_{,\mu}R_{,\nu} - g_{\mu\nu}\left(g^{tt}{R_{,t}}^2 + g^{rr}{R_{,r}}^2\right)\right],\end{aligned} \tag{4.19}$$

$$\begin{aligned}\mathcal{H} = g^{\sigma\tau}\mathcal{H}_{\sigma\tau} = 3f''(R)&\Bigg[\left({g^{tt}}_{,t} + g^{tt}\ln\sqrt{-g}_{,t}\right)R_{,t} \\ &+\left({g^{rr}}_{,r} + g^{rr}\ln\sqrt{-g}_{,r}\right)R_{,r} + g^{tt}R_{,tt} + g^{rr}R_{,rr}\Bigg] \\ &+3f'''(R)\left[g^{tt}{R_{,t}}^2 + g^{rr}{R_{,r}}^2\right].\end{aligned} \tag{4.20}$$

In these equations the derivatives of $f(R)$ with respect to $R$ are distinct from the time and spatial derivatives of $R$, a feature which will allow us to better understand the dynamical behavior of the solutions.

### *4.1.4 Solutions with constant Ricci scalar*

Let us assume that the Ricci scalar is constant, $R = R_0$. The field equations (4.17) and (4.18) with $\mathscr{H}_{\mu\nu} = 0$ are

$$f_0' R_{\mu\nu} - \frac{1}{2} f_0 g_{\mu\nu} = \kappa\, T_{\mu\nu}^{(m)}\,, \tag{4.21}$$

$$f_0' R_0 - 2 f_0 = \kappa\, T^{(m)}\,, \tag{4.22}$$

where $f_0 \equiv f(R_0)$ and $f_0' \equiv f'(R_0)$, and they can be rewritten as

$$R_{\mu\nu} + \lambda g_{\mu\nu} = q\kappa\, T_{\mu\nu}^{(m)}\,, \tag{4.23}$$

$$R_0 = q\kappa\, T - 4\lambda\,, \tag{4.24}$$

where $\lambda = -\dfrac{f_0}{2 f_0'}$ and $q^{-1} = f_0'$. We restrict to Lagrangians which reduce to the Hilbert-Einstein one as $R \to 0$ and do not contain a cosmological constant $\Lambda$,

$$f(R) \simeq R \quad \text{as} \quad R \to 0\,. \tag{4.25}$$

Then, the trace equation (4.22) indicates that *in vacuo* ($T_{\mu\nu}^{(m)} = 0$) one obtains a class of solutions with constant Ricci curvature $R = R_0$. In particular, there exist solutions with $R_0 = 0$.

Let us suppose now that the above Lagrangian density reduces to a constant for small curvature values, $\lim_{R\to 0} f = \Lambda$. Interesting features emerge again from the trace equation: using Eq. (4.22) and the definition of $f(R)$, it is seen that zero curvature solutions do not exist in this case because

$$\Psi' R - 2\Psi - \Psi_0 R - 2\Lambda = \kappa\, T^{(m)}\,. \tag{4.26}$$

Contrary to GR, even in the absence of matter there are no Ricci-flat solution of the field equations since the higher order derivatives give constant curvature solutions corresponding to a sort of effective cosmological constant. In fact, in GR, solutions with non-vanishing constant curvature occur only in the presence of matter because of the proportionality of the Ricci scalar to the trace of the matter energy-momentum tensor. A similar situation can be obtained in the presence of a cosmological constant $\Lambda$. The difference between GR and higher order gravity is that the Schwarzschild-de Sitter solution is not necessarily generated by a $\Lambda$-term, while the effect of an "effective" cosmological constant can be achieved by the higher order derivative contributions, as discussed extensively in [103, 248, 816, 817].

Let us consider now the problem of finding the general solution of Eqs. (4.21) and (4.22) for the spherically symmetric metric (4.16). The substitution of this metric into the $(t,r)$ component of (4.21) yields $\frac{\dot{B}(t,r)}{rB(t,r)} = 0$, which means that $B(t,r)$ must be time-independent, $B(t,\ r) = b(r)$. On the other hand, the $(\theta,\theta)$ component of Eq. (4.21) yields $\frac{A'(t,r)}{A(t,r)} = \zeta(r)$, where $\zeta(r)$ is a time-independent function and

$$A(t,r) = \tilde{a}(t)\,\exp\left[\int \zeta(r)dr\right] = \tilde{a}(t)\,\frac{b}{r^2}\,\exp\left[\int dr\,\frac{[2 - r^2(2\lambda + 2q\kappa\ p)]b(r)}{r}\right] \tag{4.27}$$

where $P$ is the pressure of a perfect fluid with stress-energy tensor

$$T^{(m)}_{\mu\nu} = (P + \rho)\,u_\mu u_\nu + P g_{\mu\nu}\,. \tag{4.28}$$

The function $A(t,r)$ is separable, $A(t,r) = \tilde{a}(t)a(r)$, and the line element (4.16) becomes

$$ds^2 = -\tilde{a}(t)a(r)dt^2 + b(r)dr^2 + r^2 d\Omega_2^2 \tag{4.29}$$

and is rewritten as

$$ds^2 = -a(r)d\tilde{t}^2 + b(r)dr^2 + r^2 d\Omega_2^2 \tag{4.30}$$

by redefining the time coordinate $t \rightarrow \tilde{t}$ as $d\tilde{t} = \sqrt{\tilde{a}(t)}\,dt$. From now on, the tilde will be dropped from this time coordinate.

To summarize, in a spacetime with constant scalar curvature, any spherically symmetric background is necessarily static or, the Jebsen-Birkhoff theorem holds for $f(R)$ gravity with constant curvature (cf. [582]).

A remark is in order at this point. We have assumed a spacetime with constant Ricci scalar and deduced conditions on the form of the gravitational potentials. The inverse problem can also be considered: whenever the gravitational potential $a(t,r)$ is a separable function and $b(t,r)$ is time-independent, using the definition of the Ricci scalar, it is $R = R_0 =$ const. and at the same time the solutions of the field equations will be static if spherical symmetry is invoked. For a complete analysis of this problem, one should take into account the remaining field equations contained in (4.23) and (4.24) which have to be satisfied by taking into account the expression of the Ricci scalar (4.3). One must then solve the system

$$R_{tt} + \lambda\, a(r) - q\kappa\,[\rho + P\,(1 - a(r))] = 0\,, \tag{4.31}$$

$$R_{rr} - \lambda b(r) - q\kappa\, Pb(r) = 0\,, \tag{4.32}$$

$$R_0 - q\kappa\,(\rho - 3P) + 4\lambda = 0\,, \tag{4.33}$$

$$R\,(a(r), b(r)) = R_0\,, \tag{4.34}$$

which takes the form

$$e^{\int \frac{2-r^2(2\lambda+2q\kappa P)b(r)}{r}dr}\Big\{\left[r^2(2\lambda+2q\kappa P-2)\right]^2 b(r)^4-4b(r)^3$$
$$-3r\left[r^2(2\lambda+2q\kappa P)-2\right]b'(r)b(r)^2$$
$$+2r\left[b'(r)+rb''(r)\right]b(r)-2r^2b'(r)^2\Big\}$$
$$-4r^4q\kappa(P+\rho)\,b(r)^2=0\,, \tag{4.35}$$

$$\Big\{3r\left[r^2(2\lambda+2q\kappa P)\,b'(r)-2\right]-8\Big\}b(r)^2-4\left[r^2(2\lambda+2q\kappa P)-3\right]b(r)^3$$
$$-\left[r^2(2\lambda+2q\kappa P)-2\right]b(r)^4+2r^2b'(r)^2-2rb(r)\left[rb(r)''-3b'(r)\right]=0\,, \tag{4.36}$$

$$\Big\{r^2[4\lambda+2q\kappa\,(P-\rho)]-8\Big\}b(r)^3-\Big\{3r\left[r^2(2\lambda+2q\kappa P)-2\right]b'(r)-4\Big\}b(r)^2$$
$$+\left[r^2(2\lambda+2q\kappa P)-2\right]b(r)^4-2r^2b'(r)^2+2r\left[rb''(r)-b'(r)\right]b(r)=0 \tag{4.37}$$

where, using Eq. (4.3), the only unknown potential is now $b(r)$. A general solution is found for the particular equation of state $P=-\rho$:

$$ds^2=-\left(1+\frac{k_1}{r}+\frac{q\kappa\,\rho-\lambda}{3}r^2\right)dt^2+\frac{dr^2}{1+\frac{k_1}{r}+\frac{q\kappa\,\rho-\lambda}{3}r^2}+r^2d\Omega_2^2\,. \tag{4.38}$$

In the case of constant Ricci scalar $R=R_0$, all $f(R)$ theories admit solutions with de Sitter-like behavior even in the weak-field limit. This is one of the reasons why dark energy can be replaced by $f(R)$ gravity [103, 211, 212, 218, 219, 263, 275, 851, 865].

Let us consider now $f(R)$ gravity with an analytic Lagrangian function $f(R)$, which we write as

$$f(R)=\Lambda+\Psi_0\,R+\Psi(R)\,, \tag{4.39}$$

where $\Psi_0$ is a constant, $\Lambda$ plays the role of the cosmological constant, and $\Psi(R)$ is an analytic function of $R$ satisfying the condition

$$\lim_{R\to 0}\frac{\Psi(R)}{R^2}=\Psi_1 \tag{4.40}$$

with $\Psi_1$ another constant. By neglecting the cosmological constant $\Lambda$ and setting $\Psi_0$ to zero, a new class of theories is obtained which, in the limit $R\to 0$, does

**Table 4.1** Examples of $f(R)$ models admitting constant or zero scalar curvature solutions. The powers $n$ and $m$ are integers while the $\xi_i$ are real constants.

| f(R) theory | Field equations |
|---|---|
| $R$ | $R_{\mu\nu}=0$ |
| $\xi_1 R+\xi_2 R^n$ | $\begin{cases} R_{\mu\nu}=0 & \text{with } R=0,\ \xi_1\neq 0 \\ R_{\mu\nu}+\lambda g_{\mu\nu}=0 & \text{with } R=\left[\frac{\xi_1}{(n-2)\xi_2}\right]^{\frac{1}{n-1}},\ \xi_1\neq 0,\ n\neq 2 \\ 0=0 & \text{with } R=0,\ \xi_1=0 \\ R_{\mu\nu}+\lambda g_{\mu\nu}=0 & \text{with } R=R_0,\ \xi_1=0,\ n=2 \end{cases}$ |
| $\xi_1 R+\xi_2 R^{-m}$ | $R_{\mu\nu}+\lambda g_{\mu\nu}=0$ with $R=\left[-\frac{(m+2)\xi_2}{\xi_1}\right]^{\frac{1}{m+1}}$ |
| $\xi_1 R+\xi_2 R^n+\xi_3 R^{-m}$ | $R_{\mu\nu}+\lambda g_{\mu\nu}=0$, with $R=R_0$ so that $\xi_1 R_0^{m+1}+(2-n)\xi_2 R_0^{n+m}+(m+2)\xi_3=0$ |
| $\frac{R}{\xi_1+R}$ | $\begin{cases} R_{\mu\nu}=0 & \text{with } R=0 \\ R_{\mu\nu}+\lambda g_{\mu\nu}=0 & \text{with } R=-\frac{\xi_1}{2} \end{cases}$ |
| $\frac{1}{\xi_1+R}$ | $R_{\mu\nu}+\lambda g_{\mu\nu}=0$ with $R=-\frac{2\xi_1}{3}$ |

not reproduce GR (Eq. (4.40) implies that $f(R)\sim R^2$ as $R\to 0$). In this case, analyzing the complete set of equations (4.21) and (4.22), one can observe that both zero and constant (but non-vanishing) curvature solutions are possible. In particular, if $R=R_0=0$ the field equations are solved for all forms of the gravitational potentials appearing in the spherically symmetric background (4.30), provided that the Bernoulli equation (4.5) relating these functions is satisfied for $R(r)=0$. The solutions are thus defined by the relation

$$b(r)=\frac{\exp\left[-\int dr\, h(r)\right]}{K+4\int\frac{dr\, a(r)\,\exp\left[-\int dr\, h(r)\right]}{r[a(r)+ra'(r)]}}\,. \tag{4.41}$$

Table 4.1 provides examples of $f(R)$ theories admitting solutions with constant but non-zero values of $R$ or null $R$. Each model admits Schwarzschild and Schwarzschild-de Sitter solutions, in addition to the class of solutions given by (4.41).

### 4.1.5 Solutions with $R=R(r)$

Thus far, we have discussed the behavior of $f(R)$ gravity searching for spherically symmetric solutions with constant Ricci curvature. In GR this situation is well known to give rise to the Schwarzschild ($R=0$) and the Schwarzschild-de Sitter ($R=R_0\neq 0$) solutions. The search for spherically symmetric solutions can be generalized to $f(R)$ gravity by allowing the Ricci scalar to depend on the radial coordinate. This approach is interesting because, in general, higher order theories of gravity admit naturally this kind of solution, with several examples reported in the literature [216, 217, 248, 816, 817, 1052]. In the following we approach the problem from a general point of view.

If we choose the Ricci scalar as a generic function $R(r)$ of the radial coordinate, it is possible to show that also in this case the solution of the field equations (4.17) and (4.18) is time-independent (if $T^{(m)}_{\mu\nu} = 0$). In other words, the Jebsen-Birkhoff theorem holds. As in GR, it is crucial to study the off-diagonal $(t, r)$ component of (4.17) which, for a generic $f(R)$, reads

$$\frac{d}{dr}\left[r^2 f'(R)\right]\dot{B}(t,r) = 0\,, \tag{4.42}$$

and two possibilities can occur. First, one can choose $\dot{B}(t,r) \neq 0$, implying that $f'(R) \sim 1/r^2$. In this case the remaining field equation is not satisfied and there is incompatibility. The only possible solution is then given by $\dot{B}\,(t,r) = 0$ and $B\,(t,r) = b(r)$. The $(\theta,\theta)$ equation is then used to determine that the potential $A(t,r)$ can be factorized with respect to time, the solutions are of the type (4.29), and the metric can be recast in the stationary spherically symmetric form (4.30) by a suitable coordinate transformation.

Even the more general radial-dependent case admits time-independent solutions. From the trace equation and the $(\theta,\theta)$ equation, the relation

$$a(r) = b(r)\,\frac{e^{\frac{2}{3}\int\frac{(Rf'-2f)b(r)}{R'f''}dr}}{r^4 R'^2 f''^2} \tag{4.43}$$

(with $f'' > 0$) linking $a(r)$ and $b(r)$ can be obtained, in addition to [816, 817]

$$b(r) = \frac{6\left[f'(rR'f'')' - rR'^2 f''^2\right]}{rf\,(rR'f'' - 4f') + 2f'\,[rR\,(f' - rR'f'') - 3R'f'']}\,. \tag{4.44}$$

Again, three more equations have to be satisfied in order to completely solve the system (respectively the $(t,t)$ and $(r,r)$ components of the field equations plus the Ricci scalar constraint), while the only unknown functions are $f(R)$ and the Ricci scalar $R(r)$.

If we now consider a fourth order theory described by $f(R) = R + \Phi(R)$ with $\Phi(R) \ll R$ we are able to satisfy the complete set of equations up to third order in $\Phi$. In particular, we can solve the full set of equations; the relations (4.43) and (4.44) will provide the general solution depending only on the forms of the functions $\Phi(R)$ and $R(r)$, *i.e.*,

$$a(r) = b(r)\,\frac{e^{-\frac{2}{3}\int\frac{[R+(2\Phi - R\Phi')]\,b(r)}{R'\Phi''}dr}}{r^4 R'^2 \Phi''^2}\,, \tag{4.45}$$

$$b(r) = -\frac{3(rR'\Phi'')_{,r}}{rR} \,. \tag{4.46}$$

Once the radial dependence of the scalar curvature is obtained, Eq. (4.45) allows one to obtain the solution of the field equations and the gravitational potential related to the function $a(r)$. The physical relevance of this potential can be assessed by comparison with astrophysical data (*e.g.*, [650]).

### 4.1.6 Perturbations

When the Ricci scalar $R(r)$ depends on the radial coordinate, one can find perturbative solutions. Several perturbative techniques exist that enable one to investigate higher order gravity in the weak-field limit. A general approach with analytical $f(R)$ theories begins with the assumption that the background model deviates only slightly from GR, *i.e.*, $f(R) = R+\Phi(R)$ with $\Phi(R) \ll R$. An alternative approach has the background metric considered as the zero order solution and as its starting point. Both approaches view the weak-field limit of a given higher order theory of gravity as a correction to GR, which is the zero order approximation. Both methods can provide interesting results in astrophysics. The first approach, which is based on the matching of the background metric and the GR solution is discussed in the following.

In general, the perturbative search for solutions involves the study of a perturbed metric $g_{\mu\nu} = \overset{(0)}{g}_{\mu\nu} + \overset{(1)}{g}_{\mu\nu}$, which implies that the first order field equations (4.17) and (4.18) split in two orders. The metric perturbation implies a splitting of the Ricci scalar $R$ into a background part plus a perturbation, and then the analytic function $f(R)$ can be Taylor-expanded about the background value of $R$,

$$f(R) = \sum_n \frac{f^{n(0)}}{n!}\left(R - R^{(0)}\right)^n \equiv \sum_n \frac{f^{n(0)}}{n!} R^{(1)^n} \,. \tag{4.47}$$

The zero order field equations read

$$f'^{(0)}\, R^{(0)}_{\mu\nu} - \frac{1}{2}\, g^{(0)}_{\mu\nu}\, f^{(0)} + \mathscr{H}^{(0)}_{\mu\nu} = \kappa\, T^{(0)}_{\mu\nu} \,, \tag{4.48}$$

where

$$\begin{aligned} \mathscr{H}^{(0)}_{\mu\nu} = -f''^{(0)}\Big[ R^{(0)}_{,\mu\nu} - \Gamma^{(0)\rho}{}_{\mu\nu} R^{(0)}_{,\rho} - g^{(0)}_{\mu\nu}\Big( g^{(0)\rho\sigma}{}_{,\rho} R^{(0)}_{,\sigma} + g^{(0)\rho\sigma} R^{(0)}_{,\rho\sigma} \\ + g^{(0)\rho\sigma}\left(\ln\sqrt{-g}\right)_{,\rho} R^{(0)}_{,\sigma}\Big)\Big] \\ - f'''^{(0)}\left( R^{(0)}_{,\mu} R^{(0)}_{,\nu} - g^{(0)}_{\mu\nu} g^{(0)\rho\sigma} R^{(0)}_{,\rho} R^{(0)}_{,\sigma}\right) . \end{aligned} \tag{4.49}$$

To first order it is

$$f'^{(0)}\left(R^{(1)}_{\mu\nu}-\frac{1}{2}g^{(0)}_{\mu\nu}R^{(1)}\right)+f''^{(0)}R^{(1)}R^{(0)}_{\mu\nu}-\frac{1}{2}f^{(0)}g^{(1)}_{\mu\nu}+\mathscr{H}^{(1)}_{\mu\nu} = \kappa\, T^{(1)}_{\mu\nu} \tag{4.50}$$

with

$$\begin{aligned}
\mathscr{H}^{(1)}_{\mu\nu} = -f''^{(0)}\Big\{ & R^{(1)}_{,\mu\nu} - \Gamma^{(0)\rho}{}_{\mu\nu}R^{(1)}_{,\rho} - \Gamma^{(1)\rho}{}_{\mu\nu}R^{(0)}_{,\rho} \\
& -g^{(0)}_{\mu\nu}\Big[g^{(0)\rho\sigma}{}_{,\rho}R^{(1)}_{,\sigma} + g^{(1)\rho\sigma}{}_{,\rho}R^{(0)}_{,\sigma} + g^{(0)\rho\sigma}R^{(1)}_{,\rho\sigma} + g^{(1)\rho\sigma}R^{(0)}_{,\rho\sigma} \\
& \quad +g^{(0)\rho\sigma}\left(\ln\sqrt{-g}^{(0)}_{,\rho}R^{(1)}_{,\sigma} + \ln\sqrt{-g}^{(1)}_{,\rho}R^{(0)}_{,\sigma}\right) \\
& \quad +g^{(1)\rho\sigma}\ln\sqrt{-g}^{(0)}_{,\rho}R^{(0)}_{,\sigma}\Big] \\
& -g^{(1)}_{\mu\nu}\left(g^{(0)\rho\sigma}{}_{,\rho}R^{(0)}_{,\sigma} + g^{(0)\rho\sigma}R^{(0)}_{,\rho\sigma} + g^{(0)\rho\sigma}\ln\sqrt{-g}^{(0)}_{,\rho}R^{(0)}_{,\sigma}\right)\Big\} \\
-f'''^{(0)}\Big[ & R^{(0)}_{,\mu}R^{(1)}_{,\nu} + R^{(1)}_{,\mu}R^{(0)}_{,\nu} - g^{(0)}_{\mu\nu}g^{(0)\rho\sigma}\left(R^{(0)}_{,\rho}R^{(1)}_{,\sigma} + R^{(1)}_{,\rho}R^{(0)}_{,\sigma}\right) \\
& -g^{(0)}_{\mu\nu}g^{(1)\rho\sigma}R^{(0)}_{,\rho}R^{(0)}_{,\sigma} - g^{(1)}_{\mu\nu}g^{(0)\rho\sigma}R^{(0)}_{,\rho}R^{(0)}_{,\sigma}\Big] \\
-f'''^{(0)}R^{(1)}\Big[ & R^{(0)}_{,\mu\nu} - \Gamma^{(0)\rho}{}_{\mu\nu}R^{(0)}_{,\rho} - g^{(0)}_{\mu\nu} \\
& \times\left(g^{(0)\rho\sigma}{}_{,\rho}R^{(0)}_{,\sigma} + g^{(0)\rho\sigma}R^{(0)}_{,\rho\sigma} + g^{(0)\rho\sigma}\ln\sqrt{-g}_{,\rho}R^{(0)}_{,\sigma}\right)\Big] \\
-f^{IV(0)}R^{(1)}\Big( & R^{(0)}_{,\mu}R^{(0)}_{,\nu} - g^{(0)}_{\mu\nu}g^{(0)\rho\sigma}R^{(0)}_{,\rho}R^{(0)}_{,\sigma}\Big)\,.
\end{aligned} \tag{4.51}$$

Apart from analyticity, no assumption is made on the form of the function $f(R)$. At this level, the zero order solution of Eq. (4.48) is required and, in general, this could be a GR solution. This problem can be overcome by assuming the same order of perturbation on the $f(R)$, *i.e.*,

$$f(R) = R + \Phi(R) \tag{4.52}$$

with $\Phi \ll R$. Then we have

$$f = R^{(0)} + R^{(1)} + \Phi^{(0)}\,,\quad f' = 1 + \Phi'^{(0)}\,,\quad f'' = \Phi''^{(0)}\,,\qquad f''' = \Phi'''^{(0)}\,, \tag{4.53}$$

and Eq. (4.48) reduces to

$$R^{(0)}_{\mu\nu} - \frac{1}{2}g^{(0)}_{\mu\nu}R^{(0)} = \kappa\, T^{(0)}_{\mu\nu}\,. \tag{4.54}$$

On the other hand, Eq. (4.50) reduces to

$$R^{(1)}_{\mu\nu} - \frac{1}{2}\, g^{(0)}_{\mu\nu}\, R^{(1)} - \frac{1}{2}\, g^{(1)}_{\mu\nu} R^{(0)} - \frac{1}{2}\, g^{(0)}_{\mu\nu} \Phi^{(0)} + \Phi'^{(0)} R^{(0)}_{\mu\nu} + \mathscr{H}^{(1)}_{\mu\nu} = \kappa\, T^{(1)}_{\mu\nu}\,, \quad (4.55)$$

where

$$\mathscr{H}^{(1)}_{\mu\nu} = -\Phi'''^{(0)} \left( R^{(0)}_{,\mu} R^{(0)}_{,\nu} - g^{(0)}_{\mu\nu} g^{(0)rr} R^{(0)}_{,r} R^{(0)}_{,r} \right)$$
$$-\Phi''^{(0)} \Bigg\{ R^{(0)}_{,\mu\nu} - \Gamma^{(0)r}_{\mu\nu} R^{(0)}_{,r} - g^{(0)}_{\mu\nu} \Big[ g^{(0)rr}_{\ \ \ \ ,r} R^{(0)}_{,r}$$
$$+ g^{(0)rr} R^{(0)}_{,rr} + g^{(0)rr} \ln \sqrt{-g}^{(0)}_{,r} R^{(0)}_{,r} \Big] \Bigg\} . \quad (4.56)$$

This new system of field equations is simpler than the previous one and, once the zero order solution is obtained, the first order solutions are easy to find. A list of solutions obtained with this method is presented in Table 4.2 for various $f(R)$ models.

In the case of $f(R)$ models which are manifest corrections to the Hilbert-Einstein Lagrangian, such as $f(R) = \Lambda + R + \varepsilon R \ln R$ and $f(R) = R + \varepsilon R^n$ with $|\varepsilon| \ll 1$, one obtains exact solutions for the gravitational potentials $a(r)$ and $b(r)$ related by $a(r) = b(r)^{-1}$. The first order expansion is straightforward, as in GR. If the functions $a(r)$ and $b(r)$ are not related, for $f(R) = \Lambda + R + \varepsilon R \ln R$, the first order system is solved directly without any prescription on the perturbation functions $x(r)$ and $y(r)$. This is not the case for $f(R) = R + \varepsilon R^n$ since, for this model, one obtains an explicit constraint on the perturbation function implying the possibility to deduce the form of the gravitational potential $\phi(r)$ from $a(r) = 1 + 2\phi(r)$. In such a case, no corrections are found with respect to the standard Newtonian potential. The theories $f(R) = R^n$ and $f(R) = \dfrac{R}{R+\xi}$ exhibit similar behaviors. The case $f(R) = R^2$ is degenerate and must be discussed independently.

### 4.1.7 *Spherical symmetry in $f(R)$ gravity and the Noether approach*

This subsection details the application of the Noether approach to spherical symmetry in metric $f(R)$ gravity. This is a useful method to generate exact solutions.

#### 4.1.7.1 The point-like $f(R)$ Lagrangian in spherical symmetry

Exact spherically symmetric solutions with constant Ricci scalar in $f(R)$ gravity can be found using the Noether symmetry approach presented in Chap. 2. To begin, one needs to derive a point-like Lagrangian from the action of modified gravity by imposing spherical symmetry, while enforcing the constancy of the Ricci scalar by

**Table 4.2** A list of solutions for several $f(R)$ theories in the perturbative approach. The $k_i$ are integration constants, $r_g \equiv 2GM/c^2$ is the Schwarzschild radius (in standard units), and the potentials $a(r)$ and $b(r)$ are defined in Eq. (4.30).

| | |
|---|---|
| **f(R) theory** | $\boldsymbol{\Lambda + R + \varepsilon R \ln R}$ |
| spherical potentials | $a(r) = b(r)^{-1} = 1 + \frac{k_1}{r} - \frac{\Lambda r^2}{6} + \delta x(r)$ |
| solutions | $x(r) = \frac{k_2}{r} + \frac{\varepsilon\Lambda[\ln(-2\Lambda)-1]r^2}{6\delta}$ |
| first order metric | $a(r) = 1 - \frac{\Lambda r^2}{6} + \delta x(r), \;\; b(r) = \frac{1}{1-\frac{\Lambda r^2}{6}} + \delta y(r)$ |
| solutions | $\begin{cases} x(r) = (\Lambda r^2 - 6)\left\{k_1 + \int dr \frac{4\delta(2\Lambda^2 r^4 - 15\Lambda r^2 + 18)y(r) + r\{36 r\varepsilon\Lambda[\log(-2\Lambda)-1] + \delta(\Lambda r^2-6)^2 y'(r)\}}{36 r\delta(\Lambda r^2-6)}\right\} \\ y(r) = \frac{k_2\delta - 6r^3\varepsilon\Lambda[\ln(-2\Lambda)-1]}{r\delta(r^2\Lambda-6)^2} \end{cases}$ |
| **f(R) theory** | $\mathbf{R + \varepsilon R^n}$ |
| spherical potentials | $a(r) = b(r)^{-1} = 1 + \frac{k_1}{r} + \delta x(r)$ |
| solutions | $x(r) = \frac{k_2}{r}$ |
| first order metric | $a(r) = 1 + \delta\frac{x(r)}{r}, \;\; b(r) = 1 + \delta\frac{y(r)}{r}$ |
| solutions | $x(r) = k_1 + k_2 r, \;\; y(r) = k_3$ |
| **f(R) theory** | $\mathbf{R^n}$ |
| spherical potentials | $a(r) = b(r)^{-1} = 1 + \frac{k_1}{r} + \frac{R_0 r^2}{12} + \delta x(r)$ |
| solutions | $\begin{cases} n = 2, \quad R_0 \neq 0 \text{ and } x(r) = \frac{3k_2-k_3}{3r} + \frac{k_3 r^2}{12} + \frac{k_4}{r}\int dr\, r^2 \left\{\int dr \frac{\exp\left[\frac{R_0 r_0^2 \ln(r-r_0)}{8+3R_0 r_0^2}\right]}{r^5}\right\} \\ \qquad\qquad \text{with } r_0 \text{ satisfying the condition } 6k_1 + 8r_0 + R_0 r_0^3 = 0 \\ n \geq 2, \quad \text{system solved only with } R_0 = 0 \text{ and no prescriptions on } x(r) \end{cases}$ |
| first order metric | $a(r) = 1 + \delta\frac{x(r)}{r}, \;\; b(r) = 1 + \delta\frac{y(r)}{r}$ |
| solutions | $\begin{cases} n = 2 \quad y(r) = -\frac{R_0 r^3}{6} - \frac{x(r)}{2} + \frac{1}{2}rx'(r) + k_1, \;\; R(r) = \delta R_0 \\ n \neq 2 \quad y(r) = -\frac{1}{2}\int dr\, r^2 R(r) - \frac{x(r)}{2} + \frac{1}{2}rx'(r) + k_1 \text{ with } R(r) \text{ any} \end{cases}$ |
| first order metric | $a(r) = 1 - \frac{r_g}{r} + \delta x(r), \;\; b(r) = \frac{1}{1-\frac{r_g}{r}} + \delta y(r)$ |
| solutions | $\begin{cases} n = 2 \quad y(r) = \frac{rk_1}{3r_g^2 - 7r_g r + 4r^2} + \frac{r^2 k_2}{3(3r_g^2 - 7r_g r + 4r^2)} + \frac{r_g r^2 x(r) + 2(r_g r^3 - r^4)x'(r)}{(3r_g-4r)(r_g-r)^2} \\ n \neq 2 \quad \text{any } x(r), y(r), \text{ and } R(r) \end{cases}$ |
| **f(R) theory** | $\mathbf{R/(R+\xi)}$ |
| first order metric | $a(r) = 1 + \delta\frac{x(r)}{r}, \;\; b(r) = 1 + \delta\frac{y(r)}{r}$ |
| solutions | $\begin{cases} x(r) = -\frac{4e^{-\frac{\xi^{1/2}r}{\sqrt{6}}}}{\xi}k_1 - \frac{2\sqrt{6}e^{\frac{\xi^{1/2}r}{\sqrt{6}}}}{\xi^{3/2}}k_2 + k_3 r \\ y(r) = -\frac{2e^{-\frac{\xi^{1/2}r}{\sqrt{6}}}(6\xi^{1/2}+\sqrt{6}\xi r)}{3b^{3/2}}k_1 - \frac{2e^{\frac{\xi^{1/2}r}{\sqrt{6}}}(\sqrt{6}-\xi^{1/2}r)}{\xi^{3/2}}k_2 \end{cases}$ |

means of a suitable Lagrange multiplier. With the previous considerations in mind, a spherically symmetric spacetime is described by the line element

$$ds^2 = -A(r)dt^2 + B(r)dr^2 + M(r)d\Omega_2^2 \,. \tag{4.57}$$

The Schwarzschild solution of GR is obtained if $M(r) = r^2$ and $A(r) = B^{-1}(r) = 1 - 2M/r$, with $r$ an areal radius. In the presence of spherical symmetry the action

$$S = \int dr\, L\left(A, A', B, B', M, M', R, R'\right) \tag{4.58}$$

contains only a finite number of degrees of freedom, the Ricci scalar $R$ and the potentials $A$, $B$, and $M$ defining the configuration space. The point-like Lagrangian is obtained by writing the action as

$$S = \int d^4x\sqrt{-g}\Big[f(R) - \lambda(R - \bar{R})\Big], \tag{4.59}$$

where $\lambda$ is a Lagrangian multiplier and $\bar{R}$ is the Ricci scalar of the metric (4.57)

$$\begin{aligned}\bar{R} &= \frac{A''}{AB} + \frac{2M''}{BM} + \frac{A'M'}{ABM} - \frac{A'^2}{2A^2B} - \frac{M'^2}{2BM^2} - \frac{A'B'}{2AB^2} - \frac{B'M'}{B^2M} - \frac{2}{M} \\ &\equiv R^* + \frac{A''}{AB} + \frac{2M''}{BM},\end{aligned} \tag{4.60}$$

where $R^*$ collects the terms containing first order derivatives. The Lagrange multiplier $\lambda$ is obtained by varying the action (4.59) with respect to $R$, which yields $\lambda = f_R(R)$. By expressing the metric determinant $g$ and $\bar{R}$ as functions of $A$, $B$, and $M$, Eq. (4.59) gives

$$\begin{aligned}S &= \int dr\,\sqrt{AB}\,M\left[f - f_R\left(R - R^* - \frac{A''}{AB} - \frac{2M''}{BM}\right)\right] \\ &= \int dr\left\{\sqrt{AB}\,M\Big[f - f_R\,(R - R^*)\Big] - \left(\frac{f_R M}{\sqrt{AB}}\right)' A' - \left(\frac{2\sqrt{A}}{\sqrt{B}}\,f_R\right)' M'\right\}.\end{aligned} \tag{4.61}$$

The last two integrals differ by a total divergence which can be discarded, and the point-like Lagrangian becomes

$$\begin{aligned}L = &-\frac{\sqrt{A}\,f_R}{2M\sqrt{B}}(M')^2 - \frac{f_R}{\sqrt{AB}}\,A'M' - \frac{M f_{RR}}{\sqrt{AB}}\,A'R' \\ &-\frac{2\sqrt{A}\,f_{RR}}{\sqrt{B}}\,R'M' - \sqrt{AB}\,[(2 + MR)\,f_R - Mf]\,.\end{aligned} \tag{4.62}$$

The canonical Lagrangian (4.62) is written in compact form using matrix notation as

$$L = \underline{q}'^t\,\hat{T}\,\underline{q}' + V\,, \tag{4.63}$$

where $\underline{q} = (A, B, M, R)$ and $\underline{q}' = (A', B', M', R')$ are the generalized Lagrangian coordinates and velocities. The index "$t$" denotes the transposed vector. The kinetic tensor is

$$\hat{T}_{ij} = \frac{\partial^2 L}{\partial q_i' \partial q_j'} \tag{4.64}$$

and $V(q)$ is the potential energy depending only on the generalized coordinates. The Euler-Lagrange equations are

$$\frac{d}{dr}\left(\nabla_{q'}L\right) - \nabla_q L = 2\,\frac{d}{dr}\left(\hat{T}\underline{q}'\right) - \nabla_q V - \underline{q}'^t\left(\nabla_q \hat{T}\right)\underline{q}'$$
$$= 2\hat{T}\underline{q}'' + 2\left(\underline{q}'\cdot\nabla_q\hat{T}\right)\underline{q}' - \nabla_q V - \underline{q}'^t\left(\nabla_q\hat{T}\right)\underline{q}' = 0\,. \tag{4.65}$$

$R$ obeys a constraint relating the Lagrangian coordinates. The Hessian determinant of (4.62), $\left\|\frac{\partial^2 L}{\partial q_i'\partial q_j'}\right\|$, vanishes because the point-like Lagrangian does not depend on the generalized velocity $B'$. The metric component $B$ does not contribute to the dynamics, but its equation of motion has to be taken into account as a further constraint. The definition of the energy

$$E_L = \underline{q}'\cdot\nabla_{q'}L - L \tag{4.66}$$

coincides with the Euler-Lagrangian equation for the component $B$ of the generalized coordinate $\underline{q}$. Then, the Lagrangian (4.62) contains only three degrees of freedom and not four, as expected *a priori*. Now, since the equation of motion for $B$ does not contain the derivative $B'$, it can be solved explicitly in term of $B$ as a function of the other Lagrangian coordinates:

$$B = \frac{2M^2 f_{RR}A'R' + 2Mf_R A'M' + 4AMf_{RR}M'R' + Af_R M'^2}{2AM\left[(2+MR)\,f_R - Mf\right]}\,. \tag{4.67}$$

By inserting Eq. (4.67) into the Lagrangian (4.62), we obtain a non-vanishing Hessian matrix removing the singular dynamics. The new Lagrangian reads[1]

$$L^* = \sqrt{\mathbf{L}} \tag{4.68}$$

with

$$\mathbf{L} = \underline{q}'^t\hat{\mathbf{L}}\underline{q}' = \frac{[(2+MR)f_R - fM]}{M}$$
$$\cdot\left[2M^2 f_{RR}A'R' + 2MM'(f_R A' + 2Af_{RR}R') + Af_R M'^2\right]\,. \tag{4.69}$$

[1] Lowering the dimension of the configuration space through the substitution (4.67) leaves the dynamics unaffected because $B$ is non-dynamical. In fact, if Eq. (4.67) is introduced into the set of dynamical equations (4.62), these coincide with the equation derived from (4.69).

Since $\frac{\partial \mathbf{L}}{\partial r} = 0$, $\mathbf{L}$ is canonical ($\mathbf{L}$ is a quadratic form in the generalized velocities $A'$, $M'$ and $R'$ and coincides with the Hamiltonian), hence $\mathbf{L}$ can be regarded as the new Lagrangian with three degrees of freedom. It is crucial that the Hessian determinant

$$\left\| \frac{\partial^2 \mathbf{L}}{\partial q_i' \partial q_j'} \right\| = 3AM\,[(2+MR)f_R - Mf]^3\, f_R f_{RR}^{\;2}\,. \tag{4.70}$$

now does not vanish. It is assumed that $(2+MR)f_R - Mf \neq 0$, otherwise the above definitions of $B$, and $\mathbf{L}$ (Eqs. (4.67) and (4.69)) are meaningless. Moreover, it is assumed that $f_{RR} \neq 0$ allows for a wide class of fourth order gravity models. The GR case $f(R) = R$ is special: the GR point-like Lagrangian requires a further reduction of the number of degrees of freedom and the previous results cannot be applied directly. Eq. (4.62) yields

$$L_{GR} = -\frac{\sqrt{A}}{2M\sqrt{B}}(M')^2 - \frac{1}{\sqrt{AB}}A'M' - 2\sqrt{AB} \tag{4.71}$$

which, through the Euler-Lagrange equations, provides the standard GR equations for the Schwarzschild metric. The absence of the generalized velocity $B'$ in Eq. (4.71) is evident. Again, the Hessian determinant vanishes. Nevertheless, considering again the constraint (4.67) for $B$, it is possible to obtain a Lagrangian with non-vanishing Hessian. In particular, it is

$$B_{GR} = \frac{(M')^2}{4M} + \frac{A'M'}{2A}\,, \tag{4.72}$$

$$L^*_{GR} = \sqrt{\mathbf{L}_{GR}} = \sqrt{\frac{M'(2MA' + AM')}{M}}\,, \tag{4.73}$$

and the Hessian determinant is

$$\left\| \frac{\partial^2 \mathbf{L}_{GR}}{\partial q_i' \partial q_j'} \right\| = -1\,, \tag{4.74}$$

a non-vanishing sub-matrix of the $f(R)$ Hessian matrix.

The Euler-Lagrange equations derived from Eqs. (4.72) and (4.73) yield the vacuum solutions of GR

$$A = k_4 - \frac{k_3}{r + k_1}\,, \qquad B = \frac{k_2 k_4}{A}\,, \qquad M = k_2 (r + k_1)^2\,. \tag{4.75}$$

In particular, the standard form of the Schwarzschild solution is recovered for $k_1 = 0$, $k_2 = 1$, $k_3 = 2GM/c^2$, and $k_4 = 1$.

**Table 4.3** The field equations approach and the point-like Lagrangian approach differ because spherical symmetry can be imposed either in the field equations after standard variation with respect to the metric, or directly into the Lagrangian, which then becomes point-like. The energy $E_L$ corresponds to the (0, 0) component of $H_{\mu\nu}$. The absence of $B'$ in the Lagrangian implies the proportionality between the constraint equation for $B$ and the energy function $E_L$. As a consequence, there are only three independent equations and three unknown functions. The $(\theta, \theta)$ component corresponds to the field equation for $M$. $H_{\mu\nu}$ is given in Appendix B.1.

| Field equations approach | | Point-like Lagrangian approach |
|---|---|---|
| $\delta \int d^4x \sqrt{-g}\, f = 0$ | $\leftrightarrows$ | $\delta \int dr\, L = 0$ |
| $\downarrow$ | | $\downarrow$ |
| $H_{\mu\nu} = \partial_\rho \left[\frac{\partial(\sqrt{-g} f)}{\partial_\rho g^{\mu\nu}}\right] - \frac{\partial(\sqrt{-g} f)}{\partial g^{\mu\nu}} = 0$ | | $\frac{d}{dr}\left(\nabla_{q'} L\right) - \nabla_q L = 0$ |
| | $\leftrightarrows$ | |
| $H = g^{\mu\nu} H_{\mu\nu} = 0$ | | $E_L = \underline{q}' \cdot \nabla_{q'} L - L$ |
| $\downarrow$ | | $\downarrow$ |
| $H_{00} = 0$ | $\leftrightarrows$ | $\frac{d}{dr}\left(\frac{\partial L}{\partial A'}\right) - \frac{\partial L}{\partial A} = 0$ |
| $H_{rr} = 0$ | $\leftrightarrows$ | $\frac{d}{dr}\left(\frac{\partial L}{\partial B'}\right) - \frac{\partial L}{\partial B} \propto E_L = 0$ |
| $H_{\theta\theta} = \csc^2\theta\, H_{\phi\phi} = 0$ | $\leftrightarrows$ | $\frac{d}{dr}\left(\frac{\partial L}{\partial M'}\right) - \frac{\partial L}{\partial M} = 0$ |
| $H = A^{-1} H_{00} - B^{-1} H_{rr} - 2M^{-1}\csc^2\theta\, H_{\phi\phi} = 0$ | $\leftrightarrows$ | a combination of the above equations |

Table 4.3 summarizes the field equations associated with the point-like Lagrangians and their relation with respect to the ones of the standard approach.

### 4.1.8 Noether solutions of spherically symmetric $f(R)$ gravity

In spherical symmetry, the areal radius $r$ plays the role of an affine parameter. Then, the configuration space is $\mathcal{Q} = (A, M, R)$ and the tangent space is $\mathcal{TQ} = (A, A', M, M', R, R')$. According to the Noether theorem, the existence of a symmetry for the dynamics described by the Lagrangian (4.69) implies the existence of a conserved quantity. The Lie differentiation of Eq. (4.69) yields[2]

$$\mathscr{L}_{\mathbf{X}}\mathbf{L} = \underline{\alpha} \cdot \nabla_q \mathbf{L} + \underline{\alpha}' \cdot \nabla_{q'} \mathbf{L} = \underline{q}'^t \left[\underline{\alpha} \cdot \nabla_q \hat{\mathbf{L}} + 2\left(\nabla_q \underline{\alpha}\right)^t \hat{\mathbf{L}}\right] \underline{q}' . \tag{4.76}$$

This Lie derivative vanishes if the functions $\underline{\alpha}$ satisfy the system

$$\alpha_i \frac{\partial \hat{\mathbf{L}}_{km}}{\partial q_i} + 2 \frac{\partial \alpha_i}{\partial q_k} \hat{\mathbf{L}}_{im} = 0 . \tag{4.77}$$

[2] From now on, $\underline{q}$ denotes the vector $(A, M, R)$.

Solving the system (4.77) means finding the functions $\alpha_i$ which identify the Noether vector. However the system (4.77) depends implicitly on the form of the function $f(R)$ and, by solving it, one obtains the forms of the function $f(R)$ which are compatible with spherical symmetry. Alternatively, by choosing the form of $f(R)$, (4.77) can be solved explicitly. As an example, the system (4.77) is satisfied if we choose

$$f(R) = f_0 R^s\,, \quad \underline{\alpha} = (\alpha_1, \alpha_2, \alpha_3) = \Big((3-2s)kA,\ -kM,\ kR\Big) \tag{4.78}$$

with $s$ a real number, $k$ an integration constant, and $f_0$ a dimensional coupling constant.[3] This means that for $f(R) = R^s$ there exist at least one Noether symmetry and a related conserved quantity

$$\begin{aligned}\Sigma_0 &= \underline{\alpha} \cdot \nabla_{q'} \mathbf{L} \\ &= 2skMR^{2s-3}\left[2s + (s-1)MR\right]\left[(s-2)RA' - (2s^2 - 3s + 1)AR'\right]. \end{aligned} \tag{4.79}$$

A physical interpretation of $\Sigma_0$ is possible in GR. In this case, obtained for $s = 1$, the above procedure must be applied to the Lagrangian (4.73), obtaining the solution

$$\underline{\alpha}_{GR} = (-kA,\ kM)\,. \tag{4.80}$$

The functions $A$ and $M$ provide the Schwarzschild solution (4.75), and then the constant of motion takes the form

$$\Sigma_0 = \frac{2GM}{c^2} \tag{4.81}$$

in standard units; the conserved quantity is the Schwarzschild radius (or the mass of the gravitating system).

Another solution can be found for constant Ricci scalar $R = R_0$ [816], for which the field equations reduce to

$$R_{\mu\nu} + k_0 g_{\mu\nu} = 0\,, \tag{4.82}$$

where $k_0 = -\frac{1}{2} f(R_0)/f_R(R_0)$. The general solution is

$$A(r) = \frac{1}{B(r)} = 1 + \frac{k_0}{r} + \frac{R_0}{12} r^2\,, \qquad M = r^2 \tag{4.83}$$

[3] The dimensions are given by $R^{1-s}$ in term of the Ricci scalar. For simplicity, $f_0$ is set to unity in the following.

which includes the special case

$$A(r) = \frac{1}{B(r)} = 1 + \frac{k_0}{r}, \qquad M = r^2, \qquad R = 0. \tag{4.84}$$

The solution (4.83) is the well known Schwarzschild-de Sitter metric.

In the general case $f(R) = R^s$, the Lagrangian (4.69) becomes

$$\mathbf{L} = \frac{sR^{2s-3}[2s + (s-1)MR]}{M} \cdot \left[2(s-1)M^2A'R' + 2MRM'A' + 4(s-1)AMM'R' + ARM'^2\right] \tag{4.85}$$

and the expression (4.67) of $B$ is

$$B = \frac{s\left[2(s-1)M^2A'R' + 2MRM'A' + 4(s-1)AMM'R' + ARM'^2\right]}{2AMR\left[2s + (s-1)MR\right]}, \tag{4.86}$$

and GR is recovered for $s = 1$.

Using the constant of motion (4.79), one can solve for $A$, obtaining

$$A = R^{\frac{2s^2-3s+1}{s-2}} \left\{ k_1 + \Sigma_0 \int \frac{R^{\frac{4s^2-9s+5}{2-s}}\, dr}{2ks(s-2)M\,[2s + (s-1)MR]} \right\} \tag{4.87}$$

for $s \neq 2$, where $k_1$ is an integration constant. For $s = 2$ one finds

$$A = -\frac{\Sigma_0}{12kr^2(4 + r^2R)RR'}. \tag{4.88}$$

These relations allow one to find general solutions of the field equations regulating the function $R(r)$. For example, the solution corresponding to

$$s = 5/4, \qquad M = r^2, \qquad R = 5r^{-2}, \tag{4.89}$$

is the spherically symmetric metric given by

$$ds^2 = -\frac{1}{\sqrt{5}}(k_2 + k_1 r)\, dt^2 + \frac{1}{2}\left(\frac{1}{1 + \frac{k_2}{k_1 r}}\right) dr^2 + r^2 d\Omega_2^2 \tag{4.90}$$

with $k_2 = \dfrac{32\Sigma_0}{225k}$. The value of $s$ for this solution is ruled out by Solar System experiments [96, 325, 326].

To summarize, the Noether symmetry approach provides a general method to find spherically symmetric exact solutions of ETGs, and of metric $f(R)$ gravity in particular. The procedure consists of $(i)$ obtaining the point-like $f(R)$ Lagrangian with spherical symmetry; $(ii)$ writing the Euler-Lagrange equations; $(iii)$ searching for a Noether vector field; and $(iv)$ reducing the dynamics and then integrating the equations of motion using the constants of motion. *Vice-versa*, this approach also allows one to select families of $f(R)$ models with spherical symmetry. The method can be generalized. If a symmetry exists, the Noether approach allows transformations of variables to cyclic ones, reducing the dynamics to obtain exact solutions. For example, since we know that $f(R) = R^s$ gravity admits a constant of motion, the Noether symmetry suggests the coordinate transformation

$$\mathbf{L}\left(A, M, R, A', M', R'\right) \rightarrow \widetilde{\mathbf{L}}\left(\widetilde{M}, \widetilde{R}, \widetilde{A}', \widetilde{M}', \widetilde{R}'\right) , \tag{4.91}$$

for the Lagrangian (4.69), where the conserved quantity corresponds to the cyclic variable $\widetilde{A}$. In the presence of multiple symmetries one can find multiple cyclic variables. If three Noether symmetries exist, the Lagrangian $\mathbf{L}$ can be mapped into a Lagrangian with three cyclic coordinates $\widetilde{A} = \widetilde{A}(\underline{q})$, $\widetilde{M} = \widetilde{M}(\underline{q})$ and $\widetilde{R} = \widetilde{R}(\underline{q})$ which are functions of the old generalized coordinates. These new functions must satisfy the system

$$(3-2s)\,A\,\frac{\partial \widetilde{A}}{\partial A} - M\,\frac{\partial \widetilde{A}}{\partial M} + R\,\frac{\partial \widetilde{A}}{\partial R} = 1\,, \tag{4.92}$$

$$(3-2s)\,A\,\frac{\partial \widetilde{q}_i}{\partial A} - M\,\frac{\partial \widetilde{q}_i}{\partial M} + R\,\frac{\partial \widetilde{q}_i}{\partial R} = 0\,, \tag{4.93}$$

with $i = 2, 3$ (we have set $k = 1$). A solution of (4.93) for $s \neq 3/2$ is

$$\widetilde{A} = \frac{\ln A}{(3-2s)} + F_A\left(A^{\frac{\eta_A}{3-2s}} M^{\eta_A}\, A^{\frac{\xi_A}{2s-3}} M^{\xi_A}\right) , \tag{4.94}$$

$$\widetilde{q}_i = F_i\left(A^{\frac{\eta_i}{3-2s}} M^{\eta_i}, A^{\frac{\xi_i}{2s-3}} M^{\xi_i}\right) \tag{4.95}$$

while, if $s = 3/2$,

$$\widetilde{A} = -\ln M + F_A(A) G_A(MR)\,, \tag{4.96}$$

$$\widetilde{q}_i = F_i(A) G_i(MR)\,, \tag{4.97}$$

where $F_A$, $F_i$, $G_A$ and $G_i$ are arbitrary functions and $\eta_A$, $\eta_i$, $\xi_A$, and $\xi_i$ are integration constants.

The considerations of this section make it clear once again that the Jebsen-Birkhoff theorem does not hold, in general, for metric $f(R)$ gravity.

### 4.1.9 *Non-asymptotically flat and non-static spherical solutions of metric $f(R)$ gravity*

Any solution of the vacuum field equations

$$R_{\mu\nu} = \Lambda g_{\mu\nu} \tag{4.98}$$

of GR (possibly with a cosmological constant) is also a solution of metric $f(R)$ gravity in all cases in which the (reduced) trace equation

$$f'(R)R - 2f(R) = 0 \tag{4.99}$$

admits solutions. The converse is not true. In the context of spherical symmetry, the Schwarzschild metric solves the field equations of metric $f(R)$ gravity *in vacuo* if $R = 0$. If $R$ is constant *in vacuo*, then the Schwarzschild-(anti-)de Sitter metrics are solutions. Unless the constancy of $R$ is imposed, the Schwarzschild-(anti)de Sitter metric is not the unique solution because the Jebsen-Birkhoff theorem does not hold in metric $f(R)$ gravity.

There exist several studies of exact spherically symmetric solutions of metric $f(R)$ gravity in the literature. Recent motivation for these studies arises from the need to understand the weak-field limit of metric $f(R)$ theories which are of interest in cosmology. A $1+1+2$ covariant formalism for spherically symmetric solutions in metric $f(R)$ gravity was developed in [862, 863].

In addition to vacuum solutions, non-vacuum ones have been studied; usually, in these cases, the matter source is assumed to be a perfect fluid. Fluid dynamics in metric $f(R)$ gravity was studied in [761, 807, 942, 1067] long before the recent revival of $f(R)$ gravity. Spherically symmetric solutions found in the literature include those of [173, 174, 197, 249, 321, 323, 795, 816–818, 1161].

The stability of spherically symmetric solutions was discussed in [652, 1001]. Previous stability analyses of particular modified gravity theories include that of [1161] in the theory

$$S = \int d^4x \, \frac{\sqrt{-g}}{\kappa} \left[ R - \alpha R^2 - \beta R_{\mu\nu} R^{\mu\nu} + \varepsilon \mathscr{G} \right] , \tag{4.100}$$

where $\alpha$, $\beta$, and $\varepsilon$ are constants and $\mathscr{G} = R^2 - 4R_{\mu\nu}R^{\mu\nu} + R_{\mu\nu\alpha\beta}R^{\mu\nu\alpha\beta}$ is the Gauss-Bonnet term. The Schwarzschild metric is a solution of this theory and the stability analysis of Schwarzschild black holes in [1161] led to the somehow surprising conclusion that the massive ghost graviton (or "poltergeist") that appears in this theory stabilizes Schwarzschild black holes with small mass against Hawking radiation instability (an issue discussed also in [823, 824]). If $\beta = \varepsilon = 0$ the theory (4.100) reduces to quadratic $f(R)$ gravity and the stability criterion of [1161] reduces to $\alpha < 0$, consistently with the familiar stability condition $f''(R) > 0$. For $\alpha = 0$ the theory reduces to GR, in which black holes are quantum-mechanically unstable because of Hawking radiation and of their negative specific heat, but classically stable [1139]. This feature is replicated in $f(R)$ gravity and the classical stability condition for Schwarzschild black holes is again $f''(R) \geq 0$.

Regarding black holes, all black hole solutions of GR (possibly with a cosmological constant) will also be solutions of metric and Palatini $f(R)$ gravity [77, 915]. While in the Palatini version of the theory they are *all* the black hole solutions, in metric $f(R)$ gravity other black hole solutions are in principle possible because of the failure of the Jebsen-Birkhoff theorem, adding to the richness and variety of spherically symmetric solutions.

#### 4.1.9.1 Clifton and Barrow's static solution in $f(R) = R^{1+\delta}$ gravity

The Schwarzschild-de Sitter solution is not the only known static spherically symmetric solution of the fourth order field equations of metric $f(R)$ gravity. An example is given by Clifton and Barrow's static solution [321, 326]

$$ds^2 = -A_1(r)dt^2 + \frac{dr^2}{B_1(r)} + r^2 d\Omega_2^2, \tag{4.101}$$

where

$$A_1(r) = r^{\frac{2\delta(1+2\delta)}{1-\delta}} + \frac{C_1}{r^{\frac{1-4\delta}{1-\delta}}}, \tag{4.102}$$

$$B_1(r) = \frac{(1-\delta)^2}{(1-2\delta+4\delta^2)\left[1-2\delta\,(1+\delta)\right]}\left(1+\frac{C_1}{r^{\frac{1-2\delta+4\delta^2}{1-\delta}}}\right), \tag{4.103}$$

and where $C_1$ is a constant and $r$ is the area radius. This solution is not asymptotically flat and reduces to the Schwarzschild one in the limit $\delta \to 0$ in which the theory goes over to GR. The Ricci scalar is non-constant. Perturbations of this solution were studied in [321, 326] and light deflection in this spacetime was studied in [862] using a $1+1+2$ covariant approach.

#### 4.1.9.2 A dynamical solution in $f(R) = R^{1+\delta}$ gravity

Since the $f(R)$ theories of interest for cosmology are designed to produce a time-varying effective cosmological constant in order to explain the present acceleration of the universe, black hole solutions in these theories are likely to represent central objects embedded in cosmological backgrounds. Not much is known about this kind of solution even in the context of GR, although a few GR examples are available [267, 462, 465, 468, 520, 765, 783, 784, 859, 957, 1059], and even less is known about $f(R)$ black holes and spherically symmetric solutions.

As an example of how spherical solutions of metric $f(R)$ gravity can differ from the Schwarzschild spacetime, we consider here an exact solution of $f(R) = R^{1+\delta}$ gravity found by Clifton [321] and describing a dynamical spherical spacetime which is asymptotically FLRW. This solution exhibits a peculiar behavior: it contains a strong spacetime singularity which becomes naked at late times.

Solar System experiments set the constraints $\delta = (-1.1 \pm 1.2) \cdot 10^{-5}$ on the parameter $\delta$ of $f(R) = R^{1+\delta}$ gravity [96, 321, 325, 326, 1171], while the local stability criterion requires $f''(R) \geq 0$ [396, 460] and $\delta > 0$. We only retain positive values of this parameter.

As already discussed, the fourth order field equations of vacuum metric $f(R)$ gravity can be rewritten as effective Einstein equations with geometric terms acting as a form of effective matter,

$$R_{\mu\nu} - \frac{1}{2} g_{\mu\nu} R = \frac{1}{f'(R)} \left[ \nabla_\mu \nabla_\nu f' - g_{\mu\nu} \Box f' + g_{\mu\nu} \frac{(f - Rf')}{2} \right], \quad (4.104)$$

where the effective matter spoils the Jebsen-Birkhoff theorem.

The line element is [321]

$$ds^2 = -A_2(r) dt^2 + a^2(t) B_2(r) \left( dr^2 + r^2 d\Omega_{(2)}^2 \right), \quad (4.105)$$

where

$$A_2(r) = \left( \frac{1 - C_2/r}{1 + C_2/r} \right)^{2/q}, \quad (4.106)$$

$$B_2(r) = \left( 1 + \frac{C_2}{r} \right)^4 A_2(r)^{q+2\delta-1}, \quad (4.107)$$

$$a(t) = t^{\frac{\delta(1+2\delta)}{1-\delta}}, \quad (4.108)$$

$$q^2 = 1 - 2\delta + 4\delta^2, \quad (4.109)$$

in isotropic coordinates. There are two distinct classes of solutions for a given value of $\delta$: the first has $C_2 q r > 0$ and the second has $C_2 q r < 0$. The line element (4.105) approaches the FLRW one in the limit $C_2 \to 0$. In the limit $\delta \to 0$ the theory reduces to GR while (4.105) reduces to the Schwarzschild line element in isotropic coordinates provided that $C_2 q r > 0$. Assuming that $r > 0$, $C_2 > 0$, and the positive root in the expression $q = \pm\sqrt{1 - 2\delta + 4\delta^2}$ in Eq. (4.109), it is $q \simeq 1 - \delta$ in the limit $0 < \delta \ll 1$.

The solution (4.105)–(4.109) is conformal to the Fonarev solution [498] which is conformally static [763] and, therefore, it is also conformally static, a property shared with the Sultana-Dyer solution [1059] and with certain generalized McVittie solutions of GR [465].

The metric (4.105) is conveniently recast in the Nolan gauge, in which it is straightforward to identify apparent horizons which may exist. Using first the radial coordinate

$$\tilde{r} \equiv r\left(1+\frac{C_2}{r}\right)^2 \tag{4.110}$$

and then the areal radius

$$\rho \equiv \frac{a(t)\sqrt{B_2(r)}\,\tilde{r}}{\left(1+\frac{C_2}{r}\right)^2} = a(t)\,\tilde{r}\,A_2(r)^{\frac{q+2\delta-1}{2}}\,, \tag{4.111}$$

one obtains the line element [445]

$$ds^2 = -A_2 dt^2 + a^2 A_2^{2\delta-1} d\tilde{r}^2 + \rho^2 d\Omega_2^2\,. \tag{4.112}$$

Using the fact that [445]

$$d\tilde{r} = \frac{d\rho - A_2^{\frac{q+2\delta-1}{2}}\,\dot{a}\,\tilde{r}\,dt}{a A_2^{\frac{q+2\delta-1}{2}}\,C(r)}\,, \tag{4.113}$$

where

$$C(r) = 1 + \frac{2(q+2\delta-1)}{q}\,\frac{C_2 a}{\rho}\,A_2^{\frac{2\delta-1-q}{2}}\,, \tag{4.114}$$

the metric assumes the Painlevé-Gullstrand-like form

$$ds^2 = -A_2\left[1-\frac{A_2^{2(\delta-1)}}{C^2}\,\dot{a}^2\tilde{r}^2\right]dt^2 - \frac{2A_2^{\frac{-q+2\delta-1}{2}}}{C^2}\,\dot{a}\,\tilde{r}\,dt\,d\rho$$
$$+\frac{d\rho^2}{A_2^q C^2} + \rho^2 d\Omega_2^2\,. \tag{4.115}$$

The cross-term in $dtd\rho$ is then eliminated by the use of the new time coordinate $\bar{t}$ defined by the differential equation

$$d\bar{t} = \frac{1}{F(t,\rho)}\left[dt + \beta(t,\rho)d\rho\right]\,, \tag{4.116}$$

where $F(t,\rho)$ is an integrating factor which guarantees that $d\bar{t}$ is an exact differential and satisfies

$$\frac{\partial}{\partial\rho}\left(\frac{1}{F}\right) = \frac{\partial}{\partial t}\left(\frac{\beta}{F}\right)\,. \tag{4.117}$$

By setting

$$\beta = \frac{A_2^{\frac{-q+2\delta-3}{2}}}{C^2} \frac{\dot{a}\,\tilde{r}}{1 - \frac{A_2^{2(\delta-1)}}{C^2}\dot{a}^2\tilde{r}^2} \tag{4.118}$$

the line element becomes [445]

$$ds^2 = -A_2 D F^2 d\bar{t}^2 + \frac{d\rho^2}{A_2^q C^2 D} + \rho^2 d\Omega_2^2 \tag{4.119}$$

in the Nolan gauge, where $H \equiv \dot{a}/a$ and

$$D = 1 - \frac{A_2^{-q-1}}{C^2} H^2 \rho^2 \,. \tag{4.120}$$

The apparent horizons are located at $g^{\rho\rho} = 0$, or

$$A_2^q \left( C^2 - H^2 R^2 A_2^{-q-1} \right) = 0 \,, \tag{4.121}$$

hence $g^{\rho\rho}$ vanishes if $A_2 = 0$ or $H^2 R^2 = C^2 A_2^{q+1}$. $A_2$ vanishes at $r = C_2$, which describes the Schwarzschild event horizon in the limit to GR $\delta \to 0$ and locates a spacetime singularity. In fact, the Ricci scalar is

$$R = \frac{6\left(\dot{H} + 2H^2\right)}{A_2(r)} \tag{4.122}$$

and diverges as $r \to C_2$. This quantity reduces to the familiar FLRW value $6\left(\dot{H} + 2H^2\right)$ in the limit $C_2 \to 0$. This spacetime singularity is strong in the sense of Tipler's classification [1074] since the areal radius $\rho = a\tilde{r} A_2^{\frac{q+2\delta-1}{2}}$ vanishes when $r = C_2$. This behavior is in contrast with the Schwarzschild solution of GR corresponding to $\delta = 0$ and $\rho = \tilde{r} = 4C_2$ at $r = C_2$.

Let us focus on the second root of Eq. (4.121), *i.e.*, $H^2\rho^2 = C^2 A_2^{q+1}$ or

$$H\rho = \pm \left[ 1 + \frac{2(q + 2\delta - 1)}{q} \frac{C_2 a}{\rho} A_2^{\frac{2\delta-1-q}{2}} \right] A_2^{\frac{q+1}{2}} \,, \tag{4.123}$$

where the positive sign is appropriate to an expanding universe. In the limit $\delta \ll 1$ it is

$$H\rho = \left[ 1 + \frac{2\delta C_2 a}{\rho} A_2^{-\left(1-\frac{3\delta}{2}\right)} \right] A_2^{1-\delta} \,. \tag{4.124}$$

Two limits can be studied in order to understand the properties of this solution. First, as $C_2 \to 0$ the central object disappears and the solution reduces to a FLRW universe while $r = \tilde{r}$ and $\rho$ become a comoving radius and a proper radius, respectively. Moreover, Eq. (4.123) reduces to $H\rho = 1$ which has the solution $\rho_c = 1/H$, the radius of the cosmological horizon. The second limit is the limit to GR $\delta \to 0$, in which Eq. (4.123) becomes $A_2 = 0$ or $r = C_2$ with $H \equiv 0$.

Equations (4.108) and (4.111) then allow one to compute the left hand side of Eq. (4.123) as

$$HR = \frac{\delta\,(1+2\delta)}{1-\delta}\, t^{\frac{2\delta^2+2\delta-1}{1-\delta}} \frac{C_2}{x} \frac{(1-x)^{\frac{q+2\delta-1}{q}}}{(1+x)^{\frac{-q+2\delta-1}{q}}}, \tag{4.125}$$

where $x \equiv C_2/r$. The right hand side of Eq. (4.123) is

$$\frac{(1-x)^{\frac{q+1}{q}}}{(1+x)^{\frac{q+1}{q}}} \left[1 + \frac{2\,(q+2\delta-1)}{q} \frac{x}{(1-x)^2}\right], \tag{4.126}$$

and Eq. (4.123) assumes the form

$$\left(t^{\frac{1-2\delta-2\delta^2}{1-\delta}}\right)^{-1} = \frac{(1-\delta)}{\delta\,(1+2\delta)\,C_2} \frac{x\,(1+x)^{\frac{-2q+2\delta-2}{q}}}{(1-x)^{\frac{2(\delta-1)}{q}}} \cdot \left[1 + \frac{2\,(q+2\delta-1)}{q} \frac{x}{(1-x)^2}\right], \tag{4.127}$$

where $\dfrac{1-2\delta-2\delta^2}{1-\delta}$ is positive for $0 < \delta < \dfrac{\sqrt{3}-1}{2} \simeq 0.366$.

At late times the left hand side of Eq. (4.127) vanishes, implying that $x \simeq 0$ and that there is a unique root or apparent horizon. This unique late-time horizon is identified as a cosmological horizon: in fact, $r \to \infty$ as $x = C_2/r \to 0$. The limit $x \to 0$ can also be obtained when the parameter $C_2 \to 0$ , in which case $H\rho \to 1$ and

$$r \simeq \rho \simeq H^{-1} = \frac{1-\delta}{\delta\,(1+2\delta)}\, t \tag{4.128}$$

is the radius of the cosmological horizon of the FLRW space without any central inhomogeneity. Only a cosmological apparent horizon and no black hole apparent horizons are present at late times, which means that the central singularity at $\rho = 0$ becomes naked in the late-time development of this universe.

Using the quantity $x$ as a parameter, the time $t$ and the radius $\rho$ of the apparent horizons are [445]

$$t(x) = \left\{ \frac{(1-\delta)}{\delta\,(1+2\delta)\,C_2} \frac{x\,(1+x)^{\frac{2(-q+\delta-1)}{q}}}{(1-x)^{\frac{2(\delta-1)}{q}}} \left[1 + \frac{2\,(q+2\delta-1)\,x}{q(1-x)^2}\right]\right\}^{\frac{1-\delta}{2\delta^2+2\delta-1}}, \tag{4.129}$$

$$\rho(x) = t(x)^{\frac{\delta(1+2\delta)}{1-\delta}} \frac{C_2}{x} (1-x)^{\frac{q+2\delta-1}{q}} (1+x)^{\frac{q-2\delta+1}{q}}\,. \tag{4.130}$$

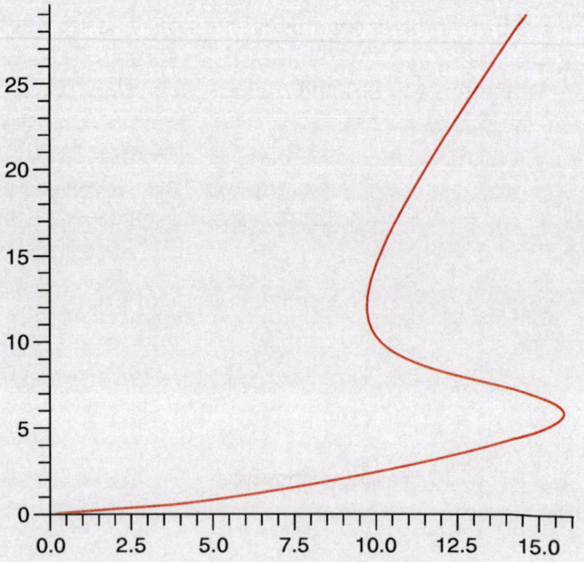


**Fig. 4.1** The radius $\rho$ of the apparent horizon as a function of time. Two inner horizons form after the Big Bang, cover the $\rho = 0$ singularity for a finite period of time, and then merge and disappear. A third horizon of cosmological nature represented by the upper branch of the curve keeps expanding. The parameter values used are $C_2 = 1$ and $\delta = 0.2$.

Figure 4.1 shows $\rho$ as a function of time. In this universe, two inner horizons develop after the Big Bang and they cover the $\rho = 0$ singularity. At a later time these apparent horizons become closer and finally merge and disappear. In the meantime a third horizon of cosmological nature keeps expanding. The $\rho = 0$ singularity becomes naked when the two inner apparent horizons disappear.

## 4.2 Spherical symmetry in scalar-tensor gravity

In the following we present spherically symmetric solutions of scalar-tensor gravity, beginning with Brans-Dicke theory and with static solutions. While it is relatively easy to find static solutions, it is considerably more difficult to find time-dependent solutions even under the assumption of spherical symmetry. Some of these are presented, although their precise significance is still uncertain – this is the case also for their counterparts in Einstein's theory.

### *4.2.1 Static solutions of Brans-Dicke theory*

The first spherically symmetric vacuum solutions of (Jordan frame) Brans-Dicke theory to be found were those forming the so-called Brans class I and given by [164]

$$ds^2 = -\left(\frac{1-\mu/r}{1+\mu/r}\right)^{2/\lambda} dt^2 + \left(1+\frac{\mu}{r}\right)^4 \left(\frac{1-\mu/r}{1+\mu/r}\right)^{\frac{2(\lambda-C-2)}{\lambda}} \left(dr^2 + r^2 d\Omega_2^2\right), \tag{4.131}$$

$$\phi(r) = \phi_0 \left(\frac{1-\mu/r}{1+\mu/r}\right)^{C/\lambda}, \tag{4.132}$$

in isotropic coordinates, where $\mu, C, \phi_0$, and $\lambda$ are constants with

$$\lambda^2 = (C+1)^2 - C\left(1 - \frac{\omega C}{2}\right) > 0\,. \tag{4.133}$$

These solutions are static and spherically symmetric, but different from the Schwarzschild metric. They solve the vacuum Brans-Dicke field equations with $V(\phi) = 0$ for $r > \mu$. Three other classes of spherically symmetric solutions were found by Brans [164], although they are not all independent from each other [134]. Usually, only positive values of the parameters $C$ and $\lambda$ occur in the literature and, for these, the scalar field diverges at the horizon (when this is present).

Various choices of the constant $C$ are possible, giving rise to a two-parameter family of solutions (assuming that $\omega$ is already fixed). The values of these parameters are not completely arbitrary, for example, the requirement that the tensor mass be positive restricts the parameter space [117]. For certain values of the parameter $C$ the solution (4.131) and (4.132) does not reduce to the Schwarzschild solution when $\omega \to \infty$, while the scalar field $\phi$ exhibits the asymptotic behavior

$$\phi = \phi_0 + \mathrm{O}\left(\frac{1}{\sqrt{\omega}}\right) \tag{4.134}$$

instead of the expected scaling (see, *e.g.*, [1153])

$$\phi = \phi_0 + \mathrm{O}\left(\frac{1}{\omega}\right) \tag{4.135}$$

as $\omega \to +\infty$ [75, 780, 976]. This issue is not unique to this specific family of solutions but is more general. The problem of the correct limit of Brans-Dicke theory and its solutions to GR occurs mostly when the matter source has vanishing trace $T^{(m)} = 0$ and it has not been completely elucidated, although some insight has been gained over the years [43, 75, 292, 293, 451, 454, 780, 885, 886, 944, 946–948, 976]. In general, the limit of a spacetime when one or more parameters vary may turn out to be ill-defined even in GR because it could depend on the coordinate system adopted and therefore it could be non-unique [535]. For example, the limit of the Schwarzschild solution as the mass parameter diverges is either the Minkowski space or a Kasner space [535]. A coordinate-independent approach to the limit of GR spacetimes based on the Cartan scalars has been developed in [885] and applied to the $\omega \to \infty$ limit of Brans-Dicke theory in [886]. It was found that the limit of Brans-Dicke solutions to GR solutions corresponding to the same stress-energy tensor is not unique, or it may not yield a GR solution at all [886].

In addition to the Brans metric (4.131), other exact solutions of Brans-Dicke theory which are static and asymptotically flat are known.

Several vacuum solutions of scalar-tensor gravity were generated from Brans-Dicke solutions in [1104]. Solutions of Barker's theory were found in [403, 921]. Barker's theory corresponds to the choice

$$\omega(\phi) = \frac{4 - 3G_0\phi}{2\,(G_0\phi - 1)} \,. \tag{4.136}$$

of the Brans-Dicke coupling function, which makes the effective gravitational coupling for spherically symmetric solutions [856]

$$G_{eff} = \frac{2\,(\omega + 2)}{2\omega + 3}\,\frac{1}{\phi} \,. \tag{4.137}$$

a constant even though the scalar $\phi$ is dynamical.

Since the Brans-Dicke-like scalar $\phi$ couples to the trace of the matter energy-momentum tensor, it is not sourced by conformally invariant matter, which plays a special role. Many electro-vacuum solutions were discovered in [72, 73, 928]. Other electro-vacuum, static, asymptotically flat solutions in the Jordan frame were found in [192, 925].

### *4.2.2 Dynamical and asymptotically FLRW solutions*

An intriguing type of solutions is that interpreted as a black hole embedded in a cosmological background. The search for such solutions in the context of GR was originally motivated by the need to understand the effect of the cosmological expansion on local systems (a problem that is not yet completely closed, see [268] for a review), and led to the McVittie solution of the Einstein equations [789]. Several recent solutions of these equations describing exact spherically symmetric cosmological black holes are still poorly understood. In the context of scalar-tensor gravity there is the possibility that the effective gravitational coupling be space-dependent, *i.e.*, that gravity can be stronger or weaker inside inhomogeneities in the matter distribution. This possibility has led to the study of exact solutions describing inhomogeneities in a FLRW universe, which are taken to be spherically symmetric for simplicity. An example is the class of separable metrics found by Clifton, Mota and Barrow [328]

$$ds^2 = -A^{2k}(r)dt^2 + a^2(t)\left(1 + \frac{C}{2kr}\right)^4 A^{\frac{2(k-1)(k+2)}{k}}(r)\left(dr^2 + r^2 d\Omega_2^2\right) , \tag{4.138}$$

$$A(r) = \frac{1 - \frac{C}{2kr}}{1 + \frac{C}{2kr}}, \tag{4.139}$$

$$a(t) = a_0 \left(\frac{t}{t_0}\right)^{\frac{2\omega(2-\gamma)+2}{3\omega\gamma(2-\gamma)+4}}, \tag{4.140}$$

$$\phi(t, r) = \phi_0 \left(\frac{t}{t_0}\right)^{\frac{2(4-3\gamma)}{3\omega\gamma(2-\gamma)+4}} \left(\frac{1 - \frac{C}{2kr}}{1 + \frac{C}{2kr}}\right)^{\frac{-2(k^2-1)}{k}}, \tag{4.141}$$

$$k = \sqrt{\frac{2(\omega + 2)}{2\omega + 3}} \tag{4.142}$$

in isotropic coordinates, where $C, a_0, t_0$, and $\phi_0$ are constants and the matter content of the universe outside the spherical inhomogeneity is a perfect fluid with constant equation of state $P = (\gamma - 1)\rho$. The energy density is

$$\rho(t, r) = \rho_0 \left(\frac{a_0}{a(t)}\right)^{3\gamma} \left(\frac{1 - \frac{C}{2kr}}{1 + \frac{C}{2kr}}\right)^{-2k}. \tag{4.143}$$

The power-law dependence of the scale factor $a(t)$ on the time coordinate $t$ is the same as the one that would occur in a FLRW universe without inhomogeneities [328].

Other solutions representing spherically symmetric inhomogeneities in an otherwise spatially homogeneous and isotropic universe in scalar-tensor gravity are obtained by mapping back from the Einstein frame to the Jordan frame certain exact solutions of GR representing cosmological black holes embedded in a FLRW background filled by a minimally coupled scalar field [328], matching FLRW regions [328]. There are also Swiss-cheese models [959] and Lemaitre-Tolman-Bondi models [573].

### 4.2.3 *Collapse to black holes in scalar-tensor theory*

We will see below that, according to a theorem by Hawking[4] [581] and other results [645], all static or stationary black hole solutions of Brans-Dicke theory coincide with those of GR provided that the (Jordan frame) Brans-Dicke scalar does not vanish or diverge on the event horizon [210, 752]. Similar results exist for more general scalar-tensor theories, such as that of a scalar field non-minimally coupled

[4] Hawking's theorem, which is proved for a Brans-Dicke scalar $\phi$ without potential, is immediately extended to the case in which it has a quadratic potential $V(\phi) = m^2\phi^2/2$ [969].

to the Ricci curvature [782, 953, 1168]. In the situation in which the Brans-Dicke scalar $\phi$ vanishes on the horizon ("cold black holes") the Hawking temperature is believed to vanish, while it should diverge when $\phi \to \infty$ on the horizon.

Assuming that gravitational collapse does not lead to a naked singularity (which seems to be a non-generic outcome in GR [1137] but should be checked in alternative gravities), the final state should presumably be a stationary (Kerr-Newman) black hole. However, it is still possible that a highly dynamical situation such as gravitational collapse leads to stationary black holes with Brans-Dicke scalar vanishing/diverging at the horizon, or that starting out with the assumption of stationarity somehow restricts one to considering eternal black holes which cannot be generated during collapse (it has been suggested that this could be the case for extremal black holes in GR). It is therefore important to check explicitly the outcome of collapse in Brans-Dicke theory, which has been done numerically by Scheel, Shapiro, and Teukolsky [975, 976] following early work by Matsuda and Nariai [781]. In [975,976] the Oppenheimer-Snyder collapse of dust in spherical symmetry is studied numerically for Brans-Dicke theory in the Jordan frame and without scalar field potential. It is found that during the collapse there are significant deviations from GR: the apparent horizon area decreases, violating the second law of black hole mechanics which states that the horizon area cannot decrease [582, 1139], and the apparent horizon crosses outside the event horizon. This behavior is attributed to the violation of the null energy condition by the Brans-Dicke scalar field regarded as an effective form of matter in the field equations, which can be rewritten as effective Einstein equations. However, once scalar radiation has been radiated away and these "transients" have decayed, the final state is found to be the Schwarzschild black hole [975, 976]. This result is in agreement with Hawking's theorem, a suggestion by Penrose [899], a study by Thorne and Dykla based on an expansion in powers of $\omega^{-1}$ [1073] in which the background is GR, and with previous numerical work by Shibata, Nakao, and Nakamura [1011]. All these studies suggest that the final state should be a Kerr (for cylindrical symmetry) or a Schwarzschild (for spherical symmetry) black hole.[5] Further numerical studies by Harada and collaborators [572] focussed on the emission of scalar gravitational waves, while Novak used a perfect fluid with pressure and paid particular attention to the phenomenon of scalarization (the possibility, analogous to magnetization, that the scalar field could peak at anomalously large amplitudes) [857].

The phenomenology found by Scheel, Shapiro and Teukolsky during the dynamical phases of the collapse is recovered by Kerimo and Kalligas [665] and Kerimo [664], who studied numerically the collapse of a dust in a tensor-two-scalar theory with massless scalars $\phi$ and $\psi$ forming a $\sigma$-model and $\omega = \omega(\phi, \psi)$, in the presence of spherical symmetry. The deviations from GR are enhanced due to the extra freedom in the coupling function $\omega(\phi, \psi)$.

[5] These studies, however, assume from the outset that the collapse does not result in a naked singularity or in a solution that does not reduce to a GR one in the $\omega \to \infty$ limit [976], and such solutions do exist in scalar-tensor gravity.

More recent work has shown that GR black holes in their quiescent state are essentially indistinguishable from solutions of a wide variety of gravity theories containing additional vector and tensor true degrees of freedom [915], although perturbations of black hole spacetimes can reveal the differences between these theories and GR [77].

## 4.3 The Jebsen-Birkhoff theorem

Spherically symmetric solutions are important to understand the weak-field limit of ETGs and to confront them with Solar System experiments (after all, the three classical tests of GR are based on the spherical Schwarzschild solution) and one would like to know which static solutions are generic. One of the first questions that arises when comparing alternative gravity with GR is whether the Jebsen-Birkhoff theorem well known from spherical symmetry in Einstein's theory can be generalized, or fails, in these extensions of GR [463].

Beginners in GR are familiar with the Birkhoff theorem stating that a spherically symmetric solution of the vacuum Einstein equations is necessarily static [142]. This result was actually discovered by Jebsen [638] two years before Birkhoff [374, 375, 642]. The Jebsen-Birkhoff theorem fails to extend to metric $f(R)$ gravity while it holds in Palatini $f(R)$ gravity [1033]. Since both metric and Palatini $f(R)$ gravities can be represented as Brans-Dicke theories with a potential, understanding the Jebsen-Birkhoff theorem in scalar-tensor gravity allows for its understanding also in $f(R)$ gravity, therefore we begin with scalar-tensor theories. Since the Brans-Dicke-like scalar is often treated as an effective form of matter for effective Einstein equations, we first formulate a generalized Jebsen-Birkhoff theorem in the presence of matter in GR and conveniently transfer the results to spherical symmetry in scalar-tensor gravity, using both the Jordan and the Einstein frames [463]. In this section the cosmological constant, if present, is regarded as an effective form of matter described by the stress-energy tensor $T_{\mu\nu}^{(\Lambda)} = -\frac{\Lambda}{8\pi G} g_{\mu\nu}$ and "vacuum" implies that $\Lambda = 0$.

### 4.3.1 *The Jebsen-Birkhoff theorem of GR*

The most general spherically symmetric line element is[6]

$$ds^2 = -A^2(t,r)dt^2 + B^2(t,r)dr^2 + r^2 d\Omega_2^2 \,, \tag{4.144}$$

[6] For convenience we use here a notation which is different from the one of the previous sections.

with $r$ an areal radius. The metric components $g_{0i}$ $(i = 1, 2, 3)$, if present, can be eliminated by redefining the coordinates $t$ and $r$ [638,705]. The $(0, 1)$, $(0, 0)$, $(1, 1)$, $(2, 2)$, and $(3, 3)$ components of the Einstein equations yield

$$\frac{\dot{B}}{B} = 4\pi G r\, T_{01}^{(m)} , \tag{4.145}$$

$$\frac{1}{r^2} + \frac{2B'}{B^3 r} - \frac{1}{B^2 r^2} = 8\pi G\, \frac{T_{00}^{(m)}}{A^2} , \tag{4.146}$$

$$\frac{2A'}{Ar} - \frac{B^2}{r^2} + \frac{1}{r^2} = 8\pi G\, T_{11}^{(m)} , \tag{4.147}$$

$$\frac{r}{B^3}\left( \frac{A'B}{A} - B' - \frac{rB^2\ddot{B}}{A^2} + \frac{r\dot{A}B^2\dot{B}}{A^3} - \frac{rA'B'}{A} + \frac{rA''B}{A} \right) = 8\pi G\, T_{22}^{(m)}, \tag{4.148}$$

$$\frac{r\sin^2\theta}{B^3}\left( \frac{A'B}{A} - B' - \frac{rB^2\ddot{B}}{A^2} + \frac{r\dot{A}B^2\dot{B}}{A^3} - \frac{rA'B'}{A} + \frac{rA''B}{A} \right)$$
$$= 8\pi G\, T_{33}^{(m)} , \tag{4.149}$$

where $T_{\mu\nu}^{(m)}$ is the matter energy-momentum tensor and, in this section, a prime and an overdot denote differentiation with respect to $r$ and $t$, respectively.

### 4.3.2 *The non-vacuum case*

Consider a timelike observer with four-velocity $u^\mu = \left(A^{-1}, 0, 0, 0\right)$ at rest in the coordinate system $(t, r, \theta, \varphi)$. The matter energy density relative to this observer is

$$\rho \equiv T_{\mu\nu}^{(m)} u^\mu u^\nu = \frac{T_{00}^{(m)}}{A^2} , \tag{4.150}$$

while the radial energy current relative to this observer is

$$J_{(r)} \equiv -T_{\mu\nu}^{(m)}\, u^\mu\, e_{(r)}^\nu = \frac{T_{01}^{(m)}}{AB} \tag{4.151}$$

(with $e_{(r)}^\mu = \left(0, B^{-1}, 0, 0\right)$ the spacelike unit vector in the radial direction), and the radial pressure is

$$P_{(r)} \equiv T_{\mu\nu}^{(m)} e_{(r)}^\mu\, e_{(r)}^\nu = \frac{T_{11}^{(m)}}{B^2} . \tag{4.152}$$

The non-radial stresses $T_{ij}^{(m)}$ with $i \neq j$ $(i, j = 1, 2, 3)$ vanish because of the Einstein equations in conjunction with the vanishing of the components $G_{ij}$ with $i \neq j$ in spherical symmetry. Equations (4.148) and (4.149) imply that $\frac{T_{33}^{(m)}}{\sin^2\theta} = T_{22}^{(m)}$.

The Einstein equations require that the matter distribution be spherically symmetric and the derivatives of $\rho$, $J_{(r)}$, and $P_{(r)}$ with respect to $\theta$ and $\varphi$ vanish. A spherically symmetric $T_{\mu\nu}^{(m)}$ is said to describe a *static matter distribution* if and only if

$$\frac{\partial \rho}{\partial t} = Au^{\nu}\nabla_{\nu}\rho = 0\,, \qquad \frac{\partial P_{(r)}}{\partial t} = Au^{\nu}\nabla_{\nu}P_{(r)} = 0\,, \qquad J_{(r)} = 0\,. \tag{4.153}$$

This is equivalent to requiring that the Lie derivatives of $T_{\mu\nu}^{(m)}$ along the directions of the two Killing vectors $t^{\mu}$ (timelike) and $\psi^{\mu}$ (spacelike) vanish [980, 981].

Equation (4.145) with the assumption $T_{01}^{(m)} = 0$ guarantees that $\dot{B} = 0$ and $B = B(r)$. Then Eq. (4.147) with the assumption $\partial P_{(r)}/\partial t = 0$ implies that $A'/A$ is time-independent. It could still be possible for $A$ to depend on time through a multiplicative factor, $A(t, r) = f(t)a(r)$, but in this case a redefinition of the time coordinate $t \to \bar{t}$ with $d\bar{t} \equiv f(t)dt$ absorbs the factor $f(t)$ into $\bar{t}$ and the metric can be written in locally static form. Then the Einstein equations (4.148) and (4.149) imply that also $T_{22}^{(m)}$ and $T_{33}^{(m)}$ and the tangential pressures

$$\begin{aligned} P_{(\theta)} &\equiv T_{\mu\nu}^{(m)} e_{\theta}^{\mu} e_{(\theta)}^{\nu} = \frac{T_{22}^{(m)}}{r^2} \\ &= P_{(\varphi)} \equiv T_{\mu\nu}^{(m)} e_{(\varphi)}^{\mu} e_{(\varphi)}^{\nu} = \frac{T_{33}^{(m)}}{r^2\sin^2\theta} \end{aligned} \tag{4.154}$$

vanish (here $e_{(\theta)}^{\mu}$ and $e_{(\varphi)}^{\mu}$ are spacelike unit vectors in the angular directions). This result leads [463] to the

**Jebsen-Birkhoff theorem (version 1):** *If a solution of the Einstein equations is spherically symmetric and the matter distribution is static (i.e.,* $\frac{\partial \rho}{\partial t} = \frac{\partial P_{(r)}}{\partial t} = 0$ *and* $J_{(r)} = 0$*), then the metric is static in a region in which* $t$ *is timelike and* $(r, \theta, \varphi)$ *are spacelike.*

The theorem is necessarily restricted to the region of the spacetime manifold in which the coordinates maintain their timelike or spacelike character, as originally remarked by Ehlers and Krasinski [413]. This restriction is not satisfied, for example, in the region inside the Schwarzschild black hole horizon or outside the de Sitter cosmological horizon, where these metrics are time-dependent. In Ehlers and Krasinki's words [413], "a spherically symmetric solution admits, besides the SO(3) generators, an additional hypersurface-orthogonal Killing vector field" (see also [511, 980, 981, 1053] for discussions of the local character of the Jebsen-Birkhoff theorem). The metric can be put in static form only where this additional Killing field stays timelike, which excludes black hole horizons where it vanishes. The theorem does not contemplate black hole horizons and does not imply that the solution of the Einstein equations is the Schwarzschild metric.

Vacuum is trivially static and gives rise to the familiar version of the Jebsen-Birkhoff theorem. The assumptions made on the matter distribution described by $T_{\mu\nu}^{(m)}$ are fairly strong because this tensor contains the metric $g_{\mu\nu}$ and the assumptions on $T_{\mu\nu}^{(m)}$ require already, indirectly that the metric is static. However, the assumption of a static matter distribution still allows for physically non-trivial situations. One example is that of a cosmological constant seen as a form of effective matter described by $T_{\mu\nu}^{(\Lambda)}$, which is spherically symmetric and static. In this case the GR solution is the Schwarzschild-(anti)de Sitter one. For example, the Schwarschild-de Sitter (or Kottler) line element admits the locally static form

$$ds^2 = -\left(1 - \frac{2GM}{r} - \frac{\Lambda r^2}{3}\right) dt^2 + \left(1 - \frac{2GM}{r} - \frac{\Lambda r^2}{3}\right)^{-1} dr^2 + r^2 d\Omega_2^2 \tag{4.155}$$

in the spacetime region comprised between the black hole and the cosmological horizons. Another non-trivial example is electro-vacuum [365]; the solution with a static electric charge $Q$ and no current is the static Reissner-Nordstrom metric

$$ds^2 = -\left(1 - \frac{2GM}{r} + \frac{Q^2}{r^2}\right) dt^2 + \left(1 - \frac{2GM}{r} + \frac{Q^2}{r^2}\right)^{-1} dr^2 + r^2 d\Omega_2^2 . \tag{4.156}$$

The absence of a radial energy current, $J_{(r)} = 0$, without the assumption of staticity of $\rho$ and $P_{(r)}$ is not sufficient to guarantee time-independence of the metric. As a counterexample consider the McVittie solution describing a spherical object embedded in a cosmological background [789] for which $J_{(r)} = 0$ and the metric is time-dependent (except for the special case in which it reduces to the Schwarzschild-(anti)de Sitter metric).

Version 1 of the Jebsen-Birkhoff theorem is more useful for the discussion of spherical symmetry in scalar-tensor gravity than the familiar textbook version because the Brans-Dicke-like scalar field acts as an effective form of matter [463].

### 4.3.3 *The vacuum case*

If vacuum ($T_{\mu\nu}^{(m)} = 0$) is regarded as a trivial form of static matter, the Jebsen-Birkhoff theorem reduces to the familiar version

**Jebsen-Birkhoff theorem (version 2):** *a spherically symmetric solution of the vacuum Einstein equations is necessarily static in a region in which $t$ is timelike and $(r, \theta, \varphi)$ are spacelike.*

"Vacuum" rules out the possibility of a cosmological constant and electro-vacuum, and the Schwarzschild-(anti)de Sitter and Reissner-Nordstrom solutions: then the solution of the Einstein equations must necessarily be the Schwarzschild metric.

"Vacuum", defined by $T_{\mu\nu}^{(m)} = 0$ is not necessarily a trivial configuration in alternative gravity. In scalar-tensor theories the Brans-Dicke-like scalar field $\phi$ describing the gravitational field together with the tensor $g_{\mu\nu}$ may be dynamical and still conspire to give a zero effective stress-energy tensor $T_{\mu\nu}^{(\phi)}$; then $\phi$ is called a "non-gravitating" or "stealth" scalar field. Two examples of massive stealth $\phi$ waves coupled non-minimally to the Ricci curvature and with a power-law potential were found in [61], with a Minkowski metric providing a non-trivial realization of version 2 of the Jebsen-Birkhoff theorem. Other examples of non-gravitating matter distributions are given in [62, 74, 371, 943, 1022]. At present, it is not clear whether these rather exotic solutions are stable, although there are indications that they may be [475].

From the physical point of view, the field content of GR consists only of a massless spin two field and the lowest order gravitational radiation is quadrupole. A spherically symmetric source cannot emit gravitational radiation and its mass-energy is conserved, therefore the spacetime around a spherically symmetric source must be static.

A well known corollary of the Jebsen-Birkhoff theorem is the extension to GR of Newton's iron sphere theorem stating that the gravitational field due to a spherically symmetric distribution of mass inside a spherical cavity vanishes. In spherical symmetry, if the energy distribution inside a cavity is static, the solution of the Einstein equations will be static, being flat space (the Schwarzschild solution with zero mass) *in vacuo* and the (anti-)de Sitter metric in the presence of a cosmological constant.

### 4.3.4 *The Jebsen-Birkhoff theorem in scalar-tensor gravity*

The validity of the Jebsen-Birkhoff theorem was investigated early on in Jordan [649] and Brans-Dicke theories [404, 699, 879, 927, 929–931, 995, 1016–1018, 1112] and it was quickly found that, in general, it is not valid. In order for the theorem to hold it is necessary to impose that the effective stress-energy tensor $T_{\mu\nu}^{(\phi)}$ of the Brans-Dicke-like scalar is time-independent.[7] The fact that the theorem fails in the presence of time-dependent scalars opens the door for new scalar-tensor phenomenology. Physically, it is expected that the Jebsen-Birkhoff theorem must be

[7] In addition to the obvious way of satisfying this requirement by imposing that $\phi$ is time-independent, also stealth scalar fields [61, 62, 74, 371, 943] satisfy it.

abandoned in scalar-tensor gravity because of the new spin zero degree of freedom that occurs in scalar-tensor gravity, which allows for scalar monopole and dipole radiation. In Einstein's theory monopole and dipole radiation are forbidden because the gravitational field is described only by a spin two field which admits only quadrupole, or higher multipole, radiation. Since in GR spherically symmetric pulsating sources cannot generate gravitational waves, a spherically symmetric metric must be static. This is no longer true in scalar-tensor gravity, in which the time-varying monopole moment of a radially pulsating spherical source generates scalar waves which transfer energy and make the metric time-dependent.

Consider the scalar-tensor action

$$S_{ST} = \frac{1}{16\pi} \int d^4x \sqrt{-g} \left[ \phi R - \frac{\omega(\phi)}{\phi} g^{\mu\nu} \nabla_\mu \phi \, \nabla_\nu \phi - V(\phi) \right] + S^{(m)} \tag{4.157}$$

and the corresponding field equations

$$R_{\mu\nu} - \frac{1}{2} g_{\mu\nu} R = \frac{8\pi}{\phi} T^{(m)}_{\mu\nu} + \frac{\omega(\phi)}{\phi^2} \left( \nabla_\mu \phi \nabla_\nu \phi - \frac{1}{2} g_{\mu\nu} \nabla^\alpha \phi \nabla_\alpha \phi \right)$$
$$+ \frac{1}{\phi} \left( \nabla_\mu \nabla_\nu \phi - g_{\mu\nu} \Box \phi \right) - \frac{V(\phi)}{2\phi} g_{\mu\nu} \equiv \frac{8\pi}{\phi} \left( T^{(m)}_{\mu\nu} + T^{(\phi)}_{\mu\nu} \right) , \tag{4.158}$$

$$(2\omega + 3) \Box \phi = 8\pi T^{(m)} - \frac{d\omega}{d\phi} \nabla^\alpha \phi \nabla_\alpha \phi + \phi \frac{dV}{d\phi} - 2V , \tag{4.159}$$

written in the form of effective Einstein equations with the scalar field $\phi$ acting as an effective form of matter (we assume $\phi > 0$ and $\omega > -3/2$ in the following). By imposing that the matter stress-energy tensor $T^{(m)}_{\mu\nu}$ vanishes, only the effective stress-energy tensor $T^{(\phi)}_{\mu\nu}$ is left, in which $T^{(\phi)}_{00}$ could depend on time and $T^{(\phi)}_{0i}$ can be non-vanishing if $\phi$ is time-dependent, spoiling the Jebsen-Birkhoff theorem. The validity of the theorem is restored only upon the assumption that $\phi$ is time-independent or does not gravitate ($T^{(\phi)}_{\mu\nu} = 0$).

### 4.3.5 *The trivial case $\phi$ = constant*

In the trivial situation $\phi = \text{const.} \equiv \phi_0 > 0$, Eq. (4.158) becomes

$$R_{\mu\nu} - \frac{1}{2} g_{\mu\nu} R = \frac{8\pi}{\phi_0} T^{(m)}_{\mu\nu} - \frac{V_0}{2\phi_0} g_{\mu\nu} , \tag{4.160}$$

where $V_0 \equiv V(\phi_0)$. The theory reduces to GR with the cosmological constant $\Lambda \equiv V_0/(2\phi_0)$. If $T^{(m)}_{\mu\nu}$ is such that the energy distribution is static (including $T^{(m)}_{\mu\nu} = 0$), then version 1 of the Jebsen-Birkhoff theorem applies and the metric is static in the region in which the coordinate gradients do not change their causal character.

### 4.3.6 *Static non-constant Brans-Dicke-like field*

Assume that the spherically symmetric solution of the field equations (4.158) and (4.159) has line element (4.144). Then the only non-vanishing Christoffel symbols are

$$\Gamma^0_{00} = \frac{\dot{A}}{A}, \qquad \Gamma^0_{01} = \Gamma^0_{10} = \frac{A'}{A}, \qquad \Gamma^0_{11} = \frac{B\dot{B}}{A^2}, \tag{4.161}$$

$$\Gamma^1_{00} = \frac{AA'}{B^2}, \qquad \Gamma^1_{01} = \Gamma^1_{10} = \frac{\dot{B}}{B}, \qquad \Gamma^1_{11} = \frac{B'}{B}, \tag{4.162}$$

$$\Gamma^1_{22} = -\frac{r}{B^2}, \qquad \Gamma^1_{33} = -\frac{r}{B^2}\sin^2\theta, \tag{4.163}$$

$$\Gamma^2_{12} = \Gamma^2_{21} = \frac{1}{r}, \qquad \Gamma^2_{33} = -\sin\theta\cos\theta, \tag{4.164}$$

$$\Gamma^3_{13} = \Gamma^3_{31} = \frac{1}{r}, \qquad \Gamma^3_{23} = \Gamma^3_{32} = \frac{\cos\theta}{\sin\theta}, \tag{4.165}$$

and

$$\nabla^\alpha\phi\nabla_\alpha\phi = -\frac{\dot{\phi}^2}{A^2} + \frac{\phi'^2}{B^2}, \tag{4.166}$$

while the d'Alembertian of the scalar $\phi$ is

$$\Box\phi = -\frac{1}{A^2}\left(\ddot{\phi} - \frac{\dot{A}}{A}\dot{\phi} - \frac{AA'}{B^2}\phi'\right) + \frac{1}{B^2}\left(\phi'' - \frac{B\dot{B}}{A^2}\dot{\phi} - \frac{B'}{B}\phi'\right) + \frac{2\phi'}{rB^2}. \tag{4.167}$$

The (0, 1), (0, 0), and (1, 1) components of the field equations (4.158) yield

$$\frac{2\dot{B}}{Br} = \frac{8\pi}{\phi}T^{(m)}_{01} + \omega\frac{\dot{\phi}\phi'}{\phi^2} + \frac{1}{\phi}\left(\dot{\phi}' - \frac{A'}{A}\dot{\phi} - \frac{\dot{B}}{B}\phi'\right), \tag{4.168}$$

$$\begin{aligned} A^2\left(\frac{1}{r^2} + \frac{2B'}{B^3 r} - \frac{1}{B^2 r^2}\right) &= \frac{8\pi}{\phi}T^{(m)}_{00} + \frac{\omega}{2\phi^2}\left(\dot{\phi}^2 + \frac{A^2}{B^2}\phi'^2\right) \\ &\quad + \frac{A^2}{B^2\phi}\left(\phi'' - \frac{B\dot{B}}{A^2}\dot{\phi} - \frac{B'}{B}\phi' + \frac{2\phi'}{r}\right) + \frac{VA^2}{2\phi}, \end{aligned} \tag{4.169}$$

$$\begin{aligned} \frac{2A'}{Ar} - \frac{B^2}{r^2} + \frac{1}{r^2} &= \frac{8\pi}{\phi}T^{(m)}_{11} + \frac{\omega}{\phi^2}\left(\phi'^2 + \frac{B^2}{A^2}\dot{\phi}^2\right) \\ &\quad + \frac{B^2}{A^2\phi}\left(\ddot{\phi} - \frac{\dot{A}}{A}\dot{\phi} - \frac{AA'}{B^2}\phi' - \frac{2A^2}{B^2 r}\phi'\right) - \frac{VB^2}{2\phi}. \end{aligned} \tag{4.170}$$

In the absence of radial energy currents ($T_{01}^{(m)} = 0$) and assuming that $\phi = \phi(r)$ only, Eq. (4.168) yields

$$\frac{\dot{B}}{B}\left(\frac{2}{r}+\frac{\phi'}{\phi}\right)=0\,, \tag{4.171}$$

hence either $\dot{B} = 0$ or $\phi(r) = C/r^2$, where $C > 0$ is a constant. If $\phi = C/r^2$ Eq. (4.169) implies that $B^2 = \dfrac{2C\,(2\omega+3)}{2C+Vr^4}$ *in vacuo* and $B = B(r)$; in both cases we reduce to $\dot{B} = 0$, and we confine our attention to this situation from now on. Equation (4.170) becomes

$$\frac{A'}{A}\left(\frac{2}{r}+\frac{\phi'}{\phi}\right)=\frac{8\pi}{\phi}\,T_{11}^{(m)}+\frac{B^2-1}{r^2}+\omega\left(\frac{\phi'}{\phi}\right)^2-\frac{2}{r}\frac{\phi'}{\phi}-\frac{B^2V}{2\phi}\,. \tag{4.172}$$

If $\partial T_{11}^{(m)}/\partial t = 0$ the right hand side is time-independent and $A$ can depend from $t$ at most through a multiplicative factor, $A(t,r) = f(t)a(r)$. Then the time-dependent factor can be absorbed by a redefinition of the time coordinate $d\bar{t} = f(t)dt$ and the metric assumes the locally static form while the other field equations imply that the radial pressures $T_{ii}^{(m)}$ ($i = 1, 2, 3$) are also static.

The fact that the scalar field $\phi$ needs to be static (or non-gravitating) in order to restore the Jebsen-Birkhoff theorem was established for particular scalar-tensor theories. Early on, Schücking [995] derived the result for Jordan's theory, Reddy [930] did it for the electro-vacuum case of the Sen-Dunn theory (Sen-Dunn gravity [1005] is a scalar-tensor theory in which the second derivative terms $\nabla_\mu\nabla_\nu\phi - g_{\mu\nu}\Box\phi$ in the effective stress-energy tensor of $\phi$ vanish identically) and for the conformally coupled scalar field [927]. Electro-vacuum Sen-Dunn gravity was revisited in [404] and errors corrected in [699], while the conformally coupled situation was reconsidered in [1017]. Electro-vacuum in more general scalar-tensor theories was studied by Venkateswarlu and Reddy [1112].

In the light of the discussion above, the Jebsen-Birkhoff theorem turns out to be not very interesting in scalar-tensor gravity because it is valid only under conditions that seem too restrictive. When the Brans-Dicke-like scalar is static but not constant, the solution of the field equations can be different from Schwarzschild-(anti)de Sitter. If this field is constant the theory reduces to GR and the familiar Jebsen-Birkhoff theorem holds together with its iron sphere corollary (the spherically symmetric solution inside an empty cavity is static if the scalar field $\phi$ is static or non-gravitating).

### 4.3.7 *The Jebsen-Birkhoff theorem in Einstein frame scalar-tensor gravity*

Let us now examine the Einstein frame description of scalar-tensor gravity. We recall, for the reader's benefit, the conformal transformation of the metric and the redefinition of the scalar field

$$g_{\mu\nu} \rightarrow \tilde{g}_{\mu\nu} = \Omega^2 \, g_{\mu\nu}, \qquad \Omega = \sqrt{G\phi}\,, \tag{4.173}$$

$$d\tilde{\phi} = \sqrt{\frac{|2\omega(\phi)+3|}{16\pi G}}\,\frac{d\phi}{\phi} \tag{4.174}$$

for $\omega \neq -3/2$. The scalar-tensor action (4.157) becomes

$$S_{ST} = \int d^4x\,\sqrt{-\tilde{g}}\left[\frac{\tilde{R}}{16\pi G} - \frac{1}{2}\,\tilde{g}^{\mu\nu}\tilde{\nabla}_\mu\tilde{\phi}\tilde{\nabla}_\nu\tilde{\phi} - U(\tilde{\phi}) + \frac{\mathcal{L}^{(m)}}{(G\phi)^2}\right], \tag{4.175}$$

with the tilde denoting rescaled quantities as usual and

$$U\left(\tilde{\phi}\right) = \frac{V\left[\phi(\tilde{\phi})\right]}{\left[G\phi(\tilde{\phi})\right]^2}\,. \tag{4.176}$$

The Einstein frame field equations are

$$\tilde{R}_{\mu\nu} - \frac{1}{2}\,\tilde{g}_{\mu\nu}\tilde{R} = \frac{8\pi G}{(G\phi)^2}\,T^{(m)}_{\mu\nu} + 8\pi G\,\tilde{T}^{(\tilde{\phi})}_{\mu\nu}\,, \tag{4.177}$$

$$\tilde{\Box}\tilde{\phi} - \frac{dU}{d\tilde{\phi}} = \frac{8\pi G T^{(m)}}{(G\phi)^2}\,. \tag{4.178}$$

Here

$$\tilde{T}^{(\tilde{\phi})}_{\mu\nu} = \tilde{\nabla}_\mu\tilde{\phi}\tilde{\nabla}_\nu\tilde{\phi} - \frac{1}{2}\,\tilde{g}_{\mu\nu}\tilde{g}^{\alpha\beta}\,\tilde{\nabla}_\alpha\tilde{\phi}\tilde{\nabla}_\beta\tilde{\phi} - \frac{U\left(\tilde{\phi}\right)}{2}\,\tilde{g}_{\mu\nu} \tag{4.179}$$

is the canonical stress-energy tensor for a scalar field minimally coupled with the Ricci curvature and satisfies the weak energy condition if $V \geq 0$. If the metric $g_{\mu\nu}$ has the spherically symmetric form (4.144) also the rescaled metric $\tilde{g}_{\mu\nu}$ has the same form with $\Omega = \Omega(\phi) = \Omega(t,r)$.

As discussed previously in this book, the Jordan and the Einstein conformal frames are physically equivalent representations at the classical level [392,473,495] as long as the conformal transformation remains well-defined, hence the previous discussion of the Jebsen-Birkhoff theorem should be easily recovered in the Einstein frame. The Einstein frame scalar field $\tilde{\phi}$ can only be constant if its Jordan frame counterpart $\phi$ is constant; in this case one obtains, in the Einstein frame, the equations of motion of GR which include a cosmological constant term if $U(\tilde{\phi}) \neq 0$ (or, equivalently, $V(\phi) \neq 0$, cf. Eq. (4.176)), and version 1 of the Jebsen-Birkhoff theorem is valid.

If instead $\phi \neq$ const. but $\phi$ is independent of the time coordinate, then $\Omega = \Omega(r)$ and $\tilde{\phi}$ given by Eq. (4.174) is also static and, introducing the rescaled four-velocity $\tilde{u}^\mu = \frac{u^\mu}{\Omega} = \left(\tilde{A}^{-1}, 0, 0, 0, \right)$, the Einstein frame quantities are

$$\tilde{\rho}[\tilde{\phi}] \equiv \tilde{T}^{(\tilde{\phi})}_{\mu\nu} \tilde{u}^{\mu} \tilde{u}^{\nu} = \frac{\tilde{\phi}'^2}{2\tilde{B}^2} + \frac{U(\tilde{\phi})}{2}, \tag{4.180}$$

$$\tilde{J}_{(r)}[\tilde{\phi}] \equiv -\tilde{T}^{(\tilde{\phi})}_{\mu\nu} \tilde{u}^{\mu} \tilde{e}^{\nu}_{(r)} = 0, \tag{4.181}$$

$$\tilde{P}_{(r)}[\tilde{\phi}] \equiv \tilde{T}^{(\tilde{\phi})}_{\mu\nu} \tilde{e}^{\mu}_{(r)} \tilde{e}^{\nu}_{(r)} = \frac{\tilde{\phi}'^2}{2\tilde{B}^2} - \frac{U(\tilde{\phi})}{2}. \tag{4.182}$$

By assuming that $\phi$ is static the effective energy distribution $\tilde{T}^{(\tilde{\phi})}_{\mu\nu}$ is static and, if $T^{(m)}_{\mu\nu}$ is static as well, version 1 of the Jebsen-Birkhoff theorem holds. Then the spherical solution $g_{\mu\nu}$ of the field equations is static in a spacetime region in which the coordinates maintain their causal character.

In the case $\omega = -3/2$ the field $\tilde{\phi}$ becomes ill-defined but the variables $(\tilde{g}_{\mu\nu}, \phi)$ can still be employed as Einstein frame variables and the action is

$$S_{(-3/2)} = \int d^4x \sqrt{-\tilde{g}} \left[ \frac{\tilde{R}}{16\pi G} + \frac{3}{2} \tilde{g}^{\mu\nu} \tilde{\nabla}_{\mu}\phi \tilde{\nabla}_{\nu}\phi - V(\phi) + \frac{\mathcal{L}^{(m)}}{(G\phi)^2} \right], \tag{4.183}$$

in which $\phi$ is a phantom field. Once again, the metric is static only if $\phi$ and $T^{(m)}_{\mu\nu}$ are static and the theory reduces again to GR with a cosmological constant if $\phi$ is constant.

The Jordan and Einstein frames are only equivalent when the conformal transformation is well-defined and this equivalence ceases to hold if $\phi \to 0^+$ or $\phi \to +\infty$, situations that may occur near a black hole horizon. For example, the black hole solutions of Brans class I [164], those of Bekenstein [118], and those of Campanelli and Lousto [210,752] are spherically symmetric and static but not Schwarzschildian. They apparently contradict a theorem by Hawking which, loosely speaking, states that stationary black holes in Brans-Dicke theory are the same as those of Einstein's theory and whose proof relies on the use of the Einstein frame.

### *4.3.8 Hawking's theorem and Jebsen-Birkhoff in Brans-Dicke gravity*

Hawking's theorem [581] states that a stationary metric containing a black hole is a solution of the vacuum Brans-Dicke field equations (with $V = 0$) if and only if it solves the Einstein equations, and therefore it must be axially symmetric or static. The theorem is usually quoted as saying that Brans-Dicke black holes are exactly the same as those of GR. From the mathematical point of view this is an overstatement and many exact solutions of scalar-tensor gravity describe black holes with a static scalar field which do not coincide with the Schwarzschild metric. A common feature of these solutions is that the scalar field either vanishes or diverges on an event or apparent horizon. Technically, this feature invalidates the proof of Hawking's theorem, as shown below. From the physical point of view a vanishing or divergent scalar field corresponds to a zero or infinite Hawking temperature (see Sect. 4.4 below), which makes these solutions highly questionable. The most well known example is probably that of Brans class I solutions (4.131) and (4.132) with divergent scalar $\phi$ on the horizon.

Let us consider again the Einstein frame used in the proof of Hawking's theorem [581]. Another theorem by Hawking [580] states that a stationary GR black hole must be axisymmetric and have spherical topology and relies on the null energy condition. Hawking's theorem extends this GR result to Brans-Dicke gravity [581] with the intention to prove that $\phi$ is static. The advantage of using the Einstein frame is that the rescaled Brans-Dicke scalar

$$\tilde{\phi} = \sqrt{\frac{|2\omega+3|}{16\pi G}}\,\ln\left(\frac{\phi}{\phi_*}\right) \tag{4.184}$$

(where $\phi_*$ is a constant) satisfies the null energy condition. The assumption that spacetime is stationary then implies that it is also axially symmetric [580] and, therefore, there exist timelike and spacelike Killing fields $t^\mu$ and $\psi^\mu$ such that the Einstein frame scalar $\tilde{\phi}$ is necessarily constant along their orbits, or else these symmetries are broken. Therefore, $\partial^\mu\tilde{\phi}$ can only be spacelike or zero outside the horizon. Consider a four-dimensional volume $\mathscr{V}$ bounded by two Cauchy hypersurfaces $\mathscr{S}$ and $\mathscr{S}'$ at two consecutive instants of time, a portion of the black hole event horizon, and spatial infinity [581]: the Einstein frame equation of motion (4.178) *in vacuo* and with $V=0$ becomes $\tilde{\Box}\tilde{\phi}=0$. Multiplying this equation by $\tilde{\phi}$, integrating over $\mathscr{V}$, and using the Gauss theorem and the identity

$$\tilde{\phi}\,\tilde{\Box}\tilde{\phi} = \tilde{\nabla}^\mu\left(\tilde{\phi}\tilde{\nabla}_\mu\tilde{\phi}\right) - \tilde{\nabla}^\mu\tilde{\phi}\,\tilde{\nabla}_\mu\tilde{\phi}\,, \tag{4.185}$$

yields [581]

$$\int_{\mathscr{V}} d^4x\;\tilde{g}^{\mu\nu}\tilde{\nabla}_\mu\tilde{\phi}\tilde{\nabla}_\nu\tilde{\phi} = \int_{\partial\mathscr{V}} ds^\alpha\left(\tilde{\phi}\,\tilde{\nabla}_\alpha\tilde{\phi}\right)\,. \tag{4.186}$$

The integral over the boundary $\partial\mathscr{V}$ on the right hand side separates into four contributions,

$$\int_{\partial\mathscr{V}} ds^\alpha\left(\tilde{\phi}\tilde{\nabla}_\alpha\tilde{\phi}\right) = \left(\int_{\mathscr{S}} + \int_{\mathscr{S}'} + \int_{r=+\infty} + \int_{horizon}\right) ds^\alpha\left(\tilde{\phi}\tilde{\nabla}_\alpha\tilde{\phi}\right)\,. \tag{4.187}$$

The two portions of the Cauchy hypersurfaces $\mathscr{S}$ and $\mathscr{S}'$ give contributions that cancel out because they have the same absolute value (due to the time symmetry) but opposite signs due to the fact that outgoing unit normals have opposite directions on these hypersurfaces. There is zero contribution from spatial infinity because $\tilde{\phi}$ vanishes there. The contribution from the integral over the portion of the horizon is declared to vanish in [581] because the projection of $\partial^\mu \tilde{\phi}$ along the null vector tangent to the horizon (which is a linear combination of $t^\mu$ and $\psi^\mu$) vanishes due to the symmetries. This argument fails when the Einstein frame scalar $\tilde{\phi}$ diverges at the horizon, which happens if the Jordan frame $\phi$ either vanishes or diverges there. In these cases the conformal transformation to the Einstein frame and its variables $(\tilde{g}_{\mu\nu}, \tilde{\phi})$ become ill-defined at the horizon. The proof cannot be completed in the Jordan frame because $\phi$ violates the null energy condition [463].

The Jordan frame scalar $\phi$ in the known solutions which violate the Hawking theorem is indeed static (which is what [581] intended to prove), but it has a radial dependence and these solutions do not coincide with the Schwarzschild metric. These solutions include those of Campanelli and Lousto [210, 752] in Brans-Dicke theory. If the Brans-Dicke scalar $\phi$ does not vanish or diverge on the horizon, Hawking's theorem applies and the solution is the Schwarzschild metric.

## 4.3.9 *The Jebsen-Birkhoff theorem in $f(R)$ gravity*

The previous results can be easily extended to metric and Palatini $f(R)$ gravity using the scalar-tensor representation of these theories.

### 4.3.9.1 Palatini $f(R)$ gravity

In the $\omega = -3/2, V \neq 0$ Brans-Dicke equivalent of Palatini $f(R)$ gravity *in vacuo*, *electro-vacuo*, or in any region in which the trace $T^{(m)}$ is constant, the d'Alembertian disappears from the field equation (4.159), which reduces to

$$8\pi G T^{(m)} + \phi \frac{dV}{d\phi} - 2V(\phi) = 0 \,. \tag{4.188}$$

This is no longer a differential equation but is instead algebraic or transcendental. If Eq. (4.188) has solutions, they are of the form $\phi = \text{const.} \equiv \phi_0$ and Eq. (4.158) becomes

$$R_{\mu\nu} - \frac{1}{2} g_{\mu\nu} R = \frac{8\pi}{\phi_0} T^{(m)}_{\mu\nu} - \frac{V(\phi_0)}{2\phi_0} g_{\mu\nu} \,, \tag{4.189}$$

describing GR with a cosmological constant $\Lambda = \frac{V(\phi_0)}{2\phi_0}$. Then the Jebsen-Birkhoff theorem is valid if the matter distribution described by $T^{(m)}_{\mu\nu}$ is static.

The fact, well known in the literature, that the Jebsen-Birkhoff theorem is valid in Palatini $f(R)$ gravity with a static matter distribution reflects once again the non-dynamical nature of the Brans-Dicke scalar in this class of theories.

#### 4.3.9.2 Metric $f(R)$ gravity

For metric $f(R)$ gravity *in vacuo* the field equation (4.159) is

$$\Box\phi = \frac{1}{3}\left[\phi \frac{dV}{d\phi} - 2V(\phi)\right] , \tag{4.190}$$

a dynamical PDE. Since $\phi$ is now dynamical and, in general, time-dependent the Jebsen-Birkhoff theorem does not hold. Currently, most studies impose that $\phi = \text{const.}$, which is equivalent to the assumption $R = \text{const.}$ This assumption is, no doubt, due to the need to simplify calculations and to compare static solutions with Solar System experiments. However, by so doing, one misses the variety of solutions with time-dependent $R$ which are certainly more generic than static ones. Moreover, metric $f(R)$ theories of current interest for cosmology are designed to produce an effective time-varying cosmological constant propelling the present acceleration of the universe without dark energy, and typical solutions are expected to be asymptotically FLRW. This brings us again to the need to understand these solutions, few of which are known [321,328,445,462,465,468,520,765,783,784,789,859,957,1059] in GR as well as in ETGs.

## 4.4 Black hole thermodynamics in extended gravity

Black hole thermodynamics [85, 1138, 1140, 1142] is a milestone of gravitational physics and, more in general, of 20th century theoretical physics. The discovery by Hawking that black holes emit semiclassical radiation made it possible to assign a temperature to black holes and it was the missing ingredient that gave meaning to the entire construction of black hole thermodynamics, making sense of the notion of entropy and of the laws of black hole thermodynamics. An important motivation for the development of black hole thermodynamics is the hope to learn about quantum gravity through the construction of the microscopic statistical mechanics underlying this macroscopic thermodynamics. To summarize (see [85,1138,1140,1142] for a full treatment), black hole thermodynamics in GR links the black hole Hawking temperature $T_H$ given by $K_B T_H = \frac{\hbar\kappa_g}{2\pi K c}$ (where $K$ and $\kappa_g$ are the Boltzmann constant and the surface gravity of the horizon, respectively), the entropy $S_{BH} = \dfrac{c^3 A}{4\hbar G} = \dfrac{A}{l_{Pl}^2}$ (where $A$ is the area of the event horizon and $l_{Pl} = \sqrt{\hbar G/c^3}$ is the Planck length), and the internal energy $Mc^2$ (where $M$ is the black hole mass). In the formulation of [85], black hole thermodynamics for stationary GR black holes comprises the:

- **Zeroth law:** the surface gravity $\kappa_g$, and hence the temperature $T_H$ of a stationary black hole are constant on the event horizon.

- **First law:** in the thermodynamical transition between stationary states in which a black hole with mass $M$, angular momentum $J$, and charge $Q$ change, it is

$$dM = K_B T_H dS_{BH} + \Omega_H dJ + \Phi_H dQ + \delta q\,, \tag{4.191}$$

  where $\Omega_H$ and $\Phi_H$ are the angular velocity and electrostatic potential on the horizon, and $\delta q$ is the change in $M$ due to a stationary matter distribution outside the black hole horizon (if present). The first law relates the quantities $M, J$, and $Q$ measured at spatial infinity with the *local* quantities $S_{BH}, T_H, A, \Omega_H$, and $\Phi_H$ on the horizon.
- **Second law:** in any classical process the horizon area $A$ cannot decrease and therefore $\Delta S_{BH} \geq 0$.
- **Third law:** it is impossible to reduce the temperature of a black hole to zero by a finite number of physical processes, or (since $T_H$ and $\kappa_g$ vanish for extremal black holes), a non-extremal black hole cannot be made extremal by a finite number of physical processes in which matter satisfying the weak energy condition falls into the horizon.

The second law is then extended to include the entropy increase of both the black hole and matter outside of it (*generalized second law* [119–121]). The generalization of black hole thermodynamics to non-stationary, dynamical black holes was begun later ([57–59, 161, 838, 1124] and references therein). The phenomenon of Hawking radiation does not depend on the field equations and hence it is present in alternative gravity as well and black hole thermodynamics extends to ETGs. If information about quantum gravity can be learned by studying black hole thermodynamics, it is necessary to understand this subject in extensions of GR given that quantum corrections, renormalization, and the formulation of effective theories unavoidably introduce extra corrections to the Hilbert-Einstein action in the form of extra fields, higher derivatives, and non-minimal couplings, as seen in Chap. 1. It has also been pointed out that the stability of black hole thermodynamics with respect to perturbations of the GR action may help us selecting physically preferred classes of theories [632].

We have already mentioned the construction of a thermodynamics of spacetime by Jacobson using local Rindler horizons and assuming the entropy-area relation $S_{BH} = \frac{A}{4G}$ [631]. A similar derivation of the field equations using the thermodynamics of local Rindler horizons has been performed also for metric $f(R)$ gravity [419]. Then, $f(R)$ corrections to the Hilbert-Einstein action seem to describe non-equilibrium thermodynamics [419] (see also [310, 424]). Similar derivations are possible also for Lovelock and Gauss-Bonnet gravity [693, 859, 880, 883, 889].

The property of the first law of relating quantities measured at infinity with local quantities on the horizon is still valid in alternative theories of gravity [632] but the expression of the entropy $S_{BH}$ must be changed in these theories, as has been known for a long time [206, 351, 753, 822, 824, 924, 1127]. Various techniques have been developed to compute black hole entropy, including Wald's Noether charge method [629, 630, 632, 1127, 1141], field redefinition techniques [632], and the Euclidean path integral approach [536]. Wald's Noether charge method is based on a Lagrangian formulation of the first law and applies to stationary black holes with bifurcate Killing horizons in any diffeomorphism-invariant theory of gravity with arbitrary spacetime dimension. The method was applied to Palatini $f(R)$ gravity, metric modified gravity, and other gravitational theories [170, 184, 1130]. We now focus on the Bekenstein-Hawking entropy in scalar-tensor and $f(R)$ gravity, following [447].

### 4.4.1 Scalar-tensor gravity

Stimulated by numerical studies showing that the collapse of dust to black holes in Brans-Dicke theory violates the area law during its dynamical phases [664,665,975, 976], Kang studied black hole entropy in Brans-Dicke theory [662]. He realized that the area law *per se* is not problematic but, rather, it is the expression of the entropy-area relation which must be corrected in Brans-Dicke gravity. The correct expression of the entropy is

$$S_{BH} = \frac{1}{4}\int_{\Sigma} d^2x \sqrt{g^{(2)}}\, \phi = \frac{\phi A}{4}, \tag{4.192}$$

where $g^{(2)}$ is the determinant of the restriction of the metric to the horizon surface $\Sigma$ and $\phi$ is, as usual, the Brans-Dicke scalar field. Equation (4.192) is understood by replacing the Newton constant $G$ with the effective gravitational coupling $G_{eff} = \phi^{-1}$ of Brans-Dicke theory [662]. The quantity $S_{BH}$ turns out to be non-decreasing. The philosophy of replacing the gravitational coupling with an effective gravitational coupling determined by rewriting the field equations as effective Einstein equations and read scalar field or, in $f(R)$ gravity, geometric terms as effective forms of matter, is useful also in more general gravitational theories. Equation (4.192) has now been derived using various procedures [629,632,1127].

An alternative viewpoint [662] uses the Einstein frame representation of Brans-Dicke gravity. The Jordan frame Brans-Dicke action

$$S_{BD} = \int d^4x \frac{\sqrt{-g}}{16\pi}\left[\phi R - \frac{\omega}{2} g^{\mu\nu}\nabla_\mu\phi\nabla_\nu\phi - V(\phi) + 16\pi\mathscr{L}^{(m)}\right] \tag{4.193}$$

assumes the Einstein frame form

$$S_{BD} = \int d^4x \sqrt{-\tilde{g}}\left[\frac{\tilde{R}}{16\pi G} - \frac{1}{2}\tilde{g}^{\mu\nu}\tilde{\nabla}_\mu\tilde{\phi}\tilde{\nabla}_\nu\tilde{\phi} - U(\tilde{\phi}) + \frac{\mathscr{L}^{(m)}}{(G\phi)^2}\right], \tag{4.194}$$

with the tilde denoting Einstein frame quantities and $U\left(\tilde{\phi}\right) = \frac{V(\phi(\tilde{\phi}))}{\left(G\phi(\tilde{\phi})\right)^2}$. A black hole event horizon, being a null surface, is invariant under the conformal rescaling but its area is changed and the change in the entropy formula

$$S_{BH} = \frac{A}{4G} \rightarrow \frac{A}{4G_{eff}} = \frac{\phi A}{4} \tag{4.195}$$

can be understood as the change in the area due to the conformal rescaling of the metric. To wit, since $\tilde{g}^{(2)}_{\mu\nu} = \Omega^2\, g^{(2)}_{\mu\nu}$, the Einstein frame area is

$$\tilde{A} = \int_\Sigma d^2x\, \sqrt{\tilde{g}^{(2)}} = \int_\Sigma d^2x\, \Omega^2\, \sqrt{g^{(2)}} = G\phi\, A \tag{4.196}$$

using the fact that the scalar field is constant on the horizon, which is necessary otherwise the surface gravity is not constant on the horizon and the zeroth law does not hold. Therefore, the relation $\tilde{S}_{BH} = \tilde{A}/4G$ is valid also in the Einstein frame. This is expected because, *in vacuo*, the theory reduces to GR with varying units of length $\tilde{l}_u \sim \Omega\, l_u$, time $\tilde{t}_u \sim \Omega\, t_u$, and mass $\tilde{m}_u = \Omega^{-1}\, m_u$ (where $t_u$, $l_u$, and $m_u$ are the constant units of time, length, and mass in the Jordan frame, respectively) [392] and the unit of area scales as $\tilde{l}_A \sim \Omega^2\, l_A = G\phi l_A$. Since $c$ and $\hbar$ do not scale because of their dimensions, the entropy is dimensionless and is not rescaled, hence the Jordan frame and Einstein frame entropies coincide [662]. The equality between Jordan frame and Einstein frame entropies is not limited to scalar-tensor gravity but extends to all theories with action $\int d^4x \sqrt{-g}\, f\left(g_{\mu\nu}, R_{\mu\nu}, \phi, \nabla_\alpha\phi\right)$ which admit an Einstein frame representation [680].

We note in passing that the equivalence between Jordan and Einstein frames with respect to black hole entropies supports the view that these two frames are physically equivalent. What is more, this equivalence is expected to hold at the classical level and break down when quantum processes are introduced. However, black hole thermodynamics is not purely classical: the phenomenon of Hawking radiation which is crucial to the understanding of black hole thermodynamics is instead semiclassical and the physical equivalence between conformal frames with respect to black hole entropy seems to extend the scope of the conformal transformation technique.

Another point worth making is that, if the scalar field vanishes on the horizon of a Brans-Dicke black hole, a zero temperature is assigned to it, seemingly violating the third law. Black holes with vanishing $\phi$ on the horizon, known as "cold black holes", have nevertheless been the subject of several studies [175–179, 210, 698, 752, 826, 1170, 1174]. Similarly, if $\phi$ diverges on the horizon of a scalar-tensor black hole, the entropy diverges. Thermodynamics seems to rule out the possibility that either $\phi \rightarrow \infty$ or $\phi \rightarrow 0$ as the horizon is approached.

For Brans-Dicke theory with zero scalar field potential, the theorem by Hawking already discussed [581] states that, unless $\phi$ vanishes or diverges on the horizon (in which cases the proof of the theorem becomes invalid), all stationary black holes are the same as those of GR, in the sense that the scalar $\phi$ becomes constant outside the horizon (then the theory reduces to GR). This theorem explains the result of the numerical studies of black hole collapse in Brans-Dicke gravity finding GR black holes as the end product [664,665,975,976]. If cold black holes and black holes with divergent $\phi$ on the horizon are discarded as pathological on the basis of thermodynamics ([170, 670], *e.g.*, advocate the use of thermodynamics to exclude similar situations), GR black holes are the only possible final state of equilibrium in Brans-Dicke theory. To date, Hawking's theorem [581] has not been generalized beyond Brans-Dicke gravity with a massless scalar field.

### 4.4.2 Metric modified gravity

The correct expression of the entropy-area relation is, in this case, [19,169,184,331, 547]

$$S_{BH} = \frac{f'(R)A}{4G} \tag{4.197}$$

This formula is proved by applying the Noether charge method [169, 184, 331]. Again (assuming a $D$-dimensional static black hole), the only correction required consists of the replacement of Newton's constant $G$ with the effective gravitational coupling which, in this case, is $G_{eff} = G/f'(R)$. This apparently heuristic justification is supported by the study of Brustein and collaborators [184] who identify $G_{eff}$ by using the matrix of coefficients of the kinetic terms of metric perturbations [184]. The metric perturbations contributing to the Noether charge in Wald's formula and its generalizations are identified with specific polarizations of the metric perturbation associated with fluctuations of the area density on the bifurcation surface $\Sigma$ of the horizon.[8] For a theory described by the action

$$S = \int d^4x\, \sqrt{-g}\, \mathscr{L}\left(g_{\mu\nu}, R_{\alpha\beta\rho\lambda}, \nabla_\sigma R_{\alpha\beta\rho\lambda}, \phi, \nabla_\alpha\phi, \ldots\right) , \tag{4.198}$$

where $\phi$ is a gravitational scalar field, the black hole entropy is then

$$S_{BH} = \frac{A}{4G_{eff}} . \tag{4.199}$$

The Noether charge is

$$S_{BH} = -2\pi \int_\Sigma \left(\frac{\delta \mathscr{L}}{\delta R_{\mu\nu ab}}\right)_{(0)} \hat{\varepsilon}_{\mu\nu}\hat{\varepsilon}_{\mu\nu} , \tag{4.200}$$

[8] The bifurcation surface $\Sigma$ is the $(D-2)$-dimensional spacelike cross-section of a Killing horizon on which the Killing field vanishes.

where the subscript (0) denotes the fact that the quantity in brackets is evaluated on solutions of the equations of motion and $\hat{\boldsymbol{\varepsilon}}_{\rho\sigma}$ is the antisymmetric binormal vector to the bifurcation surface $\Sigma$. This binormal satisfies $\nabla_\mu \chi_\nu = \hat{\boldsymbol{\varepsilon}}_{\mu\nu}$ on the bifurcation surface $\Sigma$, where $\chi^\mu$ is the Killing field vanishing on the horizon. The binormal is normalized to $\hat{\boldsymbol{\varepsilon}}^{\mu\nu}\hat{\boldsymbol{\varepsilon}}_{\mu\nu} = -2$. The effective gravitational coupling then turns out to be [184]

$$G_{\mathit{eff}} = -2\pi \left( \frac{\delta \mathscr{L}}{\delta R_{\mu\nu\rho\sigma}} \right)_{(0)} \hat{\boldsymbol{\varepsilon}}_{\mu\nu}\hat{\boldsymbol{\varepsilon}}_{\rho\sigma} \,. \tag{4.201}$$

For metric modified gravity with $\mathscr{L} = f(R)$, this prescription gives back $G_{\mathit{eff}} = G/f'(R)$ and the entropy (4.197). It is straightforward to see that this calculation is consistent with the description of metric $f(R)$ gravity as a scalar-tensor theory containing the massive dynamical degree of freedom $f'(R)$, and with the corresponding Eq. (4.192) derived in scalar-tensor gravity.

### 4.4.3 *Palatini modified gravity*

The Noether charge method of Wald was applied to Palatini modified gravity by Vollick [1130]. The entropy of a static black hole horizon corresponds again to the Noether charge

$$S_{BH} = \frac{2\pi}{\kappa_g} \int_\Sigma \mathbf{Q} \,, \tag{4.202}$$

where the $(D-2)$-form $\mathbf{Q}$ is the Noether potential corresponding to spacetime diffeomorphisms, $\Sigma$ is the bifurcation surface, and $\kappa_g$ is the surface gravity on the horizon. Vollick [1130] considered the $D$-dimensional action

$$S_{\mathit{Palatini}} = \int d^D x \sqrt{-g} \left[ \frac{f(\mathscr{R})}{16\pi G} + \mathscr{L}^{(m)} \right]; \tag{4.203}$$

since Palatini $f(\mathscr{R})$ gravity *in vacuo* is equivalent to GR with a cosmological constant, the entropy of a stationary black hole is

$$S_{BH} = \frac{f'(\mathscr{R}) A}{4G} \,. \tag{4.204}$$

In the presence of matter it is useful to consider again the trace of the field equations which, as noted above, is an algebraic (or transcendental) equation and not a differential equation due to the non-dynamical nature of the scalar $f'(\mathscr{R})$. When the trace $T^{(m)}$ is constant (in particular for conformally invariant matter), the Ricci curvature $\mathscr{R}$ is expressed in terms of $T^{(m)}$ and eliminated, becoming a constant together with $f'(\mathscr{R})$. The theory reduces again to GR with a cosmological constant and the effective gravitational coupling of the theory is identified with $G_{\mathit{eff}} = G/f'$ [1130].

Again, the black hole entropy given by the Noether charge is

$$S_{BH} = \frac{A}{4G_{eff}} = \frac{f' A}{4G} \,. \tag{4.205}$$

In the presence of matter with non-constant trace $T^{(m)}$ the situation is more complicated and the black hole entropy depends on the ratio of the effective gravitational couplings on the horizon and at spatial infinity:

$$S_{BH} = \frac{f'_{\Sigma}}{f'_{\infty}} \frac{A}{4G} \,, \tag{4.206}$$

where $f'_{\Sigma}$ is the value of $f'(\mathscr{R})$ on the horizon and $f'_{\infty}$ is the value far away from the black hole [1130].

### 4.4.4 Dilaton gravity

Dilaton gravity theories (in both the metric and Palatini formulations) with action

$$S_{dilaton} = \int d^D x \frac{\sqrt{-g}}{16\pi G} f\left(g^{\mu\nu}, R_{(\mu\nu)}\right) \tag{4.207}$$

were studied in [184, 1130]. By varying with respect to the metric $g^{\mu\nu}$ one obtains the field equations

$$\frac{\partial f}{\partial g^{\mu\nu}} - \frac{f}{2} g_{\mu\nu} = 0 \,, \tag{4.208}$$

while varying with respect to the independent connection $\Gamma$ yields

$$\bar{\nabla}_{\alpha} \left[ \sqrt{-g} \frac{\partial f}{\partial \mathscr{R}_{(\mu\nu)}} \right] = 0 \,. \tag{4.209}$$

*In vacuo* these equations describe GR with an effective cosmological constant $\lambda$ appearing in the effective matter tensor $\frac{\partial f}{\partial \mathscr{R}_{(\mu\nu)}} = \lambda g_{\mu\nu}$ [1130].

The effective gravitational coupling is $G_{eff} = G/\lambda$, and the black hole entropy is

$$S_{BH} = \frac{A}{4G_{eff}} = \frac{\lambda A}{4G} \,. \tag{4.210}$$

Before closing this section we mention that also the thermodynamics of cosmological horizons has been the subject of several works, at least in metric $f(R)$ gravity [19, 199–201, 331, 547, 806, 1146]. The Bekenstein-Hawking entropy has been studied in Lovelock gravity [198, 345, 633, 824, 889], in Gauss-Bonnet theories [693, 880, 883, 889], and in theories with Lorentz violation. Regarding the latter, there are claims that perpetual motion machines of the second kind are made possible by Lorentz violation ([398], see also [418, 634]) but this unpleasant feature seems to be impossible in tensor-vector-scalar (*TeVeS*) theories [954], and the issue requires further attention. From a general point of view, and putting together the various works available in the literature, it seems that black hole thermodynamics can be quite useful to constrain families of ETGs inspired by low-energy quantum gravity.

## 4.5 From spherical to axial symmetry: an application to $f(R)$ gravity

We now show how it is possible to obtain an axially symmetric solution starting from a spherically symmetric one, using a method developed by Newman and Janis in GR [835, 836]. This method can be applied to a static spherically symmetric metric adopted as a "seed" metric. In principle, the procedure could be applied whenever Noether symmetries are present. We apply this procedure to solutions of metric $f(R)$ gravity [229].

In general, the approach is not straightforward since, if $f(R) \neq R$, the field equations are of fourth order and the relevant existence theorems and boundary conditions are different from those of GR. However, the existence of a Noether symmetry guarantees the consistency of the chosen $f(R)$ model with the field equations.

Let us consider a spherically symmetric metric of the form

$$ds^2 = -e^{2\phi(r)}dt^2 + e^{2\lambda(r)}dr^2 + r^2 d\Omega_2^2 \,. \tag{4.211}$$

Following Newman and Janis, the line element (4.211) can be written in Eddington-Finkelstein coordinates $(u, r, \theta, \varphi)$, *i.e.*, the $g_{rr}$ component is eliminated by the coordinate change and a cross term is introduced [804]. We set $dt = du + F(r)dr$ with $F(r) = \pm e^{\lambda(r)-\phi(r)}$, turning the line element (4.211) into

$$ds^2 = -e^{2\phi(r)}du^2 \mp 2\,e^{\lambda(r)-\phi(r)}dudr + r^2 d\Omega_2^2 \,. \tag{4.212}$$

The surface $u$ = constant is a light cone with vertex in $r = 0$. The inverse metric tensor in null coordinates is

$$(g^{\mu\nu}) = \begin{pmatrix} 0 & \mp e^{-\lambda(r)-\phi(r)} & 0 & 0 \\ \mp e^{-\lambda(r)-\phi(r)} & e^{-2\lambda(r)} & 0 & 0 \\ 0 & 0 & \frac{1}{r^2} & 0 \\ 0 & 0 & 0 & \frac{1}{r^2\sin^2\theta} \end{pmatrix} . \tag{4.213}$$

The matrix (4.213) can be written in terms of a null tetrad as

$$g^{\mu\nu} = -l^{\mu}n^{\nu} - l^{\nu}n^{\mu} + m^{\mu}\bar{m}^{\nu} + m^{\nu}\bar{m}^{\mu}\,, \tag{4.214}$$

where $l^{\mu}$, $n^{\mu}$, $m^{\mu}$, and $\bar{m}^{\mu}$ satisfy the conditions

$$l_{\mu}l^{\mu} = m_{\mu}m^{\mu} = n_{\mu}n^{\mu} = 0,\ l_{\mu}n^{\mu} = -m_{\mu}\bar{m}^{\mu} = -1,\ l_{\mu}m^{\mu} = n_{\mu}m^{\mu} = 0, \tag{4.215}$$

and where an overbar denotes complex conjugation. At any spatial point, the tetrad can be chosen in the following manner: $l^{\mu}$ is the outward null vector tangent to the light cone, $n^{\mu}$ is the inward null vector pointing toward the origin, and $m^{\mu}$ and $\bar{m}^{\mu}$ are vectors tangent to the two-dimensional sphere defined by constant $r$ and $u$. For the spacetime (4.213), the null tetrad can be chosen as

$$l^{\mu} = \delta^{\mu}_{1}\,, \tag{4.216}$$

$$n^{\mu} = -\frac{1}{2}\,\mathrm{e}^{-2\lambda(r)}\,\delta^{\mu}_{1} + \mathrm{e}^{-\lambda(r)-\phi(r)}\,\delta^{\mu}_{0}\,, \tag{4.217}$$

$$m^{\mu} = \frac{1}{\sqrt{2}\,r}\left(\delta^{\mu}_{2} + \frac{i}{\sin\theta}\,\delta^{\mu}_{3}\right), \tag{4.218}$$

$$\bar{m}^{\mu} = \frac{1}{\sqrt{2}\,r}\left(\delta^{\mu}_{2} - \frac{i}{\sin\theta}\,\delta^{\mu}_{3}\right). \tag{4.219}$$

Now we extend the set of coordinates $x^{\mu} = (u, r, \theta, \phi)$ by promoting the real radius to the role of a complex variable. The null tetrad then becomes[9]

$$l^{\mu} = \delta^{\mu}_{1}\,, \tag{4.220}$$

$$n^{\mu} = -\frac{1}{2}\,\mathrm{e}^{-2\lambda(r,\bar{r})}\,\delta^{\mu}_{1} + \mathrm{e}^{-\lambda(r,\bar{r})-\phi(r,\bar{r})}\,\delta^{\mu}_{0}\,, \tag{4.221}$$

$$m^{\mu} = \frac{1}{\sqrt{2}\,\bar{r}}\left(\delta^{\mu}_{2} + \frac{i}{\sin\theta}\delta^{\mu}_{3}\right), \tag{4.222}$$

$$\bar{m}^{\mu} = \frac{1}{\sqrt{2}\,r}\left(\delta^{\mu}_{2} - \frac{i}{\sin\theta}\delta^{\mu}_{3}\right). \tag{4.223}$$

A new metric is obtained by performing the complex coordinate transformation

$$x^{\mu} \longrightarrow \tilde{x}^{\mu} = x^{\mu} + i\,y^{\mu}(x^{\sigma})\,, \tag{4.224}$$

[9] A certain degree of arbitrariness is present in the complexification of the functions $\lambda$ and $\phi$. Obviously, we must recover the metric (4.213) as soon as $r = \bar{r}$.

where $y^\mu(x^\sigma)$ are analytic functions of the real coordinates $x^\sigma$, and simultaneously letting the null tetrad $Z^\mu_a \equiv (l^\mu, n^\mu, m^\mu, \bar{m}^\mu)$ with $a = 1, 2, 3, 4$, undergo the transformation

$$Z^\mu_a \longrightarrow \tilde{Z}^\mu_a(\tilde{x}^\sigma, \bar{\tilde{x}}^\sigma) = Z^\rho_a \frac{\partial \tilde{x}^\mu}{\partial x^\rho} . \tag{4.225}$$

Obviously, one has to recover the old tetrad and metric as soon as $\tilde{x}^\sigma = \bar{\tilde{x}}^\sigma$. In summary, the effect of the "tilde transformation" (4.224) is to generate a new metric whose components are real functions of complex variables,

$$g_{\mu\nu} \longrightarrow \tilde{g}_{\mu\nu} : \tilde{\mathbf{x}} \times \bar{\tilde{\mathbf{x}}} \mapsto \mathbb{R} \tag{4.226}$$

with

$$\tilde{Z}^\mu_a(\tilde{x}^\sigma, \bar{\tilde{x}}^\sigma)|_{\mathbf{x}=\tilde{\mathbf{x}}} = Z^\mu_a(x^\sigma) . \tag{4.227}$$

For our purposes, we can make the choice

$$\tilde{x}^\mu = x^\mu + ia\left(\delta^\mu_1 - \delta^\mu_0\right)\cos\theta \longrightarrow \begin{cases} \tilde{u} = u + ia\cos\theta , \\ \tilde{r} = r - ia\cos\theta , \\ \tilde{\theta} = \theta , \\ \tilde{\phi} = \phi , \end{cases} \tag{4.228}$$

where $a$ is a constant and, with the choice $\tilde{r} = \bar{\tilde{r}}$, the null vectors (4.220)–(4.223) reduce to

$$\tilde{l}^\mu = \delta^\mu_1 , \tag{4.229}$$

$$\tilde{n}^\mu = -\frac{1}{2} e^{-2\lambda(\tilde{r},\theta)} \delta^\mu_1 + e^{-\lambda(\tilde{r},\theta)-\phi(\tilde{r},\theta)} \delta^\mu_0 , \tag{4.230}$$

$$\tilde{m}^\mu = \frac{1}{\sqrt{2}(\tilde{r} - ia\cos\theta)}\left[ia(\delta^\mu_0 - \delta^\mu_1)\sin\theta + \delta^\mu_2 + \frac{i}{\sin\theta}\delta^\mu_3\right] , \tag{4.231}$$

$$\bar{\tilde{m}}^\mu = \frac{1}{\sqrt{2}(\tilde{r} + ia\cos\theta)}\left[-ia(\delta^\mu_0 - \delta^\mu_1)\sin\theta + \delta^\mu_2 - \frac{i}{\sin\theta}\delta^\mu_3\right] . \tag{4.232}$$

A new metric is recovered from the transformed null tetrad via Eq. (4.214). With the null vectors (4.229)–(4.232) and the transformation (4.228), the new metric in coordinates $\tilde{x}^\mu = (\tilde{u}, \tilde{r}, \theta, \phi)$ is

$$(\tilde{g}^{\mu\nu}) = \begin{pmatrix} \frac{a^2\sin^2\theta}{\Sigma^2} & -e^{-\lambda(\tilde{r},\theta)-\phi(\tilde{r},\theta)} - \frac{a^2\sin^2\theta}{\Sigma^2} & 0 & \frac{a}{\Sigma^2} \\ \cdot & e^{-2\lambda(\tilde{r},\theta)} + \frac{a^2\sin^2\theta}{\Sigma^2} & 0 & -\frac{a}{\Sigma^2} \\ \cdot & \cdot & \frac{1}{\Sigma^2} & 0 \\ \cdot & \cdot & \cdot & \frac{1}{\Sigma^2\sin^2\theta} \end{pmatrix} \tag{4.233}$$

where $\Sigma = \sqrt{\tilde{r}^2 + a^2\cos^2\theta}$. The covariant metric $\tilde{g}_{\mu\nu}$ is

$$\begin{pmatrix} -e^{2\phi(\tilde{r},\theta)} & -e^{\lambda(\tilde{r},\theta)-\phi(\tilde{r},\theta)} & 0 & -ae^{\phi(\tilde{r},\theta)}\left[e^{\lambda(\tilde{r},\theta)} - e^{\phi(\tilde{r},\theta)}\right]\sin^2\theta \\ \cdot & 0 & 0 & ae^{\phi(\tilde{r},\theta)-\lambda(\tilde{r},\theta)}\sin^2\theta \\ \cdot & \cdot & \Sigma^2 & 0 \\ \cdot & \cdot & \cdot & \left[\Sigma^2 + a^2\sin^2\theta e^{\phi(\tilde{r},\theta)}(2e^{\lambda(\tilde{r},\theta)} - e^{\phi(\tilde{r},\theta)})\right]\sin^2\theta \end{pmatrix} \tag{4.234}$$

The dots in the matrix denote symmetric entries satisfying the metric symmetry $g^{\mu\nu} = g^{\nu\mu}$. The form of this metric gives the general result of the Newman-Janis algorithm starting from any spherical seed metric.

The metric (4.234) can be simplified by a further gauge transformation so that the only off-diagonal component is $g_{\phi t}$. This procedure makes it easier to compare with the standard Boyer-Lindquist form of the Kerr metric [804] and to interpret physical properties such as frame dragging. The coordinates $\tilde{u}$ and $\varphi$ can be redefined in such a way that the metric in the new coordinates has the properties described above. Explicitly, using

$$d\tilde{u} = dt + g(\tilde{r})d\tilde{r} \tag{4.235}$$

and

$$d\phi = d\phi + h(\tilde{r})d\tilde{r}\,, \tag{4.236}$$

where

$$g(\tilde{r}) = -\frac{e^{\lambda(\tilde{r},\theta)}(\Sigma^2 + a^2\sin^2\theta e^{\lambda(\tilde{r},\theta)+\phi(\tilde{r},\theta)})}{e^{\phi(\tilde{r},\theta)}(\Sigma^2 + a^2\sin^2\theta e^{2\lambda(\tilde{r},\theta)})}\,, \tag{4.237}$$

$$h(\tilde{r}) = -\frac{a\,e^{2\lambda(\tilde{r},\theta)}}{\Sigma^2 + a^2\sin^2\theta\, e^{2\lambda(\tilde{r},\theta)}}\,, \tag{4.238}$$

after algebraic manipulations the covariant metric (4.234) becomes, in coordinates $(t, \tilde{r}, \theta, \phi)$,

$$\begin{pmatrix} e^{2\phi} & 0 & 0 & a\, e^{\phi}[e^{\lambda} - e^{\phi}] \sin^2\theta \\ \cdot & -\frac{\Sigma^2}{(\Sigma^2 e^{-2\lambda} + a^2 \sin^2\theta)} & 0 & 0 \\ \cdot & \cdot & -\Sigma^2 & 0 \\ \cdot & \cdot & \cdot & -\left[\Sigma^2 + a^2 \sin^2\theta\, e^{\phi}(2e^{\lambda} - e^{\phi})\right] \sin^2\theta \end{pmatrix}, \tag{4.239}$$

where $\phi = \phi(\tilde{r}, \theta)$ and $\lambda = \lambda(\tilde{r}, \theta)$. This metric represents the complete family of metrics that may be obtained by performing the Newman-Janis algorithm on any static spherically symmetric seed metric, written in Boyer-Lindquist coordinates. These transformations require that $\Sigma^2 + a^2 \sin^2\theta\, e^{2\lambda(\tilde{r},\theta)} \neq 0$, where $e^{2\lambda(\tilde{r},\theta)} > 0$. We now show that this approach can be used to derive axially symmetric solutions also in $f(R)$ gravity.

Begin with the spherically symmetric solution (4.90), that we rewrite as

$$ds^2 = -(\alpha + \beta r)\, dt^2 + \frac{\beta r}{2(\alpha + \beta r)}\, dr^2 + r^2 d\Omega_2^2 \,, \tag{4.240}$$

where $\alpha$ is a combination of $\Sigma_0$, $k$, and $\beta = k_1$ obtained with the Noether approach. The metric tensor in Eddington-Finkelstein coordinates $(u, r, \theta, \phi)$ of the form (4.213) is

$$(g^{\mu\nu}) = \begin{pmatrix} 0 & -\sqrt{\frac{2}{\beta r}} & 0 & 0 \\ \cdot & 2 + \frac{2\alpha}{\beta r} & 0 & 0 \\ \cdot & \cdot & \frac{1}{r^2} & 0 \\ \cdot & \cdot & \cdot & \frac{1}{r^2 \sin^2\theta} \end{pmatrix}. \tag{4.241}$$

The complex null tetrad (4.220)–(4.223) is now

$$l^{\mu} = \delta_1^{\mu} \,, \tag{4.242}$$

$$n^{\mu} = -\left[1 + \frac{\alpha}{\beta}\left(\frac{1}{\bar{r}} + \frac{1}{r}\right)\right]\delta_1^{\mu} + \sqrt{\frac{2}{\beta}}\, \frac{1}{(\bar{r} r)^{1/4}}\, \delta_0^{\mu} \,, \tag{4.243}$$

$$m^{\mu} = \frac{1}{\sqrt{2}\bar{r}}\left(\delta_2^{\mu} + \frac{i}{\sin\theta}\delta_3^{\mu}\right). \tag{4.244}$$

By computing the complex coordinate transformation (4.228), the null tetrad becomes

$$\tilde{l}^\mu = \delta_1^\mu\,, \tag{4.245}$$

$$\tilde{n}^\mu = -\left[1 + \frac{\alpha}{\beta}\frac{\mathrm{Re}\{\tilde{r}\}}{\Sigma^2}\right]\delta_1^\mu + \sqrt{\frac{2}{\beta\Sigma}}\,\delta_0^\mu\,, \tag{4.246}$$

$$\tilde{m}^\mu = \frac{1}{\sqrt{2}\,(\tilde{r} + ia\cos\theta)}\left[ia\left(\delta_0^\mu - \delta_1^\mu\right)\sin\theta + \delta_2^\mu + \frac{i}{\sin\theta}\delta_3^\mu\right]. \tag{4.247}$$

By performing the same procedure as in GR, one derives an axially symmetric metric of the form (4.239) but starting from the spherically symmetric covariant metric (4.240),

$$g_{\mu\nu} = \begin{pmatrix} -\frac{r(\alpha+\beta r)+a^2\beta\cos^2\theta}{\Sigma} & 0 & 0 & \frac{\sigma_1}{2\Sigma} \\ \cdot & \frac{\beta\Sigma^2}{2\alpha r+\beta(a^2+r^2+\Sigma^2)} & 0 & 0 \\ \cdot & \cdot & \Sigma^2 & 0 \\ \cdot & \cdot & \cdot & \left(\Sigma^2 - \frac{\sigma_2}{\Sigma}\right)\sin^2\theta \end{pmatrix}. \tag{4.248}$$

where

$$\sigma_1 = a\left(2\alpha r + 2\beta\Sigma^2 - \sqrt{2\beta}\,\Sigma^{3/2}\right)\sin^2\theta\,, \tag{4.249}$$

$$\sigma_2 = a^2\left(\alpha r + \beta\Sigma^2 - \sqrt{2\beta}\,\Sigma^{3/2}\right)\sin^2\theta\,. \tag{4.250}$$

By setting $a = 0$, the metric (4.240) is immediately recovered.

The method illustrated by this example is general and can be extended to any spherically symmetric solution of $f(R)$ gravity [229].

## 4.6 Conclusions

Much insight into relativistic theories of gravity has been gained by exploring spherically symmetric solutions. A complication with respect to Einstein's theory is given by the fact that the Jebsen-Birkhoff theorem is not valid in general in ETGs. In scalar-tensor and $f(R)$ gravity, and presumably also in other theories, in order for the spherical solutions of the field equations to be static, one must impose that the distribution of effective matter corresponding to the geometric terms in the field equations other than the Einstein tensor is static. This assumption is very restrictive because, given that theories of current interest for cosmology mimic a time-dependent cosmological "constant", one expects generic solutions to be dynamical. In general, and with the exception of Palatini $f(R)$ gravity, even static solutions of the field equations do not reduce to the Schwarzschild-(anti)de Sitter metrics familiar from GR (although the latter are usually solutions of the modified field equations). This fact testifies of the larger variety of solutions allowed in the ETG when the number of degrees of freedom of GR is enlarged, for example by including the massive scalar degree of freedom of metric $f(R)$ gravity, which appears because of the fourth (as opposed to second) order of the field equations in this class of theories.

Overall, studies in the literature have focused on static solutions and we still have too few examples of truly dynamical exact solutions. The understanding of time-dependent, spherically symmetric solutions of the field equations of various ETGs including black holes, as well as the systematic search for their counterparts in GR together with the understanding of their precise significance, are identified as open problems of classical gravity. Black hole thermodynamics in ETGs has the promise of being another fruitful area of research.

# Chapter 5
# Weak-field limit

*The effort to understand the universe is one of the very few things that lifts human life a little above the level of farce, and gives it some of the grace of tragedy.*
*– Steven Weinberg*

Astrophysical applications of ETGs include the possibility of replacing dark matter in galaxy and clusters with modifications of gravity, the weak-field (Newtonian and post-Newtonian) limit, and gravitational waves. Dark matter at galactic and cluster scales is traditionally included in the realm of cosmology and is discussed in Chap. 7. Here we focus on the weak-field limit of metric $f(R)$ gravity, referring the reader to well known sources for other ETGs. We then discuss gravitational waves in ETGs.

## 5.1 The weak-field limit of extended gravity

The weak-field limit of scalar-tensor gravity is discussed in detail in many works (*e.g.*, [360, 1166, 1167]) and will not be repeated here. Due to the problems with Palatini $f(R)$ gravity already discussed, we will also omit the discussion of its weak-field limit and refer the reader to [1033] and the references therein.

At shorter (galactic and Solar System) spatial scales, ETGs exhibit gravitational potentials with non-Newtonian corrections. This feature was discovered long ago [1051], and recent interest arises from the possibility of explaining the flatness of the rotation curves of spiral galaxies without huge amounts of dark matter. In particular, the rotation curves of a wide sample of low surface brightness spiral galaxies can be fitted successfully by the corrected potentials [216, 217], and this possibility may be extended to other types of galaxies [510].

One could attempt to investigate other issues such as, for example, the Pioneer anomaly [44, 45] with the same approach [130]. A systematic analysis of ETGs at scales much smaller than the Hubble radius is then necessary. In this section we discuss the weak-field limit of $f(R)$ gravity without specifying the form of the theory and highlighting the differences and similarities with the post-Newtonian and post-Minkowskian limits of GR. The literature contains conflicting claims [14,66,96,249,250,252–254,288,325,326,389,409,605,624,816,818,830,831,868,

S. Capozziello and V. Faraoni, *Beyond Einstein Gravity: A Survey of Gravitational Theories for Cosmology and Astrophysics*, Fundamental Theories of Physics 170, DOI 10.1007/978-94-007-0165-6_5, 

871, 950, 1010, 1031, 1036, 1172], and clarity is needed in order to compare theory and experiment [1035]. Based on the scalar-tensor representation of $f(R)$ gravity with[1] $\omega = 0$, Chiba [301] originally suggested that all $f(R)$ theories are ruled out because of the experimental limit $|\omega| > 40,000$ [133]. While this constraint can be circumvented by giving the scalar degree of freedom a large mass and, therefore, making it short-ranged, it seemed that its range must be at least comparable with the Hubble radius in order to affect the dynamics of the universe. This conclusion is incorrect because of the chameleon mechanism and the weak-field limit is subtler than it appears, as will be clear below. Solar System experiments constrain the PPN parameter $\gamma$, and then the Brans-Dicke parameter $\omega$, only when the range of the scalar degree of freedom is comparable to, or larger than the spatial scale of the experiment (for the Cassini experiment providing the lower bound on $\omega$ [133], this is the size of the Solar System) [1133]. If the mass of this scalar is large, the parameter $\gamma$ is close to unity. However, the scalar does not have a fixed range but, rather, its mass depends on the energy density of its environment, so that this field becomes short-ranged and is undetectable at small (Solar System) scales, while its range is cosmological at cosmological densities $\rho$. This *chameleon mechanism* is widely used in quintessence models of dark energy [666, 667].

Again, a direct approach independent of the equivalence of metric $f(R)$ and scalar-tensor gravity is more convincing, and was first formulated for the prototype model $f(R) = R - \mu^4/R$ (which, at the time, was already ruled out by the Dolgov-Kawasaki instability [396,460]) in [430]. The weak-field limit for a general function $f(R)$ was presented in [306, 641, 869].

Weak-field experiments such as light bending, the perihelion shift of planets, and frame-dragging experiments are valuable tests of ETGs. There are sufficient theoretical predictions to state that certain higher order theories of gravity can be compatible with Newtonian and post-Newtonian experiments [25,253,301,306,326, 430, 459, 641, 849, 1031], as can be shown also by using the scalar-tensor representation of $f(R)$ gravity.

In the following we outline a formalism addressing the weak-field and small velocity limit of fourth order gravity allowing a Jordan frame systematic discussion of these limits and of spherically symmetric solutions [251]. This discussion is valid also for general higher order theories containing the invariants $R_{\mu\nu}R^{\mu\nu}$ or $R_{\alpha\beta\mu\nu}R^{\alpha\beta\mu\nu}$ [247]. The non-Newtonian corrections in the gravitational potentials could potentially explain known astrophysical phenomenology.

A preliminary step consists of concentrating on the vacuum case and then building a Newtonian and post-Newtonian formalism for $f(R)$ theories in the presence of matter. It is possible to estimate the post-Newtonian parameter $\gamma$ by considering second order solutions for the metric components *in vacuo*. For completeness, we treat the problem also by imposing the harmonic gauge on the field equations.

---

[1] Although some caution about the equivalence with scalar-tensor theory in the Newtonian and the GR limits is necessary [459, 650], the equivalence holds in the post-Newtonian limit [461].

## 5.2 The Newtonian and post-Newtonian approximations: general remarks

Certain general features must be taken into account when performing the Newtonian and post-Newtonian limits of a relativistic theory of gravity. For a virialized system of particles of total mass $M$ interacting gravitationally, the kinetic energy $\bar{M}(\bar{v})^2/2$ is approximately of the same order of magnitude as the potential energy $U = G\bar{M}^2/\bar{r}$, where $\bar{r}$ and $\bar{v}$ are typical average values of the separations and velocities of these particles. As a consequence, it is

$$\bar{v^2} \sim \frac{G\bar{M}}{\bar{r}} \tag{5.1}$$

(for instance, in Newtonian mechanics, a test particle in a circular orbit of radius $r$ about a spherically distributed mass $M$ has velocity $v$ given by $v^2 = GM/r$). The post-Newtonian approximation can be described as a method for obtaining the motion of the system beyond first (*i.e.*, Newtonian) order with respect to the quantities $G\bar{M}/\bar{r}$ and $(\bar{v})^2$, which are assumed to be small with respect to the square of the speed of light $c^2$ (this approximation is an expansion in inverse powers of $c$).

Typical values of the Newtonian gravitational potential $U$ in the Solar System are nowhere larger than $10^{-5}$ (the quantity $U/c^2$ is dimensionless). Planetary velocities satisfy the condition $(\bar{v})^2 \lesssim U$, while[2] the matter pressure $P$ inside the Sun and the planets is much smaller than the energy density $\rho U$ of matter,[3] $P/\rho \lesssim U$. Furthermore, one must consider that other forms of energy in the Solar System (stresses, radiation, thermal energy, *etc.*) have small magnitudes and their specific energy density $\Pi$ (the ratio of the energy density to the rest mass density) is related to $U$ by $\Pi \lesssim U$ ($\Pi$ is approximately $10^{-5}$ in the Sun and $10^{-9}$ in the Earth [1166, 1167]). One can consider that these quantities, as functions of velocity, give only second order contributions,

$$U \sim v^2 \sim \frac{P}{\rho} \sim \Pi \sim O(2)\,, \tag{5.2}$$

therefore the velocity $v$ contributes to order O(1), $U^2$ to order O(4), $Uv$ to order O(3), $U\Pi$ is of order O(4), *etc.* In this approximation, one has

$$\frac{\partial}{\partial x^0} \sim \mathbf{v}\cdot\nabla\,, \tag{5.3}$$

and

$$\frac{|\partial/\partial x^0|}{|\nabla|} \sim O(1)\,. \tag{5.4}$$

---

[2] Here the velocity $v$ is expressed in units of $c$.

[3] Typical values of $P/\rho$ are $10^{-5}$ in the Sun and $10^{-10}$ in the Earth [1166, 1167].

Massive test particles move along geodesics given by the equation

$$\frac{d^2x^\mu}{ds^2} + \Gamma^\mu_{\sigma\tau}\frac{dx^\sigma}{ds}\frac{dx^\tau}{ds} = 0\,, \tag{5.5}$$

or

$$\begin{aligned}\frac{d^2x^i}{dx^{0\,2}} &= -\Gamma^i_{00} - 2\Gamma^i_{0m}\frac{dx^m}{dx^0} - \Gamma^i_{mn}\frac{dx^m}{dx^0}\frac{dx^n}{dx^0} \\ &+\left(\Gamma^0_{00} + 2\Gamma^0_{0m}\frac{dx^m}{dx^0} + 2\Gamma^0_{mn}\frac{dx^m}{dx^0}\frac{dx^n}{dx^0}\right)\frac{dx^i}{dx^0}\,.\end{aligned} \tag{5.6}$$

In the small velocity approximation and retaining only first order terms in the deviations of $g_{\mu\nu}$ from the Minkowski metric $\eta_{\mu\nu}$, the particle equations of motion reduce to the Newtonian result

$$\frac{d^2x^i}{d(x^0)^2} \simeq -\Gamma^i_{00} \simeq -\frac{1}{2}\frac{\partial g_{00}}{\partial x^i}\,. \tag{5.7}$$

The quantity $(1 + g_{00})$ is of order $G\bar{M}/\bar{r}$, hence the Newtonian approximation gives $\frac{d^2x^i}{d(x^0)^2}$ to order $G\bar{M}/\bar{r}^2$, that is, to order $(\bar{v})^2/r$. As a consequence, the post-Newtonian approximation requires one to compute $\frac{d^2x^i}{d(x^0)^2}$ to order $(\bar{v})^4/\bar{r}$. According to the Equivalence Principle and the local flatness of the spacetime manifold, it is possible to find a coordinate system in which the metric tensor is nearly equal to $\eta_{\mu\nu}$, with the correction expanded in powers of $G\bar{M}/\bar{r} \sim (\bar{v})^2$,

$$g_{00}(x^0, \mathbf{x}) = -1 + g^{(2)}_{00}(x^0, \mathbf{x}) + g^{(4)}_{00}(x^0, \mathbf{x}) + \mathrm{O}(6)\,, \tag{5.8}$$

$$g_{0i}(x^0, \mathbf{x}) = g^{(3)}_{0i}(x^0, \mathbf{x}) + \mathrm{O}(5)\,, \tag{5.9}$$

$$g_{ij}(x^0, \mathbf{x}) = \delta_{ij} + g^{(2)}_{ij}(x^0, \mathbf{x}) + \mathrm{O}(4)\,, \tag{5.10}$$

and with inverse metric

$$g^{00}(x^0, \mathbf{x}) = -1 + g^{(2)00}(x^0, \mathbf{x}) + g^{(4)00}(x^0, \mathbf{x}) + \mathrm{O}(6)\,, \tag{5.11}$$

$$g^{0i}(x^0, \mathbf{x}) = g^{(3)0i}(x^0, \mathbf{x}) + \mathrm{O}(5)\,, \tag{5.12}$$

$$g^{ij}(x^0, \mathbf{x}) = \delta^{ij} + g^{(2)ij}(x^0, \mathbf{x}) + \mathrm{O}(4)\,. \tag{5.13}$$

When computing the connection coefficients $\Gamma^{\mu}_{\alpha\beta}$, one must take into account the fact that the space and time scales in the gravitational system are set by $\bar{r}$ and $\bar{r}/\bar{v}$, respectively, hence spatial and time derivatives are of order

$$\frac{\partial}{\partial x^i} \sim \frac{1}{\bar{r}}\,, \qquad \frac{\partial}{\partial x^0} \sim \frac{\bar{v}}{\bar{r}}\,. \tag{5.14}$$

Using the approximations (5.8)–(5.13), we have

$$\Gamma^{(3)}{}^{0}{}_{00} = \frac{1}{2} g^{(2),0}_{00}\,, \tag{5.15}$$

$$\Gamma^{(2)}{}^{i}{}_{00} = \frac{1}{2} g^{(2),i}_{00}\,, \tag{5.16}$$

$$\Gamma^{(2)}{}^{i}{}_{jk} = \frac{1}{2}\left( g^{(2)}{}_{jk}{}^{,i} - g^{(2)}{}^{i}{}_{j,k} - g^{(2)}{}^{i}{}_{k,j} \right), \tag{5.17}$$

$$\Gamma^{(3)}{}^{0}{}_{ij} = \frac{1}{2}\left( g^{(3)}{}^{0}{}_{i,j} + g^{(3)}{}^{0}{}_{j,i} - g^{(3)}{}_{ij}{}^{,0} \right), \tag{5.18}$$

$$\Gamma^{(3)}{}^{i}{}_{0j} = \frac{1}{2}\left( g^{(3)}{}_{0j}{}^{,i} - g^{(3)}{}^{i}{}_{0,j} - g^{(2)}{}^{i}{}_{j,0} \right), \tag{5.19}$$

$$\Gamma^{(4)}{}^{0}{}_{0i} = \frac{1}{2}\left( g^{(4)}{}^{0}{}_{0,i} + g^{(2)00} g^{(2)}_{00,i} \right), \tag{5.20}$$

$$\Gamma^{(4)}{}^{i}{}_{00} = \frac{1}{2}\left( g^{(4)}{}_{00}{}^{,i} + g^{(2)im} g^{(2)}_{00,m} - 2 g^{(3)}{}^{i}{}_{0,0} \right), \tag{5.21}$$

$$\Gamma^{(2)}{}^{0}{}_{0i} = \frac{1}{2} g^{(2)}{}^{0}{}_{0,i}\,. \tag{5.22}$$

The only non-vanishing components of the Ricci tensor are

$$R^{(2)}_{00} = \frac{1}{2}\nabla^2 g^{(2)}_{00}\,, \tag{5.23}$$

$$\begin{aligned} R^{(4)}_{00} = {} & \frac{1}{2}\nabla^2 g^{(4)}_{00} - \frac{1}{2} g^{(2)mn}{}_{,m} g^{(2)}_{00,n} - \frac{1}{2} g^{(2)mn} g^{(2)}_{00,mn} \\ & + \frac{1}{2} g^{(2)}{}^{m}{}_{m,00} - \frac{1}{4} g^{(2)}{}^{0,m}{}_{0} g^{(2)}_{00,m} - \frac{1}{4} g^{(2)}{}^{m,n}{}_{m} g^{(2)}_{00,n} - g^{(3)}{}^{m}{}_{0,m0}\,, \end{aligned} \tag{5.24}$$

$$R^{(3)}_{0i} = \frac{1}{2}\nabla^2 g^{(3)}_{0i} - \frac{1}{2} g^{(2)}{}^{m}{}_{i,m0} - \frac{1}{2} g^{(3)}{}^{m}{}_{0,mi} + \frac{1}{2} g^{(2)}{}^{m}{}_{m,0i}\,, \tag{5.25}$$

$$\begin{aligned} R^{(2)}_{ij} = {} & \frac{1}{2}\nabla^2 g^{(2)}_{ij} - \frac{1}{2} g^{(2)}{}^{m}{}_{i,mj} - \frac{1}{2} g^{(2)}{}^{m}{}_{j,mi} - \frac{1}{2} g^{(2)}{}^{0}{}_{0,ij} \\ & + \frac{1}{2} g^{(2)}{}^{m}{}_{m,ij}\,. \end{aligned} \tag{5.26}$$

In the harmonic gauge $g^{\rho\sigma}\Gamma^{\mu}_{\rho\sigma} = 0$ these expressions become (see Appendix C)

$$R^{(2)}_{00} = \frac{1}{2}\nabla^2 g^{(2)}_{00}\,, \tag{5.27}$$

$$R^{(4)}_{00} = \frac{1}{2}\nabla^2 g^{(4)}_{00} - \frac{1}{2}g^{(2)^{mn}} g^{(2)}_{00,mn} - \frac{1}{2}g^{(2)^{0}}{}_{0,00} - \frac{1}{2}\left|\nabla\nabla_\eta g^{(2)}_{00}\right|^2\,, \tag{5.28}$$

$$R^{(3)}_{0i} = \frac{1}{2}\nabla^2 g^{(3)}_{0i}\,, \tag{5.29}$$

$$R^{(2)}_{ij} = \frac{1}{2}\nabla^2 g^{(2)}_{ij}\,, \tag{5.30}$$

where $\nabla^2$ and $\boldsymbol{\nabla}$ denote the Laplacian and the gradient in flat space, respectively. The Ricci scalar in this gauge is

$$R^{(2)} = R^{(2)^{0}}{}_{0} - R^{(2)^{m}}{}_{m} = \frac{1}{2}\nabla^2 g^{(2)^{0}}{}_{0} - \frac{1}{2}\nabla^2 g^{(2)^{m}}{}_{m}\,, \tag{5.31}$$

$$\begin{aligned}R^{(4)} &= R^{(4)^{0}}{}_{0} + g^{(2)^{00}} R^{(2)}_{00} + g^{(2)^{mn}} R^{(2)}_{mn}\\ &= \frac{1}{2}\nabla^2 g^{(4)^{0}}{}_{0} - \frac{1}{2}g^{(2)^{0,0}}{}_{0,0} - \frac{1}{2}g^{(2)^{mn}}\left(g^{(2)^{0}}{}_{0,mn} - \nabla^2 g^{(2)}_{mn}\right) - \frac{1}{2}|\boldsymbol{\nabla} g^{(2)^{0}}{}_{0}|^2\\ &\quad + \frac{1}{2}g^{(2)^{00}}\nabla^2 g^{(2)}_{00}\,.\end{aligned} \tag{5.32}$$

The inverse of the metric tensor is defined by $g^{\alpha\rho}g_{\rho\beta} = \delta^{\alpha}_{\beta}$. The relations between terms of order higher than first are

$$g^{(2)00}(x^0, \mathbf{x}) = -g^{(2)}_{00}(x^0, \mathbf{x})\,, \tag{5.33}$$

$$g^{(4)00}(x^0, \mathbf{x}) = g^{(2)}_{00}(x^0, \mathbf{x})^2 - g^{(4)}_{00}(x^0, \mathbf{x})\,, \tag{5.34}$$

$$g^{(3)0i}(x^0, \mathbf{x}) = g^{(3)}_{0i}(x^0, \mathbf{x})\,, \tag{5.35}$$

$$g^{(2)ij}(x^0, \mathbf{x}) = -g^{(2)}_{ij}(x^0, \mathbf{x})\,. \tag{5.36}$$

Finally, the Lagrangian of a particle in the gravitational field is proportional to the invariant distance $ds$,

$$\begin{aligned}L &= \left(g_{\rho\sigma}\frac{dx^\rho}{dx^0}\frac{dx^\sigma}{dx^0}\right)^{1/2} = \left(g_{00} + 2g_{0m}v^m + g_{mn}v^m v^n\right)^{1/2}\\ &= \left(1 + g^{(2)}_{00} + g^{(4)}_{00} + 2g^{(3)}_{0m}v^m - \mathbf{v}^2 + g^{(2)}_{mn}v^m v^n\right)^{1/2}\,.\end{aligned} \tag{5.37}$$

To second order, this expression reduces to the Newtonian test particle Lagrangian $L_{\text{Newt}} = \left(1 + g_{00}^{(2)} - \mathbf{v}^2\right)^{1/2}$, where $v^2 = \frac{dx^m}{dx^0}\frac{dx_m}{dx^0}$. Post-Newtonian physics involves terms of order higher than fourth in the Lagrangian.

Since the odd-order perturbation terms O(1) or O(3) contain odd powers of the velocity $\mathbf{v}$ or of time derivatives, they are related to the dissipation or absorption of energy by the system. Mass-energy conservation prevents losses of energy and mass and, as a consequence, in the Newtonian limit it prevents terms of order O(1) and O(3) to appear in the Lagrangian. When contributions of order higher than O(4) are included, different theories produce different predictions. For example, due to the conservation of post-Newtonian energy, GR forbids terms of order O(5), while terms of order O(7) can appear and are related to the energy lost due to gravitational radiation.

### 5.2.1 *The Newtonian and post-Newtonian limits of metric $f(R)$ gravity with spherical symmetry*

Let us apply the formalism of the previous section to the weak-field and small velocity regime of metric $f(R)$ gravity. Assuming spherical symmetry and vacuum, we have

$$g_{tt}(t,r) = A(t,r) \simeq -1 + g_{tt}^{(2)}(t,r) + g_{tt}^{(4)}(t,r)\,, \tag{5.38}$$

$$g_{rr}(t,r) = B(t,r) \simeq 1 + g_{rr}^{(2)}(t,r)\,, \tag{5.39}$$

$$g_{\theta\theta}(t,r) = r^2\,, \tag{5.40}$$

$$g_{\phi\phi}(t,r) = r^2\sin^2\theta\,, \tag{5.41}$$

while the inverse metric components are

$$g^{tt} = A(t,r)^{-1} \simeq -1 - g_{tt}^{(2)} + g_{tt}^{(2)2} - g_{tt}^{(4)}\,, \tag{5.42}$$

$$g^{rr} = B(t,r)^{-1} \simeq 1 - g_{rr}^{(2)}\,, \tag{5.43}$$

the metric determinant is

$$g \simeq r^4\sin^2\theta\left[-1 + \left(g_{rr}^{(2)} - g_{tt}^{(2)}\right) + \left(g_{tt}^{(2)}g_{rr}^{(2)} - g_{tt}^{(4)}\right)\right], \tag{5.44}$$

and the Christoffel symbols are given by

$$\Gamma^{(3)}{}^{t}_{tt} = \frac{g_{tt,t}^{(2)}}{2}\,, \quad \Gamma^{(2)}{}^{r}_{tt} + \Gamma^{(4)}{}^{r}_{tt} = \frac{g_{tt,r}^{(2)}}{2} + \frac{g_{rr}^{(2)}g_{tt,r}^{(2)} + g_{tt,r}^{(4)}}{2}\,, \tag{5.45}$$

$$\Gamma^{(3)}{}^{r}_{tr} = -\frac{g_{rr,t}^{(2)}}{2}\,, \quad \Gamma^{(2)}{}^{t}_{tr} + \Gamma^{(4)}{}^{t}_{tr} = \frac{g_{tt,r}^{(2)}}{2} + \frac{g_{tt,r}^{(4)} - g_{tt}^{(2)}g_{tt,r}^{(2)}}{2}\,, \tag{5.46}$$

$$\Gamma^{(3)}{}^{t}_{rr} = -\frac{g^{(2)}_{rr,t}}{2}\,, \quad \Gamma^{(2)}{}^{r}_{rr} + \Gamma^{(4)}{}^{r}_{rr} = -\frac{g^{(2)}_{rr,r}}{2} - \frac{g^{(2)}_{rr}\,g^{(2)}_{rr,r}}{2}\,, \tag{5.47}$$

$$\Gamma^{r}_{\phi\phi} = \sin^2\theta\,\Gamma^{r}_{\theta\theta}\,, \quad \Gamma^{(0)}{}^{r}_{\theta\theta} + \Gamma^{(2)}{}^{r}_{\theta\theta} + \Gamma^{(4)}{}^{r}_{\theta\theta} = -r - r\,g^{(2)}_{rr} - r\,g^{(2)2}_{rr}\,. \tag{5.48}$$

The only non-vanishing components of the Ricci tensor are

$$R_{tt} = R^{(2)}_{tt} + R^{(4)}_{tt}\,, \tag{5.49}$$

$$R_{tr} = R^{(3)}_{tr}\,, \tag{5.50}$$

$$R_{rr} = R^{(2)}_{rr}\,, \tag{5.51}$$

$$R_{\theta\theta} = R^{(2)}_{\theta\theta}\,, \tag{5.52}$$

$$R_{\phi\phi} = R^{(2)}_{\theta\theta}\sin^2\theta\,, \tag{5.53}$$

where

$$R^{(2)}_{tt} = \frac{r\,g^{(2)}_{tt,rr} + 2g^{(2)}_{tt,r}}{2r}\,, \tag{5.54}$$

$$R^{(4)}_{tt} = \Big[-r(g^{(2)}_{tt,r})^2 + 4\,g^{(4)}_{tt,r} + r\,g^{(2)}_{tt,r}\,g^{(2)}_{rr,r} + 2\,g^{(2)}_{rr}\left(2g^{(2)}_{tt,r} + r\,g^{(2)}_{tt,rr}\right)$$
$$+2r\,g^{(4)}_{tt,rr} + 2r\,g^{(2)}_{rr,tt}\Big](4r)^{-1}\,, \tag{5.55}$$

$$R^{(3)}_{tr} = -\frac{g^{(2)}_{rr,t}}{r}\,, \tag{5.56}$$

$$R^{(2)}_{rr} = -\frac{r\,g^{(2)}_{tt,rr} + 2\,g^{(2)}_{rr,r}}{2r}\,, \tag{5.57}$$

$$R^{(2)}_{\theta\theta} = -\frac{2\,g^{(2)}_{rr} + r\left(g^{(2)}_{tt,r} + g^{(2)}_{rr,r}\right)}{2}\,, \tag{5.58}$$

and the post-Newtonian Ricci scalar is

$$R \simeq R^{(2)} + R^{(4)} \tag{5.59}$$

with

$$R^{(2)} = \frac{2\,g^{(2)}_{rr} + r\left(2\,g^{(2)}_{tt,r} + 2\,g^{(2)}_{rr,r} + r\,g^{(2)}_{tt,rr}\right)}{r^2}\,, \tag{5.60}$$

$$R^{(4)} = \Big\{4g^{(2)2}_{rr} + 2rg^{(2)}_{rr}\left(2g^{(2)}_{tt,r} + 4g^{(2)}_{rr,r} + rg^{(2)}_{tt,rr}\right) \tag{5.61}$$

$$+r\left[-rg^{(2)2}_{tt,r} + 4g^{(4)}_{tt,r} + rg^{(2)}_{tt,r}g^{(2)}_{rr,r} - 2g^{(2)}_{tt}\left(2g^{(2)}_{tt,r} + rg^{(2)}_{tt,rr}\right)\right.$$

$$\left.+2rg^{(4)}_{tt,rr} + 2rg^{(2)}_{rr,tt}\right]\Big\}\cdot\left(2r^2\right)^{-1}\,. \tag{5.62}$$

We restrict the discussion to functions $f(R)$ which are analytic at the value $R_0$ of the Ricci curvature,[4]

$$f(R) = \sum_{n=0}^{+\infty}\frac{f^n(R_0)}{n!}\,(R-R_0)^n = f_0 + f_1R + f_2R^2 + f_3R^3 + \ldots\,, \tag{5.63}$$

where in the last equality we have assumed $R_0 = 0$. The coefficient $f_1$ must be positive in order to have a positive gravitational coupling. The post-Newtonian formalism consists of using this expansion in the field equations, which are expanded to orders O(0), O(2), and O(4), and then solved.

The substitution of Eq. (5.63) in the vacuum field equations and their expansion to orders O(0), O(2), and O(4) yield

$$H^{(0)}_{\mu\nu} = 0\,, \quad H^{(0)} = 0\,, \tag{5.64}$$

$$H^{(2)}_{\mu\nu} = 0\,, \quad H^{(2)} = 0\,, \tag{5.65}$$

$$H^{(3)}_{\mu\nu} = 0\,, \quad H^{(3)} = 0\,, \tag{5.66}$$

$$H^{(4)}_{\mu\nu} = 0\,, \quad H^{(4)} = 0\,. \tag{5.67}$$

The order O(0) approximation gives

$$f_0 = 0\,, \tag{5.68}$$

[4] At least, the non-analytic part of $f(R)$ (if it is allowed to exist) must go to zero faster than $R^3$ as $R \to 0$.

a trivial consequence of the assumption (5.8)–(5.10) that space-time is asymptotically Minkowskian. If the Lagrangian is expandable around the zero value of the Ricci scalar ($R_0 = 0$), the cosmological constant must vanish *in vacuo*.

If we now consider the second order approximation, the system (5.64)–(5.67) *in vacuo* yields

$$f_1 r R^{(2)} - 2 f_1 g^{(2)}_{tt,r} + 8 f_2 R^{(2)}_{,r} - f_1 r g^{(2)}_{tt,rr} + 4 f_2 r R^{(2)} = 0\,, \tag{5.69}$$

$$f_1 r R^{(2)} - 2 f_1 g^{(2)}_{rr,r} + 8 f_2 R^{(2)}_{,r} - f_1 r g^{(2)}_{tt,rr} = 0\,, \tag{5.70}$$

$$2 f_1 g^{(2)}_{rr} - r\left(f_1 r R^{(2)} - f_1 g^{(2)}_{tt,r} - f_1 g^{(2)}_{rr,r} + 4 f_2 R^{(2)}_{,r} + 4 f_2 r R^{(2)}_{,rr}\right) = 0\,, \tag{5.71}$$

$$f_1 r R^{(2)} + 6 f_2 \left(2 R^{(2)}_{,r} + r R^{(2)}_{,rr}\right) = 0\,, \tag{5.72}$$

$$2 g^{(2)}_{rr} + r\left(2 g^{(2)}_{tt,r} - r R^{(2)} + 2 g^{(2)}_{rr,r} + r g^{(2)}_{tt,rr}\right) = 0\,. \tag{5.73}$$

The trace equation (5.72), in particular, is a differential equation for the Ricci scalar which allows one to solve the system (5.69)–(5.73) to order O(2) as

$$g^{(2)}_{tt} = -\delta_0 + \frac{\delta_1(t)}{3\xi r}\,\mathrm{e}^{-r\sqrt{-\xi}} - \frac{\delta_2(t)}{6(-\xi)^{3/2} r}\,\mathrm{e}^{r\sqrt{-\xi}}\,, \tag{5.74}$$

$$g^{(2)}_{rr} = -\frac{\delta_1(t)\left(r\sqrt{-\xi}+1\right)}{3\xi r}\,\mathrm{e}^{-r\sqrt{-\xi}} + \frac{\delta_2(t)\left(\xi r + \sqrt{-\xi}\right)}{6\xi^2 r}\,\mathrm{e}^{r\sqrt{-\xi}}\,, \tag{5.75}$$

$$R^{(2)} = \frac{\delta_1(t)}{r}\,\mathrm{e}^{-r\sqrt{-\xi}} - \frac{\delta_2(t)\sqrt{-\xi}}{2\xi r}\,\mathrm{e}^{r\sqrt{-\xi}}\,, \tag{5.76}$$

where

$$\xi = \frac{f_1}{6 f_2}\,, \tag{5.77}$$

and $f_1$ and $f_2$ are expansion coefficients of $f(R)$. The integration constant $\delta_0$ is dimensionless, while the two arbitrary functions of time $\delta_1(t)$ and $\delta_2(t)$ have the dimensions of an inverse length and an inverse length squared, respectively, and $\xi$ has the dimensions on an inverse length squared. These functions are completely arbitrary because the differential system (5.69)–(5.73) contains only spatial derivatives. The additive quantity $\delta_0$ can be set to zero.

The gravitational potential for a generic analytic $f(R)$ can now be obtained. Equations (5.74)–(5.76) provide the second order solution in term of the metric expansion (see the definition (5.38)–(5.41)) but, as said above, this term coincides with the gravitational potential at the Newtonian order, $g_{tt} = -1-2\phi = -1+g^{(2)}_{tt}$. The gravitational potential of a fourth order theory of gravity analytic in $R$ is

$$\phi^{(FOG)} = \frac{K_1}{3\xi r} e^{-r\sqrt{-\xi}} + \frac{K_2}{6(-\xi)^{3/2} r} e^{r\sqrt{-\xi}} \tag{5.78}$$

with $K_1 = \delta_1(t)$ and $K_2 = \delta_2(t)$.

For a given $f(R)$ theory, the structure of the potential is determined by the parameter $\xi$, which depends on the first and second derivatives of $f(R)$ at $R_0$. The potential (5.78) is valid for non-vanishing $f_2$, since we manipulated Eqs. (5.69)–(5.73) dividing by $f_2$. The Newtonian limit of GR cannot be obtained directly from the solution (5.78) but requires the field equations (5.69)–(5.73) once the appropriate expressions in terms of the constants $f_i$ are derived.

The solution (5.78) must be discussed in relation to the sign of the term under square root in the exponents. If this sign is positive (which means that $f_1$ and $f_2$ have opposite signature), the solutions (5.74)–(5.76) and (5.78) can be rewritten introducing the scale parameter $l = |\xi|^{-1/2}$. In particular, considering $\delta_0 = 0$, the functions $\delta_i(t)$ as constants, $k_1 = l\delta_1(t)/3$ and $k_2(t) = l^2\,\delta_2(t)/6$ and introducing a radial coordinate $\tilde{r}$ in units of $l$, we have

$$g_{tt}^{(2)} = -\delta_0 - \frac{\delta_1(t)l}{3}\frac{e^{-r/l}}{r/l} - \frac{\delta_2(t)l^2}{6}\frac{e^{r/l}}{r/l} = \frac{k_1}{\tilde{r}} e^{-\tilde{r}} + \frac{k_2}{\tilde{r}} e^{\tilde{r}}, \tag{5.79}$$

$$\begin{aligned} g_{rr}^{(2)} &= \frac{\delta_1(t)l}{3}\frac{(r/l+1)}{r/l} e^{-r/l} - \frac{\delta_2(t)l^2}{6}\frac{(r/l-1)}{r/l} e^{r/l} \\ &= -k_1 \frac{(\tilde{r}+1)\,e^{-\tilde{r}}}{\tilde{r}} + k_2 \frac{(\tilde{r}-1)\,e^{\tilde{r}}}{\tilde{r}}, \end{aligned} \tag{5.80}$$

$$R^{(2)} = \frac{\delta_1(t)}{l}\frac{e^{-r/l}}{r/l} + \frac{\delta_2(t)}{2}\frac{e^{r/l}}{r/l} = \frac{3}{l^2}\left(k_1 \frac{e^{-\tilde{r}}}{\tilde{r}} + k_2 \frac{e^{\tilde{r}}}{\tilde{r}}\right). \tag{5.81}$$

The gravitational potential can then be rewritten as

$$\phi^{(FOG)} = \frac{k_1}{\tilde{r}} e^{-\tilde{r}} + \frac{k_2}{\tilde{r}} e^{\tilde{r}}, \tag{5.82}$$

which is analogous to the result or [1051] derived for the theory $R + \alpha R^2 + \beta R_{\mu\nu}R^{\mu\nu}$ and consistent[5] with [920,986] discussing higher order Lagrangians such as $f(R, \Box R) = R + \sum_{k=0}^{p} a_k R\,\Box^k R$. In this last case, it was demonstrated that the number of Yukawa corrections to the gravitational potential is related to the order of the theory (this point will be reconsidered in Chap. 7 when we discuss large scale structures). However, it is straightforward to show [247] that the usual form Newton

[5] In a spatially homogeneous and isotropic spacetime manifold, the higher order curvature invariants $R_{\mu\nu}R^{\mu\nu}$ and $R_{\alpha\beta\mu\nu}R^{\alpha\beta\mu\nu}$ can be written in terms of $R^2$.

plus Yukawa is recovered in Eq. (5.82) using a coordinate change, and Eq. (5.82) assumes the form

$$\phi^{(FOG)} = -\left(\frac{GM}{f_1 r} + \frac{\delta_1(t)}{6\xi r}\, \mathrm{e}^{-r\sqrt{-\xi}}\right), \tag{5.83}$$

where $\delta_1(t)$ is again an arbitrary function of time and the parameters depend on the Taylor coefficients. An effective Newton constant $G_{eff} = G/f_1$ and a range $l = |\xi|^{-1/2}$ emerge, and depend on the form of the function $f(R)$.

The inspection of Eqs. (5.74)–(5.76) and (5.79)–(5.81) reveals that the Newtonian limit of an analytic $f(R)$ theory depends only on the first and second coefficients of the Taylor expansion of $f(R)$. The gravitational potential is always characterized by two Yukawa corrections determined only by the first two terms of the Taylor expansion.

The diverging contribution, arising from the exponentially growing mode, has to be analyzed carefully and, in particular, the physical relevance of this term must be evaluated in relation to the length scale $(-\xi)^{-1/2}$. For $r \gg (-\xi)^{-1/2}$, the weak-field approximation turns out to be unphysical and (5.74)–(5.76) no longer holds. One can obtain a modified gravitational potential which can work as a standard Newtonian one in the appropriate limit and provides interesting behavior at larger scales, even in the presence of the growing mode, once the constants in Eq. (5.78) have been suitably adjusted. Once the growing exponential term is discarded, this potential reproduces the Yukawa-like potential phenomenologically introduced in order to explain the flat rotation curves of spiral galaxies without dark matter [966].

Yukawa-like corrections to the gravitational potential have been suggested in several contexts, for example, in a model describing the gravitational interaction between dark matter and baryons. In this model the interaction suppressed at small (subgalactic) scales is described by a Yukawa contribution to the standard Newtonian potential. This behavior is suggested by observations of the inner rotation curves of low-mass galaxies and provides a natural scenario in which to interpret the cuspy profile of dark matter halos arising in $N$-body simulations [905].

The result outlined here is consistent with other calculations. Since an exponential potential is expanded in a power-law series, it is not surprising to find a power-law correction to the Newtonian potential [216, 217] when a less rigorous approach is considered in order to calculate the weak-field limit of a generic $f(R)$ theory, and perturbative calculations will provide effective potentials which can be recovered by means of an appropriate approximation from the general case (5.82).

Let us consider now a negative sign of $\xi$, when the two Yukawa corrections in (5.79)–(5.81) are complex. Using the form of $g_{tt}$, the gravitational potential (5.82) is

$$\phi^{(FOG)} = \frac{k_1}{\tilde{r}} \exp(-i\tilde{r}) + \frac{k_2}{\tilde{r}} \exp(i\tilde{r}), \tag{5.84}$$

which can be recast as

$$\phi^{(FOG)} = \frac{1}{\tilde{r}}\Big[(k_1 + k_2)\cos\tilde{r} + i\,(k_2 - k_1)\sin\tilde{r}\Big]. \tag{5.85}$$

This gravitational potential, which could *a priori* be discarded as physically irrelevant, satisfies the Helmholtz equation $\nabla^2\phi + \mathbf{k}^2\phi = 4\pi G\rho$, where $\rho$ is a real function acting both as matter and antimatter density. As discussed in [108, 109], $Re\{\phi^{(FOG)}\}$ can be seen as a classically modified Newtonian potential corrected by a Yukawa factor while $Im\{\phi^{(FOG)}\}$ could have implications for quantum mechanics. This term can provide an astrophysical origin for the puzzling decay $K_L \to \pi^+\pi^-$, whose phase is related to an imaginary potential in the kaon mass matrix. Of course, these considerations are purely speculative but it could be interesting to pursue them.

Let us consider now third order contributions in the system (5.64)–(5.67); at this order the off-diagonal equation

$$f_1\, g^{(2)}_{rr,t} + 2f_2 r\, R^{(2)}_{,tr} = 0 \tag{5.86}$$

relating the time derivatives of $R$ and $g^{(2)}_{rr}$ must be taken into account. If the Ricci scalar depends on time, also the metric components and the gravitational potential do. This result agrees with the analysis of [251] in terms of the dynamical evolution of $R$ and demonstrating that a time-independent Ricci scalar implies static spherically symmetric solutions, which is confirmed (and explained) by Eq. (5.86). In conjunction with Eqs. (5.79)–(5.81), Eq. (5.86) suggests that if one considers the problem to lower (second) order, the background metric can have static solutions according to the Jebsen-Birkhoff theorem, but this is no longer true when higher orders are considered. The validity of the Jebsen-Birkhoff theorem in higher order theories of gravity depends on the approximation order considered. This theorem holds in metric $f(R)$ gravity only when the Ricci scalar is time-independent, and to second order in a $v/c$ expansion of the metric coefficients. According to Eqs. (5.74)–(5.76) and (5.78), it is only in the limit of small velocities and weak fields that the gravitational potential is effectively time-independent. But, contrary to GR, in metric $f(R)$ gravity a spherically symmetric background can have time-dependent evolution.

The next step is the order O(4) analysis of the system (5.64)–(5.67) providing the solutions in terms of $g^{(4)}_{tt}$, the order necessary to compute the post-Newtonian parameters. Unfortunately, at this order the system is much more complicated and a general solution is not possible. One sees from Eqs. (5.64)–(5.67) that the general solution is characterized only by the first three orders of the $f(R)$ expansion, in agreement with the $f(R)$ reconstruction using the post-Newtonian parameters in the scalar-tensor representation [250, 253]. Although a complete description is difficult, an estimate of the post-Newtonian parameter $\gamma$ can be obtained from the order O(2) evaluation of the metric coefficients *in vacuo*. Since (5.74)–(5.76) suggest a non-Newtonian gravitational potential as a general solution of analytic $f(R)$ gravity, there is no reason to ask for a post-Newtonian description of these theories. In fact, as said earlier, the post-Newtonian analysis presupposes to evaluate deviations from the Newtonian potential at a higher than second order approximation in $v/c$. Thus, if the gravitational potential deduced from a given $f(R)$ theory is a general function of the radial coordinate displaying a Newtonian behavior only in a certain regime (or in a given range of the radial coordinate), it would be meaningless to develop a general post-Newtonian formalism as in GR [856, 1166, 1167]. Of course, by a proper expansion of the gravitational potential for small values of the radial coordinate, and only in this limit, one can develop an analog of the post-Newtonian limit for these theories.

In order to estimate the post-Newtonian parameter $\gamma$, one proceeds by expanding $g_{tt}$ and $g_{rr}$, obtained to second order in (5.79)–(5.81), with respect to the dimensionless coordinate $\tilde{r}$, obtaining

$$g_{tt}^{(2)} = k_2 - k_1 + \frac{k_1 + k_2}{\tilde{r}} + \frac{k_1 + k_2}{2}\tilde{r} + \mathrm{O}(2)\,, \tag{5.87}$$

$$g_{rr}^{(2)} = -\frac{k_1 + k_2}{\tilde{r}} + \frac{k_1 + k_2}{2}\tilde{r} + \mathrm{O}(2)\,, \tag{5.88}$$

where $k_1 + k_2 = GM$ and $k_1 = k_2$ in the standard case. When $\tilde{r} \to 0$ (*i.e.*, when $r \ll \sqrt{-\xi}$) the linear and successive order terms are small and the first (Newtonian) term dominates. Since the post-Newtonian parameter $\gamma$ is related to the coefficients of the $1/r$ terms in $g_{tt}$ and $g_{rr}$, one can estimate this quantity by comparing the coefficients of the Newtonian terms relative to both expressions in (5.88). Since $\gamma = 1$ in GR, the difference between these two coefficients gives the effective deviation from the GR value.

A generic fourth order gravity theory provides a post-Newtonian parameter $\gamma$ consistent with the GR prescription if $k_1 = k_2$. Conversely, deviations from this behavior can be accommodated by tuning the relation between the two integration constants $k_1$ and $k_2$. This is equivalent to adjusting the form of the $f(R)$ theory to obtain the correct GR limit first, and then the Newtonian potential. This result agrees with recovering the GR behavior from generic $f(R)$ theories in the post-Newtonian limit [1032, 1172]. This is particularly true when the $f(R)$ Lagrangian behaves, in the weak-field and small velocity regime, as the Hilbert-Einstein Lagrangian. If deviations from this regime are observed, an $f(R)$ Lagrangian which is a third order polynomial in the Ricci scalar can be more appropriate [250].

The degeneracy in the integration constants can be partially broken once a complete post-Newtonian parameterization is developed in the presence of matter. Then, the integration constants are constrained by the Boltzmann-Vlasov equation describing conservation of matter at small scales [141].

So far, no specific gauge choice has been made, however particular gauges can be considered to simplify the calculations. A natural choice consists of the conditions (5.27)–(5.30), which coincide with the standard post-Newtonian gauge

$$h_{jk},^{k} - \tfrac{1}{2}h_{,j} = \mathrm{O}(4)\,, \tag{5.89}$$

$$h_{0k},^{k} - \tfrac{1}{2}h^{k}{}_{k,0} = \mathrm{O}(5)\,, \tag{5.90}$$

where $h_{\mu\nu} \equiv g_{\mu\nu} - \eta_{\mu\nu}$. In this gauge the Ricci tensor becomes[6]

$$R_{tt|hg} = R^{(2)}_{tt|hg} + R^{(4)}_{tt|hg}\,, \tag{5.91}$$

$$R_{rr|hg} = R^{(2)}_{rr|hg}\,, \tag{5.92}$$

where

$$R^{(2)}_{tt|hg} = \frac{r\, g^{(2)}_{tt,rr} + 2 g^{(2)}_{tt,r}}{2r}\,, \tag{5.93}$$

$$R^{(4)}_{tt|hg} = \frac{r\, g^{(4)}_{tt,rr} + 2\, g^{(4)}_{tt,r} + r\left(g^{(2)}_{rr}\, g^{(2)}_{tt,rr} - g^{(2)}_{tt,tt} - g^{(2)\,2}_{tt,rr}\right)}{2r}\,, \tag{5.94}$$

$$R^{(2)}_{rr|hg} = \frac{r\, g^{(2)}_{rr,rr} + 2\, g^{(2)}_{rr,r}}{2r}\,, \tag{5.95}$$

$$R^{(2)}_{\theta\theta|hg} = R^{(2)}_{\phi\phi|hg} = 0\,, \tag{5.96}$$

while the Ricci scalar to order O(2) and O(4) is

$$R^{(2)}_{|hg} = \frac{r g^{(2)}_{tt,rr} + 2 g^{(2)}_{tt,r} - r g^{(2)}_{rr,rr} - 2 g^{(2)}_{rr,r}}{2r}\,, \tag{5.97}$$

$$R^{(4)}_{|hg} = \Big[r\, g^{(4)}_{tt,rr} + 2\, g^{(4)}_{tt,r} + r\left(g^{(2)}_{rr}\, g^{(2)}_{tt,rr} - g^{(2)}_{tt,tt} - g^{(2)\,2}_{tt,rr}\right) - g^{(2)}_{tt}\left(r\, g^{(2)}_{tt,rr} + 2\, g^{(2)}_{tt,r}\right)$$
$$- g^{(2)}_{rr}\left(r\, g^{(2)}_{rr,rr} + 2\, g^{(2)}_{rr,r}\right)\Big](2r)^{-1}\,. \tag{5.98}$$

The gauge choice does not affect the connection coefficients. The solution of the system (5.64)–(5.67) in this gauge is

$$g_{tt|hg}(t,r) = -1 + \frac{k_1}{r} + \frac{k_2}{r^2} + \frac{k_3 \log r}{r}\,, \tag{5.99}$$

$$g_{rr|hg}(t,r) = 1 + \frac{k_4}{r}\,, \tag{5.100}$$

where the constants $k_1$ and $k_4$ pertain to the order O(2), while $k_2$ and $k_3$ pertain to the order O(4). The Ricci scalar vanishes to orders O(2) and O(4).

[6] We denote harmonic gauge quantities with the subscript $hg$.

Using Eqs. (5.99) and (5.100), it is easy to check that the GR prescriptions are immediately recovered for $k_1 = k_4$ and $k_2 = k_3 = 0$. The $g_{rr}$ component contains only the second order term, as required by a GR-like behavior, while the $g_{tt}$ component exhibits also the fourth order corrections which determine the second post-Newtonian parameter $\beta$ [1166, 1167]. A full post-Newtonian formalism requires the consideration of matter in the system (5.64)–(5.67): the presence of matter links the second and fourth order contributions in the metric coefficients [1166, 1167].

## 5.2.2 Comparison with the standard formalism and the chameleon effect

The weak-field formalism developed thus far is somehow different from the standard formalism used in the literature on metric $f(R)$ gravity and we now bridge the gap between these two descriptions by reformulating the previous discussion in more standard form. Here we essentially report the work of [306, 869] (see also [470]) in order to compare the previous formalism with more standard results in the literature on metric $f(R)$ gravity. Again, the goal is writing the field equations of a general metric $f(R)$ theory in the post-Newtonian approximation, solving them, and from their solutions computing the PPN parameter $\gamma$. We consider a spherically symmetric, static, non-compact body embedded in a background de Sitter universe. Contrary to the previous section, in which the Minkowski background was assumed right away, the de Sitter space represents an adiabatic situation in which the universe evolves very slowly with respect to the dynamical time scales of local systems. A de Sitter space with $R_{\mu\nu} = R_0 g_{\mu\nu}/4$ and $R_0 = 12H_0^2$ exists if the conditions

$$f_0' R_0 - 2 f_0 = 0, \qquad H_0 = \sqrt{\frac{f_0}{6 f_0'}} \tag{5.101}$$

are satisfied. The metric is given by

$$ds^2 = -\left[1 + 2\Psi(r) - H_0^2 r^2\right] dt^2 + \left[1 + 2\Phi(r) + H_0^2 r^2\right] dr^2 + r^2 d\Omega_2^2 \tag{5.102}$$

in Schwarzschild coordinates, and where the post-Newtonian potentials $\Psi(r)$ and $\Phi(r)$ are small perturbations. The PPN parameter $\gamma$ is

$$\gamma = -\Psi/\Phi \tag{5.103}$$

and can be obtained by solving the equations of motion for the potentials $\Psi$ and $\Phi$. A linearized analysis is performed by assuming

$$|\Psi(r)|\,,|\Phi(r)| \ll 1, \qquad r \ll H_0^{-1}\,, \tag{5.104}$$

$$R(r) = R_0 + R_1(r), \tag{5.105}$$

with $R_1(r) \ll R_0$. In order to proceed, one needs the assumptions

1. $f(R)$ is analytical at $R_0$.
2. $mr \ll 1$, where $m$ is the effective mass of the scalar degree of freedom of metric $f(R)$ gravity, which must have a range larger than the size of the Solar System (or, taking into account all the present terrestrial experiments, its range must be larger than 0.2 mm [602]).
3. The pressure $P$ of the local star-like object is negligible and $T_1^{(m)} \simeq -\rho$.

By expanding $f(R)$ and $f'(R)$ around $R_0$, the trace equation simplifies to

$$3f_0''\,\Box R_1 + \left(f_0'' R_0 - f_0'\right) R_1 = \kappa\, T_1^{(m)}\,. \tag{5.106}$$

For a static spherically symmetric body it is $R_1 = R_1(r)$ and $\Box R_1 = \frac{1}{r^2}\frac{d}{dr}\left(r^2 \frac{dR_1}{dr}\right)$, and Eq. (5.106) becomes

$$\nabla^2 R_1 - m^2 R_1 = -\frac{\kappa\,\rho}{3f_0''}\,, \tag{5.107}$$

where

$$m^2 = \frac{f_0' - f_0'' R_0}{3f_0''}\,. \tag{5.108}$$

Taking into account $R_0 = 12H_0^2 = 2f_0/f_0'$, one further obtains

$$m^2 = \frac{\left(f_0'\right)^2 - 2f_0 f_0''}{3f_0' f_0''}\,. \tag{5.109}$$

This equation for the effective mass of the scalar $R$ is derived independently in various studies of perturbations of de Sitter space [472, 832, 860, 869].

The assumption that the scalar $R_1$ is very light allows one to neglect the term $m^2 R_1$ in Eq. (5.107). The Green function of the equation $\nabla^2 R_1 = -\frac{\kappa\rho}{3f_0''}$ is then $G(r) = -\frac{1}{4\pi r}$ and

$$R_1 \simeq \frac{-\kappa}{3f_0''}\int d^3\mathbf{x}'\,\rho(r')\,G(r-r')\,, \tag{5.110}$$

yielding

$$R_1 \simeq \frac{\kappa\,M}{12\pi f_0'' r} \qquad (mr \ll 1)\,. \tag{5.111}$$

The condition $m^2r^2 \ll 1$ yields

$$\frac{1}{3}\left|\frac{f_0'}{f_0''} - R_0\right| r^2 \ll 1 \tag{5.112}$$

and, using $H_0 r \ll 1$,

$$\left|\frac{f_0'}{f_0''}\right| r^2 \ll 1\,. \tag{5.113}$$

Expanding $f(R)$ and $f'(R)$ in the full field equations and using $f_0 = 6H_0^2 f_0'$, one obtains

$$\begin{aligned} &\delta^\alpha_\beta f_0'' \Box R_1 + f_0' \left(R^\alpha_\beta - 3H_0^2 \delta^\alpha_\beta\right) - \frac{f_0'}{2} R_1 \delta^\alpha_\beta \\ &- f_0'' \nabla^\alpha \nabla_\beta R_1 + f_0'' R_1 R^\alpha_\beta = \kappa\, {T^{(m)}}^\alpha_\beta\,. \end{aligned} \tag{5.114}$$

For $H_0 r \ll 1$ the d'Alembertian reduces to $\nabla^2_\eta$ and the $(0,0)$ component of the field equations yields

$$f_0' \left(R^0_0 - 3H_0^2\right) - \frac{f_0'}{2} R_1 + f_0'' R_1 R^0_0 + f_0'' \nabla^2 R_1 = -\kappa\, \rho\,. \tag{5.115}$$

Further computing $R^0_0 = 3H_0^2 - \nabla^2 \Psi(r)$ and discarding $f_0'' H_0^2 R_1 \ll f_0' \nabla^2 \Psi$ and similar terms yields

$$f_0' \nabla^2 \Psi(r) + \frac{f_0'}{2} R_1 - f_0'' \nabla^2 R_1 = \kappa\, \rho\,. \tag{5.116}$$

Since $\nabla^2 R_1 \simeq -\frac{\kappa \rho}{3 f_0''}$ for $mr \ll 1$, it is

$$f_0' \nabla^2 \Psi(r) = \frac{2\kappa\, \rho}{3} - \frac{f_0'}{2} R_1\,. \tag{5.117}$$

Equation (5.117) can be integrated from $r = 0$ to radii $r$ larger than the radius $r_0$ of the star-like object and Gauss' law then gives

$$\frac{d\Psi}{dr} = \frac{\kappa}{6\pi f_0'} \frac{\kappa M}{48\pi f_0''\, r^2} - \frac{C_1}{r^2}\,, \tag{5.118}$$

where $M(r) = 4\pi \int_0^{r_0} dr' \,(r')^2 \rho(r')$. The integration constant $C_1$ is set to zero so that the Newtonian potential is regular at $r = 0$ [306] and $\Psi(r)$ becomes

$$\Psi(r) = -\frac{\kappa\, M}{6\pi f_0'\, r} - \frac{\kappa\, M}{48\pi f_0''}\, r\,. \tag{5.119}$$

Neglecting the second term on the right hand side due to the fact that

$$\left| \frac{\frac{\kappa M r}{48\pi f_0''}}{\frac{-\kappa M}{6\pi f_0' r}} \right| = \left| \frac{f_0'}{8 f_0''} \right| r^2 \ll 1 \,, \tag{5.120}$$

one obtains [306]

$$\Psi(r) \simeq -\frac{\kappa M}{6\pi f_0' \, r} \,. \tag{5.121}$$

We now need to solve the remaining (1, 1) field equation for the potential $\Phi(r)$,

$$\begin{aligned} f_0' \left(R_1^1 - 3H_0^2\right) - \frac{f_0'}{2} R_1 - f_0'' \nabla^1 \nabla_1 R_1 \\ + f_0'' R_1 R_1^1 + f_0'' \Box R_1 = \kappa \, {T^{(m)}}_1^1 \end{aligned} \tag{5.122}$$

with ${T^{(m)}}_1^1 \simeq 0$ outside the star. We have

$$R_1^1 \simeq 3H_0^2 - \tfrac{d^2\Psi}{dr^2} + \tfrac{2}{r}\tfrac{d\Phi}{dr} \,, \tag{5.123}$$

$$g^{11} \nabla_1 \nabla_1 R_1 \simeq \tfrac{d^2 R_1}{dr^2} \,, \tag{5.124}$$

and neglecting higher order terms, one obtains [306]

$$f_0' \left( -\frac{d^2\Psi}{dr^2} + \frac{2}{r}\frac{d\Phi}{dr} \right) - \frac{f_0' R_1}{2} + \frac{2 f_0''}{r}\frac{dR_1}{dr} \simeq 0 \,. \tag{5.125}$$

Equation (5.111) for $R_1$ shows that the third term in Eq. (5.125) is negligible in comparison with the fourth one because

$$\left| \frac{\frac{f_0' R_1}{2}}{\frac{2 f_0''}{r}\frac{dR_1}{dr}} \right| \simeq \left| \frac{f_0'}{f_0''} \right| r^2 \ll 1 \,. \tag{5.126}$$

Equation (5.111) for $dR_1/dr$ and Eq. (5.121) for $\Psi(r)$, finally yield

$$\frac{d\Phi}{dr} = -\frac{\kappa M}{12\pi f_0' r^2} \,, \tag{5.127}$$

which admits the solution

$$\Phi(r) = \frac{\kappa M}{12\pi f_0' \, r} \,. \tag{5.128}$$

The post-Newtonian parameter $\gamma$ arising from the metric (5.102) which solves the field equations is then

$$\gamma = -\frac{\Phi(r)}{\Psi(r)} = \frac{1}{2}. \tag{5.129}$$

This value of $\gamma$ violates the experimental bound $|\gamma - 1| < 2.3 \cdot 10^{-5}$ [133] and coincides with the value $\frac{\omega_0 + 1}{\omega_0 + 2}$ obtained by using the equivalence between metric $f(R)$ gravity and $\omega_0 = 0$ scalar-tensor gravity [301]. The result is also confirmed by [869] in isotropic coordinates and by a study of spherically symmetric interior solutions matched to exterior solutions [650]. This would be the end of the story for metric $f(R)$ gravity if it was not for the fact that the second assumption used in the calculation is not satisfied.[7]

If the condition $mr \ll 1$ does not hold, the scalar degree of freedom of metric $f(R)$ gravity is massive and if this mass is sufficiently large, the associated range is so short that the scalar is effectively hidden from experiments probing gravity in the Solar System as is well known, for example, for the theory $f(R) = R + \alpha R^2$ incorporating renormalization-motivated corrections to GR. If $m \geq 10^{-3}$ eV (*i.e.*, the scalar field range is less than 0.2 mm), the scalar goes undetected at small scales. What saves metric $f(R)$ gravity is the *chameleon mechanism* known from scalar field models of dark energy [666, 667] and consisting of the effective mass $m$ depending on the curvature and the energy density of its environment. $m$ can be large at Solar System and terrestrial densities but small at cosmological densities. The scalar field is short-ranged in the Solar System and long-ranged at cosmological densities, still having a chance to affect the dynamics of the universe and explaining the cosmic acceleration. The chameleon effect is naturally present in metric $f(R)$ gravity [288, 478, 832, 1042]; for example, it appears in theories of the form [36, 39, 40, 275]

$$f(R) = R - (1-n)\mu^2 \left(\frac{R}{\mu^2}\right)^n \tag{5.130}$$

making them compatible with experiment in the parameter range $0 < n \leq 0.25$ if $\mu$ is sufficiently small [478]. The Cassini constraint on $\gamma$ [133] yields the upper limit [478]

$$\frac{\mu}{H_0} \leq \sqrt{3}\left[\frac{2}{n(1-n)}\right]^{\frac{1}{2(1-n)}} 10^{\frac{-6-5n}{2(1-n)}}. \tag{5.131}$$

Fifth force experiments set the limit

$$\frac{\mu}{H_0} \leq \sqrt{1-n}\left[\frac{2}{n(1-n)}\right]^{\frac{1}{2(1-n)}} 10^{\frac{-2-12n}{1-n}}, \tag{5.132}$$

[7] Other situations are possible: the function $f(R)$ may be non-analytic at $R_0$ [641].

where the stability condition $f'' > 0$ requires that $n > 0$. Admissible values of $m$ seem to be $m \simeq 10^{-50}$ eV$\sim 10^{-17} H_0$ [478]. The acceleration of the universe can be explained because, for small values of $R$, the $R^n$ correction with $n < 1$ is larger than the Hilbert-Einstein $R$-term and eventually dominates the dynamics. However, from the observational point of view, these kinds of theories are also practically indistinguishable from a cosmological constant [36,39–41,478]. From thc theoretical point of view, it is still possible to avoid the huge fine-tuning of $\Lambda$ by means of a much smaller fine-tuning of $\mu$.[8]

It is also possible to study the weak-field limit of metric $f(R)$ theories which admit a *global* Minkowski solution and linearize around this global flat background [324]. However, these theories are not relevant for the cosmology of the present epoch of the universe and may be unstable [326]. It is interesting that they contain several new post-Newtonian potentials in addition to the two discussed here [324].

## 5.3 The Post-Minkowskian approximation

We have developed a general analytic procedure to deduce the Newtonian and post-Newtonian limits of $f(R)$ gravity outside matter sources. Now we discuss a different limit of these theories, obtained when the small velocity assumption is relaxed and only the weak-field approximation is retained. Again, we assume spherical symmetry of the metric, considering gravitational potentials $A$ and $B$ of the form

$$A\,(t,r) = -1 + a\,(t,r)\;, \tag{5.133}$$

$$B\,(t,r) = 1 + b\,(t,r)\;, \tag{5.134}$$

with $|a(t,r)|, |b(t,r)| \ll 1$. Let us perturb the field equations considering again the Taylor expansion (5.63). *In vacuo* and to first order in $a$ and $b$, one obtains

$$f_0 = 0\,, \tag{5.135}$$

$$f_1\left(R^{(1)}_{\mu\nu} - \frac{1}{2}\,g^{(0)}_{\mu\nu}R^{(1)}\right) + \mathscr{H}^{(1)}_{\mu\nu} = 0\,, \tag{5.136}$$

where

$$\mathscr{H}^{(1)}_{\mu\nu} = -f_2\Big[R^{(1)}_{,\mu\nu} - \Gamma^{(0)\rho}{}_{\mu\nu}R^{(1)}_{,\rho} - g^{(0)}_{\mu\nu}\Big(g^{(0)\rho\sigma}{}_{,\rho}R^{(1)}_{,\sigma} + g^{(0)\rho\sigma}R^{(1)}_{,\rho\sigma}$$

$$+g^{(0)\rho\sigma}\ln\sqrt{-g}^{(0)}_{,\rho}\,R^{(1)}_{,\sigma}\Big)\Big]\,. \tag{5.137}$$

[8] The cosmological constant problem, of course, is still not solved.

In this approximation the Ricci scalar vanishes and the derivatives are evaluated at $R = 0$. Let us consider now the large $r$ limit far from the sources of the gravitational field: Equations (5.135) and (5.136) become

$$\frac{\partial^2 a(t,r)}{\partial r^2} - \frac{\partial^2 b(t,r)}{\partial t^2} = 0\,, \tag{5.138}$$

$$f_1\Big[a(t,r) - b(t,r)\Big] - 8f_2\Big[\frac{\partial^2 b(t,r)}{\partial r^2} + \frac{\partial^2 a(t,r)}{\partial t^2} - 2\frac{\partial^2 b(t,r)}{\partial t^2}\Big] = \Psi(t)\,, \tag{5.139}$$

where $\Psi(t)$ is a generic time-dependent function. Equations (5.138) and (5.139) are coupled wave equations for $a(t,r)$ and $b(t,r)$, therefore we search for a wave-like solution

$$a\,(t,r) = \int \frac{d\omega\, dk}{2\pi}\,\tilde{a}(\omega,k)\,\mathrm{e}^{i\,(kr\omega t)}\,, \tag{5.140}$$

$$b\,(t,r) = \int \frac{d\omega\, dk}{2\pi}\,\tilde{b}(\omega,k)\,\mathrm{e}^{i\,(kr\omega t)}\,, \tag{5.141}$$

where $k \equiv |\mathbf{k}|$, and we substitute these into Eqs. (5.138) and (5.139), setting $\Psi(t) = 0$. Equations (5.138) and (5.139) are satisfied if

$$\tilde{a}(\omega,k) = \tilde{b}(\omega,k)\,, \qquad \omega = \pm k\,, \tag{5.142}$$

$$\tilde{a}(\omega,k) = \left(1 - \frac{3\xi}{4k^2}\right)\tilde{b}(\omega,k)\,, \qquad \omega = \pm\sqrt{k^2 - \frac{3\xi}{4}} \tag{5.143}$$

where, as before, $\xi = f_1/6f_2$. In particular, for $f_1 = 0$ or $f_2 = 0$, one obtains solutions with dispersion relation $\omega = \pm k$. For $f_i \neq 0$ $(i = 1, 2)$, the dispersion relation suggests that massive modes are present. In particular, for $\xi < 0$, the mass of the scalar graviton is $m_{grav} = -3\xi/4$ and, accordingly, it is obtained for a modified real gravitational potential. A non-Newtonian gravitational potential describes a massive degree of freedom in the particle spectrum of the gravity sector with interesting perspectives for the detection and production of gravitational waves [224]. The presence of massive modes in higher order gravity is well known [1051].

If $\xi > 0$, the solution

$$a\,(\tilde{t},\tilde{r}) = (a_0 + a_1\tilde{r})\,\mathrm{e}^{\pm\frac{\sqrt{3}}{2}\tilde{t}}\,, \tag{5.144}$$

$$b\,(\tilde{t},\tilde{r}) = (b_0 + b_1\tilde{t})\cos\left(\frac{\sqrt{3}}{2}\tilde{r}\right) + (b_0' + b_1'\tilde{t})\sin\left(\frac{\sqrt{3}}{2}\tilde{r}\right) + b_0'' + b_1''\tilde{t}\,, \tag{5.145}$$

with $a_0$, $a_1$, $b_0$, $b_1$, $b'_0$, $b'_1$, $b''_0$, $b''_1$ constants is admitted. The variables $\tilde{r}$ and $\tilde{t}$ are expressed in units of $\xi^{-1/2}$. In the post-Minkowskian approximation, as expected, the gravitational field propagates via wave-like solutions. The gravitational wave content of fourth order gravity originates new phenomenology (massive modes) to be taken into account by the gravitational wave community. These massive degrees of freedom could also constitute a potential candidate for cold dark matter [399].

To conclude this section, we have a general analytic approach for the weak-field, small velocity (Newtonian) limit of a generic $f(R)$ metric theory. The scheme can be used to compute the post-Newtonian parameters of these theories without resorting to the scalar-tensor description of $f(R)$ gravity. At first sight, the scalar-tensor equivalent with Brans-Dicke parameter $\omega = 0$ seems to imply a post-Newtonian parameter $\gamma = 1/2$ incompatible with Solar System tests [301], but this is not the case because of the chameleon mechanism.

### 5.3.1 The energy-momentum pseudotensor in $f(R)$ gravity and gravitational radiation

As we have seen, higher order theories of gravity introduce extra degrees of freedom which can be described by writing the field equations as effective Einstein equations and introducing an additional curvature "effective source" in their right hand side. This quantity behaves as an effective energy-momentum tensor contributing to the energy loss of a system due to the emission of gravitational radiation. The procedure to calculate the stress-energy pseudotensor of gravitational waves in GR can be extended to more general theories and this quantity can be obtained by varying the gravitational Lagrangian. In GR this quantity is known as the Landau-Lifshitz pseudotensor [705].

Let us consider $f(R)$ gravity, for which

$$\begin{aligned}\delta \int d^4x \sqrt{-g}\, f(R) &= \delta \int d^4x \mathscr{L}\left(g_{\mu\nu}, g_{\mu\nu,\rho}, g_{\mu\nu,\rho\sigma}\right)\\ &\approx \int d^4x \left[\frac{\partial \mathscr{L}}{\partial g_{\rho\sigma}} - \partial_\lambda \left(\frac{\partial \mathscr{L}}{\partial g_{\rho\sigma,\lambda}}\right) + \partial^2_{\lambda\xi}\left(\frac{\partial \mathscr{L}}{\partial g_{\rho\sigma,\lambda\xi}}\right)\right]\delta g_{\rho\sigma}\\ &\equiv \int d^4x \sqrt{-g}\, H^{\rho\sigma}\delta g_{\rho\sigma} = 0\,. \end{aligned} \tag{5.146}$$

The Euler-Lagrange equations

$$\frac{\partial \mathscr{L}}{\partial g_{\rho\sigma}} - \partial_\lambda \left(\frac{\partial \mathscr{L}}{\partial g_{\rho\sigma,\lambda}}\right) + \partial^2_{\lambda\xi}\left(\frac{\partial \mathscr{L}}{\partial g_{\rho\sigma,\lambda\xi}}\right) = 0 \tag{5.147}$$

coincide with the vacuum field equations. Even in the case of more general theories, it is possible to define the energy-momentum pseudotensor

$$t_\alpha^\lambda = \frac{1}{\sqrt{-g}}\left\{\left[\frac{\partial \mathscr{L}}{\partial g_{\rho\sigma,\lambda}} - \partial_\xi\left(\frac{\partial \mathscr{L}}{\partial g_{\rho\sigma,\lambda\xi}}\right)\right] g_{\rho\sigma,\alpha} + \frac{\partial \mathscr{L}}{\partial g_{\rho\sigma,\lambda\xi}}\, g_{\rho\sigma,\xi\alpha} - \delta_\alpha^\lambda\, \mathscr{L}\right\}. \tag{5.148}$$

This quantity, together with the matter energy-momentum tensor $T_{\mu\nu}^{(m)}$, satisfies a conservation law as required by the contracted Bianchi identities in conjunction with the effective Einstein equations. In fact, in the presence of matter one has $H_{\mu\nu} = \frac{\kappa}{2} T_{\mu\nu}^{(m)}$ and

$$\left(\sqrt{-g}\, t_\alpha^\lambda\right)_{,\lambda} = -\sqrt{-g}\, H^{\rho\sigma}\, g_{\rho\sigma,\alpha} = -\frac{\kappa}{2}\sqrt{-g}\, T_{(m)}^{\rho\sigma} g_{\rho\sigma,\alpha} = -\kappa\left(\sqrt{-g}\, T_{(m)}{}_\alpha^\lambda\right)_{,\lambda}; \tag{5.149}$$

as a consequence,

$$\left[\sqrt{-g}\left(t_\alpha^\lambda + \kappa\, T^{(m)}{}_\alpha^\lambda\right)\right]_{,\lambda} = 0, \tag{5.150}$$

which is the conservation law given by the contracted Bianchi identities. We can now write the expression of the energy-momentum pseudotensor $t_\alpha^\lambda$ in terms of $f(R)$ and its derivatives

$$\begin{aligned} t_\alpha^\lambda = f'\left\{\left[\frac{\partial R}{\partial g_{\rho\sigma,\lambda}} - \frac{1}{\sqrt{-g}}\partial_\xi\left(\sqrt{-g}\,\frac{\partial R}{\partial g_{\rho\sigma,\lambda\xi}}\right)\right] g_{\rho\sigma,\alpha} + \frac{\partial R}{\partial g_{\rho\sigma,\lambda\xi}}\, g_{\rho\sigma,\xi\alpha}\right\} \\ -f'' R_{,\xi}\,\frac{\partial R}{\partial g_{\rho\sigma,\lambda\xi}}\, g_{\rho\sigma,\alpha} - \delta_\alpha^\lambda\, f\,. \end{aligned} \tag{5.151}$$

$t_\alpha^\lambda$ is a non-covariant quantity in GR while its generalization to fourth order gravity turns out to be covariant. This expression reduces to the Landau-Lifshitz pseudotensor of GR in the limit $f(R) \to R$, in which

$$t_\alpha^\lambda{}_{|\mathrm{GR}} = \frac{1}{\sqrt{-g}}\left(\frac{\partial \mathscr{L}_{\mathrm{GR}}}{\partial g_{\rho\sigma,\lambda}}\, g_{\rho\sigma,\alpha} - \delta_\alpha^\lambda \mathscr{L}_{\mathrm{GR}}\right) \tag{5.152}$$

and where the GR Lagrangian has been considered in its effective form containing the symmetric part of the Ricci tensor which leads to the equations of motion

$$\mathscr{L}_{\mathrm{GR}} = \sqrt{-g}\, g^{\mu\nu}\left(\Gamma_{\mu\sigma}^{\rho}\Gamma_{\rho\nu}^{\sigma} - \Gamma_{\mu\nu}^{\sigma}\Gamma_{\sigma\rho}^{\rho}\right). \tag{5.153}$$

The definitions of energy-momentum pseudotensor in GR and in $f(R)$ gravity are different. The difference is due to the fact that terms of order higher than second are present in $f(R)$ gravity and they cannot be discarded as boundary terms following integration by parts, as is done in GR. The effective Lagrangian of GR turns out to be the symmetric part of the Ricci scalar since the second order terms appearing in the expression of $R$ can be removed integrating by parts. An analytic $f(R)$ Lagrangian can be rewritten, to linear order, as $f \sim f_0' R + \mathscr{F}(R)$, where the function $\mathscr{F}$ is such that $\mathscr{F}(R) \approx R^2$ as $R \to 0$. As a consequence, one can rewrite $t_\alpha^\lambda$ as

$$t_\alpha^\lambda = f_0' t_\alpha^\lambda{}_{|\mathrm{GR}} + \mathscr{F}' \left\{ \left[ \frac{\partial R}{\partial g_{\rho\sigma,\lambda}} - \frac{1}{\sqrt{-g}} \partial_\xi \left( \sqrt{-g} \frac{\partial R}{\partial g_{\rho\sigma,\lambda\xi}} \right) \right] g_{\rho\sigma,\alpha} + \frac{\partial R}{\partial g_{\rho\sigma,\lambda\xi}} g_{\rho\sigma,\xi\alpha} \right\} - \mathscr{F}'' R_{,\xi} \frac{\partial R}{\partial g_{\rho\sigma,\lambda\xi}} g_{\rho\sigma,\alpha} - \delta_\alpha^\lambda \mathscr{F} \,. \tag{5.154}$$

The general expression of the Ricci scalar obtained by splitting its linear ($R^*$) and quadratic ($\bar{R}$) parts in the metric perturbations is

$$R = g^{\mu\nu} (\Gamma^\rho_{\mu\nu,\rho} - \Gamma^\rho_{\mu\rho,\nu}) + g^{\mu\nu} (\Gamma^\rho_{\sigma\rho} \Gamma^\sigma_{\mu\nu} - \Gamma^\sigma_{\rho\mu} \Gamma^\rho_{\nu\sigma}) = R^* + \bar{R} \tag{5.155}$$

(note that $\mathscr{L}_{\mathrm{GR}} = -\sqrt{-g}\,\bar{R}$). In the GR case $t_\alpha^\lambda{}_{|\mathrm{GR}}$, the first non-vanishing term of the Landau-Lifshitz pseudotensor is of order $h^2$ [625, 705]. A similar result can be obtained in $f(R)$ gravity: using Eq. (5.154) one obtains that, to lowest order,

$$\begin{aligned} t_\alpha^\lambda \sim t_\alpha^\lambda{}_{|h^2} &= f_0' t_\alpha^\lambda{}_{|\mathrm{GR}} + f_0'' R^* \left[ \left( -\partial_\xi \frac{\partial R^*}{\partial g_{\rho\sigma,\lambda\xi}} \right) g_{\rho\sigma,\alpha} + \frac{\partial R^*}{\partial g_{\rho\sigma,\lambda\xi}} g_{\rho\sigma,\xi\alpha} \right] \\ &\quad - f_0'' R^*_{,\xi} \frac{\partial R^*}{\partial g_{\rho\sigma,\lambda\xi}} g_{\rho\sigma,\alpha} - \frac{f_0''}{2} \delta_\alpha^\lambda R^{*2} \\ &= f_0' t_\alpha^\lambda{}_{|\mathrm{GR}} + f_0'' \left[ R^* \left( \frac{\partial R^*}{\partial g_{\rho\sigma,\lambda\xi}} g_{\rho\sigma,\xi\alpha} - \frac{1}{2} R^* \delta_\alpha^\lambda \right) \right. \\ &\quad \left. - \partial_\xi \left( R^* \frac{\partial R^*}{\partial g_{\rho\sigma,\lambda\xi}} \right) g_{\rho\sigma,\alpha} \right] . \end{aligned} \tag{5.156}$$

Using the perturbed metric we have $R^* \sim R^{(1)}$, where $R^{(1)}$ is defined by

$$R^{(1)}_{\mu\nu} = h^\sigma_{(\mu,\nu)\sigma} - \frac{1}{2} \Box h_{\mu\nu} - \frac{1}{2} h_{,\mu\nu} \,, \tag{5.157}$$

$$R^{(1)} = h_{\sigma\tau}{}^{,\sigma\tau} - \Box h \tag{5.158}$$

with $h \equiv h^\sigma{}_\sigma$. In terms of $h$ and $\eta_{\mu\nu}$, one obtains

$$\frac{\partial R^*}{\partial g_{\rho\sigma,\lambda\xi}} \sim \frac{\partial R^{(1)}}{\partial h_{\rho\sigma,\lambda\xi}} = \eta^{\rho\lambda} \eta^{\sigma\xi} - \eta^{\lambda\xi} \eta^{\rho\sigma} \,, \tag{5.159}$$

$$\frac{\partial R^*}{\partial g_{\rho\sigma,\lambda\xi}} g_{\rho\sigma,\xi\alpha} \sim h^{\lambda\xi}{}_{,\xi\alpha} - h^{,\lambda}{}_\alpha \,. \tag{5.160}$$

The first significant term in Eq. (5.156) is of second order in the perturbations. We can now write the explicit expression of the pseudotensor in terms of the perturbation $h$,

$$t^\lambda_\alpha \sim f_0' t^\lambda_{\alpha\,|\mathrm{GR}} + f_0'' \left\{ \left(h^{\rho\sigma}{}_{,\rho\sigma} - \Box h\right) \left[ h^{\lambda\xi}{}_{,\xi\alpha} - h^{,\lambda}{}_{\alpha} - \frac{1}{2}\delta^\lambda_\alpha \left(h^{\rho\sigma}{}_{,\rho\sigma} - \Box h\right)\right] \right.$$
$$\left. - h^{\rho\sigma}{}_{,\rho\sigma\xi} h^{\lambda\xi}{}_{,\alpha} + h^{\rho\sigma}{}_{,\rho\sigma}{}^{\lambda} h_{,\alpha} + h^{\lambda\xi}{}_{,\alpha} \Box h_{,\xi} - \Box h^{,\lambda} h_{,\alpha} \right\} . \tag{5.161}$$

This expression can be put in compact form using the metric perturbation $\tilde{h}_{\mu\nu}$ as

$$t^\lambda_{\alpha\,|f} = \frac{1}{2}\left[ \frac{1}{2}\tilde{h}^{,\lambda}{}_{\alpha} \Box \tilde{h} - \frac{1}{2}\tilde{h}_{,\alpha} \Box \tilde{h}^{,\lambda} - \tilde{h}^{\lambda}{}_{\sigma,\alpha} \Box \tilde{h}^{,\sigma} - \frac{1}{4}\left(\Box \tilde{h}\right)^2 \delta^\lambda_\alpha \right] . \tag{5.162}$$

The energy-momentum pseudotensor of the gravitational field describing the energy transport during propagation has a natural generalization to $f(R)$ gravity. Here we have adopted the Landau-Lifshitz construct, but many other pseudotensors can be used [814]. The general definition of $t^\lambda_\alpha$ obtained above consists of the sum of a GR contribution plus a term characteristic of $f(R)$ gravity,

$$t^\lambda_\alpha = f_0'\, t^\lambda_{\alpha\,|\mathrm{GR}} + f_0''\, t^\lambda_{\alpha\,|f} . \tag{5.163}$$

In the limit $f(R) \to R$ one obtains $t^\lambda_\alpha = t^\lambda_{\alpha\,|\mathrm{GR}}$. Massive gravitational modes are contained in $t^\lambda_{\alpha\,|f}$ since $\Box \tilde{h}$ can be considered as an effective scalar field degree of freedom evolving in a potential and $t^\lambda_\alpha$ describes the transport of energy and momentum.

## 5.4 Gravitational waves

Gravitational waves are a fundamental new prediction of relativistic theories of gravity with respect to Newton's theory, which emerge in the post-Minkowskian approximation. The Einstein field equations of GR are hyperbolic and admit solutions which describe waves of the gravitational field leaving their sources (energy currents) and propagating away at the speed of light and carrying energy and momentum. GR admits analytical solutions describing strong plane waves [1053], but these are just idealizations: realistic gravitational waves have extremely small amplitudes which make their detection very challenging [273, 1139, 1153].

Given that gravitational waves generated in the laboratory have amplitudes so small as to make their detection impossible with any means foreseeable in the near future, research has concentrated on gravitational waves generated by astrophysical sources and on those of cosmological origin forming a gravitational wave background.

Experimental efforts involving the *LIGO* [739], *VIRGO* [1123], and other giant interferometers or resonant detectors are presently operating, while more ambitious projects such as the space-based *LISA* (Laser Interferometer Space Antenna) and *BBO* (Big Bang Observer) interferometers [342, 352, 745, 904, 1098] are being designed. The gravitational wave community believes that gravitational waves will be detected within the next decade or so [998]. Since the signal to noise ratio in these experiments is low, one needs to compare the data with theoretical templates of the gravitational waveforms emitted by astrophysical objects and processes to filter out the noise. Currently, theoretical efforts in gravitational wave research focus on predicting accurate waveforms and building banks of templates for interferometric detectors. Roughly speaking, each detector is sensitive only to gravitational waves of wavelength comparable to its size, hence various types of detectors are needed to explore the entire spectrum. Resonant bar detectors are typically sensitive only to a narrow band, with giant laser interferometers being much more broad-banded. The detection of cosmological gravitational waves, which have much larger wavelengths than the typical kHz wave generated by astrophysical processes, is best left to future space-based interferometric experiments with a baseline comparable to the astronomical unit.

In the language of field theory, the gravitational waves of GR correspond to a massless spin two graviton field propagating at the speed of light with two independent polarizations. In alternative theories of gravity, there are additional degrees of freedom in addition to this massless spin two graviton, which contribute extra modes of various spin, massless or massive. In scalar-tensor and metric $f(R)$ theories of gravity there is an extra scalar mode, while in vector-tensor-scalar theories a richer spectrum of modes appear. Gravitational waves in ETGs can be classified according to the effect of their transversal and longitudinal modes on a sphere of test particles at rest before the wave arrives, resulting in the so-called *E(2)* classification scheme [407, 408, 1167]. We refer the reader to [1167] for a more comprehensive description.

To facilitate comparison with ETGs, we recall the form of the field equations of GR linearized around a Minkowski background with metric $\eta_{\mu\nu}$ in an asymptotically Cartesian coordinate system and in the weak-field limit [1139, 1153, 1167]. The metric is

$$g_{\mu\nu} = \eta_{\mu\nu} + h_{\mu\nu} \, , \tag{5.164}$$

where $|h_{\mu\nu}| \ll 1$ in these asymptotically Cartesian coordinates.[9] The linearized Einstein equations are

$$\Box_\eta h_{\mu\nu} + \partial_\mu \partial_\nu h - \partial_\mu \partial_\alpha h^\alpha_\nu - \partial_\nu \partial_\alpha h^\alpha_\mu = 0 \, , \tag{5.165}$$

where $\Box_\eta$ is the d'Alembertian of the Minkowski metric and $h \equiv \eta^{\mu\nu} h_{\mu\nu}$ is the trace of the metric perturbation $h_{\mu\nu}$ (indices are raised and lowered with

[9] Gravity is no longer weak for an observer in fast motion with respect to observers at rest in this asymptotically Cartesian coordinate system [777].

$\eta^{\mu\nu}$ and $\eta_{\mu\nu}$). By using the freedom of changing coordinates it is possible to choose the Lorentz gauge

$$\partial_\alpha h^\alpha_\mu - \frac{1}{2}\,\partial_\mu h = 0 \tag{5.166}$$

to reduce the linearized equation of motion for the metric perturbations in this gauge to

$$\Box_\eta h_{\mu\nu} = 0\,. \tag{5.167}$$

The solutions can be expanded in plane waves as

$$h_{\mu\nu} = A^{(+)}_{\mu\nu}\,\mathrm{e}^{ik_\alpha x^\alpha} + A^{(\times)}_{\mu\nu}\,\mathrm{e}^{ik_\alpha x^\alpha}\,, \tag{5.168}$$

where $A^{(+,\times)}_{\mu\nu}$ are constant polarization tensors and $\eta_{\mu\nu}k^\mu k^\nu = 0$.

### 5.4.1 Gravitational waves in scalar-tensor gravity

Let us begin by considering gravitational waves around a Minkowskian background in scalar-tensor gravity. For simplicity, we first consider the Jordan frame formulation of Brans-Dicke gravity without scalar field potential, described by the action

$$S_{BD} = \frac{1}{16\pi}\int d^4x\,\sqrt{-g}\left(\phi R - \frac{\omega}{\phi}\,g^{\mu\nu}\nabla_\mu\phi\nabla_\nu\phi\right), \tag{5.169}$$

and with the field equations written in the form

$$R_{\mu\nu} - \frac{1}{2}\,g_{\mu\nu}R = \frac{\omega}{\phi^2}\left(\nabla_\mu\phi\nabla_\nu\phi - \frac{1}{2}\,g_{\mu\nu}\nabla^\alpha\phi\nabla_\alpha\phi\right) + \frac{1}{\phi}\left(\nabla_\mu\nabla_\nu\phi - g_{\mu\nu}\Box\phi\right), \tag{5.170}$$

$$\Box\phi = \frac{1}{2\omega+3}\left(\phi\,\frac{dV}{d\phi} - 2V\right) = 0\,. \tag{5.171}$$

A flat background is antithetical to the original inspiration for Brans-Dicke theory, *i.e.*, the implementation of Mach's principle in gravity. However, this is now regarded as only one of the reasons for the study of scalar-tensor gravity, and a marginal one, and the pair $(\eta_{\mu\nu}, \phi_0)$ (where $\phi_0$ is a constant) is indeed a legitimate solution of the field equations of scalar-tensor gravity. Scalar-tensor gravitational waves are small perturbations of this space, according to

$$g_{\mu\nu} = \eta_{\mu\nu} + h_{\mu\nu}\,, \tag{5.172}$$

$$\phi = \phi_0 + \varphi\,, \tag{5.173}$$

where the scalar and tensor modes have the same order of magnitude in terms of a smallness parameter $\varepsilon$,

$$\mathrm{O}\left(\frac{\varphi}{\phi_0}\right) = \mathrm{O}\left(h_{\mu\nu}\right) = \mathrm{O}\left(\varepsilon\right) . \tag{5.174}$$

The Jordan frame linearized field equations are [1133, 1135, 1153, 1167]

$$\Box_\eta h_{\mu\nu} + \partial_\mu \partial_\nu h - \partial_\mu \partial_\alpha h^\alpha_\nu - \partial_\nu \partial_\alpha h^\alpha_\mu = -\frac{\partial_\mu \partial_\nu \varphi}{\phi_0} , \tag{5.175}$$

$$\Box_\eta \varphi = 0 . \tag{5.176}$$

It is possible to choose a gauge in which

$$\partial_\alpha h^\alpha_\mu - \frac{1}{2}\partial_\mu h - \frac{\partial_\mu \varphi}{\phi_0} = 0 . \tag{5.177}$$

In this gauge the Jordan frame linearized field equations reduce to

$$\Box_\eta h_{\mu\nu} = 0 , \tag{5.178}$$

$$\Box_\eta \varphi = 0 . \tag{5.179}$$

The solutions can be expanded in plane waves

$$h_{\mu\nu} = A^{(+)}_{\mu\nu}\, \mathrm{e}^{i k_\alpha x^\alpha} + A^{(\times)}_{\mu\nu}\, \mathrm{e}^{i k_\alpha x^\alpha} , \tag{5.180}$$

$$\varphi = \varphi_0\, \mathrm{e}^{i l_\alpha x^\alpha} , \tag{5.181}$$

where $A^{(+,\times)}_{\mu\nu}$ and $\varphi_0$ are constants and $\eta_{\mu\nu} k^\mu k^\nu = \eta_{\mu\nu} l^\mu l^\nu = 0$. Scalar modes accompany the spin two modes. Because of the assumption that the potential $V(\phi)$ is zero in the action (5.169), $l^\mu$ is a null vector, and the scalar $\varphi$-waves are massless and propagate at light speed.

If a potential $V(\phi)$ is introduced in the action (5.169), the Minkowski space $\left(\eta_{\mu\nu}, \phi_0\right)$ is a solution of the Brans-Dicke field equations only if

$$\phi_0 V'(\phi_0) - 2V(\phi_0) = 0 . \tag{5.182}$$

In this case, the linearized field equation for $\varphi$ becomes

$$\Box_\eta \varphi + \frac{V_0' - \phi_0 V_0''}{2\omega + 3}\, \varphi = 0 . \tag{5.183}$$

Unless the potential is $V(\phi) = m^2\phi^2/2$ (in which case $V(\phi)$ disappears from the equation of motion for $\phi$), the $\varphi$-waves acquire a mass. The plane wave expansion (5.181) now yields

$$\eta_{\mu\nu} l^\mu l^\nu = \frac{V_0' - \phi_0 V_0''}{2\omega + 3} \tag{5.184}$$

and $\frac{V_0' - \phi_0 V_0''}{2\omega + 3} < 0$ is required in order to preserve causality. Using $l^\mu \equiv (\omega, \mathbf{l})$, the scalar waves obey the dispersion relation

$$\omega(l) = \sqrt{\mathbf{l}^2 - \left(\frac{V_0' - \phi_0 V_0''}{2\omega + 3}\right)} \tag{5.185}$$

*in vacuo*.

Although at first sight scalar modes seem to always go hand-in-hand with tensor ones, gravitational radiation of spin two is quadrupole to lowest order and a spherically symmetric motion of a spherical distribution of mass-energy does not excite (spin two) radiation in GR, while it does excite scalar radiation in Brans-Dicke gravity.

Consider now a plane monochromatic scalar wave given by

$$\varphi = \varphi_0 \cos\left(l_\alpha x^\alpha\right) , \tag{5.186}$$

where $\varphi_0$ is a constant and $\eta_{\mu\nu} l^\mu l^\nu = 0$. The effective energy density of the waves as seen by an observer with timelike four-velocity $\xi^\mu$ is given, to first order, by the projection of the effective stress energy tensor of the scalar wave

$$T_{\mu\nu} \xi^\mu \xi^\nu = -\left(l_\mu \xi^\mu\right)^2 \frac{\varphi}{\phi_0} . \tag{5.187}$$

This quantity oscillates with the frequency of $\varphi$, violating the weak energy condition [1139]. Note that this energy density is not quadratic in the first derivatives of the field but is instead linear in its second derivatives, which implies that the energy density of the scalar is of order $O(\varepsilon)$, while the contribution of the tensor modes $h_{\mu\nu}$ is only of order $O(\varepsilon^2)$, as described by the well known Isaacson effective stress-energy tensor $T_{\mu\nu}^{(eff)}[h_{\alpha\beta}]$ [625].

The presence of negative energies for a free Brans-Dicke scalar perturbation of Minkowski space is sometimes regarded as a negative feature of (Jordan frame) scalar-tensor gravity because of the fear that negative energies cause a runaway and the system decays to a lower and lower energy state *ad infinitum*. However, this argument only applies when the total energy (including the gravitational energy of the spin two graviton) is known and is not bounded from below. It is not at all clear that this is the case here and, indeed, a covariant and gauge-invariant linear analysis with respect to inhomogeneous perturbations of Minkowski space in Brans-Dicke theory shows stability to first order [458].

Let us now discuss Brans-Dicke gravitational waves in the Einstein frame description of this theory. The weak-field metric and scalar field in the Einstein frame are

$$\tilde{g}_{\mu\nu} = \eta_{\mu\nu} + \tilde{h}_{\mu\nu} \,, \tag{5.188}$$

$$\tilde{\phi} = \tilde{\phi}_0 + \tilde{\varphi} \,, \tag{5.189}$$

where $\tilde{\phi}_0$ is constant and $\mathrm{O}(\tilde{h}_{\mu\nu}) = \mathrm{O}(\tilde{\varphi}/\tilde{\phi}_0) = \mathrm{O}(\varepsilon)$. The Einstein frame linearized field equations are

$$\tilde{R}_{\mu\nu} - \frac{1}{2}\tilde{g}_{\mu\nu}\tilde{R} = 8\pi\left(\tilde{T}_{\mu\nu}[\tilde{\varphi}] + \tilde{T}^{(eff)}_{\mu\nu}[\tilde{h}_{\mu\nu}]\right) \,, \tag{5.190}$$

$$\tilde{\Box}\tilde{\varphi} = 0 \,, \tag{5.191}$$

where

$$\tilde{T}_{\mu\nu}[\tilde{\varphi}] = \partial_\mu\tilde{\varphi}\partial_\nu\tilde{\varphi} - \frac{1}{2}\eta_{\mu\nu}\partial^\alpha\tilde{\varphi}\partial_\alpha\tilde{\varphi} \tag{5.192}$$

now has canonical form (*i.e.*, quadratic in the first derivatives of the scalar field). Again, one can consider plane monochromatic waves

$$\tilde{\varphi} = \tilde{\varphi}_0 \cos\left(l_\alpha x^\alpha\right) \,, \tag{5.193}$$

where $\tilde{\varphi}_0$ is a constant and $\eta_{\mu\nu}l^\mu l^\nu = 0$. The energy density for an observer characterized by a four-velocity $\xi^\mu$ in the Einstein frame is

$$\tilde{T}_{\mu\nu}\xi^\mu\xi^\nu = \left[l_\mu\xi^\mu\tilde{\varphi}_0 \sin\left(l_\alpha x^\alpha\right)\right]^2 + \tilde{T}^{(eff)}_{\mu\nu}[\tilde{h}_{\alpha\beta}]\,\xi^\mu\xi^\nu \,, \tag{5.194}$$

and is positive definite. In the Einstein frame, the contributions of scalar and tensor modes to the total effective energy density are both quadratic in the first derivatives of the fields and have the same order of magnitude $\mathrm{O}(\varepsilon^2)$. Contrary to the Jordan frame, the weak energy condition is satisfied in the Einstein frame. Of course, the difference is a consequence of the fact that the identification of what is matter and what is gravity is frame-dependent in ETGs [1035].

### 5.4.2 *Gravitational waves in higher order gravity*

Detecting new gravitational wave modes could be a crucial experiment able to discriminate among theories since these modes would constitute evidence that GR must be enlarged or modified [124, 236]. In general, field equations containing higher order terms describe, in addition to the massless spin two field (the standard graviton of GR), also spin zero and spin two massive modes, the latter possibly being ghosts. This result is general and can be obtained by means of a straightforward generalization of the discussion for $f(R)$ gravity and mode counting.

Let us generalize the Hilbert-Einstein action by adding curvature invariants different from the Ricci scalar,

$$S = \int d^4x \sqrt{-g}\, f(R, P, Q) \,, \tag{5.195}$$

where

$$P \equiv R_{\mu\nu} R^{\mu\nu} \,, \tag{5.196}$$

$$Q \equiv R_{\mu\nu\rho\sigma} R^{\mu\nu\rho\sigma} \,. \tag{5.197}$$

By varying the action (5.195) with respect to $g^{\mu\nu}$, one obtains the field equations [274]

$$\begin{aligned} F\, G_{\mu\nu} = {} & \frac{1}{2} g_{\mu\nu} (f - R\, F) - \left(g_{\mu\nu}\Box - \nabla_\mu \nabla_\nu\right) F \\ & -2\left(f_P R^{\alpha}_{\mu} R_{\alpha\nu} + f_Q\, R_{\alpha\beta\gamma\mu} R^{\alpha\beta\gamma}{}_{\nu}\right) \\ & -g_{\mu\nu} \nabla_\alpha \nabla_\beta \left(f_P R^{\alpha\beta}\right) - \Box\left(f_P R_{\mu\nu}\right) \\ & +2\nabla_\alpha \nabla_\beta \left(f_P\, R^{\alpha}{}_{(\mu} \delta^{\beta}{}_{\nu)} + 2 f_Q\, R^{\alpha}{}_{(\mu\nu)}{}^{\beta}\right) , \end{aligned} \tag{5.198}$$

where

$$F \equiv \frac{\partial f}{\partial R} , \qquad f_P \equiv \frac{\partial f}{\partial P} , \qquad f_Q \equiv \frac{\partial f}{\partial Q} . \tag{5.199}$$

The trace of Eq. (5.198) yields

$$\begin{aligned} & \Box\left(F + \frac{f_P}{3} R\right) \\ & = \frac{1}{3}\left\{2f - RF - 2\nabla_\alpha \nabla_\beta \left[(f_P + 2f_Q) R^{\alpha\beta}\right] - 2\left(f_P P + f_Q Q\right)\right\} . \end{aligned} \tag{5.200}$$

Expanding the third term on the right hand side of (5.200) and using the contracted Bianchi identities, one obtains

$$\Box\left(F+\frac{2}{3}(f_P+f_Q)R\right)$$
$$=\frac{1}{3}\left[2f-RF-2R^{\mu\nu}\nabla_\mu\nabla_\nu\left(f_P+2f_Q\right)-R\Box(f_P+2f_Q)\right.$$
$$\left.-2\left(f_P P+f_Q Q\right)\right]. \tag{5.201}$$

By defining

$$\Phi\equiv F+\frac{2}{3}\left(f_P+f_Q\right)R \tag{5.202}$$

and

$$\frac{dV}{d\Phi}\equiv\frac{1}{3}\left[2f-RF-2R^{\mu\nu}\nabla_\mu\nabla_\nu\left(f_P+2f_Q\right)-R\,\Box(f_P+2f_Q)\right.$$
$$\left.-2\left(f_P P+f_Q Q\right)\right], \tag{5.203}$$

the Klein-Gordon equation

$$\Box\Phi-\frac{dV}{d\Phi}=0 \tag{5.204}$$

is obtained. In order to find the modes of the gravity waves of this theory, we linearize around the Minkowski background,

$$g_{\mu\nu}=\eta_{\mu\nu}+h_{\mu\nu}, \tag{5.205}$$
$$\Phi=\Phi_0+\delta\Phi; \tag{5.206}$$

then Eq. (5.202) yields

$$\delta\Phi=\delta F+\frac{2}{3}\left(\delta f_P+\delta f_Q\right)R_0+\frac{2}{3}\left(f_{P0}+f_{Q0}\right)\delta R, \tag{5.207}$$

where $R_0\equiv R(\eta_{\mu\nu})=0$ and, similarly, $f_{P0}=\frac{\partial f}{\partial P}\,|_{\eta_{\mu\nu}}$, which is either constant or zero (a zero subscript denoting quantities evaluated with the Minkowski metric). $\delta R$ denotes the first order perturbation of the Ricci scalar which, together with the perturbed parts of the Riemann and Ricci tensors, is given by (*e.g.*, [270])

$$\delta R_{\mu\nu\rho\sigma}=\frac{1}{2}\left(\partial_\rho\partial_\nu h_{\mu\sigma}+\partial_\sigma\partial_\mu h_{\nu\rho}-\partial_\sigma\partial_\nu h_{\mu\rho}-\partial_\rho\partial_\mu h_{\nu\sigma}\right), \tag{5.208}$$

$$\delta R_{\mu\nu}=\frac{1}{2}\left(\partial_\sigma\partial_\nu h^\sigma{}_\mu+\partial_\sigma\partial_\mu h^\sigma{}_\nu-\partial_\mu\partial_\nu h-\Box h_{\mu\nu}\right), \tag{5.209}$$

$$\delta R=\partial_\mu\partial_\nu h^{\mu\nu}-\Box h, \tag{5.210}$$

where $h \equiv \eta^{\mu\nu} h_{\mu\nu}$. The first term of Eq. (5.207) is

$$\delta F = \frac{\partial F}{\partial R}\,|_0\, \delta R + \frac{\partial F}{\partial P}\,|_0\, \delta P + \frac{\partial F}{\partial Q}\,|_0\, \delta Q\,, \tag{5.211}$$

however since $\delta P$ and $\delta Q$ are second order, it is $\delta F \simeq F_{,R0}\,\delta R$ and

$$\delta\Phi = \left[F_{,R0} + \frac{2}{3}\left(f_{P0} + f_{Q0}\right)\right]\delta R\,. \tag{5.212}$$

Eq. (5.201) then yields the Klein-Gordon equation for the scalar perturbation $\delta\Phi$

$$\begin{aligned}\Box\delta\Phi &= \frac{1}{3}\,\frac{F_0}{F_{,R0} + \frac{2}{3}\left(f_{P0} + f_{Q0}\right)}\,\delta\Phi \\ &\quad -\frac{2}{3}\,\delta R^{\alpha\beta}\,\partial_\alpha\partial_\beta\left(f_{P0} + 2f_{Q0}\right) - \frac{1}{3}\,\delta R\,\Box(f_{P0} + 2f_{Q0}) \\ &= m_s^2\,\delta\Phi\,.\end{aligned} \tag{5.213}$$

The second line of Eq. (5.213) vanishes because $f_{P0}$ and $f_{Q0}$ are constant and the scalar mass is defined as

$$m_s^2 \equiv \frac{F_0}{3F_{,R0} + 2\left(f_{P0} + f_{Q0}\right)}\,. \tag{5.214}$$

Perturbing the field equations (5.198), one obtains

$$\begin{aligned}&F_0\left(\delta R_{\mu\nu} - \frac{1}{2}\eta_{\mu\nu}\delta R\right) \\ &\quad = -\left(\eta_{\mu\nu}\Box - \partial_\mu\partial_\nu\right)\left[\delta\Phi - \frac{2}{3}\left(f_{P0} + f_{Q0}\right)\delta R\right] \\ &\qquad -\eta_{\mu\nu}\partial_\alpha\partial_\beta\left(f_{P0}\delta R^{\alpha\beta}\right) - \Box\left(f_{P0}\delta R_{\mu\nu}\right) \\ &\qquad +2\,\partial_\alpha\partial_\beta\left(f_{P0}\,\delta R^{\alpha}{}_{(\mu}\delta^{\beta}{}_{\nu)} + 2f_{Q0}\,\delta R^{\alpha}{}_{(\mu\nu)}{}^{\beta}\right)\,.\end{aligned} \tag{5.215}$$

It is convenient to work in Fourier space so that, for example, $\partial_\gamma h_{\mu\nu} \to i k_\gamma h_{\mu\nu}$ and $\Box h_{\mu\nu} \to -k^2 h_{\mu\nu}$, where now $k^2 \equiv k^\mu k_\mu$. Then, Eq. (5.215) becomes

$$\begin{aligned}&F_0\left(\delta R_{\mu\nu} - \frac{1}{2}\eta_{\mu\nu}\delta R\right) \\ &\quad = \left(\eta_{\mu\nu}k^2 - k_\mu k_\nu\right)\left[\delta\Phi - \frac{2}{3}\left(f_{P0} + f_{Q0}\right)\delta R\right]\end{aligned}$$

$$+\eta_{\mu\nu}k_a k_b(f_{P0}\delta R^{ab}) + k^2(f_{P0}\delta R_{\mu\nu})$$
$$-2k_\alpha k_\beta\left(f_{P0}\,\delta R^{\alpha}{}_{(\mu}\delta^{\beta}{}_{\nu)}\right) - 4k_\alpha k_\beta\left(f_{Q0}\,\delta R^{\alpha}{}_{(\mu\nu)}{}^{\beta}\right). \tag{5.216}$$

We rewrite the metric perturbation as

$$h_{\mu\nu} = \bar{h}_{\mu\nu} - \frac{\bar{h}}{2}\,\eta_{\mu\nu} + \eta_{\mu\nu}h_f \tag{5.217}$$

and use the gauge freedom to demand that the usual conditions $\partial_\mu \bar{h}^{\mu\nu} = 0$ and $\bar{h} = 0$ hold. The first condition implies that $k_\mu \bar{h}^{\mu\nu} = 0$, while the second one gives

$$h_{\mu\nu} = \bar{h}_{\mu\nu} + \eta_{\mu\nu}h_f\,, \tag{5.218}$$
$$h = 4h_f\,. \tag{5.219}$$

With these conditions in mind, we have

$$\delta R_{\mu\nu} = \frac{1}{2}\left(2k_\mu k_\nu h_f + k^2\eta_{\mu\nu}h_f + k^2\bar{h}_{\mu\nu}\right), \tag{5.220}$$
$$\delta R = 3k^2 h_f\,, \tag{5.221}$$
$$k_\alpha k_\beta\,\delta R^{\alpha}{}_{(\mu\nu)}{}^{\beta} = -\frac{1}{2}\left[\left(k^4\eta_{\mu\nu} - k^2 k_\mu k_\nu\right)h_f + k^4\bar{h}_{\mu\nu}\right], \tag{5.222}$$
$$k_\alpha k_\beta\,\delta R^{\alpha}{}_{(\mu}\delta^{\beta}{}_{\nu)} = \frac{3}{2}k^2 k_\mu k_\nu h_f\,. \tag{5.223}$$

Using Eqs. (5.217)–(5.223) in Eq. (5.216), a little algebra yields

$$\frac{1}{2}\left(k^2 - k^4\,\frac{f_{P0} + 4f_{Q0}}{F_0}\right)\bar{h}_{\mu\nu}$$
$$= (\eta_{\mu\nu}k^2 - k_\mu k_\nu)\frac{\delta\Phi}{F_0} + (\eta_{\mu\nu}k^2 - k_\mu k_\nu)h_f\,. \tag{5.224}$$

Defining now $h_f \equiv -\delta\Phi/F_0$, we find the perturbation equation

$$k^2\left(1 + \frac{k^2}{m^2_{spin\ 2}}\right)\bar{h}_{\mu\nu} = 0\,, \tag{5.225}$$

where

$$m^2_{spin\ 2} \equiv -\frac{F_0}{f_{P0} + 4f_{Q0}}\,, \tag{5.226}$$

while Eq. (5.213) gives

$$\Box h_f = m_s^2 h_f \,. \tag{5.227}$$

It is easy to see from Eq. (5.225) that we have a modified dispersion relation corresponding to a massless spin two field ($k^2 = 0$) and a massive spin two ghost mode with

$$k^2 = \frac{2F_0}{f_{P0} + 4f_{Q0}} \equiv -m_{spin\ 2}^2 \tag{5.228}$$

with mass $m_{spin\ 2}^2$. In fact, the propagator of $\bar{h}_{\mu\nu}$ can be rewritten as

$$G(k) \propto \frac{1}{k^2} - \frac{1}{k^2 + m_{spin2}^2} \,. \tag{5.229}$$

The negative sign of the second term indicates its ghost nature, which agrees with the results found in the literature for this class of theories [300, 861, 1051]. As a check, we can see that for the Gauss-Bonnet Lagrangian density $\mathcal{G} = Q - 4P + R^2$, we have $f_{P0} = -4$ and $f_{Q0} = 1$, then Eq. (5.225) simplifies to $k^2\bar{h}_{\mu\nu} = 0$ and in this case we have no ghosts, as expected.

The solution of Eqs. (5.225) and (5.227) can be expanded in plane waves as

$$\bar{h}_{\mu\nu} = A_{\mu\nu}(\overrightarrow{p}) \exp(ik^\alpha x_\alpha) + \text{c.c.} \,, \tag{5.230}$$

$$h_f = a(\overrightarrow{p}) \exp(iq^\alpha x_\alpha) + \text{c.c.} \,, \tag{5.231}$$

where

$$k^\alpha \equiv \left(\omega_{m_{spin\ 2}}, \overrightarrow{p}\right) , \qquad \omega_{m_{spin\ 2}} = \sqrt{m_{spin\ 2}^2 + p^2} \,, \tag{5.232}$$

$$q^\alpha \equiv \left(\omega_{m_s}, \overrightarrow{p}\right) , \qquad \omega_{m_s} = \sqrt{m_s^2 + p^2} \,, \tag{5.233}$$

and where $m_{spin\ 2}$ is zero (respectively, non-zero) in the case of massless (respectively, massive) spin two modes and the polarization tensor $A_{\mu\nu}(\overrightarrow{p})$ is given by Eqs. (21)–(23) of [1102]. In Eqs. (5.225) and (5.230), the equation and the solution for the standard waves of GR [804] have been obtained while Eqs. (5.227) and (5.231) are the equation and the solution for the massive mode, respectively (see also [225]).

The fact that the dispersion law for the modes of the massive field $h_f$ is not linear has to be emphasized. The velocity of every "ordinary" (*i.e.*, arising from GR) mode $\bar{h}_{\mu\nu}$ is the light speed $c$, but the dispersion law (5.233) for the modes of $h_f$ is that of a massive field which can be discussed like a wave packet [225]. The group velocity of a wave packet of $h_f$ centered in $\overrightarrow{p}$ is

$$\vec{v_g} = \frac{\vec{p}}{\omega}, \tag{5.234}$$

which is exactly the velocity of a massive particle with mass $m$ and momentum $\vec{p}$. From Eqs. (5.233) and (5.234), it is easy to obtain

$$v_g = \frac{\sqrt{\omega^2 - m^2}}{\omega}. \tag{5.235}$$

In order for the wave packet to have constant speed, it must be [225]

$$m = \sqrt{(1 - v_g^2)}\,\omega\,. \tag{5.236}$$

Before proceeding, we discuss the phenomenological constraints on the mass of the gravitational wave field [343]. For frequencies in the range relevant for space-based and terrestrial gravitational antennas, *i.e.*, $10^{-4}$ Hz$\leq f \leq 10$ kHz [1, 17, 47, 745, 746, 1014, 1066, 1162], a strong constraint is available. For a massive gravitational wave it is [223]

$$\omega = \sqrt{m^2 + p^2}\,, \tag{5.237}$$

and then

$$0\ \text{eV} \leq m \leq 10^{-11}\ \text{eV}\,. \tag{5.238}$$

A stronger bound comes from cosmology and Solar System tests, which provide

$$0\ \text{eV} \leq m \leq 10^{-33}\ \text{eV}\,. \tag{5.239}$$

The effects of these light scalars can be discussed as those of a coherent gravitational wave.

#### 5.4.2.1 Polarization states of gravitational waves

Looking at Eq. (5.213) we see that we can have a $k^2 = 0$ mode corresponding to a massless spin two field with two independent polarizations plus a scalar mode while, if $k^2 \neq 0$, we have a massive spin two ghost mode ("poltergeist") and there are five independent polarization tensors plus a scalar mode. First, let us consider the case in which the spin two field is massless.

Taking $\vec{p}$ in the $z$-direction, a gauge in which only $A_{11}$, $A_{22}$, and $A_{12} = A_{21}$ are different from zero can be chosen. The condition $\bar{h} = 0$ gives $A_{11} = -A_{22}$. In this frame, we can take the polarization bases[10]

---

[10] These polarizations are defined in the physical three-space. The polarization vectors are orthogonal to each other and are normalized according to $e_{\mu\sigma}e^{\sigma\nu} = 2\delta^{\nu}_{\mu}$. The other modes are not traceless, in contrast to the ordinary "plus" and "cross" polarization modes of GR.

$$e_{\mu\nu}^{(+)} = \frac{1}{\sqrt{2}} \begin{pmatrix} 1 & 0 & 0 \\ 0 & -1 & 0 \\ 0 & 0 & 0 \end{pmatrix}, \quad e_{\mu\nu}^{(\times)} = \frac{1}{\sqrt{2}} \begin{pmatrix} 0 & 1 & 0 \\ 1 & 0 & 0 \\ 0 & 0 & 0 \end{pmatrix}, \tag{5.240}$$

$$e_{\mu\nu}^{(s)} = \frac{1}{\sqrt{2}} \begin{pmatrix} 0 & 0 & 0 \\ 0 & 0 & 0 \\ 0 & 0 & 1 \end{pmatrix}. \tag{5.241}$$

Substituting these expressions into Eq. (5.217), it follows that

$$h_{\mu\nu}(t,z) = A^{+}(t-z)\, e_{\mu\nu}^{(+)} + A^{\times}(t-z)\, e_{\mu\nu}^{(\times)}$$
$$+ h_s(t - v_g z)\, e_{\mu\nu}^{s}\,. \tag{5.242}$$

The terms $A^{+}(t-z)\, e_{\mu\nu}^{(+)}$ and $A^{\times}(t-z)\, e_{\mu\nu}^{(\times)}$ describe the two standard polarizations of gravitational waves which arise in GR, while the term $h_s(t - v_g z)\, \eta_{\mu\nu}$ is the massive field arising from the generic $f(R)$ theory.

When the spin two field is massive, the bases of the six polarizations are defined by

$$e_{\mu\nu}^{(+)} = \frac{1}{\sqrt{2}} \begin{pmatrix} 1 & 0 & 0 \\ 0 & -1 & 0 \\ 0 & 0 & 0 \end{pmatrix}, \quad e_{\mu\nu}^{(\times)} = \frac{1}{\sqrt{2}} \begin{pmatrix} 0 & 1 & 0 \\ 1 & 0 & 0 \\ 0 & 0 & 0 \end{pmatrix}, \tag{5.243}$$

$$e_{\mu\nu}^{(B)} = \frac{1}{\sqrt{2}} \begin{pmatrix} 0 & 0 & 1 \\ 0 & 0 & 0 \\ 1 & 0 & 0 \end{pmatrix}, \quad e_{\mu\nu}^{(C)} = \frac{1}{\sqrt{2}} \begin{pmatrix} 0 & 0 & 0 \\ 0 & 0 & 1 \\ 0 & 1 & 0 \end{pmatrix}, \tag{5.244}$$

$$e_{\mu\nu}^{(D)} = \frac{\sqrt{2}}{3} \begin{pmatrix} \frac{1}{2} & 0 & 0 \\ 0 & \frac{1}{2} & 0 \\ 0 & 0 & -1 \end{pmatrix}, \quad e_{\mu\nu}^{(s)} = \frac{1}{\sqrt{2}} \begin{pmatrix} 0 & 0 & 0 \\ 0 & 0 & 0 \\ 0 & 0 & 1 \end{pmatrix}, \tag{5.245}$$

and the amplitude can be written in terms of the six polarization states as

$$h_{\mu\nu}(t,z) = A^{+}(t - v_{g_{s2}} z)\, e_{\mu\nu}^{(+)} + A^{\times}(t - v_{g_{s2}} z)\, e_{\mu\nu}^{(\times)}$$
$$+ B^{B}(t - v_{g_{s2}} z)\, e_{\mu\nu}^{(B)} + C^{C}(t - v_{g_{s2}} z)\, e_{\mu\nu}^{(C)}$$
$$+ D^{D}(t - v_{g_{s2}} z)\, e_{\mu\nu}^{(D)} + h_s(t - v_g z)\, e_{\mu\nu}^{s}\,, \tag{5.246}$$

where

$$v_{g_{s2}} = \frac{\sqrt{\omega^2 - m_{s2}^2}}{\omega} \tag{5.247}$$

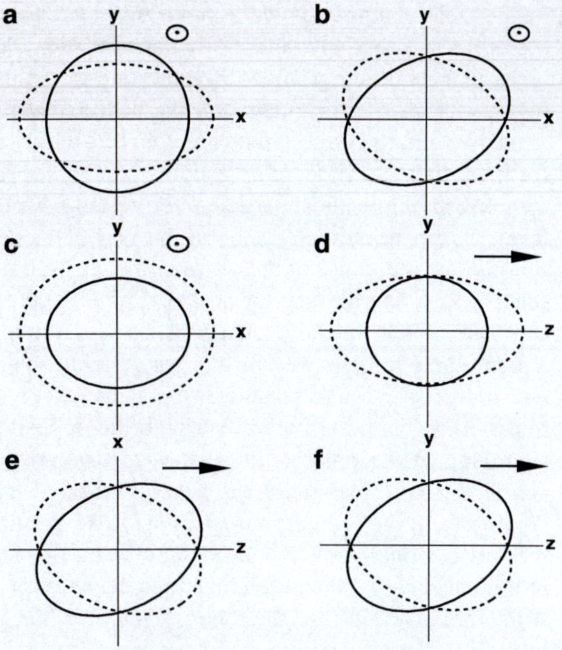


**Fig. 5.1** The six polarization modes of gravitational waves. We illustrate the displacement induced at phases spaced by $\pi$ radians by each mode on a circle of test particles at rest before the wave impinges upon them. The wave propagates out of the plane of the page in (**a**), (**b**), and (**c**) and into this plane in (**d**), (**e**), and (**f**). While (**a**) and (**b**) describe the "plus" and "cross" modes, respectively, (**c**) corresponds to the scalar mode, and (**d**), (**e**), and (**f**) to the D, B, and C modes.

is the group velocity of the massive spin two field. The first two polarizations are the same as in the massless case, inducing tidal deformations of the $(x, y)$ plane. Figure 5.1 illustrates how each gravitational wave polarization affects test masses arranged on a circle before the wave impinges on them.

From a purely quantum-mechanical point of view, the presence of the ghost mode may seem as a pathology of the theory. There are several reasons why this mode is problematic in the particle interpretation of the metric perturbations. The ghost mode can be viewed as either a particle state with positive energy and negative probability density, or as a positive probability density state with negative energy. In the first case, allowing the presence of such a particle will induce violations of unitarity, while the negative energy scenario leads to a theory without ground state and the system becomes unstable. Vacuum can decay into pairs of ordinary and ghost gravitons leading to a catastrophic instability.

A way out of these problems consists of imposing a very weak coupling of the ghost with the other particles in the theory, such that the decay rate of the vacuum becomes comparable to the inverse of the Hubble time. The present vacuum state will then appear to be sufficiently stable. This is not a viable option in our theory because the ghost state appears in the gravitational sector, which is bound to couple to all forms of matter present and it seems physically and mathematically unlikely for the ghost graviton to couple differently than the ordinary massless graviton does.

Another possibility consists of assuming that this picture does not hold up to arbitrarily high energies and that at some cutoff scale $M_{cutoff}$ the theory gets modified appropriately to ensure a ghost-free behavior and a stable ground state. This can happen, for example, if we assume that Lorentz-invariance is violated at $M_{cutoff}$, thereby restricting any potentially harmful decay [429]. However, there is no guarantee that modified gravities like the one investigated here are valid to arbitrarily high energies. Such models are plagued at the quantum level by the same problems of ordinary GR, *i.e.*, they are not renormalizable. It is, therefore, not necessary for them to be considered as genuine candidates for a quantum gravity theory and the corresponding ghost particle interpretation becomes ambiguous. At the classical level, the perturbation $h_{\mu\nu}$ should be viewed as nothing more than a tensor representing the stretching of spacetime away from flatness. A ghost mode then makes sense as just another way of propagating this perturbation of the spacetime geometry, one which, in the propagator, carries a sign opposite to that of an ordinary massive graviton. Viewed in this way, the presence of the massive ghost graviton will induce on an interferometer the same effects as an ordinary massive graviton transmitting the perturbation, but with the opposite sign of the displacement. Tidal stretching of the polarization plane by a polarized wave will turn into shrinking and *vice-versa*. Eventually, the signal will be a superposition of the displacements coming from the ordinary massless spin two graviton and the massive ghost. Since these two modes induce competing effects, their superposition will lead to a less pronounced signal than the one expected were the ghost mode absent, setting less stringent constraints on the theory. However, the presence of the new modes will also affect the total energy density carried by the gravitational waves and this may also appear as a candidate signal in stochastic gravitational wave backgrounds.

#### 5.4.2.2 Detector response

Let us consider now the possible response of a detector in the presence of gravitational waves coming from a definite direction. The detector output depends on the gravitational wave amplitude, which is determined by specific theoretical models. However, one can study the detector response to each gravitational wave polarization without specifying *a priori* the theoretical model. Following [2, 63, 158, 493, 711, 766, 1113], the angular pattern function of a detector of gravitational waves is given by

$$F_A(\hat{\Omega}) = \mathbf{D} : \mathbf{e}_A(\hat{\Omega}) , \tag{5.248}$$

$$\mathbf{D} = \frac{1}{2}\left(\hat{\mathbf{u}} \otimes \hat{\mathbf{u}} - \hat{\mathbf{v}} \otimes \hat{\mathbf{v}}\right) , \tag{5.249}$$

where $A = +, \times, B, C, D, s$ and : denotes a contraction between tensors. $\mathbf{D}$ is the *detector tensor* representing the response of a laser-interferometric detector. It maps the metric perturbation in a signal on the detector. The vectors $\hat{\mathbf{u}}$ and $\hat{\mathbf{v}}$ are unitary and orthogonal to each other, they are directed to each detector arm, and they form an orthonormal coordinate basis together with the unit vector $\hat{\mathbf{w}}$ (see Fig. 5.2). $\hat{\Omega}$ is the unit vector directed along the direction of propagation of the gravitational wave. Equation (5.248) holds only when the arm length of the detector is much smaller

**Fig. 5.2** The coordinate systems used to calculate the polarization tensors and a view of the coordinate transformation.

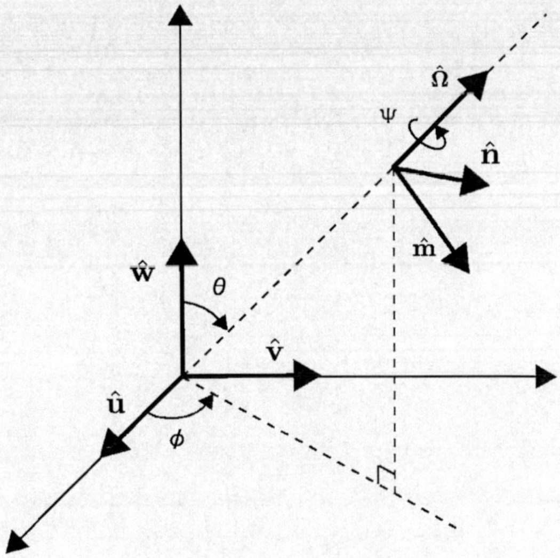


than the gravitational wave wavelength, a condition satisfied by ground-based laser interferometers but not by space interferometers such as *LISA*. A standard orthonormal coordinate system for the detector is

$$\hat{\mathbf{u}} = (1, 0, 0) , \tag{5.250}$$

$$\hat{\mathbf{v}} = (0, 1, 0) , \tag{5.251}$$

$$\hat{\mathbf{w}} = (0, 0, 1) , \tag{5.252}$$

and the coordinate system for the gravitational wave, rotated by $(\theta, \phi)$, is given by

$$\hat{\mathbf{u}}' = (\cos\theta\cos\phi, \cos\theta\sin\phi, -\sin\theta) , \tag{5.253}$$

$$\hat{\mathbf{v}}' = (-\sin\phi, \cos\phi, 0) , \tag{5.254}$$

$$\hat{\mathbf{w}}' = (\sin\theta\cos\phi, \sin\theta\sin\phi, \cos\theta) . \tag{5.255}$$

A rotation by the angle $\psi$ around the direction of propagation of the gravitational wave gives the most general choice of coordinates, that is

$$\hat{\mathbf{m}} = \hat{\mathbf{u}}'\cos\psi + \hat{\mathbf{v}}'\sin\psi , \tag{5.256}$$

$$\hat{\mathbf{n}} = -\hat{\mathbf{v}}'\sin\psi + \hat{\mathbf{u}}'\cos\psi , \tag{5.257}$$

$$\hat{\Omega} = \hat{\mathbf{w}}' . \tag{5.258}$$

The coordinates $(\hat{\mathbf{u}}, \hat{\mathbf{v}}, \hat{\mathbf{w}})$ are related to $\left(\hat{\mathbf{m}}, \hat{\mathbf{n}}, \hat{\Omega}\right)$ by the rotation angles $(\phi,\ \theta,\ \psi)$, as shown in Fig. 5.2. Using the vectors $\hat{\mathbf{m}}$, $\hat{\mathbf{n}}$, and $\hat{\Omega}$, the polarization tensors are

$$\mathbf{e}_{+} = \frac{1}{\sqrt{2}}\left(\hat{\mathbf{m}} \otimes \hat{\mathbf{m}} - \hat{\mathbf{n}} \otimes \hat{\mathbf{n}}\right), \tag{5.259}$$

$$\mathbf{e}_{\times} = \frac{1}{\sqrt{2}}\left(\hat{\mathbf{m}} \otimes \hat{\mathbf{n}} + \hat{\mathbf{n}} \otimes \hat{\mathbf{m}}\right), \tag{5.260}$$

$$\mathbf{e}_{B} = \frac{1}{\sqrt{2}}\left(\hat{\mathbf{m}} \otimes \hat{\Omega} + \hat{\Omega} \otimes \hat{\mathbf{m}}\right), \tag{5.261}$$

$$\mathbf{e}_{C} = \frac{1}{\sqrt{2}}\left(\hat{\mathbf{n}} \otimes \hat{\Omega} + \hat{\Omega} \otimes \hat{\mathbf{n}}\right). \tag{5.262}$$

$$\mathbf{e}_{D} = \frac{\sqrt{3}}{2}\left(\frac{\hat{\mathbf{m}}}{\mathbf{2}} \otimes \frac{\hat{\mathbf{m}}}{\mathbf{2}} + \frac{\hat{\mathbf{n}}}{\mathbf{2}} \otimes \frac{\hat{\mathbf{n}}}{\mathbf{2}} + \hat{\Omega} \otimes \hat{\Omega}\right), \tag{5.263}$$

$$\mathbf{e}_{s} = \frac{1}{\sqrt{2}}\left(\hat{\Omega} \otimes \hat{\Omega}\right). \tag{5.264}$$

Taking into account Eqs. (5.248) and (5.249), the angular patterns for each polarization are

$$\begin{aligned} F_{+}(\theta,\phi,\psi) &= \frac{1}{\sqrt{2}}(1+\cos^2\theta)\cos 2\phi \cos 2\psi \\ &\quad - \cos\theta \sin 2\phi \sin 2\psi\,, \end{aligned} \tag{5.265}$$

$$\begin{aligned} F_{\times}(\theta,\phi,\psi) &= -\frac{1}{\sqrt{2}}(1+\cos^2\theta)\cos 2\phi \sin 2\psi \\ &\quad - \cos\theta \sin 2\phi \cos 2\psi\,, \end{aligned} \tag{5.266}$$

$$F_{B}(\theta,\phi,\psi) = \sin\theta\,(\cos\theta\cos 2\phi\cos\psi - \sin 2\phi \sin\psi)\,, \tag{5.267}$$

$$F_{C}(\theta,\phi,\psi) = \sin\theta\,(\cos\theta\cos 2\phi\sin\psi + \sin 2\phi \cos\psi)\,, \tag{5.268}$$

$$F_{D}(\theta,\phi) = \frac{\sqrt{3}}{32}\cos 2\phi\left[6\sin^2\theta + (\cos 2\theta + 3)\cos 2\psi\right], \tag{5.269}$$

$$F_{s}(\theta,\phi) = \frac{1}{\sqrt{2}}\sin^2\theta\cos 2\phi\,. \tag{5.270}$$

The angular pattern functions for each polarization are plotted in Fig. 5.3. Even if we have considered a different model, these results are consistent, for example, with those of [2, 493, 839, 1075].

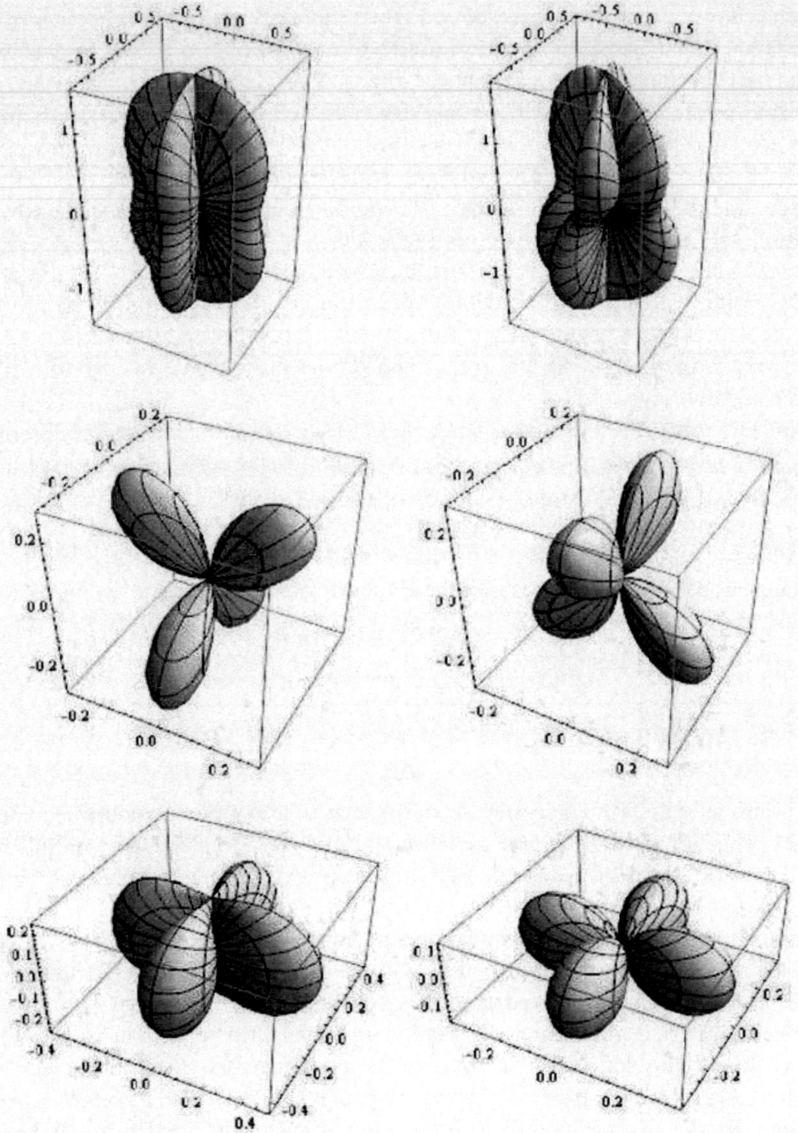

**Fig. 5.3** Angular pattern functions of an interferometric detector for the various polarizations. From left to right and from top to bottom, one sees constant level surfaces corresponding to the "plus", "cross", B, C, D, and scalar modes.

Another area of research which we do not discuss here consists of the study of the stochastic background of gravitational waves which may contain the possible signature of extra gravitational wave modes and be relevant for the detectability of these contributions to gravitational radiation.

The above analysis covers extended gravity models with a generic class of higher order Lagrangian densities and Lagrangian terms of the form $f\,(R, P, Q)$. We have linearized the field equations of these theories around a Minkowski background and found that, in addition to a massless spin two field, the theory contains also spin zero and two massive modes with the latter being, in general, ghosts. If the interferometer is directionally sensitive and we also know the orientation of the source (and, of course, if the source is coherent) the discussion is straightforward. In this case, the massive mode coming from the simplest extension of GR, *i.e.*, $f(R)$ gravity, would induce longitudinal displacements along the direction of propagation of the wave, which should be detectable, and only the scalar mode would be the detectable truly new signal [225]. But, even in this case, there could be a second scalar mode inducing a similar effect and representing a massive ghost, although with a negative sign.

For the situation considered here, massive modes are certainly of interest for the *LISA* space interferometer. It is in principle possible that massive gravitational wave modes could be produced in more significant quantities in cosmological or early astrophysical processes in alternative theories of gravity, a possibility which is still largely unexplored. This situation should be kept in mind when looking for a signature capable of distinguishing these theories from GR, and it seems to deserve further investigation.

## 5.5 Conclusions

The weak-field limit of ETGs shows new aspects of gravitation which are not present in GR. The Newtonian and post-Newtonian limits give weak-field potentials which are not of the standard Newtonian form. The corrections, in general, are Yukawa-like terms which, as we will see in the following chapters, could explain in a very natural fashion several astrophysical and cosmological observations.

The post-Minkowskian limit of ETGs exhibits new gravitational field modes which can easily be interpreted as massive gravitons.

The study of the generation, propagation, and detection of gravitational waves in the weak-field limit of a given relativistic theory of gravity is an important part of astrophysics. Primordial gravitational waves generated during the early epochs of the universe (especially during inflation) would allow, when detected, to rule out or constrain certain theories and investigate others. The detection of gravitational waves of astrophysical or cosmological origin can hardly be overemphasized because it would open a new branch of astronomy providing information which is not accessible with visible, infrared, optical, X-ray, or $\gamma$-ray astronomy. In fact, gravitational waves can be generated in regions deep inside supernovae, near black hole horizons, or very early in the history of the universe when the latter is completely opaque to photons. The study of relativistic astrophysics not related to gravitational waves in ETGs is a broad and complex subject for which we refer the reader to specialized books and review articles (*e.g.*, [1166, 1167]) and to the references therein.

# Chapter 6
# Qualitative analysis and exact solutions in cosmology

*A part of the secret of analysis is the art of using notations well.*
*– Gottfried Wilhelm von Leibniz*

In this chapter we focus again on scalar-tensor and $f(R)$ gravity. We begin by studying the phase space of spatially homogenous and isotropic scalar-tensor (and, by extension, $f(R)$) cosmology. Understanding the structure of the phase space is extremely useful when exact solutions cannot be obtained, or when one needs to know whether known solutions are generic or not. After a general discussion of the geometry of the phase space, we continue by examining particular analytical solutions of scalar-tensor and metric $f(R)$ gravity. While these rare exact solutions may be very special, they still allow us to gain insight into the properties of these ETGs.

## 6.1 The Ehlers-Geren-Sachs theorem

The identification of our universe with a FLRW space relies on the observational fact that, on a cosmological scale, the observable universe is spatially homogeneous and isotropic, and on the extrapolation of these properties to the much larger portion of the universe which is not accessible to us (Copernican principle).

In GR, as well as in ETGs, the strongest support for this identification is the high degree of isotropy of the cosmic microwave background, and the assumption that isotropy would be observed from any spatial point in the universe, *i.e.*, the Copernican principle. Mathematical relativists have something to say here: that a spacetime in which a family of observers exists who see the CMB isotropic can be identified with a FLRW space is not a trivial statement. This property is a kinematical characterization of FLRW spaces known as the Ehlers-Geren-Sachs theorem [412]. Usually, the vanishing of acceleration, shear, and vorticity,

$$\dot{u}^\mu \equiv u^\alpha \nabla_\alpha u^\mu = 0\,, \qquad \sigma_{\mu\nu} = 0\,, \qquad \omega_{\mu\nu} = 0\,, \tag{6.1}$$

for a congruence of "typical" observers with four-velocity $u^\mu$ is assumed to imply that the spacetime has FLRW line element

$$ds^2 = -dt^2 + a^2(t)\,\gamma_{ij}(x^k)\,dx^i dx^j \tag{6.2}$$

S. Capozziello and V. Faraoni, *Beyond Einstein Gravity: A Survey of Gravitational Theories for Cosmology and Astrophysics*, Fundamental Theories of Physics 170,
DOI 10.1007/978-94-007-0165-6_6, © Springer Science+Business Media B.V. 2011

where $\gamma_{ij}$ is a constant curvature three-metric. However, this result is guaranteed only if *(i)* matter is described by a perfect fluid, and *(ii)* the Einstein equations are imposed: then the Weyl tensor is guaranteed to vanish.

In its original version, the Ehlers-Geren-Sachs theorem states that, if a congruence of timelike, freely falling observers in a dust-dominated universe all see an isotropic radiation field, then the spacetime is spatially homogeneous and isotropic (and, therefore, a FLRW universe). The original Ehlers-Geren-Sachs theorem was generalized to an arbitrary perfect fluid that is geodesic and barotropic and with observers that are geodesics and irrotational [318, 428, 483] (see [320] and references therein for a discussion of inhomogeneous or anisotropic cosmological models which admit an isotropic radiation field). An "almost Ehlers-Geren-Sachs theorem" also holds: spacetimes that are "close" to satisfying the Ehlers-Geren-Sachs conditions are "close" to FLRW spaces [1054, 1067].

Since the Ehlers-Geren-Sachs theorem is so basic for cosmology, one would like to know whether it is still valid in the context of ETGs, in particular in scalar-tensor and $f(R)$ gravity. Early investigations [761, 1067] proved an Ehlers-Geren-Sachs theorem for the metric version of the theory with action

$$S = \frac{1}{2\kappa}\int d^4x\,\sqrt{-g}\left(R + \alpha R^2 + \beta R_{\mu\nu}R^{\mu\nu}\right) + S^{(m)}\,. \tag{6.3}$$

The Ehlers-Geren-Sachs theorem was later extended to general *metric* $f(R)$ gravity in [942]. In scalar-tensor gravity, the result was proved in [319], the authors of which studied theories described by the action

$$S_{ST} = \frac{1}{2\kappa}\int d^4x\,\sqrt{-g}\left[\phi R - \frac{\omega(\phi)}{\phi}\nabla^\mu\phi\nabla_\mu\phi - V(\phi)\right] + S^{(m)}\,. \tag{6.4}$$

As we have already seen, when $f'' \neq 0$ metric and Palatini $f(R)$ gravity are equivalent to Brans-Dicke theories with scalar field $\phi = f'(R)$ and with $\omega = 0$ or $\omega = -3/2$, respectively, and with a particular non-vanishing potential $V(\phi)$. This equivalence allows one to provide an independent proof of the Ehlers-Geren-Sachs theorem for metric $f(R)$ gravity, consistently with [761, 942, 1067], and also to extend the theorem to the Palatini version of these theories [448].

## 6.2 The phase space of FLRW cosmology in scalar-tensor and $f(R)$ gravity

The interest of physicists in scalar-tensor cosmology [446, 516] was revived by the extended [701, 702, 733, 1152] and hyperextended [12, 596, 702, 708, 785] scenarios of inflation in the early universe and, more recently, by models of scalar-tensor quintessence [31–33, 64, 65, 127, 132, 149, 281, 284, 297, 302, 303, 384, 420, 422, 450, 476, 477, 513, 517, 778, 847, 847, 851, 903, 934, 935, 1002, 1083, 1101]. Even for the spatially homogeneous and isotropic FLRW cosmology, the field equations of scalar-tensor gravity allow us to find only rare exact solutions and a phase space picture is very valuable in describing qualitatively the dynamics. Many phase portraits describing the dynamics of Brans-Dicke homogeneous and isotropic cosmological models can be found in the literature [594, 689, 690, 970], including the possibilities that the Brans-Dicke-like scalar $\phi$ has a non-vanishing potential $V(\phi)$, that a perfect fluid is present when $V \equiv 0$, with or without a cosmological constant $\Lambda$, and for the possible three-geometries characterized by the values $0, \pm 1$ of the curvature index $K$. Here, following [441], we discuss the geometric structure of the phase space for scalar-tensor FLRW cosmology. We try to use, as much as possible, physical quantities such as the Hubble parameter $H$ and the Brans-Dicke-like scalar $\phi$ as dynamical variables, in contrast with many works in the literature in which the price to pay for achieving simpler mathematics is the use of unphysical dynamical quantities which can be related to physical ones only by formal transformations without physical interpretation (usually, such variables mix the scale factor $a(t)$ of the FLRW metric with the scalar field $\phi$).

We adopt the FLRW line element

$$ds^2 = -dt^2 + a^2(t) \left( \frac{dr^2}{1 - Kr^2} + r^2 d\Omega_2^2 \right) \tag{6.5}$$

and the scalar-tensor action (6.4). At early times before inflation the universe may have been anisotropic, but it is widely believed that inflation caused the universe to isotropize, washing away much of the information about the pre-inflationary state (see [139, 291, 334, 798, 828, 829] for studies of the isotropization of Bianchi models in scalar-tensor cosmology). It is appropriate to study FLRW universes in ETGs during inflation, and dark energy models based on ETGs have also been studied after 1998 [31–33, 64, 65, 127, 132, 149, 281, 284, 297, 302, 303, 384, 420, 422, 450, 476, 477, 513, 517, 778, 847, 847, 851, 903, 934, 935, 1002, 1083, 1101]. It has been suggested [1042] that the universe may evolve into a chaotic regime in the future, which would make prediction impossible, and the dimensionality of the phase space is important in this regard. Chaos is impossible in a two-dimensional phase space continuum; this property is well known when the phase space is flat and it holds also when it is curved, but the proof of this statement is non-trivial for the curved phase space of FLRW cosmology [469, 561].

Many works in the literature provide complete phase space analyses for specific choices of the coupling functions and of the scalar field potential $V(\phi)$ [594, 689, 690, 970]. In this section we focus instead on the geometric structure and dimensionality of the phase space for general coupling functions $\omega(\phi)$ and (except special cases) for any potential $V(\phi)$. Obviously, a complete phase space analysis can only be performed when the theory is completely specified, and we refer the reader to the literature for details about specific models, discussing here only the general features of the phase space applicable to all models. The results that we summarize here were originally obtained for conformally or non-minimally coupled scalar fields [38, 501, 560, 562, 952] (formally included in the scalar-tensor class), and then generalized.

### 6.2.1 The dynamical system

The field equations obtained by varying the action (6.4) can be written in the form of effective Einstein equations as

$$G_{\mu\nu} = \frac{\omega(\phi)}{\phi^2}\left[\nabla_\mu\phi\,\nabla_\nu\phi - \frac{1}{2}\,g_{\mu\nu}\,\nabla^\alpha\phi\,\nabla_\alpha\phi\right] + \frac{1}{\phi}\left(\nabla_\mu\nabla_\nu\phi - g_{\mu\nu}\Box\phi\right) - \frac{V}{2\phi}\,g_{\mu\nu}$$
$$+\frac{8\pi}{\phi}\,T^{(m)}_{\mu\nu}\,, \tag{6.6}$$

$$\Box\phi = \frac{1}{2\omega+3}\left(\phi\,\frac{dV}{d\phi} - 2V - \frac{d\omega}{d\phi}\,\nabla^\alpha\phi\,\nabla_\alpha\phi + 8\pi T^{(m)}\right). \tag{6.7}$$

The combination $\phi dV/d\phi - 2V$ in Eq. (6.7) disappears if the potential is a pure mass term $V(\phi) = m^2\phi^2/2$ (e.g., [443, 970]). The field equations with the FLRW metric (6.5) are

$$H^2 = -H\left(\frac{\dot\phi}{\phi}\right) + \frac{\omega(\phi)}{6}\left(\frac{\dot\phi}{\phi}\right)^2 + \frac{V(\phi)}{6\phi} - \frac{K}{a^2} + \frac{8\pi\rho^{(m)}}{3\phi}\,, \tag{6.8}$$

$$\dot H = -\frac{\omega(\phi)}{2}\left(\frac{\dot\phi}{\phi}\right)^2 + 2H\left(\frac{\dot\phi}{\phi}\right) + \frac{1}{2\,(2\omega+3)\,\phi}\left[\phi\frac{dV}{d\phi} - 2V + \frac{d\omega}{d\phi}\,(\dot\phi)^2\right]$$
$$+\frac{K}{a^2} - \frac{8\pi}{(2\omega+3)\,\phi}\left[(\omega+2)\,\rho^{(m)} + \omega P^{(m)}\right], \tag{6.9}$$

$$\ddot\phi + \left(3H + \frac{1}{2\omega+3}\,\frac{d\omega}{d\phi}\right)\dot\phi = \frac{1}{2\omega+3}\left[2V - \phi\,\frac{dV}{d\phi} + 8\pi\left(\rho^{(m)} - 3P^{(m)}\right)\right], \tag{6.10}$$

where, as usual, an overdot denotes differentiation with respect to the comoving time. We assume that matter is a perfect fluid with barotropic equation of state

$$P^{(m)} = (\gamma-1)\,\rho^{(m)} \tag{6.11}$$

with $\gamma$ a constant. Then, the conservation equation $\dot\rho^{(m)} + 3H\left(\rho^{(m)} + 3P^{(m)}\right) = 0$ yields

$$\rho^{(m)} = \frac{\rho_0}{a^{3\gamma}} \tag{6.12}$$

with $\rho_0$ a constant.

By choosing $a$ and $\phi$ as dynamical variables, the dynamical equations (6.8) and (6.10) are of second order. Only two equations in the set (6.8)–(6.10) are independent and the Hamiltonian constraint reduces the dimensionality of the

$(a, \dot{a}, \phi, \dot{\phi})$ phase space to three [970]. A further simplification is often possible: *in vacuo* and when the universe is spatially flat ($K = 0$), the scale factor $a$ only appears in Eqs. (6.8)–(6.10) through the Hubble parameter. By choosing $H$ and $\phi$ as dynamical variables (which is a convenient choice because $H$ is a cosmological observable), the Hamiltonian constraint effectively reduces the dimensionality of the phase space $(H, \phi, \dot{\phi})$ to two, that is, the orbits of the solutions are constrained to move on an energy surface in this three-dimensional space.

Before we proceed, let us rewrite Eqs. (6.8)–(6.10) using the conformal time $\eta$ (with $dt \equiv a\, d\eta$) and the variable

$$x \equiv \frac{a'}{a} = aH \tag{6.13}$$

often used in the literature, where a prime denotes differentiation with respect to $\eta$. In terms of $x$ and $\eta$ it is

$$x^2 = -x\left(\frac{\phi'}{\phi}\right) + \frac{\omega(\phi)}{6}\left(\frac{\phi'}{\phi}\right)^2 + \frac{a^2 V(\phi)}{6\phi} - K + \frac{8\pi\rho_0}{3a^{3\gamma-2}\phi}, \tag{6.14}$$

$$\begin{aligned} x' = {} & x^2 - \frac{\omega}{2}\left(\frac{\phi'}{\phi}\right)^2 + 2x\left(\frac{\phi'}{\phi}\right) \\ & + \frac{1}{2\,(2\omega+3)\,\phi}\left[a^2\left(\phi\,\frac{dV}{d\phi} - 2V\right) + (\phi')^2\,\frac{d\omega}{d\phi}\right] \\ & + K - \frac{8\pi\rho_0\,(\omega\gamma+2)}{(2\omega+3)\,\phi\, a^{3\gamma-2}}\,. \end{aligned} \tag{6.15}$$

We now focus on various, rather general, situations of interest in scalar-tensor gravity.

#### 6.2.1.1 The phase space with vacuum, free scalar field, and any three-geometry

*In vacuo* and with no scalar field potential, Eq. (6.14) becomes the quadratic equation for $\phi'$

$$\omega\,(\phi')^2 - 6x\phi\phi' - 6\phi^2\left(K + x^2\right) = 0\,, \tag{6.16}$$

which has the reduced discriminant

$$\frac{\Delta_1}{4} = 3\phi^2\left[(2\omega+3)\,x^2 + 2\omega K\right]. \tag{6.17}$$

The solutions are

$$\phi'_{\pm}(x,\phi) = \frac{3\phi}{\omega}\left(x \pm \sqrt{\mathscr{F}_1(x,\phi)}\right), \tag{6.18}$$

$$\mathscr{F}_1(x,\phi) = \frac{1}{3}\left\{[2\omega(\phi)+3]\,x^2 + 2K\omega(\phi)\right\}. \tag{6.19}$$

If $\omega > 0$ and $K = 0$ or $+1$, it is always $\mathscr{F}_1 \geq 0$. However, if $K = -1$ or $\omega < 0$, regions corresponding to $\mathscr{F}_1 < 0$ exist which are forbidden to the dynamics. Equation (6.15) yields

$$x'_{\pm}(x,\phi) = x^2 + \frac{9}{2\omega}\left(x \pm \sqrt{\mathscr{F}_1}\right)^2 \left[\frac{\phi}{\omega(2\omega+3)}\frac{d\omega}{d\phi} - 1\right] + \frac{6x}{\omega}\left(x \pm \sqrt{\mathscr{F}_1}\right) + K. \tag{6.20}$$

We see that the values of $x'$ and $\phi'$ are determined once $(x,\phi)$ are specified. In general there are two values of both $x'$ and $\phi'$ for a given pair $(x,\phi)$, corresponding to the upper and lower sign in Eqs. (6.20) and (6.18), respectively. The two values describe the geometry of the phase space, which is a two-dimensional energy surface in the $(x, x', \phi, \phi')$ space. This surface consists of two sheets corresponding to the upper or lower sign, which will be referred to as "upper sheet" and "lower sheet", respectively. It would be misleading to regard the $(x,\phi)$ plane as the phase space: this plane is only a projection of the two-dimensional energy surface living in a three-dimensional phase space. If attention is restricted only to the $(x,\phi)$ plane, the projections of orbits onto this plane can intersect each other, which is impossible for the true orbits in the curved energy surface according to the uniqueness theorems for the solutions of the Cauchy problem of the system (6.8)–(6.10). Projections onto the $(x,\phi)$ plane of orbits passing from one sheet to the other may cross each other.

In certain scalar-tensor theories the orbits of the solutions change from one sheet to the other, and they can only do so at points where the two sheets touch each other, which are identified by $\mathscr{F}_1 = 0$. The set of points satisfying this condition forms the boundary

$$\mathscr{B} \equiv \left\{(x,\phi,\phi') \ : \ \mathscr{F}_1(x,\phi) = 0\right\} \tag{6.21}$$

of the forbidden region $\mathscr{F}$, where $x'_+ = x'_-$ and $\phi'_+ = \phi'_- = 3\phi\, x/\omega$.

The dimensional reduction of the effective phase space is achieved by using comoving time and the variables $(H,\phi)$. However, if $K \neq 0$ and comoving time is used, the scale factor cannot be eliminated from the field equations and the dimensional reduction cannot be achieved. In this case it is more convenient to use the conformal time $\eta$.

The fixed points of the system (6.18) and (6.20) correspond to $(x',\phi') = (0,0)$; Eqs. (6.18)–(6.20) then yield either $\phi = 0$ (which we reject because it corresponds to infinite gravitational coupling), or $x = 0$ with $\omega\, K = 0$. To summarize, for a spatially flat FLRW universe the only fixed point is the Minkowski space obtained for $x = aH = 0$, which lies on the boundary $\mathscr{B}$ of the forbidden region. For spatially curved universes $K = \pm 1$, the only fixed points are $(x,\phi) = (0,\phi_0)$ with $\phi_0$ a root of $\omega(\phi) = 0$. This fixed point is again a Minkowski space located on the boundary $\mathscr{B}$. If $\omega \neq 0$ everywhere, then there are no fixed points. As a consequence, there are no limit cycles (which must contain at least one fixed point).

#### 6.2.1.2 The phase space for vacuum, $V = m^2\phi^2/2$, and flat three-sections

In the absence of matter and when $V(\phi) = m^2\phi^2/2$ and the FLRW universe is spatially flat, the physical variables $H$ and $\phi$ are the most convenient. The Hamiltonian constraint (6.8) yields the quadratic equation for $\dot{\phi}$

$$\frac{\omega}{6}\left(\dot{\phi}\right)^2 - H\phi\dot{\phi} + \left(\frac{m^2}{12}\phi - H^2\right)\phi^2 = 0\,, \tag{6.22}$$

which has discriminant

$$\Delta_2\left(H,\phi\right) = \frac{\phi^2}{3}\left[(2\omega+3)\,H^2 - \frac{\omega\, m^2}{6}\,\phi\right] \tag{6.23}$$

and solutions

$$\dot{\phi}_{\pm}\left(H,\phi\right) = \frac{3}{\omega}\left(H\phi \pm \sqrt{\Delta_2}\right). \tag{6.24}$$

The phase space is again the union of two curved two-dimensional sheets in the $\left(H,\phi,\dot{\phi}\right)$ space. These two sheets join on the boundary of the forbidden region. This boundary corresponds to $\Delta_2 = 0$ or (discarding the unphysical solution $\phi = 0$),

$$H(\phi) = \pm m\sqrt{\frac{\phi\,\omega(\phi)}{6\left[2\omega(\phi)+3\right]}}\,. \tag{6.25}$$

The stationary points $(H_0, \phi_0)$, which must satisfy $H_0 = \pm m\sqrt{\phi_0/12}$, are de Sitter spaces with constant scalar field and are always located away from the boundary $\Delta_2 = 0$. Again, if $K \neq 0$ the scale factor $a$ cannot be eliminated to use $H$ as a dynamical variable. The wave equation for $\phi$ can be integrated for any value of $K$. Its solutions include the trivial solution $\phi$ = const. corresponding to GR with a cosmological constant and

$$\int d\phi\,\sqrt{2\omega(\phi)+3} = \text{const.}\int\frac{dt}{a^3(t)}\,. \tag{6.26}$$

A complete description of the phase space for Brans-Dicke theory in this case can be found in [970] in terms of the dynamical variables

$$X \equiv \sqrt{\frac{2\omega+3}{12}}\,\frac{\phi'}{\phi}\,, \tag{6.27}$$

$$Y \equiv \frac{a'}{a} + \frac{\phi'}{2\phi}\,, \tag{6.28}$$

and the conformal time $\eta$. The dimensional reduction of the phase space cannot be achieved by using these variables.

#### 6.2.1.3 The phase space in vacuo with $V \neq 0$ and spatially flat three-geometry

Now consider vacuum and a general potential $V(\phi)$ for a spatially flat FLRW universe, using $(H, \phi)$ as dynamical variables and the comoving time to write the field equations as

$$H^2 = -H\left(\frac{\dot{\phi}}{\phi}\right) + \frac{\omega(\phi)}{6}\left(\frac{\dot{\phi}}{\phi}\right)^2 + \frac{V(\phi)}{6\phi}, \tag{6.29}$$

$$\dot{H} = -\frac{\omega(\phi)}{2}\left(\frac{\dot{\phi}}{\phi}\right)^2 + 2H\left(\frac{\dot{\phi}}{\phi}\right) + \frac{1}{2(2\omega+3)\phi}\left[\phi\frac{dV}{d\phi} - 2V + \frac{d\omega}{d\phi}\left(\dot{\phi}\right)^2\right], \tag{6.30}$$

$$\ddot{\phi} + \left(3H + \frac{1}{2\omega+3}\frac{d\omega}{d\phi}\right)\dot{\phi} = \frac{1}{2\omega+3}\left(2V - \phi\frac{dV}{d\phi}\right). \tag{6.31}$$

For $\omega \neq 0$ the Hamiltonian constraint (6.29) provides again a quadratic equation for $\dot{\phi}$,

$$\omega\left(\dot{\phi}\right)^2 - 6H\phi\dot{\phi} + \left(V - 6H^2\phi\right)\phi = 0. \tag{6.32}$$

Its reduced discriminant is

$$\mathscr{F}_2(H, \phi) = \left[3(2\omega+3)H^2\phi - \omega V\right]\phi \tag{6.33}$$

and the roots are

$$\dot{\phi}_\pm(H, \phi) = \frac{1}{\omega(\phi)}\left[3H\phi \pm \sqrt{\mathscr{F}_2(H, \phi)}\right], \tag{6.34}$$

which makes the phase space again the union of two-dimensional sheets corresponding to the lower and upper signs in Eq. (6.34). Figures 6.1–6.3 illustrate the geometry of the phase space for a particular choice of $\omega(\phi)$ and $V(\phi)$. In general, there is no guarantee that $\mathscr{F}_2 \geq 0$ and there will be regions of the phase space in which $\mathscr{F}_2 < 0$: they cannot be penetrated by the orbits of the solutions.

If $K \neq 0$ the reduction of the phase space to two dimensions cannot be performed because the scale factor appears explicitly (*i.e.*, not in the combination $H = \dot{a}/a$) in the term $K/a^2$ in Eqs. (6.8) and (6.9).

The phase space is flat if $\mathscr{F}_2 = 0$, *i.e.*, if $\omega = -3/2$ in conjunction with[1] $V \equiv 0$. In this case one has

$$\dot{\phi} + 2H\phi = 0 \tag{6.35}$$

[1] Palatini $f(R)$ gravity does not correspond to this situation because it necessarily has $V \neq 0$.

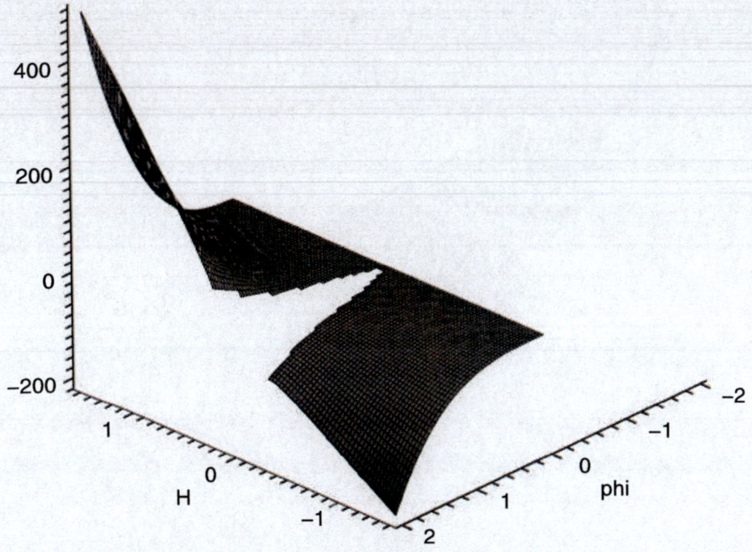


**Fig. 6.1** The upper sheet of the phase space of the scalar-tensor gravity described by $\omega(\phi) = \left(10\phi^2 - \phi + 1\right)^{-1}$ and scalar field potential $V(\phi) = \lambda\phi^4$ (for $\lambda = 1$).

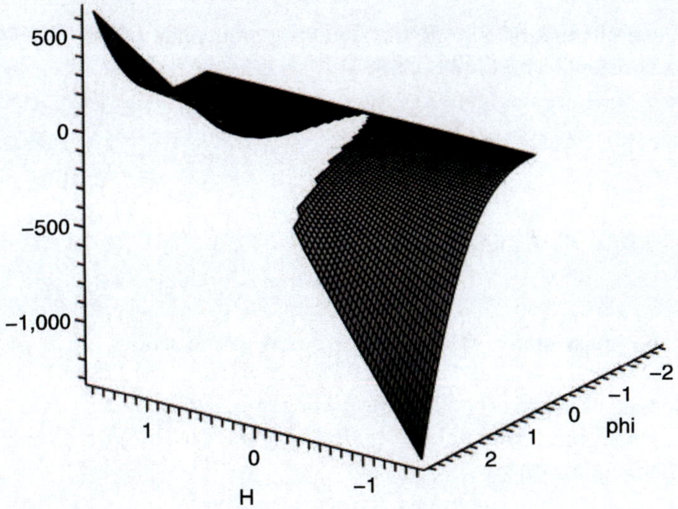


**Fig. 6.2** The lower sheet corresponding to the scalar-tensor theory described in Fig. 6.1.

and either $\phi$ is identically zero (a physically unacceptable solution) or $\phi \propto a^{-2}$. Then the Hamiltonian constraint is automatically satisfied and $\dot{H} = -H^2$, which yields the coasting universe with scale factor $a = a_0 (t - t_0)$.

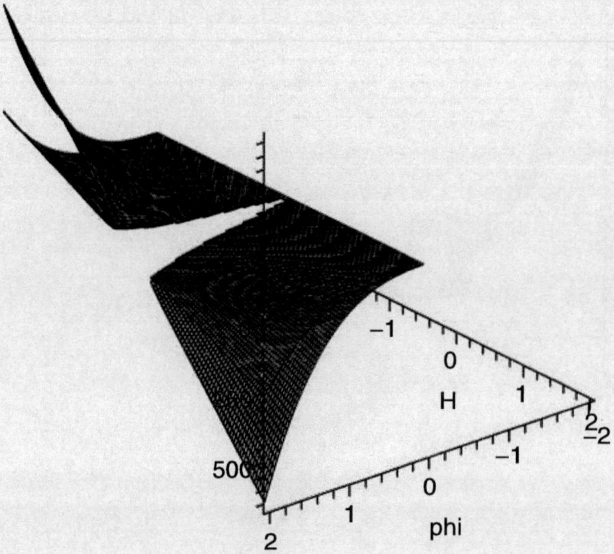


**Fig. 6.3** The energy surface of the scalar-tensor theory of Figs. 6.1 and 6.2 resulting from the two sheets of these figures joined along $\mathscr{B}$. The "hole" is the region $\mathscr{F}$ forbidden to the orbits of the dynamical system.

In the general situation $V \neq 0$, the stationary points are $(H, \phi) = (H_0, \phi_0)$ with $H_0$ and $\phi_0$ constants, and they exist if

$$H_0 = \pm\sqrt{\frac{V_0}{6\phi_0}}\,, \qquad \phi_0 V_0' - 2V_0 = 0\,, \tag{6.36}$$

where $V_0 \equiv V(\phi_0) > 0$ and $V_0' \equiv dV/d\phi\big|_{\phi_0}$.

#### 6.2.1.4 The phase space with $P = -\rho/3$ and a free scalar

Another special case corresponds to the particular equation of state $P = -\rho/3$ and a free Brans-Dicke-like scalar field ($V \equiv 0$). Using the variables $(x, \phi)$ and conformal time, the field equations are

$$x^2 = -x\left(\frac{\phi'}{\phi}\right) + \frac{\omega(\phi)}{6}\left(\frac{\phi'}{\phi}\right)^2 - K + \frac{8\pi\rho_0}{3\phi}\,, \tag{6.37}$$

$$x' = x^2 - \frac{\omega(\phi)}{2}\left(\frac{\phi'}{\phi}\right)^2 + 2x\,\frac{\phi'}{\phi} + K + \frac{(\phi')^2}{2\,(2\omega+3)\,\phi}\,\frac{d\omega}{d\phi} - \frac{A}{\phi}\,, \tag{6.38}$$

where

$$A = \frac{16\pi\rho_0\,(\omega+3)}{3\,(2\omega+3)}\,. \tag{6.39}$$

Again, Eq. (6.37) yields

$$\frac{\omega}{6}\left(\phi'\right)^2 - x\,\phi\phi' + \left(\frac{8\pi\rho_0\phi}{3} - K\phi^2 - x^2\phi^2\right) = 0\,. \tag{6.40}$$

The discriminant and roots of this quadratic equation are

$$\Delta_3\,(x,\phi) = \frac{\phi}{3}\left[(2\omega+3)\,x^2\phi - 2\omega\left(\frac{8\pi\rho_0}{3} - K\phi\right)\right], \tag{6.41}$$

$$\phi'_{\pm}\,(x,\phi) = \frac{3}{\omega}\left(x\phi \pm \sqrt{\Delta_3\,(x,\phi)}\right), \tag{6.42}$$

while Eq. (6.38) gives

$$x' = x^2 - \frac{9}{2\omega\phi^2}\left(x\phi \pm \sqrt{\Delta_3}\right)^2 + \frac{6x}{\omega\phi}\left(x\phi \pm \sqrt{\Delta_3}\right) + K$$
$$+ \frac{9}{2\omega^2\,(2\omega+3)\,\phi}\,\frac{d\omega}{d\phi}\left(x\phi \pm \sqrt{\Delta_3}\right)^2 - \frac{A}{\phi}\,. \tag{6.43}$$

The energy surface in the $(x,\phi,\phi')$ phase space consists again of two sheets attached along the boundary $\Delta_3\,(x,\phi) = 0$ of a forbidden region; there are no equilibrium points, which would correspond to constant $x_0$ and $\phi_0$ and are incompatible with Eqs. (6.37) and (6.38).

To summarize, the geometry of the phase space of FLRW scalar-tensor cosmology, in which gravity is described by the action (6.4), can be relatively complicated. In general, there are two independent equations for FLRW scalar-tensor cosmology. These are second order equations for $a$ and $\phi$ and the natural phase space seems to be the $\left(a,\dot{a},\phi,\dot{\phi}\right)$ space. The Hamiltonian constraint (6.8) confines the orbits of the solutions to an energy hypersurface, effectively reducing the dimension to three. When $K = 0$ the scale factor only enters the field equations through the Hubble parameter $H$ and it is convenient to choose $(H,\phi)$ instead of $(a,\phi)$ as dynamical variables. Then the trajectories of the solutions live in a two-dimensional hypersurface embedded in the $\left(H,\phi,\dot{\phi}\right)$ space.

If $K = \pm 1$ one can again reduce the phase space to a two-dimensional surface embedded in a three-dimensional space. The dynamics are derived without specifying the coupling function $\omega(\phi)$ and, in some cases, also for arbitrary scalar field potential $V(\phi)$. The stationary points of the dynamical system are determined in this general situation.

Usually, the phase space consists of two sheets attached to each other along the boundary $\mathscr{B}$ of a region not accessible to the orbits of the solutions. The latter can change sheet only by passing through $\mathscr{B}$, but there are scenarios in which the orbits stay in one or the other of the two sheets and never change. The $(H, \phi)$ plane is a projection of the curved energy hypersurface and projections of the orbits can possibly intersect in this plane. The equilibrium points of the dynamical system (when they exist) are de Sitter spaces, possibly degenerating into Minkowski spaces.

A complete discussion of the dynamics requires the specification of $\omega(\phi)$ and $V(\phi)$. Although many forms of $V(\phi)$ are proposed in high energy physics and in cosmology, only few proposals for $\omega(\phi)$ have been advanced, and they are dictated by mathematical convenience rather than being inspired by physics [446].

#### 6.2.1.5 The phase space of $f(R)$ gravity

As already discussed, metric and Palatini $f(R)$ gravity are equivalent to scalar-tensor theories with Brans-Dicke field $\phi = f'(R)$, Brans-Dicke couplings $\omega = 0$ and $\omega = -3/2$, respectively, and a special potential $V(\phi)$. Discarding Palatini $f(R)$ gravity which is non-dynamical and of little interest from the phase space point of view, we are left with the metric version of the theory. The general qualitative features are the same as for any scalar-tensor theory, with a simplification introduced by the fact that one needs not worry about terms in $\omega$ and its derivatives in the field equations. The phase space geometry is discussed in [386]. However, here we have only outlined what a phase space analysis is. A detailed treatment for specific models can be found in the literature (see, for example, Refs. [246, 261, 263]).

## 6.3 Analytical solutions of Brans-Dicke and scalar-tensor cosmology

In all physical theories, exact solutions are sought for in order to gain insight into the physical content and predictions of the theory, even if they describe highly idealized or oversimplified situations. Here we present a selection of analytical solutions of scalar-tensor cosmology (in the Jordan frame representation) in order to illustrate similarities and differences between scalar-tensor gravity and GR.[2]

Einstein frame solutions can be obtained from Jordan frame ones by using the conformal transformation, but in practice the reverse process is more common, *i.e.*, exact solutions of GR are mapped back into the Jordan frame to discover new scalar-tensor solutions.

[2] No attempt to be complete is made: we refer the reader to [751] for an early review, to [446] for a recent one, and to the references therein.

We adopt the line element (6.5). Vacuum solutions $(a(t), \phi(t))$ of the field equations have the gravitational scalar field as the only formal source of gravity, while non-vacuum solutions are usually derived for perfect fluids with constant equation of state

$$P^{(m)} = w\rho^{(m)} = (\gamma - 1)\,\rho^{(m)}\,, \tag{6.44}$$

with the constant $\gamma$ in the range of values $0 \leq \gamma \leq 2$ and, again we regard the cosmological constant $\Lambda$ as an effective fluid corresponding to $\gamma = 0$.

In the Jordan frame, ordinary matter is minimally coupled to the Brans-Dicke-like field $\phi$ and the stress-energy tensor $T_{\mu\nu}^{(m)}$ of the fluid is covariantly conserved. In a FLRW universe, conservation assumes the form $\frac{d\rho^{(m)}}{dt} + 3H\left(\rho^{(m)} + P^{(m)}\right) = 0$, which yields $\rho^{(m)} = C/a^{3\gamma}$, where $C$ is an integration constant. Consider a Big Bang solution with scale factor $a(t) \to 0$ as $t \to 0^{+}$: if the universe contains two barotropic fluids with indices $\gamma_1$ and $\gamma_2$ and $0 \leq \gamma_1, \gamma_2 \leq 2$, the fluid with the largest $\gamma$ will dominate the dynamics near the initial singularity.

Since the trace $T^{(m)}$ of the stress-energy tensor of ordinary matter acts as the only source for the scalar field, one finds similarities between vacuum scalar-tensor solutions and solutions corresponding to radiative matter with $T^{(m)} = 0$ [970,1145].

### 6.3.1 *Analytical solutions of Brans-Dicke cosmology*

Let us begin with spatially flat FLRW universes in Brans-Dicke theory. The field equations for arbitrary curvature index are

$$H^2 = \frac{8\pi}{3\phi}\rho^{(m)} + \frac{\omega}{6}\left(\frac{\dot{\phi}}{\phi}\right)^2 - H\frac{\dot{\phi}}{\phi} - \frac{K}{a^2} + \frac{V}{6\phi}\,, \tag{6.45}$$

$$\dot{H} = \frac{-8\pi}{(2\omega+3)\,\phi}\left[(\omega+2)\,\rho^{(m)} + \omega P^{(m)}\right] - \frac{\omega}{2}\left(\frac{\dot{\phi}}{\phi}\right)^2$$
$$+2H\frac{\dot{\phi}}{\phi} + \frac{K}{a^2} + \frac{1}{2\,(2\omega+3)\,\phi}\left(\phi\frac{dV}{d\phi} - 2V\right), \tag{6.46}$$

$$\ddot{\phi} + 3H\dot{\phi} = \frac{1}{2\omega+3}\left[8\pi\left(\rho^{(m)} - 3P^{(m)}\right) - \phi\frac{dV}{d\phi} + 2V\right]. \tag{6.47}$$

By using the new variables

$$\alpha \equiv \ln a\,, \qquad \Phi \equiv -\ln(G\phi)\,, \tag{6.48}$$

related to the scale factor and the Brans-Dicke field, the equations of Brans-Dicke cosmology are found to be symmetric under the duality transformation [735, 736]

$$\alpha \longrightarrow \left(\frac{3\omega+2}{3\omega+4}\right)\alpha - 2\left(\frac{\omega+1}{3\omega+4}\right)\Phi\,, \tag{6.49}$$

$$\Phi \longrightarrow -\left(\frac{6}{3\omega+4}\right)\alpha - \left(\frac{3\omega+2}{3\omega+4}\right)\Phi\,. \tag{6.50}$$

This transformation generalizes a scale factor duality known in the effective action of string theories [530, 542, 1087, 1111],

$$\alpha \longrightarrow -\alpha\,, \tag{6.51}$$

$$\Phi \longrightarrow \Phi - 6\alpha\,. \tag{6.52}$$

This duality is reproduced by Eqs. (6.49) and (6.50) for $\omega = -1$.

Big Bang solutions resembling those of GR were found early on, usually assuming the boundary condition near the Big Bang

$$\lim_{t\to 0}\left[a^3(t)\,\dot{\phi}(t)\right] = 0, \tag{6.53}$$

which implies that one of the four integration constants $\left(a_0, \dot{a}_0, \phi_0, \dot{\phi}_0\right)$ is eliminated, lowering the dimension of the phase space with some loss of generality of the solutions [798, 1145]. With or without the restriction (6.53), the phase space of FLRW cosmology in Brans-Dicke gravity has a higher dimension than the corresponding phase space of GR (which only requires initial conditions $(a_0, \dot{a}_0)$) and therefore scalar-tensor gravity exhibits a richer variety of solutions than GR.

Most of the exact solutions recurrent in the literature on Brans-Dicke cosmology are of the power-law type

$$a(t) \propto t^q\,, \qquad \phi(t) \propto t^s\,, \qquad 3q + s \geq 1\,. \tag{6.54}$$

In BD theory, a power-law solution plays a role analogous to that of the inflationary de Sitter attractor in GR. If a barotropic fluid is present, $V(\phi) \equiv 0$, and $a = a_0\, t^q$, then the field equation ruling the dynamics of the BD scalar reduces to

$$\ddot{\phi} + \frac{3q}{t}\dot{\phi} = \frac{8\pi\,(4-3\gamma)\,C}{(2\omega+3)\,{a_0}^{3\gamma}}\,t^{-3\gamma q}\,. \tag{6.55}$$

A particular solution of this equation is

$$\phi = \phi_1\, t^{2-3\gamma q}\,, \tag{6.56}$$

with

$$\phi_1 = \frac{8\pi C\,(4-3\gamma)}{(2\omega+3)\,a_0{}^{3\gamma}\,(2-3\gamma q)\,[1+3q\,(1-\gamma)]}, \tag{6.57}$$

and the general solution of Eq. (6.55) is therefore

$$\phi = \phi_0\, t^s + \phi_1\, t^{2-3\gamma q}\,, \tag{6.58}$$

with $s = 0$ or $3q + s = 1$.

#### 6.3.1.1 Spatially flat FLRW solutions of Brans-Dicke theory

Spatially flat solutions are, of course, more easily found than spatially curved ones. Some classic solutions are presented in the following, as well as other representatives of this class.

The **O'Hanlon and Tupper solution** [879] corresponds to vacuum, $V(\phi) \equiv 0$, and to the parameter range $\omega > -3/2$ with $\omega \neq -4/3, 0$. It is given by

$$a(t) = a_0 \left(\frac{t}{t_0}\right)^{q_\pm}, \tag{6.59}$$

$$\phi(t) = \phi_0 \left(\frac{t}{t_0}\right)^{s_\pm}, \tag{6.60}$$

with

$$q_\pm = \frac{\omega}{3\,(\omega+1) \mp \sqrt{3\,(2\omega+3)}} = \frac{1}{3\omega+4}\left(\omega + 1 \pm \sqrt{\frac{2\omega+3}{3}}\right), \tag{6.61}$$

$$s_\pm = \frac{1 \mp \sqrt{3\,(2\omega+3)}}{3\omega+4}, \tag{6.62}$$

and satisfies $3q + s = 1$. Various derivations of this solution, which exhibits a Big Bang singularity as $t \to 0$, have been given [690, 798, 798, 879]. Its limit as $\omega \to +\infty$ is

$$a(t) \propto t^{1/3}\,, \qquad \phi = \text{const.} \tag{6.63}$$

and it does not reproduce the corresponding GR solution, which is Minkowski space.

The solutions corresponding to the exponents $(q_-, s_-)$ and $(q_+, s_+)$ are called the *slow* and *fast* solution, respectively. The terminology originates in the behavior of $\phi(t)$ as $t \to 0$. The slow solution has increasing $\phi(t)$ and decreasing coupling $G_{\mathit{eff}} \simeq 1/\phi$ (for $\omega > -4/3$), while the fast solution has decreasing $\phi(t)$ and increasing $G_{eff} \simeq 1/\phi$ at early times (for $\omega > -4/3$). The fast and slow solutions are interchanged under the duality transformation (6.49) and (6.50), which yields

$$(q_{\pm}, s_{\pm}) \longrightarrow (q_{\mp}, s_{\mp}) \; . \tag{6.64}$$

The common remark that the O'Hanlon-Tupper solution goes over to the de Sitter space

$$(a(t), \phi(t)) = (a_0 \exp(H\, t)\, , \phi_0 \exp(-3H\, t))\; , \tag{6.65}$$

with $H$ = const. in the limit $\omega \to -4/3$ is not strictly speaking, true. This de Sitter space is only obtained by choosing simultaneously the values $q_+$ and $s_-$ of the exponents. However, Eq. (6.65) does describe the only de Sitter solution of the Brans-Dicke equations (6.45)–(6.47) for $K = 0$ and vacuum. This solution differs from the usual de Sitter space with constant scalar field solving the Einstein-Klein-Gordon system. Examples of de Sitter solutions with a non-constant scalar field are known in scalar-tensor theories [61, 62, 561].

The **Brans-Dicke dust solution** [165] describes the spatially flat FLRW universe containing dust with $V(\phi) = 0$ and $\omega \neq -4/3$ given by

$$(a(t), \phi(t)) = (a_0\, t^q, \phi_0\, t^s) \tag{6.66}$$

with

$$q = \frac{2(\omega+1)}{3\omega+4}\, , \qquad s = \frac{2}{3\omega+4}\, , \tag{6.67}$$

and it satisfies $3q + s = 2$, while

$$\rho^{(m)} = \frac{C}{a^{3\gamma}} = \rho_0\, t^r \tag{6.68}$$

with

$$r = -3q = \frac{-6(\omega+1)}{3\omega+4}\, , \qquad \rho_0 = \frac{C}{a_0^3}\, . \tag{6.69}$$

The **Nariai solution** [563, 827] is a particular power-law solution for a spatially flat FLRW universe with $V(\phi) \equiv 0$, $\omega \neq -4\left[3\gamma(2-\gamma)\right]^{-1} < 0$, and a perfect fluid. It is given by

$$a(t) = a_0 (1 + \delta\, t)^q\; , \tag{6.70}$$

$$\phi(t) = \phi_0 (1 + \delta\, t)^s\; , \tag{6.71}$$

where

$$q = \frac{2\,[\omega\,(2-\gamma)+1]}{3\omega\gamma\,(2-\gamma)+4}\,, \tag{6.72}$$

$$s = \frac{2\,(4-3\gamma)}{3\omega\gamma\,(2-\gamma)+4}\,, \tag{6.73}$$

and it is $s + 3\gamma q = 2$. The energy density of the fluid redshifts according to

$$\rho^{(m)}(t) = \frac{C}{a^{3\gamma}} = \rho_0\,(1+\delta\,t)^r\,, \quad r = -3\gamma q\,. \tag{6.74}$$

The exponents $q$, $s$, and $r$ are often rewritten as functions of the parameters

$$\alpha \equiv \frac{2\,(4-3\gamma)}{(2\omega+3)\,(2-\gamma)+3\gamma-4}\,, \quad A \equiv \frac{2\omega+3}{12}\,, \tag{6.75}$$

as [690]

$$q = \frac{2}{\alpha+3\gamma}\,, \quad s = \frac{2\alpha}{\alpha+3q}\,, \quad r = \frac{-6\gamma}{\alpha+3\gamma}\,, \tag{6.76}$$

and

$$\delta = \left(\frac{\alpha+3\gamma}{2}\right)\left\{\frac{8\pi\rho_0}{3\phi_0\left[(1+\alpha/2)^2 - A\alpha^2\right]}\right\}. \tag{6.77}$$

Special cases of the Nariai solution include the Brans-Dicke dust solution [165] for dust ($P^{(m)} = 0, \alpha = (\omega+1)^{-1}, \omega \neq -4/3$)

$$a(t) = a_0\,(1+\delta\,t)^{\frac{2(\omega+1)}{3\omega+4}}\,, \tag{6.78}$$

$$\phi(t) = \phi_0\,(1+\delta\,t)^{\frac{2}{3\omega+4}}\,, \tag{6.79}$$

$$\rho^{(m)}(t) = \rho_0\,(1+\delta\,t)^{\frac{-6(\omega+1)}{3\omega+4}}\,, \tag{6.80}$$

$$\delta = \left(\frac{4\pi\rho_0}{\phi_0}\,\frac{3\omega+4}{2\omega+3}\right)^{1/2}\,, \tag{6.81}$$

with $3q + s = 2$.

If $\omega = -1$ the radiation fluid and the scalar field balance each other to produce a Minkowski space with a non-trivial scalar field growing quadratically with time. The solution is instead expanding if $\omega < -4/3$ or $\omega > -1$, and pole-like if $-4/3 < \omega < -1$. In this case, there are two disconnected branches reminiscent of pre-Big Bang cosmology in string theory [738]. One of these branches expands for $t < -t_0$ and the other contracts for $t > -t_0$.

Another special case contained in the Nariai solution corresponds to a radiation fluid ($P^{(m)} = \rho^{(m)}/3$), and is given by

$$a(t) = a_0 \left(1 + \delta\, t\right)^{1/2} , \tag{6.82}$$

$$\phi(t) = \phi_0 = \text{const.} , \tag{6.83}$$

$$\rho^{(m)}(t) = \frac{C}{a^4} = \frac{\rho_0}{(1 + \delta\, t)^2} , \tag{6.84}$$

$$\delta = \left(\frac{32\pi\rho_0}{3\phi_0}\right)^{1/2} . \tag{6.85}$$

This solution does not depend on the Brans-Dicke parameter $\omega$.

Another special case of the Nariai solution is that of a FLRW universe with cosmological constant [701, 779] corresponding to $P^{(\Lambda)} = -\rho^{(\Lambda)}$, $\alpha = 4\,(2\omega + 1)^{-1}$, which results in

$$a(t) = a_0 \left(1 + \delta\, t\right)^{\omega + \frac{1}{2}} , \tag{6.86}$$

$$\phi(t) = \phi_0 \left(1 + \delta\, t\right)^2 , \tag{6.87}$$

$$\delta = \left[\frac{32\pi\rho_0}{\phi_0} \frac{1}{(6\omega + 5)\,(2\omega + 3)}\right]^{1/2} . \tag{6.88}$$

The extended inflationary scenario is based on this particular solution, which is not the only solution describing a FLRW universe dominated by a cosmological constant but is a phase space attractor. Other solutions with cosmological constant were found in [944].

**Generalized Nariai solutions** describing spatially flat FLRW universes were found in [563, 810] using the time coordinate $\tau$ defined by

$$d\tau = \frac{dt}{a^{3(\gamma-1)}} \tag{6.89}$$

(which coincides with $t$ for a dust fluid):

$$a\,(\tau) = a_0\,(\tau - \tau_-)^{q_\mp}\,(\tau - \tau_+)^{q_\pm} , \tag{6.90}$$

$$\phi\,(\tau) = \phi_0\,(\tau - \tau_-)^{s_\mp}\,(\tau - \tau_+)^{s_\pm} , \tag{6.91}$$

where

$$q_\pm = \frac{\omega}{3\left[1 + \omega\,(2 - \gamma) \mp \sqrt{\frac{2\omega+3}{3}}\right]} , \tag{6.92}$$

$$s_\pm = \frac{1 \pm \sqrt{\frac{2\omega+3}{3}}}{1 + \omega\,(2 - \gamma) \mp \sqrt{\frac{2\omega+3}{3}}} , \tag{6.93}$$

with $a_0$, $\phi_0$ and $\tau_\pm$ constants and $\omega > -3/2$. The Nariai solution is obtained when $\tau_+ = \tau_-$. If $\tau_+ \neq \tau_-$, the solution (6.90)–(6.93) approaches the O'Hanlon-Tupper vacuum solution (6.59)–(6.62) as $\tau \to \tau_\pm$.

To conclude this subject, we discuss the phase space analysis of spatially flat FLRW universes in Brans-Dicke cosmology with $V(\phi) = 0$. Dynamical system methods provide a qualitative overall view of the cosmological dynamics [333, 1040, 1136]. Allowing a cosmological constant described as an effective perfect fluid, the dynamics described are analyzed in [594, 689, 690, 944, 970]. The Brans-Dicke cosmological equations form a system of two coupled autonomous first order equations for the (rescaled) scale factor and the Brans-Dicke scalar field [594, 689, 970]. The phase space is two-dimensional, eliminating the possibility of chaos [469, 561]. The parameters $\omega$ and $\gamma$ vary in the range $(\omega, \gamma) \in (-3/2, +\infty) \times [0, 2]$. The possible behavior of the scale factor $a(t)$ is quite varied and includes bouncing models, *i.e.*, universes that contract to a minimum size and then re-expand superaccelerating with $\dot{H} > 0$; these bouncing models occur for $\omega < 0$.

A phase space analysis for spatially flat FLRW models with a perfect fluid and a linear potential can be found in [689] and the case with no fluid is studied in [945]. A linear potential $V = \Lambda\phi$ is obtained by adding a cosmological constant $\Lambda$ to the Ricci curvature multiplying the Brans-Dicke field $\phi$ in the Brans-Dicke action, which is different from the simple addition of a constant $\Lambda$ to the overall Lagrangian density as in GR [443].

For $\omega > 0$ all universes which are initially expanding approach de Sitter space at late times, irrespective of the value of the EoS parameter $w$, but this is not the case if $\omega < 0$; then, there are bouncing universes and *vacillating universes, i.e.,* solutions that expand, slow down, contract for a short time, and re-expand again.

Two de Sitter equilibrium points are always present irrespective of the value of the EoS parameter, corresponding to [100, 689, 945]

$$a^{(\pm)}(t) = a_0 \exp\left\{ \pm (\omega + 1) \left[ \frac{2\Lambda}{(2\omega + 3)(3\omega + 4)} \right]^{1/2} t \right\}, \tag{6.94}$$

$$\phi^{(\pm)}(t) = \phi_0 \exp\left\{ \pm \left[ \frac{2\Lambda}{(2\omega + 3)(3\omega + 4)} \right]^{1/2} t \right\}. \tag{6.95}$$

In general the scalar field is not constant for these de Sitter spaces, but the general relativistic solution with constant scalar is recovered in the $\omega \to \infty$ limit.

The two equilibrium points (6.94) and (6.95) corresponding to the upper or lower sign respectively, are phase space attractors. Their attraction basin is limited to most, but not all, of the solutions with expanding initial data. They reduce to Minkowski spaces with exponentially expanding or contracting Brans-Dicke scalar field if $\omega = -1$ (the parameter value given by the low-energy limit of bosonic string theory). These Minkowski spaces are stable and have analogues in string cosmology [50, 821]). There are additional equilibrium points of various nature of the dynamical system for the special values 0, 1, and 4/3 of the EoS parameter $\gamma = w + 1$ of the perfect fluid.

If $\omega > 0$, all the initially contracting solutions end in a Big Crunch singularity. If $-1 < \omega < 0$, there are non-singular bouncing universes, some of which end up near the de Sitter attractors for suitable initial conditions [336, 435, 563, 689].

A phase plane analysis of the dynamics with potential $V(\phi) = V_0\,\phi^{2n}$ can be found in [595].

#### 6.3.1.2 Spatially curved FLRW solutions with $V = 0$ and Bianchi models

Spatially closed ($K = +1$) or open ($K = -1$) FLRW universes in Brans-Dicke gravity are found in [94, 718, 750, 798, 809, 1145] and phase portraits are calculated in [594, 690, 970]. In terms of the time coordinate $\tau$ defined by

$$d\tau = \sqrt{\frac{8\pi\rho^{(m)}}{(2\omega+3)\,\phi}}\,dt \tag{6.96}$$

for $\omega > -3/2$ and of the dynamical variables

$$x \equiv \frac{1}{2\phi}\frac{d\phi}{d\tau}, \qquad y \equiv \frac{1}{a}\frac{da}{d\tau}, \tag{6.97}$$

the Brans-Dicke field equations for a FLRW model with arbitrary value of $K$ are [594]

$$x' = -x^2 - \frac{3\,(2-\gamma)}{2}\,xy + \left(2 - \frac{3\gamma}{2}\right), \tag{6.98}$$

$$y' = \frac{-2\,(1-3\alpha^2)}{3\alpha^2}\,x^2 + 3xy + \frac{3\gamma-2}{2}\,y^2 - \frac{3\,(4-3\gamma)\,\alpha^2 + 3\gamma - 2}{6\alpha^2}, \tag{6.99}$$

where $\alpha \equiv (2\omega+3)^{-1/2}$. The field equations form an autonomous system of two coupled first order equations and, using these variables, the phase space is two-dimensional. This situation is to be compared with that of GR with a single scalar field, in which the phase space is three-dimensional when $K \neq 0$ [38, 501, 561], making chaos possible [146, 155, 208, 209, 344, 1079]. In Brans-Dicke FLRW cosmology with the variables $(x, y)$ the phase space may be three-dimensional when a potential $V(\phi)$ is present [970]. The $K = 0$ FLRW universes lie on the separatrix described by the equation

$$x + y = \pm\sqrt{\frac{1+x^2}{3\alpha^2}}. \tag{6.100}$$

On either side of this separatrix there are $K = +1$ and $K = -1$ FLRW spaces and the orbits of the system cannot cross the separatrix.[3]

---

[3] This separation of the phase space in three regions was discovered for a particular scenario of inflation in GR in [122].

The phase plane is symmetrical under the time inversion $\tau \to -\tau$ together with the reflection about the origin $(x, y) \to (-x, -y)$. The equilibrium points $(x', y') = (0, 0)$ of the dynamical system (6.98) and (6.99) are

$$\left(x_{\pm}^{(1)}, y_{\pm}^{(1)}\right) = \frac{\pm 1}{\sqrt{3\,(2-\gamma)^2 - \alpha^2\,(4-3\gamma)^2}} \cdot \left[\alpha\,(4-3\gamma)\,,\, 2-\gamma-\alpha^2\,(4-3\gamma)\right], \tag{6.101}$$

$$\left(x_{\pm}^{(2)}, y_{\pm}^{(2)}\right) = \pm\left(\frac{\sqrt{2-3\gamma}}{2}\,,\, \frac{1}{\sqrt{2-3\gamma}}\right), \tag{6.102}$$

where the solutions denoted by a '+' or a '−' are expanding or contracting, respectively. Because of this duality we consider only the '+' solutions. The critical points $\left(x_{\pm}^{(1)}, y_{\pm}^{(1)}\right)$ exist only for $K = 0$ and lie on the separatrix. When mapped back to the original variables, the equilibrium points correspond to the power-law solutions

$$(a(t), \phi(t)) = \left(a_0\, t^q, \phi_0\, t^s\right)\,, \tag{6.103}$$

where

$$q^{(i)} = \frac{2y^{(i)}}{2x^{(i)} + 3\,\gamma\, y^{(i)}}\,, \tag{6.104}$$

$$s = \frac{4x^{(i)}}{2x^{(i)} + 3\,\gamma\, y^{(i)}}\,. \tag{6.105}$$

The critical point $\left(x_{+}^{(1)}, y_{+}^{(1)}\right)$ is the Nariai solution; it exists if $\alpha < \frac{\sqrt{3}(2-\gamma)}{|4-3\gamma|}$. In the limiting case $\omega = -4/\left[3\gamma\,(2-\gamma)\right]$, the equilibrium point becomes a de Sitter space. If $\gamma = 0$, the equilibrium point reduces to the solution (6.86)–(6.88) of extended inflation.

Mapping back the $(x, y)$ variables to the original ones, the equilibrium point $\left(x_{+}^{(2)}, y_{+}^{(2)}\right)$ corresponds to the coasting universe $(a(t), \phi(t)) = \left(a_0\, t, \phi_0\, t^{2-3/\gamma}\right)$ and it exists if $\gamma < 2/3$ irrespective of $\omega$. In GR, instead, this solution only appears for $\gamma = 2/3$. If

$$\omega > \omega_c \equiv \frac{2}{(2-\gamma)\,(2-3\gamma)}\,, \tag{6.106}$$

this equilibrium point is located in the $K > 0$ region; if $\omega < \omega_c$ it appears in the $K < 0$ region, while if $\omega = \omega_c$, it lies on the $K = 0$ separatrix.

The stability analyses of [594, 690, 970] show that:

- The equilibrium point $\left(x_{+}^{(1)}, y_{+}^{(1)}\right)$ (the expanding Nariai solution) is a late-time attractor if $\omega > \omega_c$ and a saddle point if $\omega < \omega_c$. It corresponds to power-law inflation and it is an attractor in $\omega > \omega_c$ spatially open models.

- The other equilibrium point $\left(x_+^{(2)}, y_+^{(2)}\right)$ is an attractor if $\omega < \omega_c$ and a saddle point if $\omega > \omega_c$.
- The two fixed points coincide and lie on the $K = 0$ separatrix if $\omega = \omega_c$. In this case, the equilibrium point is an attractor if $K \leq 0$ and a saddle point if $K > 0$.

The fixed points at infinity can be found by using a Poincaré projection compactifying the phase plane and are located by the values of the polar angle

$$\theta^{(3)} = \tan^{-1}\left(-1 + \frac{1}{\sqrt{3\alpha}}\right), \tag{6.107}$$

$$\theta^{(4)} = \tan^{-1}\left(-1 - \frac{1}{\sqrt{3\alpha}}\right), \tag{6.108}$$

$$\theta^{(5)} = \frac{\pi}{2}. \tag{6.109}$$

In addition, there are equilibrium points given by the symmetric angles obtained by time reversal. $\theta^{(3)}$ and $\theta^{(4)}$ lie on the $K = 0$ separatrix. Using the original variables, these equilibrium points describe the spatially flat O'Hanlon-Tupper solutions [879] $a \propto t^q$, $\phi \propto \phi^s$ with

$$q = \frac{\sin\theta^{(i)}}{F\left(\theta^{(i)}\right)}, \qquad s = \frac{2\cos\theta^{(i)}}{F\left(\theta^{(i)}\right)}, \tag{6.110}$$

$$F(\theta) = 2\cos\theta + \left[1 - 2\sin(2\theta) + \frac{2}{3\alpha^2}\cos^2\theta\right]\sin\theta. \tag{6.111}$$

Here $\theta^{(4)}$ is the *fast* (expanding) solution and $\theta^{(3)}$ is the *slow* solution, which expands if $\omega > 0$ and contracts if $\omega < 0$. $\theta^{(5)}$ describes the Milne universe of GR $(a, \phi) = (a_0\, t, \text{const.})$.

Stability analyses of the stationary points at infinity [594, 690, 970] show that:

- The fixed point $\theta^{(5)}$ (expanding Milne universe) is a late time attractor if $\gamma > 2/3$.
- Whenever the critical point $\left(x^{(1)}, y^{(1)}\right)$ lies at a finite distance from the origin, the stationary points at infinity $\theta^{(3)}$ and $\theta^{(4)}$ are nodes.
- When the critical point $\left(x^{(1)}, y^{(1)}\right)$ disappears, the nature of the critical points at infinity depends on the value of $\gamma$. If $\gamma < 4/3$ then $\theta^{(3)}$ is a saddle point, and it is also an attractor for $K = 0$ solutions lying on the separatrix, while $\theta^{(4)}$ is a node. This behavior is reversed if $\gamma > 4/3$.

If $K = +1$ certain solutions end in a Big Crunch and there are also *hesitating universes* emerging from a Big Bang, slowing down, and re-expanding again. There are coasting universes, and also vacillating universes emerging from a Big Bang, contracting, and then re-expanding. Other solutions begin contracting, stop, expand, and then reach an accelerated regime.

If $K = -1$ there are bouncing universes and solutions which emerge from a Big Bang and approach the $\theta^{(5)}$ coasting universe attractor at late times. For both $K = \pm 1$, there are solutions dominated by a cosmological constant but which do not approach a $K = 0$ de Sitter solution at late times [970]. They show that the cosmic no-hair theorems of GR stating that the inflationary de Sitter space is a late time attractor for the cosmic dynamics, and that cosmological inflation generically leads to a flat universe, do not hold in general in Brans-Dicke (or scalar-tensor) gravity. This property is related to the fact that the spatially flat solution of Brans-Dicke cosmology corresponding to pure cosmological constant is not de Sitter space but a power-law solution instead.

#### 6.3.1.3 Phase space for $V = m^2\phi^2/2$ and any three-geometry

When $V(\phi) = m^2\phi^2/2$, $K$ is arbitrary and with a barotropic fluid, the use of conformal time $\eta$ and of the variables

$$X \equiv \sqrt{\frac{2\omega+3}{12}}\frac{\phi'}{\phi}, \tag{6.112}$$

$$Y \equiv \frac{a'}{a} + \frac{\phi'}{2\phi}, \tag{6.113}$$

$$Z \equiv \frac{m^2}{2}a^2\phi \tag{6.114}$$

(where $' \equiv d/d\eta$), reduces the dynamical system to the three first order coupled equations [970]

$$X' = -2XY + \frac{4-3\gamma}{4A}\left(Y^2 - X^2 - \frac{Z}{6} + K\right), \tag{6.115}$$

$$Y' = \frac{2-3\gamma}{2}\left(Y^2 - X^2 + K\right) - 2X^2 + \frac{\gamma}{4}Z, \tag{6.116}$$

$$Z' = 2ZY, \tag{6.117}$$

where

$$A \equiv \frac{1}{2}\sqrt{\frac{2\omega+3}{3}} \tag{6.118}$$

and $\omega > -3/2$.

In general, the phase space is three-dimensional, contrary to the cases $V = 0$ [690] and $V = \lambda\phi$ [689, 945]. However, *in vacuo*, the variable $Z$ can be eliminated and the phase space becomes two-dimensional. If there is only radiation ($P^{(m)} = \rho^{(m)}/3$) the variable $X$ can be eliminated and the phase space becomes again two-dimensional [970].

Whether equilibrium points exist or not depends on the three-geometry and on the value of $\gamma$. For vacuum and for radiation there exist power-law contracting and expanding attractors and also expanding de Sitter attractors [970].

The **Dehnen-Obregon solution** [369] is a $K = +1$ coasting universe filled with dust, with zero scalar field potential and $\omega < -2$ given by

$$a(t) = \sqrt{\frac{-2}{2+\omega}}\, t \,, \tag{6.119}$$

$$\phi(t) = \frac{-8\pi}{2\omega+3}\, \rho(t)\, t^2 = \frac{\phi_0}{t} \,, \tag{6.120}$$

with $2\pi^2 a^3(t)\, \rho(t) = M$ and

$$\phi_0 = \frac{-\sqrt{2}\, M\, |\omega+2|^{3/2}}{\pi\, (2\omega+3)} \,, \tag{6.121}$$

where the constant $M$ is the mass of this closed universe and

$$\rho(t) = \frac{M\, |\omega+2|^{3/2}}{4\sqrt{2}\,\pi^2}\, \frac{1}{t^3} \,. \tag{6.122}$$

Other FLRW solutions have been derived using conformal time and a different choice of variables than $(a, \phi)$ [94, 533, 751, 798, 874, 1084, 1145].

Homogeneous and anisotropic **Bianchi universes** have been found, with special attention devoted to the initial singularity [828, 829], usually with variables different from $(a, \phi)$, which complicates their interpretation. The isotropization of Bianchi I, V, and IX models in Brans-Dicke cosmology (without potential $V$) and a perfect fluid is studied in [291, 828, 829]. Bianchi V solutions with metric

$$ds^2 = -dt^2 + a_1^2\, dx^2 + a_2^2\, e^{-2x}\, dy^2 + a_3^2\, e^{-2x}\, dz^2 \tag{6.123}$$

in synchronous coordinates are found in [289].

## 6.3.2 *Exact scalar-tensor cosmologies*

Many cosmological solutions of scalar-tensor theories with varying coupling $\omega(\phi)$ have been found [11, 15, 16, 94, 100, 101, 104, 118, 136, 230, 232, 370, 479, 480, 512, 533, 658, 659, 798, 874, 906, 1080, 1084, 1145] with specialized techniques originally developed in [94, 100] for vacuum- or radiation-dominated FLRW universes. Spatially flat FLRW universes with a perfect fluid obeying $0 \leq \gamma \leq 4/3$ in general scalar-tensor theories are studied in [101]. A solution-generating technique for FLRW universes with arbitrary three-geometry, arbitrary coupling $\omega(\phi)$, and vacuum or radiation, can be found in [104]. A solution-generating method valid for any value of the perfect fluid EoS and spatially flat FLRW models is also developed there.

Bianchi models in general scalar-tensor gravity are discussed in [798, 1145]. The isotropization of Bianchi cosmologies in scalar-tensor theories without scalar field potential and with a perfect fluid is studied in [798]. Vacuum with a scalar field potential satisfying the special condition[4]

$$\frac{\phi}{V}\frac{dV}{d\phi} = 2 \pm \alpha\sqrt{\frac{3}{2} + \omega(\phi)}\,, \tag{6.124}$$

where $\alpha$ is constant, is studied in [139, 333, 334]. With the exception of certain Bianchi type IX models, all Bianchi universes isotropize if $0 < \alpha^2 \leq 2$ and they inflate if $\alpha^2 < 2$. Bianchi VIIb models with $\alpha > 0$ isotropize but do not inflate [334].

Point-like Noether symmetries leading to first integrals of motion as conserved charges in anisotropic scalar-tensor cosmologies are studied in [735]. In the presence of a cosmological constant, only Brans-Dicke gravity admits Noether symmetries. These symmetries are still present in the $\Lambda \to 0$ limit, but other scalar-tensor theories admit Noether symmetries for $\Lambda = 0$ [735].

## 6.4 Analytical solutions of metric $f(R)$ cosmology by the Noether approach

We will now focus on the application of the Noether symmetry approach to the search for exact solutions in FLRW cosmology, concentrating on metric $f(R)$ gravity [226]. Since metric $f(R)$ gravity is equivalent to an $\omega = 0$ scalar-tensor theory with a special potential determined by the form of the function $f(R)$, this discussion continues the study of scalar-tensor cosmology begun in the previous sections. It provides insight on the nature of a particular correction to the Hilbert-Einstein action, which can however be regarded as the prototype of higher order additions motivated by renormalization and low-energy string theory. The Noether symmetry approach has been applied to scalar-tensor gravity in [234].

### *6.4.1 Point-like $f(R)$ cosmology*

The equations of FLRW cosmology in $f(R)$ gravity can be derived from a canonical point-like Lagrangian $L\left(a, \dot{a}, R, \dot{R}\right)$ where $Q = (a, R)$ is the configuration space and $TQ = \left(a, \dot{a}, R, \dot{R}\right)$ is the corresponding tangent bundle on which $L$ is

[4] This special choice of the potential makes the corresponding Einstein frame potential an exponential.

defined. As done for spherical symmetry, Lagrange multipliers can be used to turn the expression of $R$ in terms of the scale factor and its derivatives into a constraint on the dynamics. By selecting a suitable Lagrange multiplier and integrating by parts, the Lagrangian $L$ becomes canonical. Specifically, the action is

$$S = 2\pi^2 \int dt\, a^3 \left\{ f(R) - \lambda \left[ R - 6\left( \frac{\ddot{a}}{a} + \frac{\dot{a}^2}{a^2} + \frac{K}{a^2} \right) \right] - \frac{\rho_{m0}}{a^3} - \frac{\rho_{r0}}{a^4} \right\}, \tag{6.125}$$

where $a$ is the scale factor normalized to unity today (*i.e.*, $a_0 \equiv a(t_0) = 1$) and $\rho_{m0}$ and $\rho_{r0}$ are the current values of the energy densities of dust and radiation, respectively.[5]

The variation of the action term containing the Lagrange multiplier with respect to $R$ gives $\lambda = f_R$ and Eq. (6.125) becomes

$$S = 2\pi^2 \int dt\, a^3 \left\{ f - f_R \left[ R - 6\left( \frac{\ddot{a}}{a} + \frac{\dot{a}^2}{a^2} + \frac{K}{a^2} \right) \right] - \frac{\rho_{m0}}{a^3} - \frac{\rho_{r0}}{a^4} \right\}. \tag{6.126}$$

Integrating by parts, the point-like FLRW Lagrangian is written as

$$L = a^3 (f - f_R\, R) - 6a^2\, f_{RR}\, \dot{R}\, \dot{a} + 6\, f_R\, a\, \dot{a}^2 + 6K\, f_R\, a - \rho_{m0} - \frac{\rho_{r0}}{a}, \tag{6.127}$$

a canonical function of the (time-dependent) Lagrangian coordinates $(a, R)$. The total energy $E_L$ and the time-time component of the field equations obey

$$E_L = 6\, f_{RR}\, a^2\, \dot{a}\, \dot{R} + 6\, f_R\, a\, \dot{a}^2 - a^3 (f - f_R\, R) - 6K\, f_R\, a + \rho_{m0} + \frac{\rho_{r0}}{a} = 0. \tag{6.128}$$

In the following it will be convenient to search for solutions in the parametric form $[H(a), f(R(a))]$ so that $f_R = f'/R'$, where a prime now denotes differentiation with respect to the parameter $a$. Moreover, if $R$ is not constant, $f_{RR}\, \dot{R} = df_R/dt = a\, H\, f_R' = a\, H \left[ f''/R' - f'\, R''/R'^2 \right]$ and the Friedmann equation can be rewritten as

$$f - 6a \left( \frac{f''}{R'} - \frac{f'\, R''}{R'^2} \right) H^2 - \frac{6f'\, H^2}{R'} - \left( \frac{6K}{a^2} + R \right) \frac{f'}{R'} = \frac{\rho_{0m}}{a^3} + \frac{\rho_{0r}}{a^4}. \tag{6.129}$$

The equations of motion for $a$ and $R$ are

$$f_{RR} \left[ R - 6\left( H^2 + \frac{\ddot{a}}{a} + \frac{K}{a^2} \right) \right] = 0, \tag{6.130}$$

[5] With this choice, $a$ is dimensionless while the curvature index $K$ carries the dimensions of an inverse length squared and $f(R)$ those of a mass to the fourth power.

$$6\, f_{RRR}\, \dot{R}^2 + 6\, f_{RR}\, \ddot{R} + 6\, f_R\, H^2 + 12\, f_R\, \frac{\ddot{a}}{a}$$
$$= 3\,(f - f_R\, R) - 12\, f_{RR}\, H\, \dot{R} - 6\, f_R\, \frac{K}{a^2} + \frac{\rho_{r0}}{a^4}\,. \tag{6.131}$$

Considering $R$ and $a$ as independent Lagrangian coordinates, for consistency $R$ must coincide with $6\left(\frac{\ddot{a}}{a} + \frac{\dot{a}^2}{a^2} + \frac{K}{a^2}\right)$ (unless $f_{RR} = 0$), which is the Euler constraint on the dynamics. As shown below, exact solutions of the system (6.128)–(6.131) and the corresponding form of the function $f(R)$ can be obtained by imposing the existence of Noether symmetries. The existence of these symmetries guarantees the reduction of the dynamics and the integrability of the system.

### 6.4.2 *Noether symmetries in metric $f(R)$ cosmology*

The condition $\mathscr{L}_{\mathbf{X}} L = 0$ for the existence of a symmetry gives rise to the system of linear partial differential equations for $\alpha$ and $\beta$

$$f_R\,(\alpha + 2a\,\partial_a \alpha) + a\, f_{RR}\,(\beta + a\,\partial_a \beta) = 0\,, \tag{6.132}$$

$$a^2\, f_{RR}\,\partial_R \alpha = 0\,, \tag{6.133}$$

$$2\, f_R\,\partial_R \alpha + f_{RR}\,(2\alpha + a\,\partial_a \alpha + a\,\partial_R \beta) + a\,\beta\, f_{RRR} = 0\,, \tag{6.134}$$

$$3\alpha\,(f - R\, f_R) - a\,\beta\, R\, f_{RR} - \frac{6K}{a^2}\,(\alpha f_R + a\,\beta\, f_{RR}) + \frac{\rho_{r0}\,\alpha}{a^4} = 0\,, \tag{6.135}$$

obtained by equating to zero the coefficients of the quadratic terms containing $\dot{a}^2$, $\dot{R}^2$, and $\dot{a}\dot{R}$. The last equation is a relation between $a$ and $R$. A solution is found by determining explicit forms of $\alpha$ and $\beta$. If at least one of these is different from zero, then a Noether symmetry exists, which fixes the form of the function $f(R)$.

If $f_{RR} \neq 0$, Eq. (6.133) can be immediately integrated and $\alpha = \alpha(a)$ depends only on $a$ (GR corresponds to the trivial case $f_{RR} = 0$). One can write Eqs. (6.132) and (6.134) as

$$f_R\left(\alpha + 2a\,\frac{d\alpha}{da}\right) + a\, f_{RR}\,(\beta + a\,\partial_a \beta) = 0\,, \tag{6.136}$$

$$f_{RR}\left(2\alpha + a\,\frac{d\alpha}{da} + a\,\partial_R \beta\right) + a\,\beta\, f_{RRR} = 0. \tag{6.137}$$

Using separation of variables and searching for solutions of the form

$$\beta = A(a)\, B(R)\,, \tag{6.138}$$

it must be

$$f_R\left(\alpha+2a\,\frac{d\alpha}{da}\right)+a\,B\,f_{RR}\,\frac{d}{da}\,(a\,A)=0\,, \tag{6.139}$$

$$f_{RR}\left(2\,\alpha+a\,\frac{d\alpha}{da}\right)+a\,A\,\frac{d}{dR}\,(f_{RR}\,B)=0\,. \tag{6.140}$$

Equation (6.140) implies

$$\frac{1}{f_{RR}}\,\frac{d}{dR}\,(f_{RR}B)=\nu_0=\text{const.}, \tag{6.141}$$

with solution

$$B=\frac{\nu_1+\nu_0\,f_R}{f_{RR}}\,, \tag{6.142}$$

where $\nu_1$ is another constant, and we also have

$$2\alpha+a\,\frac{d\alpha}{da}=-\nu_0\,a\,A\,. \tag{6.143}$$

Since $f_{RR}\neq 0$, Eq. (6.139) implies that

$$\nu_1=0 \qquad \text{or} \qquad B=\nu_0\,\frac{f_R}{f_{RR}} \tag{6.144}$$

and

$$\alpha+2\,a\,\frac{d\alpha}{da}+a\,\frac{d}{da}\,(\nu_0\,a\,A)=0 \tag{6.145}$$

which becomes, using Eq. (6.143),

$$\alpha-a^2\,\frac{d^2\alpha}{da^2}-a\,\frac{d\alpha}{da}=0\,, \tag{6.146}$$

with general solution

$$\alpha=c_1\,a+\frac{c_2}{a}\,. \tag{6.147}$$

Since the scale factor is dimensionless, $c_1$ and $c_2$ have the same dimensions. We can further fix $\alpha$ to be dimensionless, which determines the dimensions of $\beta$ to be that of a mass squared. Finally, one obtains

$$\beta=-\left(3\,c_1+\frac{c_2}{a^2}\right)\,\frac{f_R}{f_{RR}}\,. \tag{6.148}$$

Using the solutions for $\alpha$ and $\beta$, Eq. (6.135) is rewritten as

$$\left(c_1\, a^2 + c_2\right)\left(3\, a^4\, f + \rho_{r0}\right) + 2\, a^4\, f_R\, (6\, K\, c_1 - c_2\, R) = 0\,, \tag{6.149}$$

which is equivalent to

$$f_R = \frac{\left(c_1\, a^2 + c_2\right)\left(\rho_{r0} + 3\, a^4\, f\right)}{2\, a^4\, (c_2\, R - 6K\, c_1)} \tag{6.150}$$

if $c_2\, R - 6K\, c_1 \neq 0$. It is clear that, once a form of $f$ is chosen, this equation provides $R$ as a function of $a$. Since, on shell, $R$ is a function of $H$ and its derivatives, this Noether condition constitutes a dynamical constraint. It is convenient to look for solutions in the parametric form $\left[H(a),\, f\left(R(a)\right)\right]$. In this case, since $f_R = f'/R'$, the Noether condition corresponds to the ordinary differential equation

$$\frac{f'(a)}{R'(a)} = \frac{\left(c_1 a^2 + c_2\right)\left[3 f(a) a^4 + \rho_{0r}\right]}{2a^4\left[c_2 R(a) - 6 c_1 K\right]}\,. \tag{6.151}$$

A Noether solution $\left[H(a),\, f(R(a))\right]$ must solve simultaneously the Friedmann equation (6.129) and the Noether constraint (6.151). Equation (6.150) can also be rewritten as

$$c_1\, a^2\left(\rho_{r0} + 3\, a^4\, f + 12K\, a^2\, f_R\right) + c_2\left[\rho_{r0} + a^4\,(3\, f - 2\, R\, f_R)\right] = 0. \tag{6.152}$$

There is a family of solutions that provides a class of $f(R)$ models with Noether symmetry. The symmetry implies the existence of the constant of motion

$$\alpha\left(6\, f_{RR}\, a^2\, \dot{R} + 12\, f_R\, a\, \dot{a}\right) + 6\beta\, f_{RR}\, a^2\, \dot{a} = 6\, \mu_0^3 = \text{const.}\,, \tag{6.153}$$

where $\mu_0$ carries the dimensions of a mass. Equation (6.153) can be recast as

$$\frac{df_R}{dt} = f_{RR}\, \dot{R} = \frac{\mu_0^3}{a\left(c_1\, a^2 + c_2\right)} + \frac{c_1\, a^2 - c_2}{c_1\, a^2 + c_2}\, f_R\, H \tag{6.154}$$

or, in terms of the parameter $a$,

$$a\, H(a)\left[\frac{f''(a)}{R'(a)} - \frac{f'(a) R''(a)}{R'(a)^2}\right] - \frac{\left(a^2 c_1 - c_2\right) H(a) f'(a)}{\left(c_1 a^2 + c_2\right) R'(a)} = \frac{\mu_0^3}{a\left(c_1 a^2 + c_2\right)}\,. \tag{6.155}$$

Once Eq. (6.151) is solved, because the Noether constraint is satisfied, the solution $\left[H(a),\, f\left(R(a)\right)\right]$ automatically solves also Eq. (6.155) for a particular $\mu_0$. Equation (6.153) can then be used to reduce the order of the Friedmann equation. In fact, writing Eq. (6.128) as

$$f - 6\, f_{RR}\, \dot{R}\, H - 6\, f_R\, H^2 - f_R\left(R + \frac{6K}{a^2}\right) - \frac{\rho_{m0}}{a^3} - \frac{\rho_{r0}}{a^4} = 0, \tag{6.156}$$

it is

$$f - \frac{12\,c_1\,a^2}{c_1\,a^2 + c_2}\,f_R\,H^2 - f_R\left(R + \frac{6K}{a^2}\right) = \frac{6\,\mu_0^3\,H}{a\,(c_1\,a^2 + c_2)} + \frac{\rho_{m0}}{a^3} + \frac{\rho_{r0}}{a^4} \quad (6.157)$$

with $f_R$ given by (6.150). We will use this relation to search for analytical cosmological solutions depending on the constant of motion $\mu_0$ determined by the Noether symmetry.

### 6.4.3 *Exact cosmologies*

We now discuss the Noether condition (6.152) and the dynamical system (6.128), (6.130), and (6.131) and, in particular, the values of the integration constants $c_{1,2}$, the structural parameters $K$, $\rho_{r0}$, $\rho_{m0}$, and the Noether charge $\mu_0$.

#### 6.4.3.1 $c_1 = 0$

In this case, the Noether condition (6.152) reduces to

$$2\,R\,f_R - 3\,f = \frac{\rho_{r0}}{a^4}\,. \quad (6.158)$$

- **Vacuum and pure dust**
  *In vacuo* or in the presence of pure dust ($\rho_{r0} = 0$), we find

$$f = f_0\left(\frac{R}{R_0}\right)^{3/2}\,. \quad (6.159)$$

For vacuum, this solution was presented in [239]. To avoid ghosts, it must be $f_R > 0$, *i.e.*, $f_0 > 0$. With pure dust, Eq. (6.159) can be substituted into Eq. (6.157) with $\rho_{m0} \neq 0$ and $\rho_{r0} = 0$ to obtain

$$\left(\frac{R}{R_0}\right)^{3/2} + \frac{18K}{a^2\,R_0}\left(\frac{R}{R_0}\right)^{1/2} = -\frac{12\mu_0^3\,H}{c_2\,a\,f_0} - \frac{2\rho_{m0}}{a^3\,f_0}\,. \quad (6.160)$$

1. $K = 0$: the right hand side of Eq. (6.160) must be positive. If $\mu_0 = 0$ (in this case analytical solutions exist) this is impossible because $f_0 > 0$, hence it is impossible to have ghost-free solutions. For the more general case $\mu_0^3/c_2 < 0$, there could in principle be a physical solution, which can be found numerically.
2. $K \neq 0$: the Ricci scalar can be found by solving Eq. (6.160). If $\mu_0 = 0$ we have a cubic equation for $\sqrt{R/R_0}$ which always admits a real root, although this may not be positive. As follows from Eq. (6.160), there are ghosts if $\mu_0 = 0$, $K = -1$ since $f_0 < 0$. If $\mu_0 = 0, K = +1$ there are no solutions because $x^3 + \frac{18K}{a^2\,R_0}x + \frac{2\rho_{m0}}{a^3\,f_0} > 0$.

- **Mixture of dust and radiation**
  For a mixture of dust and radiation we have

$$f_R = \frac{3}{2}\frac{f}{R} + \frac{\rho_{r0}}{2a^4 R}. \tag{6.161}$$

To keep $f_R > 0$ it must be

$$f > -\frac{\rho_{r0}}{3a^4}. \tag{6.162}$$

Substituting $f_R$ into the reduced Friedmann equation (6.157) yields

$$f = -\frac{12\,\mu_0^3\,a\,H\,R}{c_2\,(R\,a^2 + 18K)} - \frac{6K\,\rho_{r0}}{a^4\,(Ra^2 + 18K)} - \frac{3\rho_{r0}R}{a^2\,(Ra^2 + 18K)} - \frac{2\rho_{m0}R}{a\,(Ra^2 + 18K)}, \tag{6.163}$$

which gives $f$ as a function of $a$ via $R = R(a)$. It must be $c_2 \neq 0$ otherwise the Noether condition becomes trivial. This expression can be inserted into Eq. (6.161). Assuming that $R(a)$ is a monotonic function of $a$ and using $f_R = (df/da)/(dR/da)$, Eq. (6.158) becomes the differential equation for $R(a)$

$$\begin{aligned} R' = & \frac{6}{a^3\left(18a^3H\mu_0^3 + 4c_2\rho_{r0} + 3ac_2\rho_{m0}\right)(Ra^2 + 6K)} \\ & \times \left\{-R^2[2a^3(H - aH')\mu_0^3 + c_2(2\rho_{r0} + a\rho_{m0})]a^4 + 6KR \right. \\ & \left. \times \left[6a^3\mu_0^3\,(H + aH') - c_2(4\rho_{r0} + a\rho_{m0})\right]a^2 - 72c_2K^2\rho_{r0}\right\}. \end{aligned} \tag{6.164}$$

Equation (6.164) can be further rewritten as a second order differential equation for $H(a)$ by using the familiar expression of $R$ in terms of $H$ and its derivatives

$$R = 6\left(a\,HH' + 2H^2 + \frac{K}{a^2}\right). \tag{6.165}$$

Substituting Eq. (6.165) into Eq. (6.164) yields

$$\begin{aligned} H'' = & -\left[a^4H^2\left(18a^3H\mu_0^3 + 4c_2\rho_{r0} + 3ac_2\rho_{m0}\right)\right]^{-1} \\ & \cdot\left\{24aK^2\mu_0^3 + H\left[a^2\left\{a^2\left(6a^3H\mu_0^3 + 4c_2\rho_{r0} + 3ac_2\rho_{m0}\right)H'^2\right.\right.\right. \\ & + a\left[12aK\mu_0^3 + H\left(78a^3H\mu_0^3 + 32c_2\rho_{r0} + 21ac_2\rho_{m0}\right)\right]H' \\ & \left.\left.\left. + 12H\left[2aK\mu_0^3 + H\left(2a^3H\mu_0^3 + 2c_2\rho_{r0} + ac_2\rho_{m0}\right)\right]\right\} - 8c_2K\rho_{r0}\right]\right\}. \end{aligned} \tag{6.166}$$

This differential equation identifies functions $f(R)$ which satisfy simultaneously the Friedmann equation and the Noether condition. Having chosen $a$ as the evolution parameter, finding the functions $H(a)$ which solve Eq. (6.166) determines uniquely the metric. Hence, $H(a)$ represents an exact solution of the field equations. Of course, to know the dependence $a(t)$ of the scale factor on the proper time, the integral $t = \int da/(aH)$ must be computed.

Analytical solutions can be found in the special case $\mu_0 = 0$, in which the differential equation becomes linear in $H^2$. Its solution is a family

$$H = H\left(a, d_1, d_2, c_2, K, \rho_{r0}, \rho_{m0}\right) , \tag{6.167}$$

where the constants $d_{1,2}$ come from the integration of Eq. (6.166). Equation (6.165) makes it possible to define a function $R\left(a, d_1, d_2, c_2, \rho_{r0}, \rho_{m0}\right)$ which can then be substituted into Eq. (6.163) to find the explicit parametric form $f = f(a, d_1, d_2, c_2, \rho_{r0}, \rho_{m0})$ of the function $f(R)$. In other words, we find the explicit parametric form of $f(R)$ using the scale factor $a$ as a parameter.

Two situations can be distinguished:

1. $K = 0$, $\mu_0 = 0$: Eq. (6.166) can be integrated yielding

$$H^2 = \frac{d_2\left(d_1 + 8\,a\,\rho_{r0} + 3\,\rho_{m0}\,a^2\right)}{a^4} , \tag{6.168}$$

   where $d_{1,2}$ are integration constants with dimensions of a mass to the fourth power and of an inverse mass squared, respectively. This expression for $H(a)$, together with Eqs. (6.163) and (6.165) forms a solution of the system (6.129) and (6.151), so that Eq. (6.155) is satisfied by setting $\mu_0 = 0$. This solution is physically unacceptable because it corresponds to a negative gravitational coupling: Eq. (6.162) cannot be satisfied by Eq. (6.163) if $K = 0$ and $\mu_0 = 0$ (however, the non-linear case $\mu_0/c_2 < 0$ could still lead to physical solutions). Similarly, we reject the case $K = -1$, $\mu_0 = 0$.
2. $K > 0$, $\mu_0 = 0$: as long as $R > 18K/a^2$, the second term on the left hand side of Eq. (6.163) is positive, allowing for the possibility of a physical solution. The integration of Eq. (6.166) leads to

$$H^2 = \left(\sqrt{2}\,d_1 - \frac{32\,\rho_{r0}^2 K}{9\rho_{0m}^2}\right)\frac{1}{a^4} + \left(8\,d_2\,\rho_{r0} - \frac{16K\rho_{r0}}{3\rho_{m0}}\right)\frac{1}{a^3} + \frac{3\,d_2\,\rho_{m0}}{a^2} , \tag{6.169}$$

   where $d_1$ and $d_2$ have the dimensions of a mass squared and an inverse mass squared, respectively. In order to find $d_1$ and $d_2$, one fits this formula with the Friedmann equation of GR for a FLRW universe dominated by matter, radiation, or curvature, considering

$$\sqrt{2}\,d_1 - \frac{32\,\rho_{r0}^2\,K}{9\rho_{0m}^2} = H_0^2\,\Omega_{r0}^{(\mathrm{eff})} , \tag{6.170}$$

$$8\,d_2\,\rho_{r0} - \frac{16\rho_{r0}\,K}{3\rho_{m0}} = H_0^2\,\Omega_{m0}^{(\mathrm{eff})} , \tag{6.171}$$

$$3\,d_2\,\rho_{m0} = H_0^2\,\Omega_{k0}^{(\mathrm{eff})} . \tag{6.172}$$

This system does not admit physical solutions since, using present data [1039], one finds

$$K - \frac{1}{2} H_0^2 \Omega_{k0}^{(\text{eff})} - \frac{3}{16} \frac{\rho_{m0}}{\rho_{r0}} H_0^2 \Omega_{m0}^{(\text{eff})} < 0 . \tag{6.173}$$

#### 6.4.3.2 $c_2 = 0$

In this case, the Noether condition (6.152) reduces to

$$\rho_{r0} + 3a^4 f + 12K a^2 f_R = 0 . \tag{6.174}$$

- **Cosmological constant and dust**
  If only dust and a cosmological constant are present, $\rho_{r0} = 0$ and a flat universe cannot be a solution of the field equations because it would imply that $f = 0$. If $K \neq 0$ one finds

$$f_R = \frac{a^2 f}{4K} . \tag{6.175}$$

Since $f_R > 0$, $f$ is positive when $K$ is positive and *vice-versa*. Taking this fact into account in the Friedmann equation, it is

$$a^3 c_1 \left[ \left(12H^2 + R\right) a^2 + 10K \right] f = 4K \left( 6H\mu_0^3 + c_1 \rho_{m0} \right) . \tag{6.176}$$

Restricting ourselves to the study of the simplest linear case corresponding to $\mu_0 = 0$, two situations are possible:

1. $\rho_{m0} = 0, \mu_0 = 0$: then one needs to impose

$$R = 6 \left( 2H^2 + \frac{K}{a^2} \right) \tag{6.177}$$

which, together with the definition of $R$, yields

$$H^2 = 2 \left( d_1 - \frac{K}{3a^2} \right) \tag{6.178}$$

where $d_1$ is an integration constant with the dimensions of a mass squared. Equation (6.174) can now be solved for $f(a)$, obtaining

$$f = \frac{d_2}{a} = d_2 \left( - \frac{R + 24d_1}{2K} \right)^{1/2} \tag{6.179}$$

with $d_2$ an integration constant carrying the dimensions of a mass to the fourth power.

2. $\rho_{m0} \neq 0, \mu_0 = 0$: the Friedmann equation and Eq. (6.175) give

$$f = - \frac{4K \rho_{m0}}{(12H^2 + R) a^5 + 10K a^3} . \tag{6.180}$$

Substituting this expression into Eq. (6.175) and using the expression of $R$ in terms of $H(a)$, one finds a linear second order differential equation for $H^2(a)$ with solution

$$H^2 = \frac{d_1}{2a^4} + 2d_2 - \frac{2K}{3a^2}, \tag{6.181}$$

where $d_{1,2}$ are again integration constants with the dimensions of a mass squared and

$$R = -24d_2 - \frac{2K}{a^2}, \tag{6.182}$$

$$f = -\frac{2K\,\rho_{m0}}{3\,a\,d_1}. \tag{6.183}$$

- **Radiation and dust** There are three possibilities, corresponding to the possible signs of $K$.

1. $K = 0$: in this case it is

$$f = \frac{\rho_{r0}}{3a^4} \tag{6.184}$$

and

$$f_R = -\frac{f'}{R'} = \frac{4}{3}\frac{\rho_{r0}}{a^5\,R'}. \tag{6.185}$$

A well-behaved background evolution requires $R' > 0$, hence $f_R < 0$ and the effective gravitational coupling is negative, which makes this solution physically unacceptable.

2. $K \neq 0$: using Eq. (6.174) one finds

$$f_R = \frac{\rho_{r0}}{12Ka^2} + \frac{f\,a^2}{4K} \tag{6.186}$$

and, using the Friedmann equation (6.157) and solving for $f$, one obtains

$$f = \frac{-c_1\left(12H^2 + R\right)\rho_{r0}a^2 + 12K\left(6H\mu_0^3 + c_1\rho_{m0}\right)a + 6c_1K\rho_{r0}}{3a^4c_1\left[\left(12H^2 + R\right)a^2 + 10K\right]} \tag{6.187}$$

which, inserted into the Noether condition (6.174) in conjunction with the expression of $R$ in terms of $(H, H', a)$, leads to the differential equation for $H(a)$

$$\begin{aligned} H'' = {} & \left[a^5H^2\left(18aH\mu_0^3 + 3ac_1\rho_{m0} + 4c_1\rho_{r0}\right)\right]^{-1} \\ & \cdot\Big\{aH\Big[-\left(18aH\mu_0^3 + 3ac_1\rho_{m0} + 4c_1\rho_{r0}\right)H'^2a^4 \\ & - 3\left(aH\left(30aH\mu_0^3 + 5ac_1\rho_{m0} + 8c_1\rho_{r0}\right) - 4K\mu_0^3\right)H'\,a^2 \\ & + 4K\left(6aH\mu_0^3 + ac_1\rho_{m0} + 2c_1\rho_{r0}\right)\Big] - 8K^2\mu_0^3\Big\}. \end{aligned} \tag{6.188}$$

In the case $\mu_0 = 0$, $\rho_{m0} \neq 0$, Eq. (6.188) is integrated to give

$$H^2 = \frac{256K\rho_{r0}^3}{405a^5\rho_{m0}^3} + \frac{16K\rho_{r0}^2}{27a^4\rho_{m0}^2} + \frac{8d_1\rho_{r0}}{5a^5} - \frac{2K}{3a^2} + \frac{3\rho_{m0}d_1}{2a^4} + 2d_2\,, \quad (6.189)$$

where $d_{1,2}$ are integration constants with the dimensions of an inverse mass squared and a mass squared, respectively. A new cosmological term scaling as $a^{-5}$ appears in this Friedmann equation, analogous to a matter term with equation of state $P = 2\rho/3$.

If $(\mu_0, \rho_{m0}) = (0, 0)$ (*i.e.*, for a radiation-dominated universe), Eq. (6.188) admits the solution

$$H^2 = 2d_2 + \frac{2d_1}{5a^5} - \frac{2K}{3a^2}\,, \quad (6.190)$$

where both $d_1$ and $d_2$ have the dimensions of a mass squared.

### 6.4.4 $c_1, c_2 \neq 0$

Dividing Eq. (6.152) by $c_1$ one obtains

$$f_R = \frac{(a^2 + c_3)\left(\rho_{r0} + 3\,a^4\,f\right)}{2\,a^4\,(c_3\,R - 6K)} \quad (6.191)$$

where $c_3 = c_2/c_1 \neq 0$, which implies that

$$f_{RR}\,\dot{R} = \frac{\tilde{\mu}_0^3}{a\,(a^2 + c_3)} + \frac{a^2 - c_3}{a^2 + c_3}\,f_R\,H \quad (6.192)$$

with $\tilde{\mu}_0^3 = \mu_0^3/c_1$. The Friedmann equation (6.157) can be rewritten as

$$f - \frac{12\,a^2}{a^2 + c_3}\,f_R\,H^2 - f_R\left(R + \frac{6K}{a^2}\right) = \frac{6\tilde{\mu}_0^3\,H}{a\,(a^2 + c_3)} + \frac{\rho_{m0}}{a^3} + \frac{\rho_{r0}}{a^4}\,. \quad (6.193)$$

By substituting Eq. (6.191) into Eq. (6.193) and solving for $f$, one finds

$$f = \frac{12\,\tilde{\mu}_0^3\,a^5\,H\,(6K - c_3\,R)}{a^4\,(a^2 + c_3)\,[3\,(12H^2 + R)\,a^4 + (30K + c_3 R)\,a^2 + 18c_3 K]}$$
$$- \frac{\rho_{r0}\left(12H^2 + R\right)a^4 + 2\rho_{m0}\,(c_3 R - 6K)\,a^3 + 3\rho_{r0}\,(c_3 R - 2K)\,a^2 + 6c_3\,K\rho_{r0}}{a^4\,[3\,(12H^2 + R)\,a^4 + (30K + c_3 R)\,a^2 + 18c_3 K]}\,, \quad (6.194)$$

where $R = R(a)$ and $H = H(a)$. Again, one can proceed by regarding $f$ as an implicit function of $a$ into the Noether condition (6.191). Since $f = f(R(a))$, it is

$$f_R = \frac{da}{dR}\frac{df}{da} = \frac{f'}{R'} . \tag{6.195}$$

The substitution of Eqs. (6.194) and (6.195) into Eq. (6.191) yields the second order equation for $H$

$$\begin{aligned} H'' = & \frac{1}{a^4(a^2+c_3)(3a^2+c_3)H^2\left[18\tilde{\mu}_0^3 Ha^3 + (a^2+c_3)(4\rho_{r0}+3a\rho_{m0})\right]} \\ & \cdot\Big\{-24c_3(3a^2+c_3)\tilde{\mu}_0^3 H^4 a^5 - 24(a^2+c_3)^2 K^2 \tilde{\mu}_0^3 a \\ & - H^2\Big[6(3a^2+c_3)^2\tilde{\mu}_0^3 H'^2 a^4 + 24(-3a^4 - 2c_3 a^2 + c_3^2)K\tilde{\mu}_0^3 \\ & + (a^2+c_3)^2(45\rho_{m0}a^3 + 72\rho_{r0}a^2 + 21c_3\rho_{m0}a + 32c_3\rho_{r0})H'\Big]a^3 \\ & -6H^3\left[(3a^2+c_3)(15a^2+13c_3)\tilde{\mu}_0^3 H' a^4\right. \\ & \qquad \left. +2c_3(a^2+c_3)^2(2\rho_{r0}+a\rho_{m0})\right]a^2 \\ & -(a^2+c_3)H\Big[a^4 H'\left[12(c_3-3a^2)K\tilde{\mu_0}^3\right. \\ & \left. + (a^2+c_3)(3a^2+c_3)(4\rho_{r0}+3a\rho_{m0})H'\right] \\ & - 4(a^2+c_3)K\left(3\rho_{m0}a^3 + 6\rho_{r0}a^2 + 2c_3\rho_{r0}\right)\Big]\Big\} . \end{aligned} \tag{6.196}$$

Equation (6.196) rules the dynamics of the Noether solutions for choices of $f(R)$ compatible with the Noether symmetry. There is a free parameter $c_3$ which, together with the initial conditions $(H_0, H_0')$, uniquely specify the dynamics. This non-linear ODE is still of second order in $H(a)$, as the time-time component of the field equations is in any $f(R)$ theory. However, this equation is independent of the explicit form of the function $f(R)$ and contains $c_3$ and $\mu_0$ (the Noether charge) as the only unknown parameters. Then, there is a solution of Eq. (6.196) for any value of the Noether charge and the solutions spanned as $c_3$ and $\mu_0$ vary represent the whole set of Noether-charged cosmological solutions of $f(R)$ theories.

#### 6.4.4.1 Cosmological constant and dust

In the presence of dust and a cosmological constant, Eq. (6.191) reduces to

$$f_R = \frac{3f\,(a^2+c_3)}{2\,(R\,c_3 - 6K)} , \tag{6.197}$$

while $f$ can be written as

$$f = \frac{2\left(6K - Rc_3\right)\left[\left(6H\mu_0^3 + \rho_{m0}\right)a^2 + \rho_{m0}c_3\right]}{a\left(a^2 + c_3\right)\left[3\left(12H^2 + R\right)a^4 + \left(30K + Rc_3\right)a^2 + 18Kc_3\right]}. \tag{6.198}$$

If $(\rho_{m0}, \mu_0) = (0, 0)$ there are no solutions, hence we only discuss $\mu_0 = 0, \rho_{m0} \neq 0$, in which case $f$ can be written in the form

$$f = 3\left(12H^2 + R\right)a^4 + \left(30K + Rc_3\right)a^2 + 18Kc_3. \tag{6.199}$$

Inserting this relation into Eq. (6.197) together with the definition of $R$, one finds

$$H'' = \frac{-4c_3H^2 - a\left(15a^2 + 7c_3\right)H'H - a^2\left(3a^2 + c_3\right)H'^2 + 4K}{a^2\left(3a^2 + c_3\right)H}, \tag{6.200}$$

whose general solution is

$$H^2 = -\frac{c_3K}{9a^4} - \frac{2K}{3a^2} + \frac{2d_1}{a^4} + \frac{2c_3d_2}{a^2} + 3d_2. \tag{6.201}$$

- **Pure radiation**
  Once again, Eq. (6.196) with $(\mu_0, \rho_{m0}) = (0, 0)$ yields

$$\left(H^2\right)'' = -\frac{18a^2 + 8c_3}{a\left(3a^2 + c_3\right)}\left(H^2\right)' - \frac{12\,c_3\,H^2}{a^2\left(3a^2 + c_3\right)} + \frac{2k(6a^2 + 2c_3)}{a^4\left(3a^2 + c_3\right)}. \tag{6.202}$$

The general solution for positive values of $c_3$ is

$$\begin{aligned} H^2 = {} & \frac{3c_3d_1}{a^4} + \frac{27d_1}{c_3} + \frac{18d_1}{a^2} + \frac{5\sqrt{3}\sqrt{c_3}d_2}{a^3} + \frac{9\sqrt{3}d_2}{a\sqrt{c_3}} + \frac{4K}{c_3} + \frac{2K}{a^2} \\ & + \frac{3c_3d_2\tan^{-1}\left(\frac{\sqrt{3}a}{\sqrt{c_3}}\right)}{a^4} + \frac{27d_2\tan^{-1}\left(\frac{\sqrt{3}a}{\sqrt{c_3}}\right)}{c_3} + \frac{18d_2\tan^{-1}\left(\frac{\sqrt{3}a}{\sqrt{c_3}}\right)}{a^2}. \end{aligned} \tag{6.203}$$

For negative values of $c_3$ one finds

$$\begin{aligned} H^2 = {} & \frac{3c_3d_1}{a^4} + \frac{27d_1}{c_3} + \frac{18d_1}{a^2} - \frac{5\sqrt{3}\sqrt{c_3}d_2}{a^3} + \frac{9\sqrt{3}d_2}{a\sqrt{-c_3}} + \frac{4K}{c_3} + \frac{2K}{a^2} \\ & + \frac{3c_3d_2\tanh^{-1}\left(\frac{\sqrt{3}a}{\sqrt{-c_3}}\right)}{a^4} + \frac{27d_2\tanh^{-1}\left(\frac{\sqrt{3}a}{\sqrt{-c_3}}\right)}{c_3} + \frac{18d_2\tanh^{-1}\left(\frac{\sqrt{3}a}{\sqrt{-c_3}}\right)}{a^2}. \end{aligned} \tag{6.204}$$

Both expressions of $H(a)$ together with Eqs. (6.194) and (6.165) form a solution of (6.129) and (6.151) with vanishing Noether charge $\mu_0$.

- **Dust and radiation**

Let us restrict ourselves to the case $\tilde{\mu} = 0$, which admits analytical solutions. Equation (6.196) reduces to

$$(H^2)'' = -\frac{(45\rho_{m0}a^3 + 72\rho_{r0}a^2 + 21c_3\rho_{m0}a + 32c_3\rho_{r0})}{a\,(3a^2 + c_3)\,(4\rho_{r0} + 3a\rho_{m0})}\,(H^2)' - \frac{24\,c_3\,(\rho_{m0}a + 2\rho_{r0})\,H^2}{a^2\,(3a^2 + c_3)\,(4\rho_{r0} + 3a\rho_{m0})} + \frac{8k(3\rho_{m0}a^3 + 6\rho_{r0}a^2 + 2c_3\rho_{r0})}{a^4\,(3a^2 + c_3)\,(4\rho_{r0} + 3a\rho_{m0})}\,. \tag{6.205}$$

Remarkably, this ODE is linear in $H^2$ and analytical solutions can be obtained for any value of $K$.

1. $K = 0$: the solution of Eq. (6.205) is

$$H^2 = \frac{4d_1d_2\,c_3^{9/2}}{a^4} + \frac{24d_1d_2\,c_3^{7/2}}{a^2} - \frac{\rho_{0m}\,d_2c_3^{5/2}}{a^4} + 36\,d_1d_2c_3^{5/2} + \frac{2\sqrt{3}\rho_{r0}\tan^{-1}\left(\frac{\sqrt{3}a}{\sqrt{c_3}}\right)d_2c_3^2}{a^4} + \frac{10\rho_{r0}d_2c_3^{3/2}}{a^3} + \frac{12\sqrt{3}\,\rho_{r0}\tan^{-1}\left(\frac{\sqrt{3}a}{\sqrt{c_3}}\right)d_2c_3}{a^2} + \frac{18\rho_{r0}d_2\sqrt{c_3}}{a} + 18\sqrt{3}\,\rho_{r0}\tan^{-1}\left(\frac{\sqrt{3}\,a}{\sqrt{c_3}}\right)d_2\,, \tag{6.206}$$

where $d_{1,2}$ are integration constants with the dimensions of a mass to the fourth power and of an inverse mass squared, respectively. This solution deviates from the corresponding GR solution because it contains an $1/a$ term which, if dominant, leads to acceleration. Furthermore, there are terms involving $\rho_{r0}$, which include the inverse tangent of $a$ ($c_3$ is assumed to be positive). These terms have different behavior at low and high redshift. In fact, since $\tan^{-1}(a) \sim a$ at high redshifts $a \to 0$, these terms behave as dust with energy densities scaling as $1/a$ and $a$ respectively, and are subdominant with respect to radiation. Moreover, since $\tan^{-1}(a) \sim \pi/2$ for large and positive $a$, these terms behave as radiation, curvature, and cosmological constant, respectively. In order to have a true dust component at late times, it must be

$$\rho_{r0}\,d_2\,c_3^{3/2} = \frac{4\pi G}{15}\,\rho_{m0}\,. \tag{6.207}$$

This fact means that $\rho_{r0}$ behaves as the material source in this modified Friedmann equation. A cosmological constant term is also present and it is determined by the integration constants related to the Noether condition.
As for the case $c_3 < 0$, the solution of Eq. (6.205) can be written as (see Appendix B.2)

$$\begin{aligned} H^2 = &-\frac{4d_1d_2(-c_3)^{9/2}}{a^4} + \frac{24d_1d_2(-c_3)^{7/2}}{a^2} \\ &+\frac{\rho_{0m}d_2(-c_3)^{5/2}}{a^4} - 36d_1d_2(-c_3)^{5/2} \\ &+\frac{2\sqrt{3}\,\rho_{r0}\tanh^{-1}\left(\frac{\sqrt{3}\,a}{\sqrt{-c_3}}\right)d_2c_3^2}{a^4} + \frac{10\rho_{r0}d_2(-c_3)^{3/2}}{a^3} \\ &+\frac{12\sqrt{3}\,\rho_{r0}\tanh^{-1}\left(\frac{\sqrt{3}\,a}{\sqrt{-c_3}}\right)d_2c_3}{a^2} \\ &-\frac{18\rho_{r0}d_2\sqrt{-c_3}}{a} + 18\sqrt{3}\,\rho_{r0}\tanh^{-1}\left(\frac{\sqrt{3}\,a}{\sqrt{-c_3}}\right)d_2\,. \end{aligned} \tag{6.208}$$

2. $K \neq 0$: the general solution is

$$\begin{aligned} H^2 = &-\frac{32K\tan^{-1}\left(\frac{\sqrt{3}a}{\sqrt{c_3}}\right)\rho_{r0}^3}{9\sqrt{3}a^4\rho_{m0}^3\sqrt{c_3}} - \frac{160K\rho_{r0}^3}{27a^3\rho_{m0}^3c_3} - \frac{64K\tan^{-1}\left(\frac{\sqrt{3}a}{\sqrt{c_3}}\right)\rho_{r0}^3}{3\sqrt{3}a^2\rho_{m0}^3c_3^{3/2}} \\ &-\frac{32K\rho_{r0}^3}{3a\rho_{m0}^3c_3^2} - \frac{32K\tan^{-1}\left(\frac{\sqrt{3}a}{\sqrt{c_3}}\right)\rho_{r0}^3}{\sqrt{3}\rho_{m0}^3c_3^{5/2}} - \frac{16K\rho_{r0}^2}{3a^2\rho_{m0}^2c_3} - \frac{8K\rho_{r0}^2}{27a^4\rho_{m0}^2} \\ &-\frac{8K\rho_{r0}^2}{\rho_{m0}^2c_3^2} + \frac{\sqrt{3}\tan^{-1}\left(\frac{\sqrt{3}a}{\sqrt{c_3}}\right)d_2\rho_{r0}}{2a^4c_3^{5/2}} + \frac{5d_2\rho_{r0}}{2a^3c_3^3} + \frac{3\sqrt{3}\tan^{-1}\left(\frac{\sqrt{3}a}{\sqrt{c_3}}\right)d_2\rho_{r0}}{a^2c_3^{7/2}} \\ &+\frac{9d_2\rho_{r0}}{2ac_3^4} + \frac{9\sqrt{3}\tan^{-1}\left(\frac{\sqrt{3}a}{\sqrt{c_3}}\right)d_2\rho_{r0}}{2c_3^{9/2}} - \frac{2K\tan^{-1}\left(\frac{\sqrt{3}a}{\sqrt{c_3}}\right)\sqrt{c_3}\rho_{r0}}{\sqrt{3}a^4\rho_{m0}} \\ &-\frac{4\sqrt{3}K\tan^{-1}\left(\frac{\sqrt{3}a}{\sqrt{c_3}}\right)\rho_{r0}}{a^2\rho_{m0}\sqrt{c_3}} - \frac{4\sqrt{3}K\tan^{-1}\left(\frac{\sqrt{3}a}{\sqrt{c_3}}\right)\rho_{r0}}{a^2\rho_{m0}\sqrt{c_3}} \\ &-\frac{10K\rho_{r0}}{3a^3\rho_{m0}} - \frac{6K\rho_{r0}}{a\rho_{m0}c_3} - \frac{6\sqrt{3}K\tan^{-1}\left(\frac{\sqrt{3}a}{\sqrt{c_3}}\right)\rho_{r0}}{\rho_{m0}c_3^{3/2}} - \frac{2K}{3a^2} - \frac{Kc_3}{9a^4} + \frac{6d_1}{a^2c_3} \\ &+\frac{9d_1}{c_3^2} + \frac{d_1}{a^4} - \frac{\rho_{m0}d_2}{4a^4c_3^2}\,. \end{aligned} \tag{6.209}$$

Also in these cases we have interesting behaviors matching the main cosmological eras. The integration constants $d_{1,2}$ have the dimensions of a mass squared and an inverse mass squared, respectively.
For negative values of $c_3$, Eq. (6.205) has the solution

$$
\begin{aligned}
H^2 = &\frac{32K\tanh^{-1}\left(\frac{\sqrt{3}a}{\sqrt{-c_3}}\right)\rho_{r0}^3}{9\sqrt{3}a^4\rho_{m0}^3\sqrt{-c_3}} - \frac{160K\rho_{r0}^3}{27a^3\rho_{m0}^3c_3} - \frac{64K\tanh^{-1}\left(\frac{\sqrt{3}a}{\sqrt{-c_3}}\right)\rho_{r0}^3}{3\sqrt{3}a^2\rho_{m0}^3(-c_3)^{3/2}}\\
&-\frac{32K\rho_{r0}^3}{3a\rho_{m0}^3c_3^2} + \frac{32K\tanh^{-1}\left(\frac{\sqrt{3}a}{\sqrt{-c_3}}\right)\rho_{r0}^3}{\sqrt{3}\rho_{m0}^3(-c_3)^{5/2}} - \frac{16K\rho_{r0}^2}{3a^2\rho_{m0}^2c_3}\\
&-\frac{8K\rho_{r0}^2}{27a^4\rho_{m0}^2} - \frac{8K\rho_{r0}^2}{\rho_{m0}^2c_3^2} - \frac{\sqrt{3}\tanh^{-1}\left(\frac{\sqrt{3}a}{\sqrt{-c_3}}\right)d_2\rho_{r0}}{2a^4(-c_3)^{5/2}} + \frac{5d_2\rho_{r0}}{2a^3c_3^3}\\
&+\frac{3\sqrt{3}\tanh^{-1}\left(\frac{\sqrt{3}a}{\sqrt{-c_3}}\right)d_2\rho_{r0}}{a^2(-c_3)^{7/2}} + \frac{9d_2\rho_{r0}}{2ac_3^4}\\
&-\frac{9\sqrt{3}\tanh^{-1}\left(\frac{\sqrt{3}a}{\sqrt{-c_3}}\right)d_2\rho_{r0}}{2(-c_3)^{9/2}} - \frac{2K\tanh^{-1}\left(\frac{\sqrt{3}a}{\sqrt{-c_3}}\right)\sqrt{-c_3}\rho_{r0}}{\sqrt{3}a^4\rho_{m0}}\\
&+\frac{4\sqrt{3}K\tanh^{-1}\left(\frac{\sqrt{3}a}{\sqrt{-c_3}}\right)\rho_{r0}}{a^2\rho_{m0}\sqrt{-c_3}} - \frac{10K\rho_{r0}}{3a^3\rho_{m0}}\\
&-\frac{6K\rho_{r0}}{a\rho_{m0}c_3} - \frac{6\sqrt{3}K\tanh^{-1}\left(\frac{\sqrt{3}a}{\sqrt{-c_3}}\right)\rho_{r0}}{\rho_{m0}(-c_3)^{3/2}} - \frac{2K}{3a^2} - \frac{Kc_3}{9a^4} + \frac{6d_1}{a^2c_3}\\
&+\frac{9d_1}{c_3^2} + \frac{d_1}{a^4} - \frac{\rho_{m0}d_2}{4a^4c_3^2}\,. \qquad (6.210)
\end{aligned}
$$

Once the free parameters (if there are any) are constrained by the data, one can select physically interesting $f(R)$ models as in [218].

- **The non-linear case $\tilde{\mu}_0 \neq 0$**
  For general (*i.e.*, non-vanishing) values of the parameter $\tilde{\mu}_0$, Eq. (6.196) cannot be written as a linear differential equation for $H^2$ and analytical solutions are unknown. However, assigning an initial condition for $H$ and suitable values for the parameters, one can solve this equation numerically. In turn, the initial conditions determine the functions $f(R)$ and $H(a)$.

#### 6.4.4.2 Non-Noether solutions

In general it is not possible to find a solution of the Friedmann equations which is also a Noether symmetry: such symmetries do not exist for all $f(R)$ theories.

A solution of the cosmological equations is generally incompatible with the condition $\mathscr{L}_{\mathbf{X}}L = 0$. Both equations are satisfied only in peculiar situations occurring if Noether charges are present in the structure of the theory. For example, imposing a power-law solution $a \propto t^p$ defines a function $R = R(a)$ which can be inserted in the Noether symmetry equations in order to find $f(R(a))$. Finally one can substitute the expressions of $f(a)$, $R(a)$, and $H$ in the Friedmann equations. It is easy to show that, for $K = 0$, there are no simple power-law solutions compatible with a Noether charge. The method discussed above allows one to discriminate theories which admit or do not admit cosmological solutions compatible with a Noether charge. Power-law solutions exist in general $f(R)$ models and they can be found using different methods [24, 211, 212, 219, 275, 331, 846, 865]. Assuming a power-law $H(a)$, one finds $R$ as a function of $a$ and then, in principle, determines $f(R(a))$. It is therefore possible to write the field equation as a second order differential equation for $f(a)$, with $H$ and $R$ given functions of $a$. This argument holds also if the redshift $z$ is employed [218]. For example, let us rewrite the Friedmann equation (6.128) as

$$f - 6\, f_{RR}\, \dot{R} H - 6\, f_R\, H^2 - f_R \left( R + \frac{6K}{a^2} \right) = \frac{\rho_{m0}}{a^3} + \frac{\rho_{r0}}{a^4} \tag{6.211}$$

and let us consider $H = \bar{H}(a)$ and $R = \bar{R}(a)$ as given functions, where

$$\bar{R} = 6 \left( 2\bar{H}^2 + a \bar{H} \bar{H}' + \frac{K}{a^2} \right) . \tag{6.212}$$

The Friedmann equation can be written as

$$f'' + \left[ \frac{1}{a} - \frac{\bar{R}''}{\bar{R}'} + \frac{1}{6a\, \bar{H}^2} \left( \bar{R} + \frac{6K}{a^2} \right) \right] f' - \frac{\bar{R}'}{6a\, \bar{H}^2}\, f = - \frac{\rho_{m0}\, a + \rho_{r0}}{6\, a^5\, \bar{H}^2}\, \bar{R}' . \tag{6.213}$$

The general solution of this second order linear equation in $f$ depends on the parameters $f_0$ and $f_0'$ and is a linear combination of two independent solutions of the homogeneous equation plus a particular solution. It is then clear that more than one $f(R)$ model can have the same $H(a)$ behavior, *i.e.*, multiple theories share the same cosmological evolution due to the fourth order of the field equations. The singular points of this differential equation occur as either $\bar{H}$ or $d\bar{R}/da$ vanish. Interesting classes of solutions can be found.

- **Radiation solutions**

  Let us search all the $f(R)$ choices admitting the particular solution $a = \sqrt{t/t_0}$, for which

$$\bar{H} = \frac{1}{2\, t_0\, a^2} = \frac{H_0}{a^2} \tag{6.214}$$

so that $\bar{R} = 6K/a^2$, where $H_0 \equiv (2\,t_0)^{-1}$. We have three interesting cases:

1. If $K = 0$ then it is $R = 0$, leading to the Friedmann equation

$$f(0) - 6\, f_R(0)\, \bar{H}^2 = \frac{\rho_{m0}}{a^3} + \frac{\rho_{r0}}{a^4} \tag{6.215}$$

which, if $\rho_{m0} \neq 0$, cannot be solved for $\bar{H} \sim a^{-2}$ since $f(0)$ and $f'(0)$ must be constant and cannot depend on $a$. If $\rho_{m0} = 0$, standard GR is recovered.

2. If $K < 0$ we have the differential equation

$$f'' + \frac{4}{a}\, f' + \frac{2K}{H_0^2}\, f = \frac{2K\,(\rho_{r0} + a\rho_{m0})}{H_0^2\, a^4} \tag{6.216}$$

whose general solution can be written as

$$R = \frac{6K}{a^2}\,, \tag{6.217}$$

$$\begin{aligned} f = {} & \frac{\sqrt{\frac{a\sqrt{-K}}{H_0}}\, d_2 \cos\left(\frac{a\sqrt{-2K}}{H_0}\right) H_0^2}{\sqrt[4]{2}\, a^{7/2} K \sqrt{\pi}} - \frac{\sqrt{\frac{a\sqrt{-K}}{H_0}}\, d_1 \sin\left(\frac{a\sqrt{-2K}}{H_0}\right) H_0^2}{\sqrt[4]{2}\, a^{7/2} K \sqrt{\pi}} \\ & - \frac{\sqrt[4]{2}\sqrt{\frac{a\sqrt{-K}}{H_0}}\, d_1 \cos\left(\frac{a\sqrt{-2K}}{H_0}\right) H_0}{a^{5/2}\sqrt{-K}\sqrt{\pi}} \\ & - \frac{\sqrt[4]{2}\sqrt{\frac{a\sqrt{-K}}{H_0}}\, d_2 \sin\left(\frac{a\sqrt{-2K}}{H_0}\right) H_0}{a^{5/2}\sqrt{-K}\sqrt{\pi}} + \frac{\rho_{m0}}{a^3} \\ & + \frac{\rho_{r0}\sqrt{-K}\,\mathrm{Ci}\left(\frac{\sqrt{2}a\sqrt{-K}}{H_0}\right) \sin\left(\frac{a\sqrt{-2K}}{H_0}\right)}{\sqrt{2}a^3 H_0} \\ & - \frac{\rho_{r0}\sqrt{-K}\cos\left(\frac{a\sqrt{-2K}}{H_0}\right) \mathrm{Si}\left(\frac{a\sqrt{-2K}}{H_0}\right)}{\sqrt{2}a^3 H_0} + \frac{\rho_{r0} K \cos\left(\frac{a\sqrt{-2K}}{H_0}\right) \mathrm{Ci}\left(\frac{a\sqrt{-2K}}{H_0}\right)}{a^2 H_0^2} \\ & + \frac{\rho_{r0} K \sin\left(\frac{a\sqrt{-2K}}{H_0}\right) \mathrm{Si}\left(\frac{a\sqrt{-2K}}{H_0}\right)}{a^2 H_0^2}\,, \end{aligned} \tag{6.218}$$

where the SinIntegral and CosIntegral functions are defined by

$$\mathrm{Si}(x) \equiv \int_0^x \frac{\sin t}{t}\, dt\,, \qquad \mathrm{Ci}(x) \equiv -\int_x^\infty \frac{\cos t}{t}\, dt\,, \tag{6.219}$$

respectively. Both integration constants $d_{1,2}$ have the dimensions of a mass to the fourth power.

3. Using a similar procedure for $K > 0$, one finds the solution

$$R = \frac{6K}{a^2}, \tag{6.220}$$

$$\begin{aligned} f = &\frac{\sqrt{\frac{a\sqrt{K}}{H_0}}\, d_1 \cosh\left(\frac{\sqrt{2K}\,a}{H_0}\right) H_0^2}{\sqrt[4]{2}\, a^{7/2} K \sqrt{\pi}} + \frac{\sqrt{\frac{a\sqrt{K}}{H_0}}\, d_1 \sinh\left(\frac{\sqrt{2K}\,a}{H_0}\right) H_0^2}{\sqrt[4]{2}\, a^{7/2} K \sqrt{\pi}} \\ &- \frac{\sqrt[4]{2}\sqrt{\frac{a\sqrt{K}}{H_0}}\, d_1 \cosh\left(\frac{\sqrt{2K}\,a}{H_0}\right) H_0}{a^{5/2}\sqrt{\pi K}} \\ &- \frac{\sqrt[4]{2}\sqrt{\frac{a\sqrt{K}}{H_0}}\, d_1 \sinh\left(\frac{\sqrt{2K}a}{H_0}\right) H_0}{a^{5/2}\sqrt{\pi K}} + \frac{\rho_{m0}}{a^3} \\ &- \frac{\rho_{r0}\sqrt{K}\,\mathrm{Chi}\left(\frac{\sqrt{2K}a}{H_0}\right)\sinh\left(\frac{\sqrt{2K}a}{H_0}\right)}{\sqrt{2}\,a^3 H_0} \\ &+ \frac{\rho_{r0}\sqrt{K}\cosh\left(\frac{\sqrt{2K}\,a}{H_0}\right)\mathrm{Shi}\left(\frac{\sqrt{2K}\,a}{H_0}\right)}{\sqrt{2}a^3 H_0} - \frac{\rho_{r0} K \cosh\left(\frac{\sqrt{2K}\,a}{H_0}\right)\mathrm{Chi}\left(\frac{\sqrt{2K}a}{H_0}\right)}{a^2 H_0^2} \\ &+ \frac{\rho_{r0} K \sinh\left(\frac{\sqrt{2K}a}{H_0}\right)\mathrm{Shi}\left(\frac{\sqrt{2K}a}{H_0}\right)}{a^2 H_0^2}, \end{aligned} \tag{6.221}$$

where

$$\mathrm{Shi}(x) \equiv \int_0^x \frac{\sinh t}{t}\, dt\,, \qquad \mathrm{Chi}(x) \equiv \gamma_{E,M} + \ln(x) + \int_0^x \frac{\cosh t - 1}{t}\, dt\,, \tag{6.222}$$

respectively, and $\gamma_{EM} \approx 0.577$ is the Euler-Mascheroni constant. Both $d_1$ and $d_2$ are integration constants with the dimensions of a mass to the fourth power.

- **Dust solutions**
  We now search for $f(R)$ models with dust-like behavior $a(t) = (t/t_0)^{2/3}$, and

$$\bar{H} = \frac{2}{3\,t_0\,a^{3/2}} = \frac{H_0}{a^{3/2}}, \qquad \bar{R} = \frac{2(2/t_0^2 + 9\,K\,a)}{3\,a^3}, \tag{6.223}$$

where $H_0 \equiv 2/(3\,t_0)$. For spatially flat FLRW universes there is the two-parameter family of solutions

$$R = \frac{4}{3\,t_0^2\,a^3}, \tag{6.224}$$

$$f(a) = a^{-(7+\sqrt{73})/4}\left(d_1\,a^{\sqrt{73}/2} + d_2\right) + \frac{\rho_{m0}\,a - 6\rho_{r0}}{2\,a^4} \tag{6.225}$$

depending on the integration constants $d_{1,2}$ both with dimensions of a mass to the fourth power. The Hilbert-Einstein Lagrangian $f(R) = R$, obtained when $d_1$, $d_2$, and $\rho_{r0}$ vanish, belongs to this family.

- **Exponential solutions**
  By imposing that

$$\bar{H} = H_0 = \text{const.}, \qquad \bar{R} = 6\left(2H_0^2 + \frac{K}{a^2}\right), \tag{6.226}$$

there are again three situations corresponding to the possible signs of the curvature index.

1. $K = 0$: $H$ and $R$ are constant with $R = R_0 \equiv 12\, H_0^2$ and the Friedmann equation reduces to

$$f(R_0) - \frac{1}{2}\, f_R(R_0)\, R_0 = \frac{\rho_{m0}}{a^3} + \frac{\rho_{r0}}{a^4}. \tag{6.227}$$

   Solutions exist only if $\rho_{m0} = \rho_{r0} = 0$ [103].
2. $K > 0$: $H$ is still constant but $R$ is not and one finds

$$R = 6\left(H_0^2 + \frac{K}{a^2}\right), \tag{6.228}$$

$$f = d_1 \cosh\left(\frac{\sqrt{2K}}{H_0\, a}\right) + d_2 \sinh\left(\frac{\sqrt{2K}}{H_0\, a}\right) + \frac{6\rho_{r0}\, H_0^4}{K^2} + \frac{3\rho_{m0}\, H_0^2}{a K} + \frac{6\rho_{r0}\, H_0^2}{a^2\, K} + \frac{\rho_{r0}}{a^4} + \frac{\rho_{m0}}{a^3}. \tag{6.229}$$

3. $K < 0$: the solution is

$$R = 6\left(H_0^2 + \frac{K}{a^2}\right), \tag{6.230}$$

$$f = d_1 \cos\left(\frac{\sqrt{-2K}}{H_0\, a}\right) + d_2 \sin\left(\frac{\sqrt{-2K}}{H_0\, a}\right) + \frac{6\rho_{r0}\, H_0^4}{K^2} + \frac{3\rho_{m0}\, H_0^2}{a\, K} + \frac{6\rho_{r0}\, H_0^2}{a^2\, K} + \frac{\rho_{r0}}{a^4} + \frac{\rho_{m0}}{a^3}. \tag{6.231}$$

- **$\Lambda$CDM solutions**
  Let us now look for $f(R)$ models compatible with the $\Lambda$CDM solutions of the Friedmann equations of GR, which are relevant in the confrontation of theoretical $f(R)$ models with observations (*e.g.*, [243]). Defining

$$\bar{H}^2 \equiv H_0^2\left(\frac{\Omega_{m0}}{a^3} + \frac{\Omega_{r0}}{a^4} + 1 - \Omega_{m0} - \Omega_{r0}\right) \tag{6.232}$$

with the present observational constraint $\Omega_{m0} + \Omega_{r0} = 1$, the differential equation to solve becomes

$$f'' + \left[\frac{6\Omega_{m0} H_0^2}{3\Omega_{m0} H_0^2 + 4ak} - \frac{4(\Omega_{m0} + \Omega_{r0} - 1)a^4 - 7\Omega_{m0}a - 8\Omega_{r0}}{-(\Omega_{m0} + \Omega_{r0} - 1)a^4 + \Omega_{m0}a + \Omega_{r0}}\right] \frac{f'}{2a}$$
$$- \frac{3\Omega_{m0} H_0^2 + 4ak}{2a\left[-(\Omega_{m0} + \Omega_{r0} - 1)a^4 + \Omega_{m0}a + \Omega_{r0}\right] H_0^2} f$$
$$= -\frac{\left(3\Omega_{m0} H_0^2 + 4ak\right)(\rho_{r0} + a\rho_{m0})}{2a^5\left[-(\Omega_{m0} + \Omega_{r0} - 1)a^4 + \Omega_{m0}a + \Omega_{r0}\right] H_0^2}, \tag{6.233}$$

which must be integrated numerically upon providing initial conditions $\left(f_0, f_0'\right)$.

## 6.5 Analytical cosmological solutions of $f\left(R, \Box R, \ldots, \Box^k R\right)$ gravity

We now apply the Noether symmetry approach to the search for exact solutions of a higher order gravity theory with more degrees of freedom than metric $f(R)$ gravity. The results show the extreme generality of the method.

### *6.5.1 Higher order point-like Lagrangians for cosmology*

A generic higher order theory in four dimensions is described by the action

$$S = \int d^4x \sqrt{-g}\, f(R, \Box R, \Box^2 R, \ldots, \Box^k R) \tag{6.234}$$

in units $\kappa = c = 1$. The field equations are [191, 237, 985]

$$G_{\mu\nu} = \frac{1}{D_\Sigma}\left\{\frac{1}{2} g_{\mu\nu}(f - D_\Sigma R) + \left(g_{\mu\lambda} g_{\nu\sigma} - g_{\mu\nu} g_{\lambda\sigma}\right) D_\Sigma^{;\lambda\sigma}\right.$$
$$+ \frac{1}{2}\sum_{i=1}^{k}\sum_{j=1}^{i}\left(g_{\mu\nu} g_{\lambda\sigma} + g_{\mu\lambda} g_{\nu\sigma}\right)\left(\Box^{j-i}\right)^{;\sigma}\left(\Box^{i-j}\frac{\partial f}{\partial \Box^i R}\right)^{;\lambda}$$
$$\left. - g_{\mu\nu} g_{\lambda\sigma}\left[\left(\Box^{j-1}\right)^{;\sigma} \Box^{i-j}\frac{\partial f}{\partial\left(\Box^i R\right)}\right]\right\}, \tag{6.235}$$

where

$$D_\Sigma = \sum_{j=0}^{k} \Box^j\left(\frac{\partial f}{\partial \Box^j R}\right). \tag{6.236}$$

These are vacuum field equations of order $(2k+4)$. Matter forms studied in the literature include a (non-)minimally coupled scalar field [233, 1144].

Let us restrict, for simplicity, to the Lagrangian density $f(R, \Box R)$. Then, we have eight order field equations which become sixth order equations if the function $f$ is linear in $\Box R$. In order to apply the Noether symmetry approach we use the point-like FLRW Lagrangian

$$L = L\left(a, \dot{a}, R, \dot{R}, \Box R, (\dot{\Box R})\right), \tag{6.237}$$

where $R$ and $\Box R$ are regarded as independent variables and time derivatives of order higher than one are eliminated using the method of Lagrange multipliers (see, *e.g.*, [1119] for the fourth order case). The action becomes

$$S = 2\pi^2 \int dt \left\{ a^3 f - \lambda_1 \left[ R - 6\left(\frac{\ddot{a}}{a} + \left(\frac{\dot{a}}{a}\right)^2 + \frac{K}{a^2}\right)\right] - \lambda_2 \left[ \Box R + \ddot{R} + 3\left(\frac{\dot{a}}{a}\right)\dot{R}\right]\right\}. \tag{6.238}$$

The Lagrange multipliers $\lambda_{1,2}$ are obtained by varying the action with respect to $R$ and $\Box R$,

$$\lambda_1 = a^3 \frac{\partial f}{\partial R}, \qquad \lambda_2 = a^3 \frac{\partial f}{\partial(\Box R)}. \tag{6.239}$$

Integrating by parts, the point-like Lagrangian is

$$L = 6a\dot{a}^2 \frac{\partial f}{\partial R} + 6a^2\dot{a}\frac{d}{dt}\left(\frac{\partial f}{\partial R}\right) - a^3\dot{R}\frac{d}{dt}\left(\frac{\partial f}{\partial(\Box R)}\right) + a^3\left[f - \left(R + \frac{6K}{a^2}\right)\frac{\partial f}{\partial R} - \Box R\frac{\partial f}{\partial(\Box R)}\right]. \tag{6.240}$$

Note that, alternatively, one could consider

$$\lambda_1 = a^3\left[\frac{\partial f}{\partial R} + \Box\left(\frac{\partial f}{\partial(\Box R)}\right)\right] \tag{6.241}$$

as a Lagrange multiplier [233], obtaining an alternative Lagrangian density differing from (6.240) only by a term vanishing on the constraint,

$$\tilde{L} = L - a^3\left\{R - 6\left[\frac{\ddot{a}}{a} + \left(\frac{\dot{a}}{a}\right)^2 + \frac{K}{a^2}\right]\right\}\Box\left(\frac{\partial f}{\partial(\Box R)}\right). \tag{6.242}$$

The Lagrangians $L$ and $\tilde{L}$ are equivalent.

Let us now derive the Euler-Lagrange equations from (6.240). The equation

$$\frac{d}{dt}\left(\frac{\partial L}{\partial \dot{a}}\right) - \frac{\partial L}{\partial a} = 0 \tag{6.243}$$

yields

$$\begin{aligned}
&R\,\frac{\partial f}{\partial R} + \Box R\,\frac{\partial f}{\partial(\Box R)} - f + 2\left(3H^2 + 2\dot{H} + \frac{K}{a^2}\right)\frac{\partial f}{\partial R} \\
&+2\left(\Box R - H\dot{R}\right)\frac{\partial^2 f}{\partial R^2} + \dot{R}(\dot{\Box R})\frac{\partial^2 f}{\partial(\Box R)^2} \\
&+\left[2\Box^2 R - 2H(\dot{\Box R}) + \dot{R}^2\right]\frac{\partial^2 f}{\partial R\partial(\Box R)} \\
&+2\dot{R}^2\,\frac{\partial^3 f}{\partial R^3} + 2(\dot{\Box R})^2\frac{\partial^3 f}{\partial R\partial(\Box R)^2} + 4\dot{R}(\dot{\Box R})\,\frac{\partial^3 f}{\partial R^2\partial(\Box R)} = 0\,,
\end{aligned} \tag{6.244}$$

while the other Euler-Lagrange equation

$$\frac{d}{dt}\left(\frac{\partial L}{\partial \dot{R}}\right) - \frac{\partial L}{\partial R} = 0 \tag{6.245}$$

gives

$$\Box\left(\frac{\partial f}{\partial(\Box R)}\right) = 0\,. \tag{6.246}$$

Finally,

$$\frac{d}{dt}\left(\frac{\partial L}{\partial(\dot{\Box R})}\right) - \frac{\partial L}{\partial \Box R} = 0 \tag{6.247}$$

coincides with the Lagrangian constraints

$$\Box R + \left(\ddot{R} + 3H\dot{R}\right) = 0\,, \tag{6.248}$$

$$R - 6\left(\dot{H} + 2H^2 + \frac{K}{a^2}\right) = 0\,. \tag{6.249}$$

The Hamiltonian constraint (time-time component of the field equations)

$$E_L \equiv \dot{a}\,\frac{\partial L}{\partial \dot{a}} + \dot{R}\,\frac{\partial L}{\partial \dot{R}} + (\dot{\Box R})\frac{\partial L}{\partial(\dot{\Box R})} - L = 0 \tag{6.250}$$

yields

$$H^2\left(\frac{\partial f}{\partial R}\right)+H\,\frac{d}{dt}\left(\frac{\partial f}{\partial R}\right)+\frac{\Gamma}{6}=0\,, \tag{6.251}$$

where

$$\Gamma=\left(R+\frac{6K}{a^2}\right)\frac{\partial f}{\partial R}+\Box R\,\frac{\partial f}{\partial(\Box R)}-f-\dot{R}\,\frac{d}{dt}\left(\frac{\partial f}{\partial(\Box R)}\right) \tag{6.252}$$

plays the role of an effective density [233].

### 6.5.2 The Noether symmetry approach for higher order gravities

As seen before, a Noether symmetry for the Lagrangian (6.240) exists if $\mathscr{L}_X L=0$ [230–232, 234, 235, 241]. In our case, the tangent space is

$$TQ=(a,\dot{a},R,\dot{R},\Box R,(\dot{\Box R})) \tag{6.253}$$

and the symmetry generator is

$$X=\alpha\,\frac{\partial}{\partial a}+\beta\,\frac{\partial}{\partial R}+\gamma\,\frac{\partial}{\partial(\Box R)}+\dot{\alpha}\,\frac{\partial}{\partial\dot{a}}+\dot{\beta}\,\frac{\partial}{\partial\dot{R}}+\dot{\gamma}\,\frac{\partial}{\partial(\dot{\Box R})}\,, \tag{6.254}$$

where $\alpha,\beta$, and $\gamma$ are functions of $(a,R,\Box R)$. A Noether symmetry exists if at least one of these functions is different from zero. Their analytic forms can be found by writing explicitly the condition $\mathscr{L}_X L=0$, which corresponds to a set of $1+\dfrac{n(n+1)}{2}$ partial differential equations obtained by equating to zero the quadratic terms containing $\dot{a}^2$, $\dot{R}^2$, $(\dot{\Box R})^2$, $\dot{a}\dot{R}$, *etc.* In our case, $n=3$ and one obtains the PDE system

$$\frac{\partial f}{\partial R}\left(\alpha+2a\,\frac{\partial\alpha}{\partial a}\right)+a\,\frac{\partial^2 f}{\partial R^2}\left(\beta+a\,\frac{\partial\beta}{\partial a}\right)+a\,\frac{\partial^2 f}{\partial R\partial(\Box R)}\left(\gamma+a\,\frac{\partial\gamma}{\partial a}\right)=0\,, \tag{6.255}$$

$$-6\,\frac{\partial^2 f}{\partial R^2}\frac{\partial\alpha}{\partial R}+\frac{\partial^2 f}{\partial R\partial(\Box R)}\left(3\alpha+2a\,\frac{\partial\beta}{\partial R}\right)+\beta a\,\frac{\partial^3 f}{\partial R^2\partial(\Box R)}+\gamma a\,\frac{\partial^3 f}{\partial R\partial(\Box R)^2}$$
$$+a\,\frac{\partial\gamma}{\partial R}\frac{\partial^2 f}{\partial(\Box R)^2}=0\,, \tag{6.256}$$

$$6\,\frac{\partial^2 f}{\partial R\partial(\Box R)}\,\frac{\partial\alpha}{\partial(\Box R)}-a\,\frac{\partial^2 f}{\partial(\Box R)^2}\,\frac{\partial\beta}{\partial(\Box R)}=0\,, \tag{6.257}$$

$$12\,\frac{\partial f}{\partial R}\,\frac{\partial \alpha}{\partial R}+6\,\frac{\partial^2 f}{\partial R^2}\left(2\alpha+a\frac{\partial \alpha}{\partial a}+a\,\frac{\partial \beta}{\partial R}\right)+a\,\frac{\partial^2 f}{\partial R\partial(\Box R)}\left(6\,\frac{\partial \gamma}{\partial R}-2a\,\frac{\partial \beta}{\partial a}\right)$$

$$-a^2\,\frac{\partial \gamma}{\partial a}\,\frac{\partial^2 f}{\partial(\Box R)^2}+6\beta a\,\frac{\partial^3 f}{\partial R^3}+6\gamma a\,\frac{\partial^3 f}{\partial R^2\partial(\Box R)}=0\,, \tag{6.258}$$

$$12\,\frac{\partial f}{\partial R}\,\frac{\partial \alpha}{\partial(\Box R)}+\frac{\partial^2 f}{\partial R\partial(\Box R)}\left(12\alpha+6a\,\frac{\partial \alpha}{\partial a}+6a\,\frac{\partial \gamma}{\partial(\Box R)}\right)+6a\,\frac{\partial^2 f}{\partial R^2}\,\frac{\partial \beta}{\partial(\Box R)}$$

$$-a^2\,\frac{\partial^2 f}{\partial(\Box R)^2}\,\frac{\partial \beta}{\partial a}+6\gamma a\,\frac{\partial^3 f}{\partial R\partial(\Box R)^2}+6\beta a\,\frac{\partial^3 f}{\partial R^2\partial(\Box R)}=0\,, \tag{6.259}$$

$$\frac{\partial^2 f}{\partial(\Box R)^2}\left(3\alpha+a\frac{\partial \gamma}{\partial(\Box R)}+a\frac{\partial \beta}{\partial R}\right)+\frac{\partial^2 f}{\partial R\partial(\Box R)}\left(2a\frac{\partial \beta}{\partial(\Box R)}-6\frac{\partial \alpha}{\partial R}\right)-6\frac{\partial^2 f}{\partial R^2}\,\frac{\partial \alpha}{\partial \Box R}$$

$$+\beta a\,\frac{\partial^3 f}{\partial R\partial(\Box R)^2}+\gamma a\,\frac{\partial^3 f}{\partial(\Box R)^3}=0\,, \tag{6.260}$$

$$3\alpha\left(f-R\frac{\partial F}{\partial R}\right)-\beta a R\,\frac{\partial^2 f}{\partial R^2}-3\alpha\Box R\,\frac{\partial f}{\partial(\Box R)}-\gamma a\Box R\,\frac{\partial^2 f}{\partial(\Box R)^2}$$

$$-a\,\frac{\partial^2 f}{\partial R\partial(\Box R)}(\beta\Box R+\gamma R)-\frac{6K}{a^2}\left[\alpha\frac{\partial f}{\partial R}+\beta a\,\frac{\partial^2 f}{\partial R^2}+\gamma a\,\frac{\partial^2 f}{\partial R\partial(\Box R)}\right]=0\,. \tag{6.261}$$

This system is overdetermined and, if it admits solutions, enables one to assign $\alpha$, $\beta, \gamma$, and $f\,(R, \Box R)$. Then, one can transform the Lagrangian (6.240) so that

$$L\left(a, \dot{a}, R, \dot{R}, \Box R, (\dot{\Box R})\right) \rightarrow L\,(u, \dot{u}, w, \dot{w}, \dot{z})\,, \tag{6.262}$$

where $z$ is a cyclical variable and the dynamics are thus simplified. This change of variables can be easily obtained by the conditions

$$i_X dz=\alpha\,\frac{\partial z}{\partial a}+\beta\,\frac{\partial z}{\partial R}+\gamma\,\frac{\partial z}{\partial \Box R}=1\,, \tag{6.263}$$

$$i_X dw=\alpha\,\frac{\partial w}{\partial a}+\beta\,\frac{\partial w}{\partial R}+\gamma\,\frac{\partial w}{\partial \Box R}=0\,, \tag{6.264}$$

$$i_X du=\alpha\,\frac{\partial u}{\partial a}+\beta\,\frac{\partial u}{\partial R}+\gamma\,\frac{\partial u}{\partial \Box R}=0\,. \tag{6.265}$$

Once the dynamics are solved using the variables $(z, w, u)$, the inverse transformation

$$(z(t), w(t), u(t)) \rightarrow (a(t), R(t), \Box R(t)) \tag{6.266}$$

allows its description in terms of the original variables. Note, however, that we are considering constrained dynamics since the variables $a$, $R$, $\Box R$ are related.

If $f(R, \Box R)$ depends only linearly on $\Box R$ the theory is of sixth order, otherwise it is of eighth order. A sixth-order solution of the Noether system (6.255)–(6.261) is recovered if

$$\alpha = \frac{\alpha_0}{\sqrt{a}}, \qquad \text{any } (\beta, \gamma), \qquad f(R, \Box R) = f_1 R + f_2 \Box R, \quad K = 0. \quad (6.267)$$

According to the previous discussion [675,1144], this theory reduces to GR in which standard cosmological solutions are recovered.

If the function $f$ depends on powers of $\Box R$, Noether symmetries are given by

$$\alpha = 0, \quad \beta = \beta_0, \quad \gamma = 0, \quad f(R, \Box R) = f_1 R + f_2 (\Box R)^n, \quad n \geq 2. \quad (6.268)$$

However, the theory can assume different forms following integration by parts [675, 985, 1144]. The equations of motion are

$$4 f_1 \dot{H} + 6 f_1 H^2 + \frac{2 K f_1}{a^2} - f_2 (1-n)(\Box R)^n - \left(\frac{\dot{R}}{a}\right) \Sigma_0 = 0, \quad (6.269)$$

$$-f_2 n(n-1) a^3 (\Box R)^{n-2} (\dot{\Box R}) = \Sigma_0, \quad (6.270)$$

$$\Box R + \ddot{R} + 3 H \dot{R} = 0, \quad (6.271)$$

$$6 f_1 \left(H^2 + \frac{K}{a^2}\right) - f_2 (1-n)(\Box R)^n + \dot{R} \Sigma_0 = 0, \quad (6.272)$$

where $\Sigma_0$ is the Noether charge and $R$ is the cyclic variable. The standard gravitational coupling is recovered, as usual, for $f_1 = 1/2$. A solution for this system is

$$a(t) = a_0 t \quad (6.273)$$

for $K = -1$, $\Sigma_0 = 0$, and for arbitrary $n$ and $f_2$. Another solution is

$$a(t) = a_0 \sqrt{t} \quad (6.274)$$

for $K = 0$, $\Sigma_0 = 0$, $f_1 = 0$ and for arbitrary $n$ and $f_2$. Finally, one obtains

$$a(t) = a_0 \exp(k_0 t) \quad (6.275)$$

for $K = 0$, $\Sigma_0 = 0$, and $f_1 = 0$.

The radiative solution $a(t) = a_0 \sqrt{t}$ is obtained for the cases

$$\alpha = 0, \quad \beta = \frac{\beta_0}{a}, \quad \text{any } \gamma, \quad f(R, \Box R) = f_1 R + f_2 R^2 + f_3 \Box R, \quad (6.276)$$

and

$$\alpha = 0\,, \quad \beta = 0\,, \quad \gamma = \frac{\gamma_0}{a}\,, \quad f\left(R, \Box R\right) = f_1 R + f_2 R^2 + f_3 R \Box R\,, \tag{6.277}$$

with $\gamma_0 = \text{const}$. The second situation is physically interesting and the related cosmological models have been the subject of attention [129, 550].

Another Noether symmetry is obtained for

$$\alpha = 0\,, \quad \beta = \beta_0\,, \quad \gamma = \beta_0 \frac{\Box R}{R}\,, \quad f\left(R, \Box R\right) = f_1 R + f_2 \sqrt{R \Box R} \tag{6.278}$$

or simply for

$$f\left(R, \Box R\right) = f_2 \sqrt{R \Box R}\,. \tag{6.279}$$

This case deserves attention because $\sqrt{R \Box R}$ is a part of the $a_3$-anomaly [550] occurring in the analysis of first loop corrections to the gravitational action [538, 539, 1176].

The straightforward change of variables (6.263)–(6.265) yields

$$z = R\,, \quad u = \sqrt{\frac{\Box R}{R}}\,, \quad w = a\,. \tag{6.280}$$

By selecting the standard Einstein coupling $f_1 = 1/2$, the Lagrangian (6.240) becomes

$$L = 3\left(w\dot{w}^2 - Kw\right) - f_2\left(3w\dot{w}^2 u + 3w^2 \dot{w}\dot{u} + \frac{w^3 \dot{z}\dot{u}}{2u^2} - 3Kwu\right)\,, \tag{6.281}$$

where $z$ (and therefore $R$) is the cyclic variable. The equations of motion are derived from (6.281). Again, the particular solutions

$$a(t) = a_0 t\,, \quad a(t) = a_0 \sqrt{t}\,, \quad a(t) = a_0 \exp(k_0 t)\,, \tag{6.282}$$

which depend on the set of parameters $(\Sigma_0, K, f_2)$ are obtained. A phase space and conformal analysis [129] provides the conditions for the onset and the duration of inflation, which depend on the signs and values of $f_2$ and $\Sigma_0$, restricting the set of initial conditions providing an adequate amount of inflation [129, 550]. Table 6.1 summarizes the results.

**Table 6.1** Symmetries in models of order higher than fourth.

| $f\left(R, \Box R\right)$ | $\alpha$ | $\beta$ | $\gamma$ |
|---|---|---|---|
| $f_1 R + f_2 (\Box R)^n \; (n \neq 1)$ | 0 | $\beta_0$ | 0 |
| $f_1 R + f_2 R^2 + f_3 \Box R$ | 0 | $\beta_0 a^{-1}$ | $\gamma(a, R, \Box R)$ |
| $f_0 R + f_1 R^2 + f_2 R \Box R$ | 0 | 0 | $\gamma_0\, a^{-1}$ |
| $f_2 \sqrt{R (\Box R)}$ | 0 | $\beta_0$ | $\beta_0 R^{-1} \Box R$ |
| $f_1 R + f_2 \sqrt{R (\Box R)}$ | 0 | $\beta_0$ | $\beta_0 R^{-1} \Box R$ |

## 6.6 Conclusions

The exact solutions of scalar-tensor, metric $f(R)$, and $f\left(R, \Box R, \ldots, \Box^k R\right)$ gravity presented give a flavour of the particular features of these classes of theories, and of what can be obtained by allowing the gravitational theory to include extra degrees of freedom. When analytical solutions cannot be found, or reveal themselves to be too special for the task at hand, a phase space analysis can be extremely useful when special symmetries are present, such as in FLRW cosmology or with spherical symmetry. The geometry of the phase space of scalar-tensor FLRW cosmology is rather involved.

In the presence of symmetries, a theory lends itself to the Noether approach for the search of analytical solutions. This approach has been amply illustrated in this chapter. The Noether technique requires higher order terms, such as $R^{3/2}$ or $\sqrt{R\Box R}$, which are physically relevant because they are related to first loop or trace anomaly contributions to the effective action of quantum gravity. For illustration, the PDE system (6.255)–(6.261) can have several solutions. We have not presented an exhaustive list of possible Noether symmetries in higher order theories, but we have restricted to examples in fourth, sixth, and eighth order models. Of course, obtaining higher derivative terms in the effective Lagrangian of gravity does not, automatically make the theory renormalizable [107].

The presence of a Noether symmetry makes the analysis of a given cosmology easier, however the existence of an abstract symmetry is not, by itself, a criterion for preferring a particular theory over another. The situation becomes far more interesting when the theory is physically relevant *per se* and is also the only one selected by the Noether approach [230, 235, 805].

# Chapter 7
# Cosmology

*Entia non sunt multiplicanda praeter necessitatem.*
*– William of Ockham*

The standard Big Bang model of the universe is a very successful description of the universe around us. Its success is supported by three major pieces of evidence: (1) the expansion of the universe, which constitutes the most natural interpretation of the redshift of galaxies; (2) the existence of a cosmic background of microwaves, a relic of the primordial state of high density, pressure and temperature (the predicted spectrum of the cosmic microwave background is that of a blackbody with temperature of 2.73 K, in remarkable agreement with the observations), and (3) primordial nucleosynthesis: the relative abundances of hydrogen, deuterium, helium, and a few other light elements are predicted and are in agreement with observations.

If GR is replaced by an ETG in the standard Big Bang model, one can still find many spatially homogeneous and isotropic solutions that are in qualitative and quantitative agreement with the observed recession of galaxies, spectrum and temperature of the cosmic microwave background, and primordial nucleosynthesis [390, 554, 555, 1153]. Primordial nucleosynthesis imposes constraints on alternative gravity, but these can be satisfied by many scenarios.

A consistent cosmological theory necessarily requires a relativistic theory of gravitation, and cosmology was developed after Einstein's formulation of GR. Following the 1998 discovery of the acceleration of the cosmic expansion, modified theories of gravity at large scales were proposed as explanations alternative to dark energy. An important motivation for the study of ETGS, therefore, is provided by cosmological observations. In this chapter we recall the standard equations of Big Bang and inflationary cosmology in GR, which will then be compared with the corresponding field equations of ETGs (for comprehensive introductions to cosmology we refer the reader to standard textbooks [688, 728, 1139, 1153]). We focus on the present accelerated epoch, discuss cosmography as a phenomenological formalism able to parameterize a wide spectrum of gravitational theories, and we apply it to metric $f(R)$ gravity. Then we proceed to discuss galaxy clusters and show how modified gravity can in principle replace dark matter at the cluster scale.

S. Capozziello and V. Faraoni, *Beyond Einstein Gravity: A Survey of Gravitational Theories for Cosmology and Astrophysics*, Fundamental Theories of Physics 170,
DOI 10.1007/978-94-007-0165-6_7, © Springer Science+Business Media B.V. 2011

## 7.1 Big Bang, inflationary, and late-time cosmology in GR

The FLRW line element

$$ds^2 = -dt^2 + a^2(t)\left[\frac{dr^2}{1-Kr^2} + r^2\left(d\theta^2 + \sin^2\theta\, d\varphi^2\right)\right] \tag{7.1}$$

describes spatially homogeneous and isotropic universes in comoving time $t$ and polar coordinates $(r, \theta, \varphi)$ and contains the normalized curvature index $K$ which can assume the values $0, +1$, or $-1$ corresponding to flat, positively, or negatively curved spatial sections, respectively. Thanks to these symmetries, and assuming that the universe is filled with a perfect fluid described by the stress-energy tensor $T_{\mu\nu}^{(m)} = (P+\rho)\, u_\mu u_\nu + P g_{\mu\nu}$, the Einstein equations for this metric reduce to the ordinary differential equations for the scale factor $a(t)$

$$H^2 = \frac{\kappa}{3}\,\rho - \frac{K}{a^2}, \tag{7.2}$$

$$\frac{\ddot{a}}{a} = \dot{H} + H^2 = -\frac{\kappa}{6}\,(\rho + 3P)\,, \tag{7.3}$$

where an overdot denotes differentiation with respect to the comoving time $t$, $H \equiv \dot{a}/a$ is the Hubble parameter, and $\rho$ and $P$ are the energy density and pressure of the cosmic fluid, respectively. A cosmological constant $\Lambda$, if present, will be described as a perfect fluid with energy density $\rho^{(\Lambda)}$ and pressure $P^{(\Lambda)}$ related by

$$P^{(\Lambda)} = -\rho^{(\Lambda)} = \frac{\Lambda}{\kappa}. \tag{7.4}$$

The energy density and pressure of the cosmic fluid satisfy the conservation equation

$$\dot{\rho} + 3H\,(P+\rho) = 0, \tag{7.5}$$

which follows from the covariant conservation $\nabla^\nu T_{\mu\nu}^{(m)} = 0$ for the energy-momentum tensor of the cosmic fluid. A critically open ($K = 0$), spatially flat, universe has energy density

$$\rho_c = \frac{3H^2}{\kappa} \tag{7.6}$$

(*critical density*). If many fluids are present in the universe, the energy density $\rho^{(i)}$ of the $i$-th component of this multi-component fluid can be expressed in units of the critical density by introducing the corresponding dimensionless *density parameter*

$$\Omega^{(i)} \equiv \frac{\rho^{(i)}}{\rho_c}. \tag{7.7}$$

Since $\rho = \sum_i \rho^{(i)}$, the Hamiltonian constraint (7.2) in a $K = 0$ universe is written as

$$1 = \sum_i \Omega^{(i)}. \tag{7.8}$$

### 7.1.1 *The standard Big Bang model*

By assuming reasonable equations of state for the cosmic fluid in the early universe, such as the radiation equation of state $P = \rho/3$, or even the dust equation of state $P = 0$, it is found that $\ddot{a} < 0$, which leads to a singularity at a finite time in the past (Big Bang) and, for $K = +1$ universes, to a maximum size of the universe which is reached and followed by collapse to a Big Crunch singularity [688,728,1139,1153]. The standard Big Bang model makes three fundamental predictions: the expansion of the universe and the redshift of galaxies, the existence, temperature, and blackbody spectrum of the cosmic microwave background (CMB), and the relative abundances of light elements produced during primordial nucleosynthesis in the radiation-dominated era. The observation of these phenomena established the Big Bang model as a highly successful description of the universe.

### 7.1.2 *Inflation in the early universe*

Beginning in the 1970s, certain shortcomings of the Big Bang model started being noticed, most notably the *flatness problem* (why is the universe so close to being spatially flat today given that any initial departure from a $K = 0$ model gets amplified during the dynamical evolution?), the *horizon problem* (why photons of the CMB coming from opposite directions today have the same temperature to high precision while the size of causally connected regions at the last scattering is at most one degree?), and the *monopole problem* (why monopoles predicted so profusely in grand unified theories are not observed, and why similar relics from early epochs failed to dominate the dynamics of the universe given that they are so massive?) [688, 728]. In the context of the standard Big Bang model, these three problems can only be "solved" by imposing extremely fine-tuned initial conditions, which is hardly acceptable in a physical theory. In 1980 the idea was advanced that a brief and accelerated ($\ddot{a} > 0$) expansion of the universe by a factor of approximately $e^{60}$ can solve all these problems [564,741,742,1044]. In the most common inflationary models, the universe is dominated by a scalar field $\phi$ minimally coupled to the Ricci curvature and with potential energy density $V(\phi)$, as in the action

$$S = \int d^4x \sqrt{-g} \left[ \frac{R}{2\kappa} - \frac{1}{2} g^{\mu\nu} \nabla_\mu\phi \nabla_\nu\phi - V(\phi) \right]. \tag{7.9}$$

We assume here that $\phi$ dominates the dynamics of the universe during inflation and we do not include other forms of matter in the picture. The stress-energy tensor of $\phi$ has the canonical form

$$T_{\mu\nu} = \nabla_\mu \phi \, \nabla_\nu \phi - \frac{1}{2} g_{\mu\nu} \nabla^\alpha \phi \, \nabla_\alpha \phi - g_{\mu\nu} V(\phi). \tag{7.10}$$

As it is well known, this energy-momentum tensor can be cast in the form of a perfect fluid stress-energy tensor with four-velocity $u^\mu = \frac{\nabla^\mu \phi}{\sqrt{|\nabla^\alpha \phi \nabla_\alpha \phi|}}$ and with energy density and pressure

$$\rho_\phi = T_{\mu\nu} u^\mu u^\nu = \frac{\dot{\phi}^2}{2} + V(\phi), \tag{7.11}$$

$$P_\phi = T_{\mu\nu} h^{\mu\nu} = \frac{T_{ii}}{g_{ii}} = \frac{\dot{\phi}^2}{2} - V(\phi), \tag{7.12}$$

where $h_\mu{}^\nu$ is the projection operator on the three-dimensional spatial sections defined by $h_{\mu\nu} \equiv g_{\mu\nu} + u_\mu u_\nu$. The scalar field potential $V(\phi)$ is usually taken to be positive because during inflation, in which $\dot{\phi} \simeq 0$ and $\rho \simeq V(\phi)$, the universe is dominated by the potential of the scalar and must be non-negative. The Friedmann equations (7.2) and (7.3) then assume the form

$$H^2 = \frac{\kappa}{3} \left[ \frac{\dot{\phi}^2}{2} + V \right] - \frac{K}{a^2}, \tag{7.13}$$

$$\dot{H} = -H^2 - \frac{\kappa}{3} \left[ \dot{\phi}^2 - V \right], \tag{7.14}$$

while the scalar $\phi$ satisfies the Klein-Gordon equation

$$\ddot{\phi} + 3H\dot{\phi} + \frac{dV}{d\phi} = 0. \tag{7.15}$$

The curvature term can be omitted from eq. (7.13) because, even starting from anisotropic initial conditions, the universe evolves dynamically to a state extremely close to a spatially flat FLRW universe during inflation (*cosmic no-hair*). With a few exceptions, $K = 0$ is a robust prediction of inflation [688, 728].

Only two of the three equations (7.13)–(7.15) are independent. In fact, if $\dot{\phi} \neq 0$, the Klein-Gordon equation (7.15) can be derived from the conservation equation (7.5) or, equivalently, from Eqs. (7.13) and (7.14).

The equations of inflation (7.13)–(7.15) and those satisfied by cosmological perturbations are usually solved in the *slow-roll approximation*

$$\frac{\dot{\phi}^2}{2} \ll V, \qquad \ddot{\phi} \ll H\dot{\phi}, \tag{7.16}$$

in which Eqs. (7.13)–(7.15) reduce to

$$H^2 \simeq \frac{\kappa V}{3}, \tag{7.17}$$

$$3H\dot{\phi} \simeq -\frac{dV}{d\phi}. \tag{7.18}$$

The slow-roll approximation assumes that the solution $\phi(t)$ of the field equations rolls slowly, *i.e.*, with negligible "friction" $\dot{\phi}$ over a shallow segment of its potential $V(\phi)$. The latter then mimics the effect of a cosmological constant $\Lambda \simeq V(\phi)$ making the cosmic expansion almost de Sitter,

$$a(t) = a_0 \exp[H(t)\,t], \tag{7.19}$$

where

$$H(t) = H_0 + H_1 t + \cdots \tag{7.20}$$

and the constant term $H_0$ dominates in the expression of $H(t)$. Generally speaking, there exist de Sitter solutions $H$ = const., $\phi$ = const. which are attractors in the phase space of the dynamical system (7.13)–(7.15), and this fact justifies the use of the slow-roll approximation, which turns out to be a generic phenomenon. Inflation stops when the potential $V(\phi)$ ends its plateau and quickly decreases to a zero minimum. Then, $\phi$ quickly accelerates toward this minimum, overshoots it, and oscillates around it. These oscillations are damped by particle creation due to the explicit coupling of $\phi$ to other fields (or even to the Ricci curvature if $\phi$ is non-minimally coupled [111]), a phenomenon called *reheating* which dissipates the kinetic energy of $\phi$ and raises the temperature of the universe which has dropped during the inflationary expansion. This superluminal expansion, if it lasts for approximately 60 e-folds of expansion of the scale factor $a(t)$, solves the flatness, horizon and monopole problems of the standard Big Bang model. What is more, it provides a mechanism for the generation of density perturbations via quantum fluctuations of the scalar field (which are accompanied by gravitational waves, tensor mode fluctuations of the metric tensor) and correspond to scalar density perturbations which will later seed the formation of large scale structures. Cosmic structures begin to grow after the end of the inflationary and radiation epochs. For practical purposes, inflation ends when the slow-roll parameters

$$\varepsilon \equiv \frac{1}{2\kappa}\left(\frac{V'}{V}\right)^2, \qquad \eta \equiv \frac{1}{\kappa}\frac{V''}{V} \tag{7.21}$$

(where a prime denotes differentiation with respect to $\phi$) become of order unity. The subject of inflation will be reconsidered in Chap. 8 in connection with ETGs.

### 7.1.3 *The present-day acceleration*

As already mentioned, it was discovered in 1998 that the present expansion of the universe is accelerated, a finding that shocked the community of cosmologists.

Evidence for the cosmic acceleration came from the observation of SNeIa (SNeIa) [936,982], which were adopted as standard candles because of their high intrinsic luminosity and because they possess a characteristic period-luminosity relation which makes it possible to assess their absolute magnitude, and hence their luminosity distance, from their light curve [163,548]. A systematic search for SNeIa was launched after a long time of steady progress in the study of such objects and of their light curves. The search revealed that these supernovae are fainter than it was expected at that time, which is interpreted as an effect of the recent expansion history of the universe (accelerated versus decelerated). At a fixed redshift, these supernovae are further away than they would be in a decelerating universe. For moderate redshifts $z$ the luminosity distance ($D_L$)-redshift relation is

$$H_0\, D_L = z + \frac{1}{2}\,(1 - q_0)\, z^2 + \cdots\,, \tag{7.22}$$

where $q \equiv -\ddot{a}a/(\dot{a})^2$ is the deceleration parameter. In an accelerating universe with $q < 0$, $D_L$ is larger than it would be in a decelerating universe with $q > 0$. Since the universe has accelerated its expansion in the past, light from a supernova has travelled a larger distance to reach us and the supernova looks dimmer than it would be in a decelerated universe.

The cosmic acceleration has also the advantage that it helps reconciling the age of the universe with that of globular clusters, which has been a problem for theorists in the past.

The data, first from the *BOOMERANG* [377,706,791,797], *MAXIMA* [570] and similar experiments, and then from the *WMAP* satellite [1038], provide the picture of a spatially flat universe with total energy density (in units of the critical density)

$$\Omega = \Omega^{(m)} + \Omega^{(q)} = 1 \tag{7.23}$$

where $\Omega^{(m)} \simeq 0.24$ is the matter energy density, and the rest is dark matter, and $\Omega^{(q)} \simeq 0.76$ describes an unknown form of energy called *dark energy* and unclustered [411,491,902,936–939,982]. The acceleration equation of GR

$$\frac{\ddot{a}}{a} = -\frac{\kappa}{6}\,(\rho + 3P) \tag{7.24}$$

implies that an accelerated expansion is possible if and only if the effective equation of state of the dominant cosmic fluid is such that $P < -\rho/3$. The analysis of the available data implies that $P \simeq -\rho$. This fact is at odds with the $P = 0$ equation of state of dark matter and constitutes a very exotic property of dark energy.

In the early days of research on dark energy, observational reports used to quote the constraints on the effective equation of state parameter of dark energy $w \equiv P/\rho$ as $-1 \leq w \leq -1/3$ (with the upper bound decreasing as the observations were getting better and better). It was then realized that values $w < -1$ were also compatible with the data, and it was even argued that the range of values $w \leq -1$ is favored [202,356,358,359,502,1177,1178].

A large amount of theoretical work went into building models of dark energy. An obvious candidate for dark energy is the cosmological constant $\Lambda$, which is characterized by the constant equation of state $P_\Lambda = -\rho_\Lambda$ and, as will be discussed below, fits the data better than most other models [140, 271, 955, 956]. However, explaining the cosmic acceleration with the cosmological constant brings back in a more severe form the old cosmological constant problem [1154, 1155], *i.e.*, the value of the vacuum energy density $\rho_\Lambda$ is approximately 120 orders of magnitude smaller than its natural Planck scale value $\Lambda \sim c^3/(\hbar G)$), and 40 orders of magnitude smaller than the value predicted if a cutoff at the QCD scale is introduced. Assuming that the cosmological constant is cancelled almost exactly by an unknown mechanism which nevertheless leaves behind an extremely tiny residual just sufficient to explain the cosmic acceleration, corresponds to extreme fine-tuning. Most authors are more inclined to believe that the cosmological constant is exactly cancelled, although a plausible mechanism to achieve this cancellation is unknown, and that the cosmic acceleration is due to some other form of energy, thus avoiding the need to leave the extremely small residual.

A second problem with the cosmological constant is posed by the fact that the dark energy just begins to dominate the cosmic dynamics at redshift $z \sim 1$, following the radiation- and matter-dominated epochs. In order for this to happen around the present time, the energy density of the cosmological constant must be fine-tuned: dark energy must be subdominant at the time of primordial nucleosynthesis, or it would affect the expansion rate of the universe and modify the observed relative abundances of light elements. Dark energy must also be negligible during the matter-dominated era or else the growth of density perturbations would be compromised. Then the problem arises of why the dark energy begins to dominate the dynamics of the universe only now when there are observers to notice this fact (*cosmic coincidence problem*). In other words, the densities of matter and dark energy were very different in the past and will be very different in the future, so why is it that they are approximately equal in the short epoch of the cosmic history in which we live?

Rejecting the cosmological constant explanation, most authors prefer dynamical models in which dark energy is time-dependent and begins to dominate during the matter era. Many models of dark energy have been proposed: we refer the reader to [743] for a resource letter on dark energy with a detailed list of references and to [971] for a recent review. Most models of dark energy are based on a minimally coupled scalar field (*quintessence* field) rolling in a potential, in the theoretical framework of GR. This is not surprising because it is well known from inflationary theories that a scalar field slowly rolling on a flat section of its potential can accelerate the expansion of the universe and is equivalent to a perfect fluid. The fact that the scalar field is dynamical provides a time-dependent effective equation of state, which is a necessary feature to solve the cosmic coincidence problem.

In the simplest models based on a single minimally coupled scalar field, if the potential $V(\phi)$ satisfies the condition [1050]

$$\Gamma \equiv \frac{V \frac{d^2V}{d\phi^2}}{\left(\frac{dV}{d\phi}\right)^2} \geq 1, \tag{7.25}$$

the phase space of the field equations contains late time attractors with a very large attraction basin. The condition (7.25) is useful because, for a given potential, one can sometimes test for the presence of these attractors by checking the condition $\Gamma > 1$ without solving explicitly the field equations. These attractor solutions are time-dependent (not fixed points in the $(H, \phi)$ phase space) and they exist for a wide class of potentials used in the literature. They are known as *tracking solutions* [732, 1050, 1180] and they were studied before the discovery of the cosmic acceleration as mechanisms to implement an effective time-dependent cosmological "constant" [487,488,509,923,1156] which otherwise would have had to be assumed *ad hoc* by assigning $\Lambda$ as a function of time or of the scale factor. The idea of a time-dependent vacuum energy is not new (see [875] for a review): the original reason to pursue it, which has now only historical importance, was the attempt to reconcile a low value of the matter density $\Omega^{(m)}$ observed in the 1990s with the inflationary prediction $\Omega^{(m)} + \Omega^{(\Lambda)} = 1$.

The solution of a tracking model converges to its attractor before the present era for a very large set of initial conditions spanning even 150 orders of magnitude, quintessence begins to dominate the dynamics of the universe after the matter era, and equipartition $\Omega^{(q)} = \Omega^{(m)}$ occurs at redshifts $z \sim 1$ (before that time, there is some evidence of a decelerated expansion of the universe from the supernova SN1997ff at $z = 1.7$ [540, 939] and at least two other supernovae at $z = 1.2$ [1077, 1093]). The cosmic coincidence problem is then, at least in principle, resolved, contrary to what happens with a cosmological constant which has constant effective energy density.

The energy density of quintessence $\rho_q$ redshifts more slowly than the energy densities of ordinary matter and radiation and comes to dominate late in the matter era, even if its numerical value was initially negligible because the ratio $\rho_{(m)}/\rho_{(q)}$ decreases during the cosmic expansion, until it becomes less than unity and $\rho_{(q)}$ begins to dominate. The equipartition time is determined by the energy scale of the quintessence potential, which is fixed by the requirement that $\Omega_0 \simeq 1$ today. This scale is very small on natural particle physics scales, and generally needs to be fine-tuned in order to reproduce the cosmic dynamics that we know. Hence, it is fair to say that all models of quintessence suffer from some degree of fine-tuning.

During the radiation- and matter-dominated epochs, quintessence tracks the dominant form of energy (radiation or dust) and emerges only later. The radiation energy density $\rho_{(r)} \propto a^{-4}$ initially dominates, but it redshifts faster than the matter energy density $\rho_{(m)} \propto a^{-3}$. In an analogous way, the quintessence energy density $\rho_{(q)}$ is initially much smaller than $\rho_{(m)}$ but redshifts more slowly and eventually comes to dominate.

The nature of the suggested quintessence field is completely speculative: some attempts have been made to relate this field to known particle physics, *e.g.*, [316], to identify quintessence with a supergravity field [166, 167, 339, 949], an axion, the string dilaton, or moduli fields in string theories [115, 314, 315, 503, 504, 528, 529, 585, 588, 669, 837, 1085]. None of these attempts is convincing at this time.

The available data do not exclude that dark energy could be of the very exotic form called *phantom energy*. The constraints on the dark energy equation of state are in agreement with values of the parameter $w < -1$ [358, 571, 792, 974]. The possibility that $w < -1$ was investigated by several authors [202,305,505,723,726, 792,872,997,1177,1178] using models based on a scalar field with the "wrong" sign of the kinetic energy term of the scalar (*phantom field*), or other modifications of the Lagrangian. It has also been argued that stringy matter requires $w_q < -1$ [504]. Dark energy with the extreme equation of state $w_q < -1$ is known as *phantom energy*. A fundamental phantom field is extremely unstable from both the classical and quantum points of view [276,287] and, at best, could be admissible only as part of an effective theory. However, a model of phantom energy not relying on Einstein gravity could be more plausible because of the following considerations. In GR the inequality $P < -\rho$ is equivalent to $\dot{H} > 0$, and the Hubble parameter satisfies the equations

$$H^2 = \frac{\kappa}{3}\rho, \tag{7.26}$$

$$\frac{\ddot{a}}{a} = \dot{H} + H^2 = -\frac{\kappa}{6}\left(\rho + 3P\right). \tag{7.27}$$

Assuming for simplicity that a single quintessence fluid dominates the cosmic expansion, Eqs. (7.26) and (7.27) imply that

$$\dot{H} = -\frac{\kappa}{2}\left(\rho_{(q)} + P_{(q)}\right) \tag{7.28}$$

and therefore $P_{(q)} < -\rho_{(q)}$ (or $w_q < -1$) is equivalent to $\dot{H} > 0$. If the matter component has a non-negligible energy density, the pressure of quintessence must be even more negative. A regime with $\dot{H} > 0$ (originally investigated in the context of inflationary models [754]) is known as *superinflation* or, in the context of the present-day epoch of the universe, *superacceleration* [446,450,455,934]. If our universe is truly superaccelerating today, models of quintessence based on a single scalar field minimally coupled to gravity would not be able to explain this superacceleration. In fact, the scalar field $\phi(t)$ has energy density and pressure

$$\rho = \frac{\dot{\phi}^2}{2} + V(\phi), \tag{7.29}$$

$$P = \frac{\dot{\phi}^2}{2} - V(\phi). \tag{7.30}$$

Then, Eq. (7.28) becomes

$$\dot{H} = -\frac{\kappa}{2}\dot{\phi}^2 \tag{7.31}$$

and $\dot{H} \leq 0$ in any minimally coupled GR model. The case $\dot{H} = 0$ corresponds to a de Sitter solution with scale factor $a(t) = a_0\, \mathrm{e}^{Ht}$. For any potential $V(\phi)$, the effective equation of state of a FLRW universe dominated by a minimally coupled scalar field is

$$\frac{P}{\rho} = \frac{\dot{\phi}^2 - 2V}{\dot{\phi}^2 + 2V} \equiv w(x), \tag{7.32}$$

where $x \equiv \dot{\phi}^2/2V$ is the ratio between the kinetic and the potential energy densities of the scalar. Assuming that $V \geq 0$, which guarantees positivity of the energy density of $\phi$,

$$w(x) = \frac{x-1}{x+1} \tag{7.33}$$

increases monotonically from its minimum $w_{min} = -1$ attained at $x = 0$ to its horizontal asymptote approached as $x \to +\infty$, corresponding to $V = 0$. At best, the effective equation of state parameter of a minimally coupled scalar field spans the range

$$-1 \leq w \leq 1. \tag{7.34}$$

If the effective equation of state of the cosmic fluid in the present era is such that $w < -1$, it cannot be explained by this canonical scalar field model for quintessence unless one resorts to the problematic and unnatural phantom field.

Superacceleration regimes known in pre-1998 literature consist of pole-like inflation with scale factor

$$a(t) = \frac{a_0}{t - t_0}, \tag{7.35}$$

a special form of superacceleration studied in early inflationary theories [754,911], in pre-big bang cosmology [738], and in Brans-Dicke theory [348].

From a more general point of view, negative or even phantom effective pressures appear in ETGs, as will be discussed in the rest of this book (they were noted in higher derivative gravity in [910] and in induced gravity in [911]). They appear also due to semiclassical particle production causing bulk viscosity [91, 604, 1175, 1179], in dissipative fluids in the presence of a quintessence field [308, 309], with quantum fields violating the weak energy condition [7, 872], and in models with non-linear kinetic terms [202,305,792,891,997,1177,1178]. A form of quintessence without potential $V$ and kinetic energy density non-linear in $\nabla^\mu\phi\nabla_\mu\phi$ is known as kinetically driven quintessence or *k-essence*.

The possibility has been pointed out that a time-dependent and positive equation of state parameter $w$, when interpreted as a constant, give rise to $w^{(eff)} < -1$ or, *vice-versa*, that assuming $w =$ const. and $w \geq -1$ in a likelihood analysis may incorrectly disguise an equation of state $w_q < -1$ [773].

Let us examine the possible consequences of dark energy taking the form of phantom energy. A fluid with effective equation of state parameter $w < -1$ violates the weak and dominant energy conditions and, in principle, opens the door to causality violations, wormholes and time machines. These are not mandatory, at least if the $w$-parameter is time-dependent [383,787,788]: the speed of sound does not necessarily exceed the speed of light because $P = w\rho$ describes an *effective* equation of state, not a true one. Acoustic waves (space-dependent perturbations of the exact model) do not obey the effective equation of state satisfied by the unperturbed $P$ and $\rho$.

A curious potential consequence of superacceleration is that, if it continues, the universe may end in a finite time with the scale factor $a(t)$ diverging as $t \to t_*$, with $t_*$ finite. Then the energy density and pressure also diverge as $t \to t_*$ (the energy density grows during superacceleration) in what is called a *Big Rip* singularity [202, 203, 456, 505, 787, 788, 1042]. From the mathematical point of view, it is easy to understand the Big Rip as the manifestation of a common phenomenon in the theory of ordinary differential equations. Specifically, an ordinary differential equation of the form

$$\frac{dy(t)}{dt} = f(y), \tag{7.36}$$

where the function $f$ is defined over a real interval $I$, admits maximally extendable solutions if the function $f(y)$ satisfies the Lipschitz condition, *i.e.*, if there exists a constant $M$ such that

$$\left| f(y') - f(y'') \right| \leq M \left| y' - y'' \right| \tag{7.37}$$

for any $y', y''$ in $I$. Consider, for example, the equation

$$\frac{dy(t)}{dt} = y^2, \tag{7.38}$$

the solutions of which do not admit a maximal extension. The solution is

$$y(t) = \frac{1}{t_0 - t}, \tag{7.39}$$

where $t_0$ is an integration constant corresponding to the initial condition $y(0) = 1/t_0$. For $t_0 > 0$, consider the branch $t < t_0$: the solution explodes as $t \to t_0$ from below and cannot be extended beyond this barrier. The qualitative behavior of this solution is caused by its rapid growth: while $y$ grows, its derivative grows even faster (as $y^2$) and the solution quickly explodes. A similar situation may occur for a FLRW universe in a superacceleration regime: the acceleration equation can be written as

$$\dot{H} = -H^2 - \frac{\kappa}{6}(3w + 1)\rho \tag{7.40}$$

for a fluid with equation of state $P = w\rho$, while the integration of the energy conservation equation $\dot{\rho} + 3H(w + 1)\rho = 0$ yields $\rho = \rho_0 a^{-3(w+1)}$. For a phantom fluid with $w < -1$, $\rho$ *increases* during the cosmic expansion and $\dot{H} \propto a^{|3(w+1)|}$ as the expansion proceeds, creating a situation similar to the one of the example above. When the equation of state is constant, the scale factor is easily found to be

$$a(t) = \frac{a_0}{|t - t_0|^{\frac{2}{3|1+w|}}}. \tag{7.41}$$

Realistic models of dark and phantom energy have a dynamical effective equation of state and $w$ can in principle change and cross the barrier $w = -1$ (known as the *phantom divide*). However, it has been proved difficult to construct explicit models that exhibit this behavior.

If the inequality $w < -1$ is supported by the future reconstruction of the equation of state of dark energy, the more conventional quintessence models based on minimally coupled scalars have to be abandoned in favor of alternative ones. Among these are non-minimally coupled models in ETGs. Even a very small amount of phantom energy leads to a fast (on a cosmological timescale) growth of its energy density and may lead to a Big Rip singularity.

Finally, let us discuss the prospects of testing the idea of quintessence (scalar field dark energy) with observations. Proposals to verify or falsify the idea and scenarios of quintessence exist, suggesting the reconstruction of the effective potential $V(\phi)$ and of the effective equation of state of quintessence from the luminosity distance-redshift relation $D_L(z)$ of SNeIa [149, 304, 476, 609, 646, 825, 958, 1043]. A minimally coupled scalar field and its potential can be expressed in terms of the variable $x \equiv 1 + z$ as [958]

$$\frac{V(x)}{\rho_c^{(0)}} = \frac{H^2}{H_0^2} - \frac{x}{6H_0^2}\frac{d\left(H^2\right)}{dx} - \frac{\Omega^{(m)}}{2}x^3, \tag{7.42}$$

$$\frac{1}{\rho_c^{(0)}}\left(\frac{d\phi}{dx}\right)^2 = \frac{2}{3H_0^2 x}\frac{d\left(\ln H\right)}{dx} - \frac{\Omega^{(m)}}{H^2}x, \tag{7.43}$$

and the equation of state parameter is

$$w_q(x) = \frac{2x\,\frac{d\left(\ln H\right)}{dx} - 3}{3\left[1 - \left(\frac{H_0}{H}\right)^2 \Omega^{(m)} x^3\right]}, \tag{7.44}$$

where $\rho_c^{(0)} = 3H_0^2/\kappa$ is the present value of the critical density. The Hubble parameter as a function of redshift $z$ can be derived from the kinematic relation with the luminosity distance $D_L(z)$ in a $K = 0$ FLRW universe

$$H = \left\{\frac{d}{dz}\left[\frac{D_L(z)}{1+z}\right]\right\}^{-1}. \tag{7.45}$$

If the luminosity distance $D_L(z)$ is determined from observational data, then Eqs. (7.42) and (7.44) allow the reconstruction of the potential $V$ and the parameter $w_q$ [149, 958].

Another proposal consists of testing the effective equation of state of the universe by using weak gravitational lensing [125]. Finally, for a minimally coupled scalar in GR, the relation

$$\frac{d^2 H(z)}{dz^2} \geq 3\,\Omega_0^{(m)}\,H_0\,(1+z)^2 \tag{7.46}$$

must be satisfied [956]. If the observations show a violation of this inequality, we would have evidence for non-Einsteinian gravity.

## 7.2 Using cosmography to map the structure of the universe

Both dark energy and modified gravity models exist which agree with observational data on the expansion history of the universe. As a consequence, one cannot discriminate between these competing approaches unless high precision probes of the expansion rate and the growth of cosmological structures become available. This situation suggests a conservative approach to the problem of the cosmic acceleration which relies on the smallest possible number of model-dependent quantities. A potential solution consists of going back to cosmography [1153] instead of finding and then testing solutions of the Friedmann equations. The cosmographic parameters, which are completely defined by the time derivatives of the scale factor, make it possible to fit the data on the distance-redshift relation without *a priori* assumptions on the underlying cosmological model. The only assumption is that the metric is a FLRW one, which is a solution of the field equations of many gravitational theories.

More than eighty years after Hubble's discovery of the cosmic expansion we can now, in principle, extend cosmography well beyond the purpose of Hubble's quest for the value of $H_0$. The Hubble diagram of SNeIa extends up to $z = 1.7$ making it necessary to Taylor-expand the scale factor at least to fifth order in order to have a reliable approximation of the distance-redshift relation, and variables other than the redshift $z$ are necessary when data at $z \geq 1$ are considered. It is then possible, at least in principle, to estimate up to five cosmographic parameters, although in practice the available data set is still too small to allow for a precise and realistic determination of all of them. Once these five quantities are determined, they can be used to constrain theoretical models. This approach reverses the conventional prediction of the cosmographic parameters in the context of a given theory. In cosmography, instead, the model is described by characterizing its quantities as functions of the cosmographic parameters. Such a program is particularly suited for the study of modified $f(R)$ gravity. Due to the fourth order of the field equations in these theories, it is difficult to obtain analytical expressions for the scale factor (which clearly depends on the form of the function $f(R)$) and hence predict the values of the cosmographic parameters. In the following we derive useful relations between the cosmographic parameters and the present-day values of $f^{(n)}(R) \equiv d^n f/dR^n$, with $n = 0, 1, 2, 3$ for any function $f(R)$, under rather general assumptions.[1]

The cosmographic parameters are defined in a FLRW spacetime. It is difficult to estimate *a priori* the extent to which the fifth order expansion provides a sufficiently accurate description of the quantities of interest. The number of cosmographic parameters to be used depends on the problem at hand: here we are concerned only with the Hubble diagram of SNeIa, hence we have to check that the distance modulus $\mu_{cp}(z)$ obtained using the fifth order expansion of the scale factor coincides, within experimental error, with the one ($\mu_{DE}(z)$) of the underlying physical model. Since such a model is, of course, unknown one can adopt a

[1] We only study metric $f(R)$ theories here. The reader is referred to [912, 913] for a similar approach in Palatini $f(R)$ gravity.

phenomenological parameterization of the dark energy equation of state[2] (EoS) and examine the percent deviation $\Delta\mu/\mu_{DE}$ as function of the EoS parameters. Here this exercise is carried out using the Chevallier-Polarski-Linder (CPL) model which we introduce below and it is verified that $\Delta\mu/\mu_{DE}$ is an increasing function of the redshift $z$, as expected, but remains smaller than 2% up to $z \sim 2$ over a wide region of the CPL parameter space. Moreover, truncating the Taylor expansion to a lower order may introduce sufficient deviations for $z > 1$ to potentially bias the analysis if the experimental errors are as small as those predicted for future surveys of SNeIa. There is a certain degree of confidence that the fifth order expansion presented below is both sufficient to get an accurate distance modulus over the redshift range probed by these supernovae and necessary to avoid dangerous biases.

### 7.2.1 *The cosmographic apparatus*

The key ingredient of cosmography is the Taylor expansion of the scale factor $a(t)$. It is convenient to introduce the functions

$$H(t) \equiv \frac{1}{a}\frac{da}{dt}, \tag{7.47}$$

$$q(t) \equiv -\frac{1}{aH^2}\frac{d^2a}{dt^2}, \tag{7.48}$$

$$j(t) \equiv \frac{1}{aH^3}\frac{d^3a}{dt^3}, \tag{7.49}$$

$$s(t) \equiv \frac{1}{aH^4}\frac{d^4a}{dt^4}, \tag{7.50}$$

$$l(t) \equiv \frac{1}{aH^5}\frac{d^5a}{dt^5}, \tag{7.51}$$

which are referred to as the *Hubble*, *deceleration*, *jerk*, *snap*, and *lerk* parameters, respectively. Straightforward algebra yields the relations

$$\dot{H} = -H^2(1+q), \tag{7.52}$$

$$\ddot{H} = H^3(j + 3q + 2), \tag{7.53}$$

---

[2] One can always use a phenomenological dark energy model to obtain a reliable estimate of the evolution of the scale factor even if the correct theory involves modified gravity instead of dark energy.

$$\frac{d^3H}{dt^3} = H^4\left[s - 4j - 3q(q+4) - 6\right], \tag{7.54}$$

$$\frac{d^4H}{dt^4} = H^5\left[l - 5s + 10(q+2)j + 30(q+2)q + 24\right]. \tag{7.55}$$

Equations (7.52)–(7.55) relate the derivatives of the Hubble parameter to the other cosmographic parameters. The distance-redshift relation may then be obtained using the Taylor expansion of $a(t)$ [285, 1126, 1149].

#### 7.2.1.1 The scale factor series

With these definitions in mind, the Taylor expansion of the scale factor to fifth order is

$$\frac{a(t)}{a(t_0)} = 1 + H_0(t-t_0) - \frac{q_0}{2}H_0^2(t-t_0)^2 + \frac{j_0}{3!}H_0^3(t-t_0)^3 + \frac{s_0}{4!}H_0^4(t-t_0)^4$$
$$+ \frac{l_0}{5!}H_0^5(t-t_0)^5 + \mathrm{O}\left[(t-t_0)^6\right], \tag{7.56}$$

which is the inverse of the characterization of the redshift factor in a FLRW universe

$$1 + z = \frac{a(t_0)}{a(t)}. \tag{7.57}$$

The physical distance travelled by a photon emitted at time $t_*$ and absorbed at the current epoch $t_0$ is

$$D = c\int_{t_*}^{t_0} dt = c(t_0 - t_*). \tag{7.58}$$

Assuming that $t_* = t_0 - D/c$ and inserting this value into Eq. (7.56), one obtains

$$1 + z = \frac{a(t_0)}{a(t_0 - \frac{D}{c})}$$
$$= \left[1 - \frac{H_0}{c}D - \frac{q_0}{2}\left(\frac{H_0}{c}\right)^2 D^2 - \frac{j_0}{6}\left(\frac{H_0}{c}\right)^3 D^3 + \frac{s_0}{24}\left(\frac{H_0}{c}\right)^4 D^4\right.$$
$$\left. - \frac{l_0}{120}\left(\frac{H_0}{c}\right)^5 D^5 + \mathrm{O}\left[\left(\frac{H_0 D}{c}\right)^6\right]\right]^{-1}. \tag{7.59}$$

The inverse of this expression is

$$1+z=1+\frac{H_0}{c}D+\left(1+\frac{q_0}{2}\right)\left(\frac{H_0}{c}\right)^2 D^2+\left(1+q_0+\frac{j_0}{6}\right)\left(\frac{H_0}{c}\right)^3 D^3$$
$$+\left(1+\frac{3}{2}q_0+\frac{q_0^2}{4}+\frac{j_0}{3}-\frac{s_0}{24}\right)\left(\frac{H_0}{c}\right)^4 D^4$$
$$+\left(1+2q_0+\frac{3}{4}q_0^2+\frac{q_0 j_0}{6}+\frac{j_0}{2}-\frac{s}{12}+l_0\right)\left(\frac{H_0}{c}\right)^5 D^5$$
$$+\mathrm{O}\left[\left(\frac{H_0 D}{c}\right)^6\right]. \tag{7.60}$$

Inverting the series $z(D)$ to obtain $D(z)$ yields the proper distance $D$ as a function of redshift

$$z(D)=\mathscr{Z}_D^{(1)}\left(\frac{H_0 D}{c}\right)+\mathscr{Z}_D^{(2)}\left(\frac{H_0 D}{c}\right)^2+\mathscr{Z}_D^{(3)}\left(\frac{H_0 D}{c}\right)^3+\mathscr{Z}_D^{(4)}\left(\frac{H_0 D}{c}\right)^4$$
$$+\mathscr{Z}_D^{(5)}\left(\frac{H_0 D}{c}\right)^5+\mathrm{O}\left[\left(\frac{H_0 D}{c}\right)^6\right], \tag{7.61}$$

where

$$\mathscr{Z}_D^{(1)}=1, \tag{7.62}$$

$$\mathscr{Z}_D^{(2)}=1+\frac{q_0}{2}, \tag{7.63}$$

$$\mathscr{Z}_D^{(3)}=1+q_0+\frac{j_0}{6}, \tag{7.64}$$

$$\mathscr{Z}_D^{(4)}=1+\frac{3q_0}{2}+\frac{q_0^2}{4}+\frac{j_0}{3}-\frac{s_0}{24}, \tag{7.65}$$

$$\mathscr{Z}_D^{(5)}=1+2q_0+\frac{3q_0^2}{4}+\frac{q_0 j_0}{6}+\frac{j_0}{2}-\frac{s}{12}+l_0, \tag{7.66}$$

from which it follows that

$$D(z)=\frac{cz}{H_0}\left[\mathscr{D}_z^{(0)}+\mathscr{D}_z^{(1)}z+\mathscr{D}_z^{(2)}z^2+\mathscr{D}_z^{(3)}z^3+\mathscr{D}_z^{(4)}z^4+\mathrm{O}(z^5)\right] \tag{7.67}$$

with

$$\mathscr{D}_z^{(0)} - 1, \tag{7.68}$$

$$\mathscr{D}_z^{(1)} = -\left(1 + \frac{q_0}{2}\right), \tag{7.69}$$

$$\mathscr{D}_z^{(2)} = 1 + q_0 + \frac{q_0^2}{2} - \frac{j_0}{6}, \tag{7.70}$$

$$\mathscr{D}_z^{(3)} = -\left(1 + \frac{3q_0}{2} + \frac{3q_0^2}{2} + \frac{5q_0^3}{8} - \frac{j_0}{2} - \frac{5q_0 j_0}{12} - \frac{s_0}{24}\right), \tag{7.71}$$

$$\mathscr{D}_z^{(4)} = 1 + 2q_0 + 3q_0^2 + \frac{5q_0^3}{2} + \frac{7q_0^4}{2} - \frac{5q_0 j_0}{3} - \frac{7q_0^2 j_0}{8} - \frac{q_0 s_0}{8} - j_0$$
$$+ \frac{j_0^2}{12} - \frac{s_0}{6} - \frac{l_0}{120}. \tag{7.72}$$

In astronomy it is not the proper (physical) distance $D(z)$ that is relevant. but rather the luminosity distance

$$D_L = \frac{a(t_0)}{a(t_0 - \frac{D}{c})} a(t_0) r_0 \tag{7.73}$$

and the angular diameter distance

$$D_A = \frac{a(t_0 - \frac{D}{c})}{a(t_0)} a(t_0) r_0, \tag{7.74}$$

where

$$r_0(D) = \begin{cases} \sin\left(\int_{t_0 - \frac{D}{c}}^{t_0} \frac{c\,dt}{a(t)}\right) & (K = +1) \\ \int_{t_0 - \frac{D}{c}}^{t_0} \frac{c\,dt}{a(t)} & (K = 0) \\ \sinh\left(\int_{t_0 - \frac{D}{c}}^{t_0} \frac{c\,dt}{a(t)}\right) & (K = -1). \end{cases} \tag{7.75}$$

For short distances, the series expansion of $a(t)$ in $r_0(D)$ yields

$$r_0(D) = \int_{t_0 - \frac{D}{c}}^{t_0} \frac{c\,dt}{a(t)} = \int_{t_0 - \frac{D}{c}}^{t_0} \frac{c\,dt}{a_0} \left\{ 1 + H_0(t_0 - t) + \left(1 + \frac{q_0}{2}\right) H_0^2 (t_0 - t)^2 \right.$$
$$+ \left(1 + q_0 + \frac{j_0}{6}\right) H_0^3 (t_0 - t)^3 + \left(1 + \frac{3q_0}{2} + \frac{q_0^2}{4} + \frac{j_0}{3} - \frac{s_0}{24}\right) H_0^4 (t_0 - t)^4$$
$$\left. + \left(1 + 2q_0 + \frac{3q_0^2}{4} + \frac{q_0 j_0}{6} + \frac{j_0}{2} - \frac{s}{12} + l_0\right) H_0^5 (t_0 - t)^5 + O[(t_0 - t)^6] \right\}$$

$$= \frac{D}{a_0} \left\{ 1 + \frac{1}{2} \frac{H_0 D}{c} + \left( \frac{2 + q_0}{6} \right) \left( \frac{H_0 D}{c} \right)^2 + \left( \frac{6 + 6q_0 + j_0}{24} \right) \left( \frac{H_0 D}{c} \right)^3 \right.$$

$$+ \left( \frac{24 + 36q_0 + 6q_0^2 + 8j_0 - s_0}{120} \right) \left( \frac{H_0 D}{c} \right)^4$$

$$+ \left( \frac{12 + 24q_0 + 9q_0^2 + 2q_0 j_0 + 6j_0 - s_0 + 12l_0}{72} \right) \left( \frac{H_0 D}{c} \right)^5$$

$$\left. + \mathrm{O}\left[ \left( \frac{H_0 D}{c} \right)^6 \right] \right\} . \tag{7.76}$$

To convert from the proper distance travelled to the $r$-coordinate the Taylor expansions of the sin and sinh functions are used, obtaining

$$r_0(D) = \left[ \int_{t_0 - \frac{D}{c}}^{t_0} \frac{c\, \mathrm{d}t}{a(t)} \right] - \frac{k}{3!} \left[ \int_{t_0 - \frac{D}{c}}^{t_0} \frac{c\, \mathrm{d}t}{a(t)} \right]^3 + \mathrm{O}\left( \left[ \int_{t_0 - \frac{D}{c}}^{t_0} \frac{c\, \mathrm{d}t}{a(t)} \right]^5 \right), \tag{7.77}$$

so that Eq. (7.56) with the curvature term becomes

$$r_0(D) = \frac{D}{a_0} \left\{ \mathscr{R}_D^{(0)} + \mathscr{R}_D^{(1)} \frac{H_0 D}{c} + \mathscr{R}_D^{(2)} \left( \frac{H_0 D}{c} \right)^2 + \mathscr{R}_D^{(3)} \left( \frac{H_0 D}{c} \right)^3 \right.$$

$$\left. + \mathscr{R}_D^{(4)} \left( \frac{H_0 D}{c} \right)^4 + \mathscr{R}_D^{(5)} \left( \frac{H_0 D}{c} \right)^5 + \mathrm{O}\left[ \left( \frac{H_0 D}{c} \right)^6 \right] \right\} , \tag{7.78}$$

where

$$\mathscr{R}_D^{(0)} = 1, \tag{7.79}$$

$$\mathscr{R}_D^{(1)} = \frac{1}{2}, \tag{7.80}$$

$$\mathscr{R}_D^{(2)} = \frac{1}{6} \left( 2 + q_0 - \frac{kc^2}{H_0^2 a_0^2} \right), \tag{7.81}$$

$$\mathscr{R}_D^{(3)} = \frac{1}{24} \left( 6 + 6q_0 + j_0 - 6\frac{kc^2}{H_0^2 a_0^2} \right), \tag{7.82}$$

$$\mathscr{R}_D^{(4)} = \frac{1}{120} \left( 24 + 36q_0 + 6q_0^2 + 8j_0 - s_0 - \frac{5kc^2(7 + 2q_0)}{a_0^2 H_0^2} \right), \tag{7.83}$$

$$\mathscr{R}_D^{(5)} = \frac{1}{144} \left[ 24 + 48q_0 + 18q_0^2 + 4q_0 j_0 + 12j_0 - 2s_0 + 24l_0 \right.$$

$$\left. - \frac{3kc^2(15 + 10q_0 + j_0)}{a_0^2 H_0^2} \right]. \tag{7.84}$$

The luminosity distance is then expressed as

$$D_L(z) = \frac{cz}{H_0}\left[\mathscr{D}_L^{(0)} + \mathscr{D}_L^{(1)}\, z + \mathscr{D}_L^{(2)}\, z^2 + \mathscr{D}_L^{(3)}\, z^3 + \mathscr{D}_L^{(4)}\, z^4 + \mathrm{O}(z^5)\right] \quad (7.85)$$

with

$$\mathscr{D}_L^{(0)} = 1, \quad (7.86)$$

$$\mathscr{D}_L^{(1)} = -\frac{1}{2}\,(-1 + q_0)\,, \quad (7.87)$$

$$\mathscr{D}_L^{(2)} = -\frac{1}{6}\left(1 - q_0 - 3q_0^2 + j_0 + \frac{kc^2}{H_0^2 a_0^2}\right), \quad (7.88)$$

$$\mathscr{D}_L^{(3)} = \frac{1}{24}\left[2 - 2q_0 - 15q_0^2 - 15q_0^3 + 5j_0 + 10q_0 j_0 + s_0 + \frac{2kc^2(1+3q_0)}{H_0^2 a_0^2}\right], \quad (7.89)$$

$$\mathscr{D}_L^{(4)} = \frac{1}{120}\Bigg[-6 + 6q_0 + 81q_0^2 + 165q_0^3 + 105q_0^4 - 110q_0 j_0 - 105q_0^2 j_0$$
$$- 15q_0 s_0 27 j_0 + 10j^2 - 11s_0 - l_0 - \frac{5kc^2(1 + 8q_0 + 9q_0^2 - 2j_0)}{a_0^2 H_0^2}\Bigg]. \quad (7.90)$$

The angular diameter distance used in number counts is

$$D_A(z) = \frac{cz}{H_0}\left[\mathscr{D}_A^{(0)} + \mathscr{D}_A^{(1)}\, z + \mathscr{D}_A^{(2)}\, z^2 + \mathscr{D}_A^{(3)}\, z^3 + \mathscr{D}_A^{(4)}\, z^4 + \mathrm{O}(z^5)\right], \quad (7.91)$$

where

$$\mathscr{D}_A^{(0)} = 1, \quad (7.92)$$

$$\mathscr{D}_A^{(1)} = -\frac{1}{2}\,(3 + q_0)\,, \quad (7.93)$$

$$\mathscr{D}_A^{(2)} = \frac{1}{6}\left(11 + 7q_0 + 3q_0^2 - j_0 - \frac{kc^2}{H_0^2 a_0^2}\right), \quad (7.94)$$

$$\mathscr{D}_A^{(3)} = -\frac{1}{24}\left[50 + 46q_0 + 39q_0^2 + 15q_0^3 - 13j_0 - 10q_0 j_0 - s_0 - \frac{2kc^2(5+3q_0)}{H_0^2 a_0^2}\right], \quad (7.95)$$

$$\mathscr{D}_A^{(4)} = \frac{1}{120}\Bigg[274 + 326q_0 + 411q_0^2 + 315q_0^3 + 105q_0^4 - 210q_0 j_0 - 105q_0^2 j_0$$
$$- 15q_0 s_0 137 j_0 + 10j^2 - 21s_0 - l_0 - \frac{5kc^2(17 + 20q_0 + 9q_0^2 - 2j_0)}{a_0^2 H_0^2}\Bigg]. \quad (7.96)$$

Following the notation of [285], we define

$$\Omega_0 = 1 + \frac{kc^2}{H_0^2 a_0^2}, \tag{7.97}$$

which can be regarded as a purely cosmographic parameter, or

$$\Omega_0 = 1 - \Omega_k = \Omega_{m,0} + \Omega_{r,0} + \Omega_{X,0}. \tag{7.98}$$

Using this density parameter, Eqs. (26)–(28) become

$$\mathscr{D}^{(0)}_{L,y} = 1, \tag{7.99}$$

$$\mathscr{D}^{1}_{L,y} = -\frac{1}{2}\left(-3 + q_0\right), \tag{7.100}$$

$$\mathscr{D}^{(2)}_{L,y} = -\frac{1}{6}\left(12 - 5q_0 + 3q_0^2 - j_0 - \Omega_0\right), \tag{7.101}$$

$$\mathscr{D}^{(3)}_{L,y} = \frac{1}{24}\left[52 - 20q_0 + 21q_0^2 - 15q_0^3 - 7j_0 + 10q_0 j_0 + s_0 - 2\Omega_0(1 + 3q_0)\right], \tag{7.102}$$

$$\mathscr{D}^{(4)}_{L,y} = \frac{1}{120}\left[359 - 184q_0 + 186q_0^2 - 135q_0^3 + 105q_0^4 + 90q_0 j_0 - 105q_0^2 j_0 \right.$$
$$\left. - 15q_0 s_0 57j_0 + 10j^2 + 9s_0 - l_0 - 5\Omega_0(17 - 6q_0 + 9q_0^2 - 2j_0)\right] \tag{7.103}$$

and

$$\mathscr{D}^{(0)}_{A,y} = 1, \tag{7.104}$$

$$\mathscr{D}^{(1)}_{A,y} = -\frac{1}{2}\left(1 + q_0\right), \tag{7.105}$$

$$\mathscr{D}^{(2)}_{A,y} = -\frac{1}{6}\left[-q_0 - 3q_0^2 + j_0 + \Omega_0\right], \tag{7.106}$$

$$\mathscr{D}^{(3)}_{A,y} = -\frac{1}{24}\left[-2q_0 + 3q_0^2 + 15q_0^3 - j_0 - 10q_0 j_0 - s_0 + 2\Omega_0\right], \tag{7.107}$$

$$\mathscr{D}^{(4)}_{A,y} = -\frac{1}{120}\left[1 - 6q_0 + 9q_0^2 - 15q_0^3 - 105q_0^4 + 10q_0 j_0 + 105q_0^2 j_0 + 15q_0 s_0 \right.$$
$$\left. - 3j_0 - 10j^2 + s_0 + l_0 + 5\Omega_0\right]. \tag{7.108}$$

The relations occurring earlier in this section are valid for arbitrary values of the curvature index $K$ but in the following we assume a spatially flat universe and use these relations for $K = 0$. Since we are going to use supernova data, the Taylor expansion of the luminosity distance as it appears in the distance modulus $\mu(z) = 5\log_{10} D_L(z)$ used in astronomy will also be useful:

$$\mu(z) = \frac{5}{\log 10}\left(\log z + \mathscr{M}^{(1)}z + \mathscr{M}^{(2)}z^2 + \mathscr{M}^{(3)}z^3 + \mathscr{M}^{(4)}z^4\right), \tag{7.109}$$

where

$$\mathscr{M}^{(1)} = -\frac{1}{2}(-1 + q_0), \tag{7.110}$$

$$\mathscr{M}^{(2)} = -\frac{1}{24}(7 - 10q_0 - 9q_0^2 + 4j_0), \tag{7.111}$$

$$\mathscr{M}^{(3)} = \frac{1}{24}(5 - 9q_0 - 16q_0^2 - 10q_0^3 + 7j_0 + 8q_0 j_0 + s_0), \tag{7.112}$$

$$\mathscr{M}^{(4)} = \frac{1}{2880}(-469 + 1004q_0 + 2654q_0^2 + 3300q_0^3 + 1575q_0^4 + 200j_0^2 - 1148j_0 - 2620q_0 j_0 - 1800q_0^2 j_0 - 300q_0 s_0 - 324s_0 - 24l_0). \tag{7.113}$$

#### 7.2.1.2 Cosmography and extended gravity

In metric $f(R)$ gravity and in a spatially flat universe the Hubble parameter obeys (using units such that $\kappa \equiv 8\pi G = 1$)

$$H^2 = \frac{1}{3}\left[\frac{\rho_m}{f'(R)} + \rho_{curv}\right], \tag{7.114}$$

where the prime denotes differentiation with respect to $R$ and

$$\rho_{curv} = \frac{1}{f'(R)}\left\{\frac{1}{2}\left[f(R) - Rf'(R)\right] - 3H\dot{R}f''(R)\right\} \tag{7.115}$$

is the energy density of an effective curvature fluid.[3] Assuming that matter does not couple directly to this curvature fluid, the continuity equation for matter yields the usual scaling

$$\rho_M = \frac{\rho_M(t_0)}{a^3} = \frac{3H_0^2\Omega_M}{a^3}, \tag{7.116}$$

[3] The name *curvature fluid* does not refer to the curvature index of the FLRW metric but denotes the fact that the field equations have been re-arranged in the form of effective Einstein equations.

where $\Omega_M$ is the present-day density parameter of matter. The continuity equation for $\rho_{curv}$ reads

$$\dot{\rho}_{curv} + 3H(1 + w_{curv})\rho_{curv} = \frac{3H_0^2 \Omega_M \dot{R} f''(R)}{[f'(R)]^2 a^3}, \tag{7.117}$$

where the effective equation of state parameter of the curvature fluid is

$$w_{curv} = -1 + \frac{\ddot{R} f''(R) + \dot{R}\left[\dot{R} f'''(R) - H f''(R)\right]}{\left[f(R) - R f'(R)\right]/2 - 3H\dot{R} f''(R)}. \tag{7.118}$$

The curvature fluid quantities $\rho_{curv}$ and $w_{curv}$ depend only on $f(R)$ and its derivatives up to third order. As a consequence, if one considers only their present-day values (naively, replacing $R$ with $R_0$), two $f(R)$ theories with the same values of $f(R_0)$, $f'(R_0)$, $f''(R_0)$, and $f'''(R_0)$ will be degenerate.[4]

By combining Eq. (7.117) with Eq. (7.114) the master equation for the Hubble parameter

$$\dot{H} = -\frac{1}{2f'(R)} \left\{3H_0^2 \Omega_M a^{-3} + \ddot{R} f''(R) + \dot{R}\left[\dot{R} f'''(R) - H f''(R)\right]\right\} \tag{7.119}$$

is finally obtained. Using the expression of the Ricci curvature in a $K = 0$ universe

$$R = 6\left(\dot{H} + 2H^2\right) \tag{7.120}$$

and inserting the result into Eq. (7.119), one obtains a fourth order non-linear ODE for the scale factor $a(t)$, which cannot be solved easily even in simple situations such as, *e.g.*, $f(R) \propto R^n$. Its numerical solution, although technically feasible, is plagued by the uncertainty on the boundary conditions, *i.e.*, on the present-day values of the scale factor and its derivatives up to third order.

#### 7.2.1.3 Cosmography and the derivatives of $f(R)$

We approach the problem from a different point of view: rather than choosing a parametrized expression for $f(R)$ and then numerically solving Eq. (7.119) for given boundary conditions, we relate the present-day values of the derivatives of

[4] It can be argued that this statement is not strictly true because different $f(R)$ theories will lead to different expansion rates $H(t)$ and, therefore, different present-day values of $R$ and its derivatives. However, it is likely that two functions $f(R)$ that exactly match up to third order today will give rise to the same $H(t)$ at least for $t \simeq t_0$, so that $(R_0, \dot{R}_0, \ddot{R}_0)$ will be almost the same.

$f(R)$ to the cosmographic parameters $(q_0, j_0, s_0, l_0)$ and we constrain them in a model-independent way, obtaining indications on the kind of function $f(R)$ which is able to fit the observed Hubble diagram (see [900] for a similar analysis motivated by effective energy conditions).

As a preliminary step, it is useful to consider again the constraint equation (7.120). Differentiation with respect to $t$ yields

$$\dot{R}=6\left(\ddot{H}+4H\dot{H}\right), \tag{7.121}$$

$$\ddot{R}=6\left(\frac{d^3H}{dt^3}+4H\ddot{H}+4\dot{H}^2\right), \tag{7.122}$$

$$\frac{d^3R}{dt^3}=6\left(\frac{d^4H}{dt^4}+4H\frac{d^3H}{dt^3}+12\dot{H}\ddot{H}\right). \tag{7.123}$$

Evaluating these derivatives at the present time and using Eqs. (7.52)–(7.55), one obtains

$$R_0=6H_0^2(1-q_0), \tag{7.124}$$

$$\dot{R}_0=6H_0^3(j_0-q_0-2), \tag{7.125}$$

$$\ddot{R}_0=6H_0^4\left(s_0+q_0^2+8q_0+6\right), \tag{7.126}$$

$$\frac{d^3R_0}{dt^3}=6H_0^5\left[l_0-s_0+2(q_0+4)j_0-6(3q_0+8)q_0-24\right], \tag{7.127}$$

which will be useful in the following.

Let us now come back to the expansion rate (7.114) and the master equation (7.119), which hold during the entire history of the universe. At the present time $t=t_0$ they give

$$H_0^2=\frac{H_0^2\Omega_M}{f'(R_0)}+\frac{f(R_0)-R_0f'(R_0)-6H_0\dot{R}_0f''(R_0)}{6f'(R_0)}, \tag{7.128}$$

$$-\dot{H}_0=\frac{3H_0^2\Omega_M}{2f'(R_0)}+\frac{\dot{R}_0^2f'''(R_0)+\left(\ddot{R}_0-H_0\dot{R}_0\right)f''(R_0)}{2f'(R_0)}. \tag{7.129}$$

Using Eqs. (7.52)–(7.55) and (7.124)–(7.127), Eqs. (7.128) and (7.129) can be rearranged in the form of relations between $H_0$ and the cosmographic parameters $(q_0, j_0, s_0)$, or as relations between the present-day values of $f(R)$ and its derivatives up to third order. However, two more relations are needed in order to close the system and determine the four unknown quantities $f(R_0)$, $f'(R_0)$, $f''(R_0)$, and $f'''(R_0)$. The first relation is obtained by noting that (restoring $\kappa\equiv 8\pi G$) the Hamiltonian constraint reads

$$H^2=\frac{8\pi G}{3f'(R)}\left[\rho_m+\rho_{curv}f'(R)\right] \tag{7.130}$$

where $G_{eff} = G/f'(R)$ is as an effective gravitational coupling with present-day value equal to the measured Newton constant $G_{eff}(z=0) = G$, hence

$$f'(R_0) = 1. \tag{7.131}$$

The fourth relation needed to close the system can be obtained by first differentiating Eq. (7.119) with respect to $t$,

$$\begin{aligned}\ddot{H} =& \frac{\dot{R}^2 f'''(R) + (\ddot{R} - H\dot{R})\, f''(R) + 3H_0^2 \Omega_M a^{-3}}{2\left[\dot{R} f''(R)\right]^{-1} [f'(R)]^2} \\ &- \frac{\dot{R}^3 f^{(iv)}(R) + (3\dot{R}\ddot{R} - H\dot{R}^2)\, f'''(R)}{2f'(R)} \\ &- \frac{(d^3R/dt^3 - H\ddot{R} + \dot{H}\dot{R})\, f''(R) - 9H_0^2 \Omega_M H a^{-3}}{2f'(R)}\end{aligned} \tag{7.132}$$

(with $f^{(iv)}(R) \equiv d^4 f/dR^4$), and then expanding $f(R)$ to third order,

$$f(R) \simeq f(R_0) + f'(R_0)(R - R_0) + \frac{f''(R_0)}{2}(R - R_0)^2 + \frac{f'''(R_0)}{6}(R - R_0)^3. \tag{7.133}$$

Evaluating Eq. (7.132) at the present day yields

$$\begin{aligned}\ddot{H}_0 =& \frac{\dot{R}_0^2 f'''(R_0) + (\ddot{R}_0 - H_0\dot{R}_0)\, f''(R_0) + 3H_0^2 \Omega_M}{2\left[\dot{R}_0 f''(R_0)\right]^{-1} [f'(R_0)]^2} - \frac{(3\dot{R}_0\ddot{R}_0 - H\dot{R}_0^2)\, f'''(R_0)}{2f'(R_0)} \\ &- \frac{(d^3R_0/dt^3 - H_0\ddot{R}_0 + \dot{H}_0\dot{R}_0)\, f''(R_0) - 9H_0^3 \Omega_M}{2f'(R_0)}.\end{aligned} \tag{7.134}$$

We can now schematically proceed as follows: we evaluate Eqs. (7.52)–(7.55) at $z = 0$ and insert them in the left hand sides of Eqs. (7.128), (7.129), and (7.134). Then we insert Eqs. (7.124)–(7.127) in the right hand sides of these same equations so that only the cosmographic parameters $(q_0, j_0, s_0, l_0)$ and the $f(R)$-related quantities appear. Finally, we solve the resulting equations subject to the constraint (7.131) with respect to the present-day values of $f(R)$ and its derivatives up to third order. The result is

$$\frac{f(R_0)}{6H_0^2} = -\frac{\mathscr{P}_0(q_0, j_0, s_0, l_0)\Omega_M + \mathscr{Q}_0(q_0, j_0, s_0, l_0)}{\mathscr{R}(q_0, j_0, s_0, l_0)}, \tag{7.135}$$

$$f'(R_0) = 1, \tag{7.136}$$

$$6H_0^2 f''(R_0) = -\frac{\mathscr{P}_2(q_0, j_0, s_0)\Omega_M + \mathscr{Q}_2(q_0, j_0, s_0)}{\mathscr{R}(q_0, j_0, s_0, l_0)}, \tag{7.137}$$

$$(6H_0^2)^2 f'''(R_0) = -\frac{\mathscr{P}_3(q_0, j_0, s_0, l_0)\Omega_M + \mathscr{Q}_3(q_0, j_0, s_0, l_0)}{(j_0 - q_0 - 2)\mathscr{R}(q_0, j_0, s_0, l_0)}, \tag{7.138}$$

where

$$\begin{aligned}\mathscr{P}_0 \equiv & (j_0 - q_0 - 2)l_0 - (3s_0 + 7j_0 + 6q_0^2 + 41q_0 + 22)s_0 \\ & - \left[(3q_0 + 16)j_0 + 20q_0^2 + 64q_0 + 12\right] j_0 - \left(3q_0^4 + 25q_0^3 + 96q_0^2 + 72q_0 + 20\right),\end{aligned} \tag{7.139}$$

$$\begin{aligned}\mathscr{Q}_0 \equiv & (q_0^2 - j_0 q_0 + 2q_0)l_0 + \left[3q_0 s_0 + (4q_0 + 6)j_0 + 6q_0^3 + 44q_0^2 + 22q_0 - 12\right] s_0 \\ & + \left[2j_0^2 + (3q_0^2 + 10q_0 - 6)j_0 + 17q_0^3 + 52q_0^2 + 54q_0 + 36\right] j_0 + 3q_0^5 + 28q_0^4 \\ & + 118q_0^3 + 72q_0^2 - 76q_0 - 64,\end{aligned} \tag{7.140}$$

$$\mathscr{P}_2 \equiv 9s_0 + 6j_0 + 9q_0^2 + 66q_0 + 42, \tag{7.141}$$

$$\mathscr{Q}_2 \equiv -\left\{6(q_0 + 1)s_0 + [2j_0 - 2(1 - q_0)]\, j_0 + 6q_0^3 + 50q_0^2 + 74q_0 + 32\right\}, \tag{7.142}$$

$$\mathscr{P}_3 \equiv 3l_0 + 3s_0 - 9(q_0 + 4)j_0 - (45q_0^2 + 78q_0 + 12), \tag{7.143}$$

$$\begin{aligned}\mathscr{Q}_3 \equiv & -\left\{2(1 + q_0)l_0 + 2(j_0 + q_0)s_0 - (2j_0 + 4q_0^2 + 12q_0 + 6)j_0 \right. \\ & \left. - (30q_0^3 + 84q_0^2 + 78q_0 + 24)\right\},\end{aligned} \tag{7.144}$$

$$\begin{aligned}\mathscr{R} \equiv & (j_0 - q_0 - 2)l_0 - (3s_0 - 2j_0 + 6q_0^2 + 50q_0 + 40)s_0 \\ & + \left[(3q_0 + 10)j_0 + 11q_0^2 + 4q_0 - 18\right] j_0 - (3q_0^4 + 34q_0^3 + 246q_0 + 104).\end{aligned} \tag{7.145}$$

Equations (7.135)–(7.145) make it possible to estimate the present-day values of $f(R)$ and its first three derivatives in terms of the Hubble parameter $H_0$ and the cosmographic parameters $(q_0, j_0, s_0, l_0)$, given the value of the matter density parameter $\Omega_M$. This is in principle problematic: while the cosmographic parameters may be determined in a model-independent way, the fiducial value of $\Omega_M$ is usually obtained by fitting a given dataset in the framework of an assumed dark energy scenario. However, different models all converge to the concordance value $\Omega_M \simeq 0.25$ which is also in agreement with astrophysical, model-independent estimates based on the mass fraction of gas in galaxy clusters. It has been proposed that $f(R)$ theories may avoid the need for dark matter in galaxies and galaxy clusters [148, 213, 216, 217, 510, 793, 1021]: in this case, the total matter content of the universe is essentially equal to the baryonic one. According to the primordial elements abundance and Big Bang nucleosynthesis, we would then get $\Omega_M \simeq \omega_b/h^2$ with $\omega_b = \Omega_b h^2 \simeq 0.0214$ [672] and $h$ the Hubble constant in units of 100 km/(s · Mpc). Setting $h = 0.72$ in agreement with the results of the Hubble Space Telescope Key Project [506] yields $\Omega_M = 0.041$ for a baryon-only universe. In the following we consider both cases when numerical estimates are needed.

$H_0$ only plays the role of a scaling parameter giving the correct physical dimensions to $f(R)$ and its derivatives. As such, it is not surprising that we need four cosmographic parameters $(q_0, j_0, s_0, l_0)$ in order to determine the four quantities $f(R_0)$, $f'(R_0)$, $f''(R_0)$, and $f'''(R_0)$. Moreover, Eqs. (7.135)–(7.138) are linear in these $f$-quantities and $(q_0, j_0, s_0, l_0)$ uniquely determine the previous ones. On the contrary, inverting them to obtain the cosmographic parameters in terms of the $f(R)$ ones does not produce linear relations. Indeed, the field equations in $f(R)$ theories are non-linear fourth order ODEs for the scale factor $a(t)$ and fixing the derivatives of $f(R)$ up to third order makes it possible to find a *class* of solutions, not a single solution. Each of these solutions is characterized by a different set of cosmographic parameters, explaining why the inversion of Eqs. (7.135)–(7.145) does not provide a unique result for $(q_0, j_0, s_0, l_0)$.

Let us discuss the assumptions leading to the relations above. While Eqs. (7.128) and (7.129) are exact consequences of the field equations, Eq. (7.134) relies heavily on the approximation of $f(R)$ with its third order expansion (7.133). When this approximation breaks down, the system is no longer closed because a fifth unknown parameter $f^{(iv)}(R_0)$ enters the game. Replacing $f(R)$ with its expansion is not possible for all $f(R)$ theories. By truncating the expansion to third order, one implicitly assumes that higher order terms are negligible over the redshift range probed by the data, *i.e.*,

$$f^{(n)}(R_0)(R - R_0)^n \ll \sum_{m=0}^{3} \frac{f^{(m)}(R_0)}{m!}(R - R_0)^m \qquad \text{for } n \geq 4 \tag{7.146}$$

in this range. It is impossible to check the validity of this assumption without explicitly solving the field equations. However, one can estimate the order of magnitude of the relevant quantities considering that, for all viable models, the background dynamics should not differ much from those of the $\Lambda$CDM model to at least $z \simeq 2$. Using the expression of $H(z)$ for the $\Lambda$CDM model, it is seen that $R/R_0$ is a rapidly increasing function of redshift so that, in order for Eq. (7.146) to hold, it must be $f^{(n)}(R_0) \ll f'''(R_0)$ for $n \geq 4$. This condition is easier to check for many analytical $f(R)$ models; once it is verified, we still have to worry about Eq. (7.131) relying on the assumption that the *cosmological* gravitational coupling is exactly the same as the local one. Although reasonable, this requirement is not automatic. The numerical value usually adopted for the Newton constant $G_N$ is obtained from laboratory experiments in settings that can hardly be considered homogenous and isotropic. The spacetime metric in such conditions has little to do with the cosmological metric at large scales and matching the two values of $G$ is an extrapolation. Although commonly accepted and quite reasonable, the condition $G_{local} = G_{cosmological}$ could, at least, in principle, be violated and then Eq. (7.131) would have to be reconsidered. As we will see, the condition $f'(R_0) = 1$ may not be verified in certain modified gravity models popular in the literature. However, it is reasonable to assume that $G_{eff}(z = 0) = G(1 + \varepsilon)$ with $\varepsilon \ll 1$. One should repeat the derivation of Eqs. (7.135)–(7.138) with the condition $f'(R_0) = (1 + \varepsilon)^{-1}$. By linearizing in $\varepsilon$ and comparing with the equations derived earlier, we can estimate the error induced by assuming that $\varepsilon = 0$. The resulting expressions are long and will not be reported here; they depend in a complicated way on the values of the matter density parameter $\Omega_M$, the cosmographic parameters $(q_0, j_0, s_0, l_0)$, and $\varepsilon$. However, it can be checked numerically that the error induced on $f(R_0)$, $f''(R_0)$, and $f'''(R_0)$ is much lower than 10% for values of $\varepsilon$ as large as 0.1, which is unrealistically generous.

#### 7.2.1.4 $f(R)$ gravity and the CPL model

Determining $f(R)$ and its derivatives in terms of the cosmographic parameters requires a model-independent estimate of the latter from data. Unfortunately, even in the current era hailed as the era of "precision cosmology", such a program is still too ambitious to provide useful constraints on the derivatives of $f(R)$. The cosmographic parameters may also be expressed in terms of the dark energy density and the EoS parameters in such a way that the present-day values of $f(R)$ and its derivatives generating given $(q_0, j_0, s_0, l_0)$ can be obtained for a specified dark energy model. To this end, it is convenient to adopt a parametrized expression for the dark energy EoS in order to reduce the dependence of the results on the theoretical model. In agreement with a prescription by the Dark Energy Task Force [20], in the following we use the Chevallier-Polarski-Linder (CPL) parameterization for the EoS setting [299, 744]

$$w = w_0 + w_a(1 - a) = w_0 + w_a \frac{z}{1 + z} \tag{7.147}$$

so that, in a spatially flat universe filled with dust and dark energy, the dimensionless Hubble parameter $E(z) \equiv H/H_0$ is given by

$$E^2(z) = \Omega_M(1 + z)^3 + \Omega_X(1 + z)^{3(1+w_0+w_a)} \, e^{-\frac{3w_a z}{1+z}} \tag{7.148}$$

with $\Omega_X = 1 - \Omega_M$ because of spatial flatness. In order to determine the cosmographic parameters for this model, we avoid integrating $H(z)$ to obtain $a(t)$ by noting that $d/dt = -(1 + z)H(z)d/dz$ and we use this relation to evaluate $(\dot{H}, \ddot{H}, d^3H/dt^3, d^4H/dt^4)$ and then solve Eqs. (7.52)–(7.55) evaluated at $z = 0$ with respect to the parameters of interest. Straightforward algebra yields

$$q_0 = \frac{1}{2} + \frac{3}{2}(1 - \Omega_M) w_0, \tag{7.149}$$

$$j_0 = 1 + \frac{3}{2}(1 - \Omega_M)\left[3w_0(1 + w_0) + w_a\right], \tag{7.150}$$

$$s_0 = -\frac{7}{2} - \frac{33}{4}(1 - \Omega_M) w_a - \frac{9}{4}(1 - \Omega_M)\left[9 + (7 - \Omega_M) w_a\right] w_0$$
$$-\frac{9}{4}(1 - \Omega_M)(16 - 3\Omega_M) w_0^2 - \frac{27}{4}(1 - \Omega_M)(3 - \Omega_M) w_0^3, \tag{7.151}$$

$$l_0 = \frac{35}{2} + \frac{1 - \Omega_M}{4}\left[213 + (7 - \Omega_M) w_a\right] w_a + \frac{(1 - \Omega_M)}{4}\left[489 + 9(82 - 21\Omega_M) w_a\right] w_0$$
$$+\frac{9}{2}(1 - \Omega_M)\left[67 - 21\Omega_M + \frac{3}{2}(23 - 11\Omega_M) w_a\right] w_0^2$$
$$+\frac{27}{4}(1 - \Omega_M)(47 - 24\Omega_M) w_0^3 + \frac{81}{2}(1 - \Omega_M)(3 - 2\Omega_M) w_0^4. \tag{7.152}$$

Inserting Eqs. (7.149)–(7.152) into Eqs. (7.135)–(7.145) one obtains lengthy expressions, not reported here, for the present-day values of $f(R)$ and its first three derivatives in terms of $(\Omega_M, w_0, w_a)$. The $f(R)$ model thus obtained is not dynamically equivalent to the starting CPL one: the two models have the same cosmographic parameters only today. As such, for instance, the scale factors coincide in the two theories only during the time period in which the fifth order Taylor expansion constitutes a good approximation of the actual $a(t)$. Such a procedure does not select a unique $f(R)$ model but rather a class of theories with the same third order expansion of $f(R)$.

#### 7.2.1.5 The ΛCDM model

With this *caveat* in mind, we consider first the ΛCDM model obtained by setting $(w_0, w_a) = (-1, 0)$, with

$$q_0 = \frac{1}{2} - \frac{3}{2}(1 - \Omega_M), \tag{7.153}$$

$$j_0 = 1, \tag{7.154}$$

$$s_0 = 1 - \frac{9}{2}\Omega_M, \tag{7.155}$$

$$l_0 = 1 + 3\Omega_M + \frac{27}{2}\Omega_M^2. \tag{7.156}$$

When inserted into the expressions for the $f(R)$ quantities, these relations give

$$f(R_0) = R_0 - 2\Lambda, \quad f''(R_0) = f'''(R_0) = 0, \tag{7.157}$$

and the only metric $f(R)$ theory having exactly the same cosmographic parameters as the $\Lambda$CDM model is described by $f(R) \propto R$, *i.e.*, GR. This is a consequence of the values of $(q_0, j_0)$ of the $\Lambda$CDM model: had we left $(s_0, l_0)$ unspecified and fixed $(q_0, j_0)$ to be the values in Eqs. (7.153)–(7.156), we would have obtained Eq. (7.157) again. Since the $\Lambda$CDM model fits well a large set of different data, we expect the actual values of $(q_0, j_0, s_0, l_0)$ to differ little from the $\Lambda$CDM ones. Therefore, we substitute into Eqs. (7.135)–(7.145) the expressions

$$q_0 = q_0^\Lambda (1 + \varepsilon_q), \tag{7.158}$$

$$j_0 = j_0^\Lambda (1 + \varepsilon_j), \tag{7.159}$$

$$s_0 = s_0^\Lambda (1 + \varepsilon_s), \tag{7.160}$$

$$l_0 = l_0^\Lambda (1 + \varepsilon_l), \tag{7.161}$$

with $\left(q_0^\Lambda, j_0^\Lambda, s_0^\Lambda, l_0^\Lambda\right)$ given by Eqs. (7.153)–(7.156) and $\left(\varepsilon_q, \varepsilon_j, \varepsilon_s, \varepsilon_l\right)$ quantifying the deviations from the $\Lambda$CDM values allowed by the data. A numerical estimate of these quantities can be obtained, for example, from a Markov chain analysis but this is outside the scope of the present discussion. We prefer to consider an idealized situation in which the four quantities above share the same value $\varepsilon \ll 1$. In this case we can easily investigate how much the corresponding function $f(R)$ deviates from GR by considering the ratios $f''(R_0)/f(R_0)$ and $f'''(R_0)/f(R_0)$. Inserting the previous expressions for the cosmographic parameters into the exact formulae for $f(R_0)$, $f''(R_0)$ and $f'''(R_0)$, taking their ratios and then expanding to first order in $\varepsilon$, one obtains

$$\eta_{20} = \frac{64 - 6\Omega_M(9\Omega_M + 8)}{[3(9\Omega_M + 74)\Omega_M - 556]\,\Omega_M^2 + 16}\,\frac{\varepsilon}{27}, \tag{7.162}$$

$$\eta_{30} = \frac{6\,[(81\Omega_M - 110)\Omega_M + 40]\,\Omega_M + 16}{[3(9\Omega_M + 74)\Omega_M - 556]\,\Omega_M^2 + 16}\,\frac{\varepsilon}{243\Omega_M^2}, \tag{7.163}$$

having defined the dimensionless quantities

$$\eta_{20} \equiv \frac{f''(R_0)}{f(R_0)}\,H_0^4, \tag{7.164}$$

$$\eta_{30} = \frac{f'''(R_0)}{f(R_0)}\,H_0^6, \tag{7.165}$$

which are more convenient for estimating the order of magnitude of the different terms. Inserting our fiducial values for $\Omega_M$, we obtain

$$\eta_{20} \simeq 0.15\,\varepsilon \qquad \text{for } \Omega_M = 0.041, \tag{7.166}$$

$$\eta_{20} \simeq -0.12\,\varepsilon \qquad \text{for } \Omega_M = 0.250, \tag{7.167}$$

$$\eta_{30} \simeq 4\,\varepsilon \quad \text{for } \Omega_M = 0.041, \tag{7.168}$$

$$\eta_{30} \simeq -0.18\,\varepsilon \quad \text{for } \Omega_M = 0.250. \tag{7.169}$$

For values of $\varepsilon$ up to 0.1, the above relations show that the second and third derivatives are at most two orders of magnitude smaller than the zero order term $f(R_0)$. The value of $\eta_{30}$ for a baryon-only model corresponding to $\Omega_M = 0.04$ seems to attribute a greater importance to the third order term. However, numerical checks show that the above relations approximate well the exact expressions up to $\varepsilon \simeq 0.1$, the accuracy depending on the value of $\Omega_M$ and being a decreasing function of this parameter. Using the exact expressions for $\eta_{20}$ and $\eta_{30}$, the conclusion that the effect of second and third order derivatives is negligible is significantly strengthened. This result is valid under the assumption that the narrower the constraints on the validity of the $\Lambda$CDM model, the less the cosmographic parameters deviate from their $\Lambda$CDM values. It is possible to show that this is indeed the case for the CPL parametrization used here. We have also assumed that the deviations $(\varepsilon_q, \varepsilon_j, \varepsilon_s, \varepsilon_l)$ take on the same values. Although this assumption is rather *ad hoc*, the main results are not affected by relaxing it. One can still assume that all of them are very small so that Taylor-expanding to first order leads to additional terms in Eqs. (7.162)–(7.163) which are likely of the same order of magnitude. We may therefore conclude that, if the observations confirm that the values of the cosmographic parameters agree within ~10% with those predicted in the $\Lambda$CDM model, then the deviation of the function $f(R)$ from the GR choice $f(R) \propto R$ must be small. This conclusion is justified only for the $f(R)$ models which satisfy the constraint (7.146). It is possible to construct a counterexample with $f(R_0) \propto R_0$, $f''(R_0) = f'''(R_0) = 0$ but $f^{(n)}(R_0) \neq 0$ for some $n \geq 4$. For this counterexample, Eq. (7.146) is not satisfied and the cosmographic parameters must be evaluated using the solution of the field equations and it is not excluded that the resulting $(q_0, j_0, s_0, l_0)$ are within 10% of the $\Lambda$CDM ones.

#### 7.2.1.6 The constant EoS model

Let us now take into account the condition $w = -1$ but retain $w_a = 0$, which produces the so-called *quiessence* models. In such a case, problems arise because both terms $(j_0 - q_0 - 2)$ and $\mathscr{R}$ may vanish for some combinations of the two parameters $(\Omega_M, w_0)$ of the model. For instance, we find that $j_0 - q_0 - 2 = 0$ for $w_0 = w_1$ or $w_2$, with

$$w_1 = \frac{1}{1 - \Omega_M + \sqrt{(1-\Omega_M)(4-\Omega_M)}}, \tag{7.170}$$

$$w_2 = -\frac{1}{3}\left[1 + \frac{4-\Omega_M}{\sqrt{(1-\Omega_M)(4-\Omega_M)}}\right]. \tag{7.171}$$

The equation $\mathscr{R}(\Omega_M, w_0) = 0$ may have different real roots $w$ depending on the value of $\Omega_M$ adopted. Denoting collectively with $\mathbf{w}_{null}$ the values of $w_0$ which, for a given $\Omega_M$, make $(j_0 - q_0 - 2)\mathscr{R}(\Omega_M, w_0)$ vanish, we identify a set of quiessence models whose cosmographic parameters give rise to divergent values of $f(R_0)$, $f''(R_0)$, and $f'''(R_0)$. For these models $f(R)$ is clearly not defined so that we have to exclude these cases from further consideration. It is still possible to construct an $f(R)$ theory reproducing the same background dynamics, but a different route has to be used.

Since both $q_0$ and $j_0$ now deviate from the $\Lambda$CDM values, it is not surprising that both $f''(R_0)$ and $f'''(R_0)$ assume finite non-zero values. However, to investigate the deviations of $f(R)$ from GR, it is more instructive to study the quantities $\eta_{20}$ and $\eta_{30}$. These are plotted in Figs. 7.1 and 7.2 for two fiducial values of $\Omega_M$. The range of $w_0$ in these plots has been chosen in such a way as to avoid divergences but the lessons we draw are valid for more general values of $w_0$. Even in this case, $f''(R_0)$ and $f'''(R_0)$ are two to three orders of magnitude smaller that the zero-th order term $f(R_0)$. This fact could be guessed from the previous discussion for the $\Lambda$CDM case: relaxing the assumption $w_0 = -1$ is equivalent to allowing the cosmographic parameters to deviate from their $\Lambda$CDM values. Although one cannot map directly the two cases into each other, one could argue in favor of such a relation, making these plots less surprising. Nevertheless, while in the $\Lambda$CDM case $\eta_{20}$ and $\eta_{30}$ always have opposite signs, this is not true for quiessence models with $w > -1$. Depending on the value of $\Omega_M$, we can have $f(R)$ theories with both $\eta_{20}$ and $\eta_{30}$

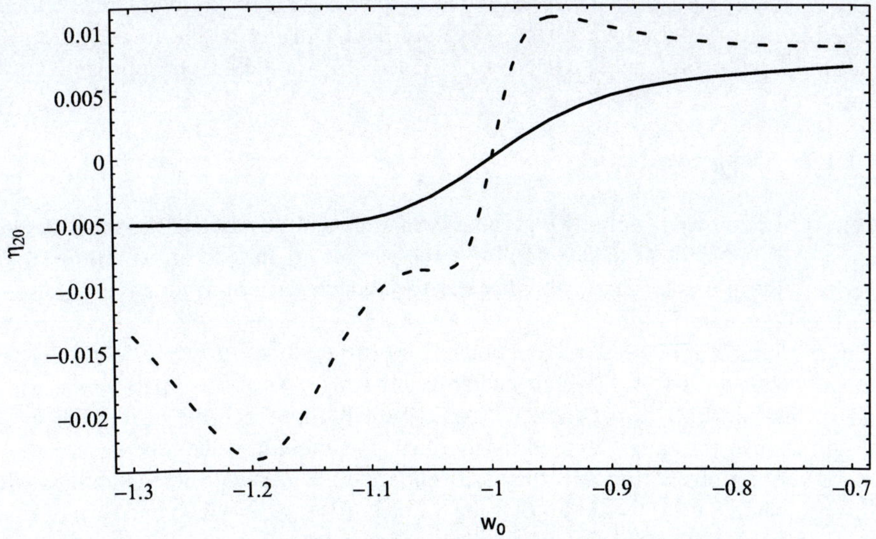


**Fig. 7.1** The dimensionless ratio between the present-day values of $f''(R)$ and $f(R)$ as a function of the constant EoS parameter $w_0$ of the corresponding quiessence model. The dashed and solid curves describe models with $\Omega_M = 0.041$ and 0.250, respectively.

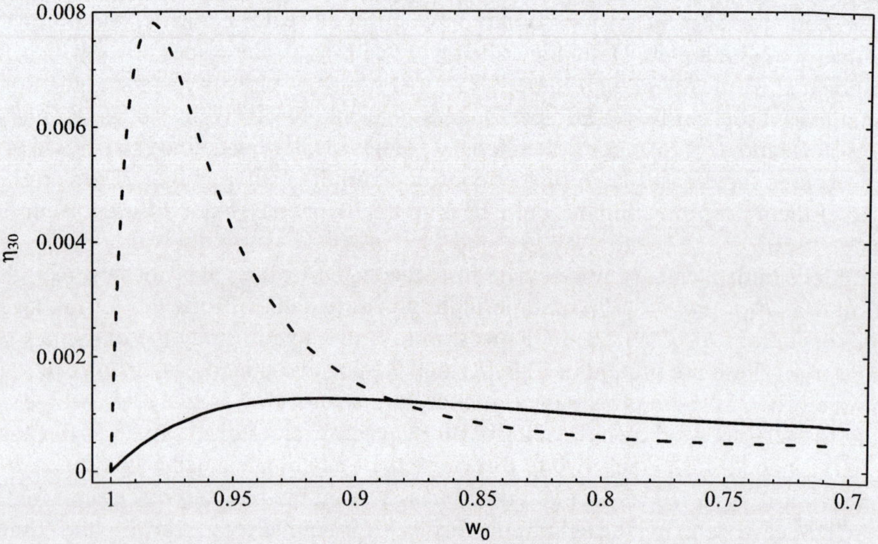


**Fig. 7.2** The dimensionless ratio between the present-day values of $f'''(R)$ and $f(R)$ as functions of the constant EoS parameter $w_0$ of the corresponding quiessence model. The dashed and solid curves refer to models with $\Omega_M = 0.041$ and 0.250, respectively.

positive. Moreover, the lower the value of $\Omega_M$, the higher the ratios $\eta_{20}$ and $\eta_{30}$ for a given value of $w_0$, which can be explained qualitatively by noting that for a lower $\Omega_M$ the density parameter of the curvature fluid must be larger, requiring higher values of the second and third derivatives (see [214] for a different approach).

#### 7.2.1.7 The general case

Finally, we consider evolving dark energy models with dynamical equation of state and $w_a \neq 0$. Needless to say, varying three parameters instead of two allows for a wider range of models that will not be discussed in detail here. We focus on evolving dark energy models with $w_0 = -1$, a value in agreement with recent analyses. The resulting $\eta_{20}$ and $\eta_{30}$ as functions of $w_a$ are reported in Figs. 7.3 and 7.4 for models with positive $w_a$, which guarantees that $w(z) \approx w_0 + w_a > w_0$ as $z \to \infty$ so that the EoS of dark energy can eventually approach the dust value $w = 0$; this is also the range favored by the data. We exclude values such that $\eta_{20}$ or $\eta_{30}$ diverge. Considering how they are defined, it is clear that these quantities will diverge when $f(R_0) = 0$ and that the values of $(w_0, w_a)$ which make $(\eta_{20}, \eta_{30})$ diverge are obtained by solving

$$\mathscr{P}_0(w_0, w_a)\Omega_M + \mathscr{Q}_0(w_0, w_a) = 0, \tag{7.172}$$

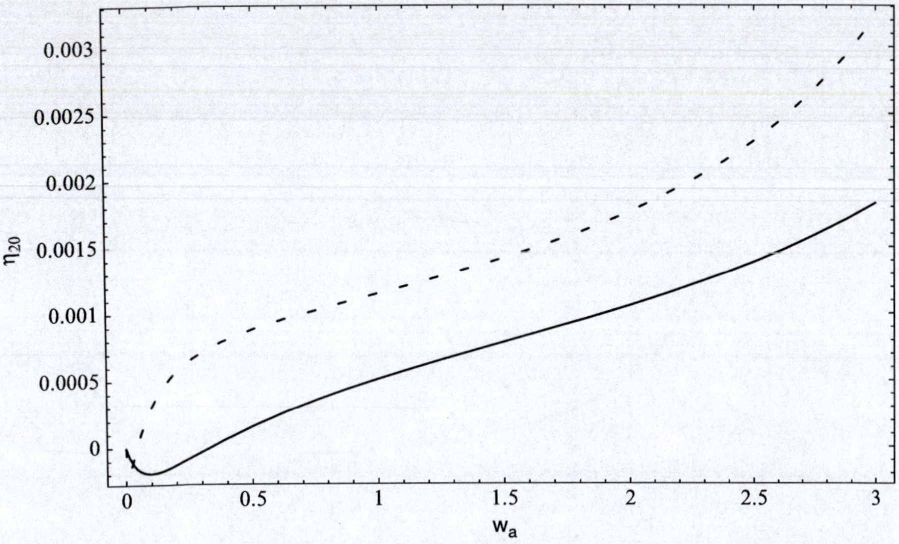


**Fig. 7.3** The dimensionless ratio between the present-day values of $f''(R)$ and $f(R)$ as functions of the $w_a$ parameter for models with $w_0 = -1$. The dashed and solid curves describe models with $\Omega_M = 0.041$ and 0.250, respectively.

where $\mathscr{P}_0(w_0, w_a)$ and $\mathscr{Q}_0(w_0, w_a)$ are obtained by inserting Eqs. (7.149)–(7.152) in the definitions (7.139)–(7.140). For these CPL models, no two $f(R)$ models have the same values of the cosmographic parameters while simultaneously satisfying all the criteria for the validity of the procedure followed here. If $f(R_0) = 0$, the condition (7.146) is likely to be violated and higher than third order derivatives must be included in the Taylor expansion of $f(R)$, invalidating the derivation of Eqs. (7.135)–(7.138).

Subject to this *caveat*, Figs. 7.3 and 7.4 show that allowing the dark energy EoS to evolve dynamically does not change significantly our conclusions. The second and third derivatives, although non-vanishing, are negligible with respect to the zero-th order term, favoring a function $f(R)$ that is only mildly non-linear which, in some sense, is expected. Eqs. (7.149) and (7.150) suggest that, having set $w_0 = -1$, $q_0$ is the same as in the $\Lambda$CDM model while $j_0$ becomes $j_0^\Lambda + (3/2)(1 - \Omega_M)w_a$. The Hilbert-Einstein Lagrangian density $f(R) = R - 2\Lambda$ is recovered for $(q_0, j_0) = (q_0^\Lambda, j_0^\Lambda)$ for all values of $(s_0, l_0)$ and introducing a $w_a \neq 0$ makes $(s_0, l_0)$ different from the $\Lambda$CDM values but the first two cosmographic parameters are only mildly affected. These deviations are then partially washed out by the complicated way in which they enter the determination of the present-day values of $f(R)$ and its first three derivatives.

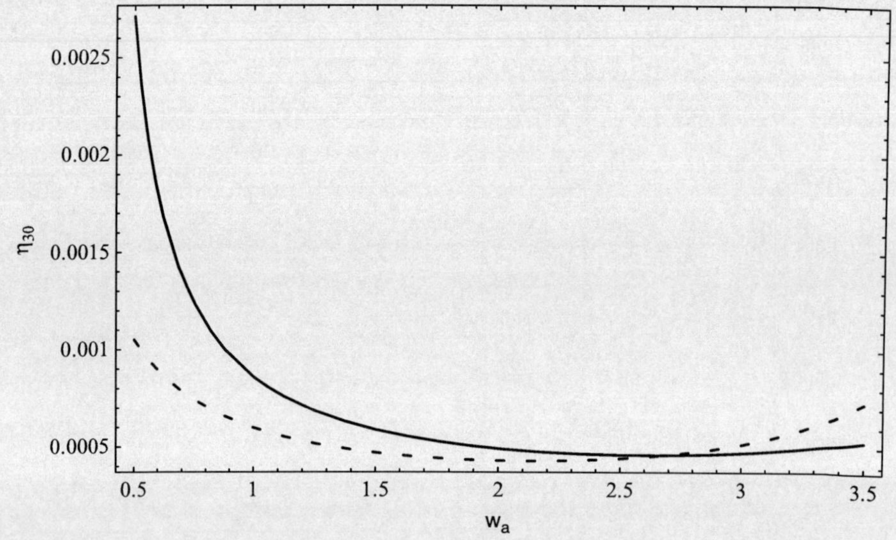


**Fig. 7.4** The dimensionless ratio between the present-day values of $f'''(R)$ and $f(R)$ as a function of the $w_a$ parameter for models with $w_0 = -1$. The dashed and solid curves describe models with $\Omega_M = 0.041$ and 0.250, respectively.

#### 7.2.1.8 Constraining the $f(R)$ parameters

We have now devised an alternative method to estimate $f(R_0)$, $f''(R_0)$, and $f'''(R_0)$ by resorting to a model-independent parameterization of the dark energy EoS. Ideally, the cosmographic parameters would be estimated directly from the data and Eqs. (7.135)–(7.145) would then be used to infer the values of the quantities related to the function $f(R)$. The latter would then be used to constrain the parameters entering a modified gravity theory with a specified function $f(R)$ characterized by a set of parameters $\mathbf{p} = (p_1, \ldots, p_n)$, provided that the assumptions underlying the derivation of Eqs. (7.135)–(7.145) are satisfied. In the following we present two examples highlighting the potentiality and the limitations of this analysis.

#### 7.2.1.9 A double power-law action

The first example, which is physically motivated in [842], is given by the choice

$$f(R) = R\left(1 + \alpha R^n + \beta R^{-m}\right) \tag{7.173}$$

with $n, m > 0$, and

$$f(R_0) = R_0\left(1 + \alpha R_0^n + \beta R_0^{-m}\right), \tag{7.174}$$

$$f'(R_0)=1+\alpha(n+1)R_0^n-\beta(m-1)R_0^{-m}, \tag{7.175}$$

$$f''(R_0)=\alpha n(n+1)R_0^{n-1}+\beta m(m-1)R_0^{-(1+m)}, \tag{7.176}$$

$$f'''(R_0)=\alpha n(n+1)(n-1)R_0^{n-2}$$
$$-\beta m(m+1)(m-1)R_0^{-(2+m)}. \tag{7.177}$$

Denoting with $\phi_i$ $(i=0,1,2,3)$ the values of $f^{(i)}(R_0)$ determined by Eqs. (7.135)–(7.145), one can solve the system

$$f(R_0)=\phi_0, \tag{7.178}$$

$$f'(R_0)=\phi_1, \tag{7.179}$$

$$f''(R_0)=\phi_2, \tag{7.180}$$

$$f'''(R_0)=\phi_3, \tag{7.181}$$

for the four unknowns $(\alpha,\beta,n,m)$ as follows. The first and second equation are solved with respect to $(\alpha,\beta)$ obtaining

$$\alpha=\frac{1-m}{n+m}\left(1-\frac{\phi_0}{R_0}\right)R_0^{-n}, \tag{7.182}$$

$$\beta=-\frac{1+n}{n+m}\left(1-\frac{\phi_0}{R_0}\right)R_0^{m}, \tag{7.183}$$

while the solution of the third and fourth equations yields

$$\alpha=\frac{\phi_2 R_0^{1-n}\left[1+m+(\phi_3/\phi_2)R_0\right]}{n(n+1)(n+m)}, \tag{7.184}$$

$$\beta=\frac{\phi_2 R_0^{1+n}\left[1-n+(\phi_3/\phi_2)R_0\right]}{m(1-m)(n+m)}. \tag{7.185}$$

By equating the two solutions one obtains the system

$$\frac{n(n+1)(1-m)\,(1-\phi_0/R_0)}{\phi_2 R_0\left[1+m+(\phi_3/\phi_2)R_0\right]}=1, \tag{7.186}$$

$$\frac{m(n+1)(m-1)\,(1-\phi_0/R_0)}{\phi_2 R_0\left[1-n+(\phi_3/\phi_2)R_0\right]}=1, \tag{7.187}$$

for the unknowns $(n, m)$. Solving this system with respect to $m$ provides two solutions: $m = -n$, which is discarded because it corresponds to divergent $(\alpha, \beta)$, and

$$m = -\left(1 - n + \frac{\phi_3}{\phi_2} R_0\right) \tag{7.188}$$

which, inserted back into the system, leads to a quadratic equation for $n$ with roots

$$n = \frac{1}{2}\left[1 + \frac{\phi_3}{\phi_2} R_0 \pm \frac{\sqrt{\mathcal{N}(\phi_0, \phi_2, \phi_3)}}{\phi_2 R_0 (1 + \phi_0/R_0)}\right] \tag{7.189}$$

where

$$\begin{aligned} \mathcal{N}(\phi_0, \phi_2, \phi_3) = & \left(R_0^2\phi_0^2 - 2R_0^3\phi_0 + R_0^4\right)\phi_3^2 \\ & + 6\left(R_0\phi_0^2 - 2R_0^2\phi_0 + R_0^3\right)\phi_2\phi_3 \\ & + 9\left(\phi_0^2 - 2R_0\phi_0 + R_0^2\right)\phi_2^2 + 4\left(R_0^2\phi_0 - R_0^3\right)\phi_2^3. \end{aligned} \tag{7.190}$$

Depending on the values of $(q_0, j_0, s_0, l_0)$, Eq. (7.189) may lead to one, two, or any number of acceptable solutions, *i.e.*, positive values of $n$. This solution has then to be inserted back into Eq. (7.188) to determine $m$ and then into Eq. (7.183) or (7.185) to estimate $(\alpha, \beta)$. If the final values of $(\alpha, \beta, n, m)$ are physically acceptable, we then conclude that the model (7.173) agrees with the data giving the cosmographic parameters inferred. The complete analytical exploration of the region of the $(q_0, j_0, s_0, l_0)$ parameter space leading to acceptable solutions $(\alpha, \beta, n, m)$ is a challenge that we will not pursue here.

#### 7.2.1.10 The Hu and Sawicki model

One of the most pressing problems of $f(R)$ theories is the need to escape the severe constraints imposed by Solar System experiments. A successful model proposed by Hu and Sawicki [605] is based on the choice[5]

$$f(R) = R - R_c \frac{\alpha (R/R_c)^n}{1 + \beta (R/R_c)^n}. \tag{7.191}$$

As in the model previously discussed, there are four parameters which can be expressed in terms of the cosmographic parameters $(q_0, j_0, s_0, l_0)$. First, it is

[5] This model does not pass the matter instability test but viable generalizations of it have been proposed [330, 848, 852].

$$f(R_0)=R_0 - R_c \frac{\alpha R_{0c}^n}{1+\beta R_{0c}^n}, \tag{7.192}$$

$$f'(R_0)=1 - \frac{\alpha n R_c R_{0c}^n}{R_0(1+\beta R_{0c}^n)^2}, \tag{7.193}$$

$$f''(R_0)=\frac{\alpha n R_c R_{0c}^n \left[(1-n)+\beta(1+n)R_{0c}^n\right]}{R_0^2(1+\beta R_{0c}^n)^3}, \tag{7.194}$$

$$f'''(R_0)=\frac{\alpha n R_c R_{0c}^n (An^2+Bn+C)}{R_0^3(1+\beta R_{0c}^n)^4}, \tag{7.195}$$

with $R_{0c} = R_0/R_c$ and

$$A=-\beta^2 R_{0c}^{2n} + 4\beta R_{0c}^n - 1, \tag{7.196}$$

$$B=3(1-\beta^2 R_{0c}^{2n}), \tag{7.197}$$

$$C=-2(1-\beta R_{0c}^n)^2. \tag{7.198}$$

Equating Eqs. (7.192)–(7.195) to the quantities $(\phi_0, \phi_1, \phi_2, \phi_3)$ defined above, one can in principle solve this system to obtain $(\alpha, \beta, R_c, n)$ in terms of $(\phi_0, \phi_1, \phi_2, \phi_3)$ and then, using Eqs. (7.135)–(7.145), express them as functions of the cosmographic parameters. However, setting $\phi_1 = 1$ as required by Eq. (7.136) gives only the trivial solution $\alpha n R_c = 0$ so that the Hu-Sawicki model reduces to the Hilbert-Einstein Lagrangian $f(R) = R$. To circumvent this problem one relaxes the condition $f'(R_0) = 1$ to $f'(R_0) = (1+\varepsilon)^{-1}$, which is equivalent to assuming that the present-day effective gravitational coupling $G_{eff,0} = G_N/f'(R_0)$ differs only slightly from its Newtonian value. Then, it is possible to solve analytically for $(\alpha, \beta, R_c, n)$ in terms of $(\phi_0, \varepsilon, \phi_2, \phi_3)$. The actual values of $(\phi_0, \phi_2, \phi_3)$ are no longer given by Eqs. (7.135)–(7.138) but it can be checked that they deviate from these expressions[6] much less than 10% for values of $\varepsilon$ reaching 10%, which are well above realistic expectations.

With this *caveat* in mind, we first solve the equation

$$f(R_0) = \phi_0, \quad f''(R_0) = \frac{1}{1+\varepsilon} \tag{7.199}$$

obtaining

$$\alpha=\frac{n(1+\varepsilon)}{\varepsilon}\left(\frac{R_0}{R_c}\right)^{1-n}\left(1-\frac{\phi_0}{R_0}\right)^2, \tag{7.200}$$

$$\beta=\frac{n(1+\varepsilon)}{\varepsilon}\left(\frac{R_0}{R_c}\right)^{-n}\left[1-\frac{\phi_0}{R_0}-\frac{\varepsilon}{n(1+\varepsilon)}\right], \tag{7.201}$$

[6] The correct expressions of $(\phi_0, \phi_2, \phi_3)$ can still be written formally as Eqs. (7.135)–(7.138) but the polynomials entering these equations are now different and depend also on powers of $\varepsilon$.

and then insert these expressions in Eqs. (7.192)–(7.195). $R_c$ drops out and its value can no longer be determined, which is expected because Eq. (7.191) can trivially be rewritten as

$$f(R) = R - \frac{\tilde{\alpha}\, R^n}{1 + \tilde{\beta} R^n} \tag{7.202}$$

where $\tilde{\alpha} \equiv \alpha\, R_c^{1-n}$ and $\tilde{\beta} \equiv \beta R_c^{-n}$ are determined by the above expressions for $(\alpha, \beta)$. Reversing the discussion, the present-day values of $f^{(i)}(R)$ depend on $(\alpha, \beta, R_c)$ only through the parameters $(\tilde{\alpha}, \tilde{\beta})$ and the use of the cosmographic parameters is unable to break this degeneracy. However, since $R_c$ only plays the role of a scaling parameter, one can arbitrarily choose its value without loss of generality. This degeneracy allows us to get a consistency relation for the Hu-Sawicki model. By solving the equation $f''(R_0) = \phi_2$, one obtains

$$n = \frac{(\phi_0/R_0) + [(1+\varepsilon)/\varepsilon](1 - \phi_2 R_0) - (1-\varepsilon)/(1+\varepsilon)}{1 - \phi_0/R_0}, \tag{7.203}$$

which can then be inserted into $f'''(R_0) = \phi_3$ to obtain a complicated relation between $\phi_0, \phi_2$, and $\phi_3$. By solving this relation with respect to $\phi_3/\phi_0$ and expanding to first order in $\varepsilon$, the constraint obtained is

$$\frac{\phi_3}{\phi_0} \simeq -\frac{1+\varepsilon}{\varepsilon}\frac{\phi_2}{R_0}\left[R_0\left(\frac{\phi_2}{\phi_0}\right) + \frac{\varepsilon\phi_0^{-1}}{1+\varepsilon}\left(1 - \frac{2\varepsilon}{1-\phi_0/R_0}\right)\right]. \tag{7.204}$$

If the cosmographic parameters $(q_0, j_0, s_0, l_0)$ are known with sufficient accuracy, one can compute the values of $(R_0, \phi_0, \phi_2.\phi_3)$ for a given $\varepsilon$ (eventually using the expressions obtained for $\varepsilon = 0$) and then check if they satisfy this relation. If this is not the case, the Hu-Sawicki model can be rejected without solving the field equations and fitting the data. In practice, the errors in the cosmographic parameters are still so large that this test is left for the future. However, the Hu-Sawicki model passes other tests [605] and is consistent with cosmography.

#### 7.2.1.11 Observational constraints on the derivatives of $f(R)$

Equations (7.135)–(7.145) relate the present-day values of $f(R)$ and its first three derivatives to the cosmographic parameters $(q_0, j_0, s_0, l_0)$ and the matter density parameter $\Omega_M$. In principle, therefore, a measurement of these quantities constrains the $f^{(i)}(R_0)$ and the parameters of a given fourth order theory through the method discussed above. The errors in the cosmographic parameters propagate through the $f(R)$ quantities; the covariance matrix for the cosmographic parameters is non-diagonal and care must be taken in the estimate of the errors on $f^{(i)}(R_0)$. A similar discussion applies to the errors in the dimensionless ratios $\eta_{20}$ and $\eta_{30}$. As a general rule, the uncertainty on a $f(R)$-related quantity $g(\Omega_M, \mathbf{p})$ which depends on $\Omega_M$ and the cosmographic parameters $\mathbf{p}$ is

$$\sigma_g^2 = \left|\frac{\partial g}{\partial \Omega_M}\right|^2 \sigma_M^2 + \sum_{i=1}^{4}\left|\frac{\partial g}{\partial p_i}\right|^2 \sigma_{p_i}^2 + 2\sum_{i \neq j} \frac{\partial g}{\partial p_i}\frac{\partial g}{\partial p_j} C_{ij}, \tag{7.205}$$

where $C_{ij}$ are the elements of the covariance matrix $C_{ii} = \sigma_{p_i}^2$, $(p_1, p_2, p_3, p_4) = (q_0, j_0, s_0, l_0)$ and it is assumed that the error $\sigma_M$ on $\Omega_M$ is uncorrelated with the errors on $\mathbf{p}$. This assumption is satisfied exactly only if the matter density parameter is estimated using an astrophysical method, for example, obtaining the total energy density of the universe from the estimated halo mass function. Alternatively, $\Omega_M$ can be constrained by CMB experiments. Since the latter probe the very high redshift universe ($z \simeq z_{lss} \simeq 1089$) while the cosmographic parameters pertain to the present-day epoch, one can argue that the determination of $\Omega_M$ is not affected by the details of the model adopted for describing the late universe. It is reasonable to assume that, whatever dark energy or $f(R)$ candidate is considered, the era probed by the CMB is well approximated by standard GR with dust only. We make the simplifying but physically well-motivated assumption that $\sigma_M$ is small and is uncorrelated with the cosmographic parameters. With this assumption, the problem of estimating the errors on $g(\Omega_M, \mathbf{p})$ reduces to that of estimating the covariance matrix for the cosmographic parameters, given the details of the data set used. We address this issue by computing the Fisher information matrix ([1071] and references therein) defined as

$$F_{ij} = \left\langle \frac{\partial^2 L}{\partial \theta_i \partial \theta_j} \right\rangle \tag{7.206}$$

where $L = -2\ln \mathscr{L}(\theta_1, \ldots, \theta_n)$, $\mathscr{L}(\theta_1, \ldots, \theta_n)$ is the likelihood of the experiment, $(\theta_1, \ldots, \theta_n)$ is the set of parameters to be constrained, and $\langle \ldots \rangle$ denotes an expectation value computed by evaluating the Fisher matrix elements for fiducial values of the model parameters $(\theta_1, \ldots, \theta_n)$, while the covariance matrix $\mathbf{C}$ is finally obtained as the inverse of $\mathbf{F}$.

A key ingredient in the computation of $\mathbf{F}$ is the definition of the likelihood which depends, of course, on the experimental constraint used. The present analysis is based on a fifth order Taylor expansion of the scale factor $a(t)$, hence we can only rely on observational tests probing quantities which are described well by this truncation. Moreover, since we do not assume a particular model, we can only characterize the background evolution of the universe and not the evolution of perturbations which unavoidably require the specification of a physical model. As a result, the Hubble diagram of SNeIa is ideal in order to constrain the cosmographic parameters. We define the likelihood as

$$\mathscr{L}(H_0, \mathbf{p}) \propto \exp\left[-\frac{\chi^2}{2}(H_0, \mathbf{p})\right], \tag{7.207}$$

$$\chi^2(H_0, \mathbf{p}) = \sum_{n=1}^{\mathscr{N}_{SNeIa}} \left[\frac{\mu_{obs}(z_i) - \mu_{th}(z_n, H_0, \mathbf{p})}{\sigma_i(z_i)}\right]^2, \tag{7.208}$$

where the distance modulus is given by

$$\mu_{th}(z, H_0, \mathbf{p}) = 25 + 5\log\left(\frac{c}{H_0}\right) + 5\log D_L(z, \mathbf{p}) \tag{7.209}$$

and

$$D_L(z) = (1+z)\int_0^z \frac{dz}{H(z)/H_0} \tag{7.210}$$

is the luminosity distance. Using the fifth order expansion of the scale factor, we obtain an analytical expression for $D_L(z, \mathbf{p})$ and the computation of $F_{ij}$ does not require numerical integrations, making the estimate faster. The final ingredient consists of the specification of the details of the SNeIa survey, *i.e.*, the redshift distribution of the sample and the error in each measurement. Following [671], we adopt[7]

$$\sigma(z) = \sqrt{\sigma_{sys}^2 + \left(\frac{z}{z_{max}}\right)^2 \sigma_m^2}, \tag{7.211}$$

where $z_{max}$ is the maximum redshift of the survey, $\sigma_{sys}$ is an irreducible scatter in the SNeIa distance modulus, and $\sigma_m$ must be assigned depending on the photometric accuracy.

In order to compute the Fisher matrix one chooses the $\Lambda$CDM predictions for the cosmographic parameters as fiducial model. For $\Omega_M = 0.3$ and $h = 0.72$ (where $h$ is the Hubble parameter in units of $100\,\mathrm{km/s \cdot Mpc}$), we obtain

$$(q_0, j_0, s_0, l_0) = (-0.55, 1.0, -0.35, 3.11). \tag{7.212}$$

As a first consistency check, we compute the Fisher matrix for a survey mimicking the recent database of [367], setting $(\mathcal{N}_{SNeIa}, \sigma_m) = (192, 0.33)$. After marginalizing over $h$ (which is fully degenerate with the SNeIa absolute magnitude $\mathcal{M}$), we obtain the uncertainties

$$(\sigma_1, \sigma_2, \sigma_3, \sigma_4) = (0.38, 5.4, 28.1, 74.0), \tag{7.213}$$

where we use the indexing introduced above for the cosmographic parameters. These values compare reasonably well with those obtained from a cosmographic fitting of the SNeIa Gold Dataset[8] [643, 644]

---

7 The authors of [671] assume that the data are binned in redshift and the error is $\sigma^2 = \sigma_{sys}^2/\mathcal{N}_{bin} + \mathcal{N}_{bin}(z/z_{max})^2\sigma_m^2$, with $\mathcal{N}_{bin}$ the number of supernovae in a bin. We prefer not to bin the data so that $\mathcal{N}_{bin} = 1$.

8 These estimates are obtained by computing the mean and standard deviation from the marginalized likelihoods of the cosmographic parameters. Then, the central values do not represent exactly the best-fit model, while the standard deviations do not give a rigorous description of the error because the marginalized likelihoods are manifestly non-Gaussian. Since we are interested in an order of magnitude estimate we do not pay attention to these statistical details.

$$q_0 = -0.90 \pm 0.65, \quad j_0 = 2.7 \pm 6.7, \tag{7.214}$$

$$s_0 = 36.5 \pm 52.9, \quad l_0 = 142.7 \pm 320. \tag{7.215}$$

Because of the Gaussian assumptions they rely on, Fisher matrix forecasts are known to be lower limits to the accuracy a given experiment can attain in the determination of a set of parameters. This is indeed the case with the comparison suggesting that our predictions are quite optimistic. However, the analysis in [643, 644] used the Gold SNeIa dataset which is poorer in high redshift supernovae than the one in [367] that we mimic here, hence larger errors on the higher order parameters $(s_0, l_0)$ are expected.

Rather than computing the errors on $f(R_0)$ and its first three derivatives, it is more interesting to look at the precision attainable on the dimensionless ratios $\eta_{20}$ and $\eta_{30}$ because they quantify the deviations from linearity. For the fiducial model adopted, both $\eta_{20}$ and $\eta_{30}$ vanish while, using the covariance matrix for a present-day survey and setting $\sigma_M / \Omega_M \simeq 10\%$, their uncertainties are

$$(\sigma_{20}, \sigma_{30}) = (0.04, 0.04). \tag{7.216}$$

As an application, Figs. 7.1 and 7.2 show how $\eta_{20}$ and $\eta_{30}$ depend on the present-day EoS parameter $w_0$ for $f(R)$ models with the same cosmographic parameters of a dark energy model with constant EoS. As is clear from these figures, also restricted to the $1\sigma$ range, the full region plotted is allowed by such large constraints on $(\eta_{20}, \eta_{30})$, meaning that the full class of corresponding $f(R)$ theories is experimentally viable. One may therefore conclude that the current SNeIa data are unable to discriminate between a $\Lambda$-dominated universe and this class of metric modified gravities.

As the next step, we consider a SuperNova Anisotropy Probe (SNAP)-like survey [23] setting $(\mathcal{N}_{SNeIa}, \sigma_m) = (2000, 0.02)$. We use the redshift distribution of Table 1 of [671] and add 300 nearby SNeIa in the redshift range $(0.03, 0.08)$. The Fisher matrix calculation gives the uncertainties on the cosmographic parameters

$$(\sigma_1, \sigma_2, \sigma_3, \sigma_4) = (0.08, 1.0, 4.8, 13.7). \tag{7.217}$$

The significant improvement of the accuracy in the determination of $(q_0, j_0, s_0, l_0)$ translates in a reduction of the errors on $(\eta_{20}, \eta_{30})$, which now read

$$(\sigma_{20}, \sigma_{30}) = (0.007, 0.008) \tag{7.218}$$

having assumed that, when SNAP data will be available, the matter density parameter $\Omega_M$ will be known with a precision $\sigma_M / \Omega_M \sim 1\%$. Looking again at Figs. 7.1 and 7.2, it is clear that the situation is improved: the constraints on $\eta_{20}$ make it possible to narrow the range of allowed models with low matter content (dashed curve) while models with typical values of $\Omega_M$ are still viable for $w_0$ spanning almost the entire horizontal axis. The constraint on $\eta_{30}$ is still too weak so that almost the full region plotted is allowed.

Finally, we consider an hypothetical future SNeIa survey working at the same photometric accuracy as SNAP and with the same redshift distribution, but increasing the number of SNeIa to $\mathcal{N}_{SNeIa} = 6 \cdot 10^4$ as expected from *DES* [5], *PanSTARRS* [655], and *SKYMAPPER* [983], while still larger numbers may potentially be achieved by *ALPACA* [341] and *LSST* [1096]. Such a survey could achieve

$$(\sigma_1, \sigma_2, \sigma_3, \sigma_4) = (0.02, 0.2, 0.9, 2.7) \tag{7.219}$$

giving (with $\sigma_M/\Omega_M \sim 0.1\%$),

$$(\sigma_{20}, \sigma_{30}) = (0.0015, 0.0016). \tag{7.220}$$

Figure 7.1 shows that, with such a precision on $\eta_{20}$, the allowed region for $w_0$ essentially reduces to the $\Lambda$CDM value while Fig. 7.2 shows that the constraint on $\eta_{30}$ definitely excludes models with low matter content, further reducing the range of $w_0$ values to small deviations from $w_0 = -1$. We can therefore conclude that such a survey will be able to discriminate between the concordance $\Lambda$CDM model and all the $f(R)$ theories giving the same cosmographic parameters as quiessence models other than the $\Lambda$CDM itself.

A similar discussion may be repeated for $f(R)$ models with the same $(q_0, j_0, s_0, l_0)$ values as the CPL model even if grasping the efficacy of the survey is less intuitive due to the fact that the parameter space is multi-valued. For the same reason we do not explore the accuracy on the double power-law or Hu-Sawicki models, even if this is technically feasible.

#### 7.2.1.12 What does cosmography teach us after all?

An unprecedented amount of high quality data have provided new input for cosmology. As it often happens in science, new and better data lead to unexpected discoveries, such as the nowadays accepted evidence for the cosmic acceleration. The equally impressive amount of more or less viable theoretical models proposed in order to explain this acceleration has also generated confusion and model-independent analyses are valuable. From this point of view, cosmography is particularly useful and preferable to assuming *ad hoc* theoretical solutions. Current and future SNeIa surveys have renewed interest in the determination of the cosmographic parameters, motivating the investigation of how these quantities can constrain cosmological models.

We have already derived the expressions of the present-day values of $f(R)$ and its first three derivatives as functions of the matter density parameter $\Omega_M$, the Hubble constant $H_0$, and the cosmographic parameters $(q_0, j_0, s_0, l_0)$ in metric $f(R)$ gravity. A third order expansion of $f(R)$ was required and we have shown that these relations hold for a large class of models and can help in the search for viable $f(R)$ models without the need to solve the field equations.

The constraints on $(q_0, j_0, s_0, l_0)$ are still too weak to efficiently apply the program outlined above. We have shown how it is possible to link the popular CPL parameterization of the dark energy equation of state and the derivatives of $f(R)$ by imposing that they produce the same cosmographic parameters. This analysis has led to the conclusion that the only $f(R)$ function able to produce the same values of $(q_0, j_0, s_0, l_0)$ as the $\Lambda$CDM model is $f(R) = R - 2\Lambda$. If future observations tell us that the cosmographic parameters are those of the $\Lambda$CDM model with high precision, we can rule out all the metric $f(R)$ theories which satisfy the assumptions used in the derivation of Eqs. (7.135)–(7.138). This result should not be seen as a no-go theorem for higher order gravity: one could still construct models with vanishing $f''(R_0)$ and $f'''(R_0)$ as required, but with non-vanishing higher order derivatives. One could argue against such models by invoking Occam's razor, but nothing compels their rejection if they turn out to be theoretically well motivated and in agreement with the data.

If new SNeIa surveys determine the cosmographic parameters with sufficient accuracy, acceptable constraints on the dimensionless quantities $\eta_{20} \propto f''(R_0)/f(R_0)$ and $\eta_{30} \propto f'''(R_0)/f(R_0)$ could be obtained, allowing for the possibility of discriminating competing $f(R)$ theories from each other. To assess the feasibility of such a program we have outlined a Fisher matrix approach forecasting the accuracy in the cosmographic parameters that can be achieved by future SNeIa surveys. A SuperNova Anisotropy Probe (SNAP)-like survey may begin giving interesting (yet still weak) constraints enabling us to reject $f(R)$ models with low matter content, while a definitive improvement is achievable with future SNeIa surveys observing $\sim 10^4$ objects and making it possible to discriminate between $\Lambda$CDM and a large class of fourth order theories. However, a necessary ingredient is the measurement of $\Omega_M$ by an independent method such as the mass fraction of gas in galaxy clusters which, at present, is still far from the necessary 1% precision. One can also rely on the $\Omega_M$ estimate from the anisotropy and polarization spectra of the CMB even if this requires assuming that the physics at recombination is strictly described by GR, which restricts the scope to $f(R)$ models reducing to linear $f(R)$ at recombination. This assumption does not seem too restrictive because in the $f(R)$ models in the literature the modifications of gravity only play a role either during early universe inflation or late in the matter-dominated era.

Regarding the kind of data useful to constrain the cosmographic parameters, the use of a fifth order expansion of the scale factor makes it possible to bypass the specification of the underlying physical model and rely on the minimal assumption that the universe is described by the spatially flat FLRW metric. While useful, this generality severely limits the useful datasets. Only observational tests which depend exclusively on the background evolution can be used, so that the range of astrophysical probes reduces to standard candles such as SNeIa and, possibly, gamma-ray bursts (if they turn out to be reliable standard candles [238]) and standard rods, *e.g.*, the angular size-redshift relation for compact radio sources. Moreover, pushing the Hubble diagram to redshifts $z \sim 2$ may rise the question of the impact of gravitational lensing amplification on the apparent magnitude of the standard candles adopted. The magnification probability distribution function depends on the growth history of perturbations [338,508,598,599,608] and one should worry about the underlying physical model in order to estimate whether this effect biases the estimate of the cosmographic parameters. It has been shown [559,647,854,941,972] that the amplification due to gravitational lensing does not alter significantly the measured distance modulus for $z \sim 1$ SNeIa. Although these studies are based on GR, one can argue that in any viable $f(R)$ model the growth of perturbations eventually leads to a distribution of structures along the line of sight close to the one observed and, therefore, the lensing amplification is approximately the same. The systematic error introduced by neglecting lensing magnification should then be lower than the statistical errors expected by future SNeIa surveys. One can also try to further reduce this possible bias using the method of flux averaging [1147] (in this case the Fisher matrix calculation should be repeated). The constraints on the cosmographic parameters may be tightened by imposing physically motivated priors in parameter space. For instance, one can impose that the Hubble parameter $H(z)$ stays always positive over the full range probed by the data, or that the transition from past deceleration to present acceleration occurs over the range probed by the data so that we can detect it. Such priors should be included in the likelihood definition and the Fisher matrix should then be re-computed.

Although the current data are still too limited to efficiently discriminate between competing $f(R)$ theories, an aggressive strategy aiming at a precise determination of the cosmographic parameters could offer stringent constraints on higher order gravity without the need to solve the field equations or address the complicated problems related to the growth of perturbations. More than eighty years after the pioneering work of Hubble, the old cosmographic approach constitutes a precious observational tool to investigate new developments in cosmology.

## 7.3 Large scale structure and galaxy clusters

Changing the gravity sector of our theories of the physical world has consequences not only at cosmological scales but also at the smaller scales of galaxies and clusters and, therefore, we now apply the $f(R)$ gravity approach to galaxy clusters. This issue is the subject of a lively debate with results arguing in favor [25,253,288,389, 830,1031] or against [301,396,868] such models at local scales. As a rule of thumb, higher order theories of gravity cause the gravitational potential to deviate from the Newtonian $1/r$ law [326,675,793,987,1021,1051], although the deviations may be small.

Consider, for illustration, power-law theories $f(R) = f_0 R^n$; their Newtonian limit was investigated in [217] assuming that the metric in the limit $\Phi/c^2 \ll 1$ is Schwarzschild-like. It turns out that a power-law term $(r/r_c)^\beta$ must be added to the Newtonian $1/r$ term in order to obtain the correct gravitational potential. While the parameter $\beta$ may be expressed analytically as a function of the slope $n$ of $f(R)$, $r_c$ sets the scale on which the correction term begins being significant. A particular range of values of $n$ such that the correction is an increasing function of the radius $r$ has been investigated: this correction causes an increase of the rotation curve with respect to the Newtonian situation and, in principle, offers the prospect of fitting the galactic rotation curves without dark matter.

A set of low surface brightness (LSB) galaxies with extended and well-determined rotation curves was considered in [378, 379]. These systems are supposed to be dominated by dark matter and fitting the relevant data successfully without dark matter supports the modified gravity approach (see [510] for an independent analysis of a different sample of galaxies). In conjunction with the hints from large scale cosmological applications there is, in principle, the possibility to address both the dark energy and dark matter problems within the context of a single fundamental theory [148, 216, 682, 747]. The simple power-law $f(R)$ considered here is just a toy model which fails to account for all the cosmological phenomena contemplated in a comprehensive theory, including the early universe, large scale structures, and the late time acceleration [216, 682].

A fundamental issue is related to galaxy clusters and superclusters, structures with size intermediate between galaxies and the universe as a whole. Like galaxies, clusters and superclusters appear to be dominated by dark matter, but this dark matter component seems to be clustered and organized in a way that is quite different from galaxies. It seems that the dark matter distribution depends on the scale and that also its fundamental nature may depend on it (see [67] for a review).

In the (metric) modified gravity approach, the task is to reconstruct the mass profile of clusters without dark matter, *i.e.*, to find corrections to the Newtonian potential which produce the same dynamics as dark matter. $f(R)$ gravity could be the paradigm to interpret both dark energy and dark matter as curvature effects acting on scales at which GR is not tested. Let us discuss now how cosmography and then galaxy clusters could help implementing this program.

### 7.3.1 *The weak-field limit of $f(R)$ gravity and galaxy clusters*

As discussed in Chap. 5, the gravitational potential of an analytic $f(R)$ theory is

$$\phi = -\frac{GM}{f_1 r} - \frac{\delta_1(t)\,\mathrm{e}^{-r\sqrt{-\xi}}}{6\xi r}. \tag{7.221}$$

Among the possible analytic $f(R)$ models, consider those for which the cosmological term $f_0$ and terms higher than second have been discarded. Rewriting the $f(R)$ Lagrangian as

$$f(R) = a_1 R + a_2 R^2 + \cdots \tag{7.222}$$

and specifying the gravitational potential (7.221) generated by a point-like matter distribution as

$$\phi(r) = -\frac{3GM}{4a_1 r}\left(1 + \frac{\mathrm{e}^{-r/L}}{3}\right), \tag{7.223}$$

where

$$L(a_1, a_2) \equiv \left(-\frac{6a_2}{a_1}\right)^{1/2} \tag{7.224}$$

(not a free parameter in the fitting procedure but a function of the coefficients $a_{1,2}$) is an *interaction length* of the problem introduced by the correction to the Newtonian potential. The change in notations reflects the specific choice in the much wider class of potentials (7.221), but the following considerations are general.

### 7.3.2 Extended systems

The gravitational potential (7.223) describes a point-like mass and it is generalized to extended galaxy clusters described as spherically symmetric systems consisting of infinitesimal mass elements *dm*, each contributing a point-like gravitational potential. The total potential is obtained by integrating all these contributions over a sphere. We must then compute the integral

$$\Phi(r) = \int_0^\infty r^2 dr \int_0^\pi \sin\theta\, d\theta \int_0^{2\pi} d\varphi\, \phi(r). \tag{7.225}$$

The point-like potential (7.223) can be split into two terms: the Newtonian component is

$$\phi_N(r) = -\frac{3GM}{4a_1 r}, \tag{7.226}$$

the extended integral of which is well known (apart from the factor $\frac{3}{4a_1}$) and is

$$\Phi_N(r) = -\frac{3}{4a_1}\frac{GM(<r)}{r}, \tag{7.227}$$

where $M(<r)$ is the mass enclosed in a sphere of radius $r$. The correction

$$\phi_C(r) = -\frac{GM}{4a_1}\frac{e^{-r/L}}{r} \tag{7.228}$$

gives

$$\Phi_C(r) = -\frac{\pi GL}{2}\int_0^\infty dr' r' \rho(r')\, \frac{\mathrm{e}^{-\frac{|r-r'|}{L}} - \mathrm{e}^{-\frac{|r+r'|}{L}}}{r}. \tag{7.229}$$

The radial integral is estimated numerically once the mass density is given. A fundamental difference between this term and the Newtonian one is that, while in the latter the matter outside a spherical shell of radius $r$ does not contribute to the potential, for the former external matter contributes to the integral. Therefore, we split the corrective potential as

$$\begin{aligned}\Phi_{C,int}(r) &= -\frac{\pi GL}{2}\int_0^r dr' r' \rho(r')\, \frac{\mathrm{e}^{-\frac{|r-r'|}{L}} - \mathrm{e}^{-\frac{|r+r'|}{L}}}{r} \\ &= -\frac{\pi GL}{2}\int_0^r dr' r' \rho(r')\, \mathrm{e}^{-\frac{r+r'}{L}}\left(\frac{-1+\mathrm{e}^{\frac{2r'}{L}}}{r}\right)\end{aligned} \tag{7.230}$$

if $r' < r$, and

$$\Phi_{C,ext}(r) = -\frac{\pi GL}{2}\int_r^{+\infty} dr' r' \rho(r') \frac{e^{-\frac{|r-r'|}{L}} - e^{-\frac{|r+r'|}{L}}}{r}$$
$$= -\frac{\pi GL}{2}\int_r^{+\infty} dr' r' \rho(r')\, e^{-\frac{r+r'}{L}} \left(\frac{-1 + e^{2r/L}}{r}\right) \quad (7.231)$$

if $r' > r$, and the total potential of the spherical mass distribution is

$$\Phi(r) = \Phi_N(r) + \Phi_{C,int}(r) + \Phi_{C,ext}(r). \quad (7.232)$$

As shown below, we need the radial derivative of the gravitational potential; it may not be possible to evaluate the two derivatives analytically, but they can be estimated numerically once an expression for the total mass density $\rho(r)$ is given. While the Newtonian term yields the simple expression

$$\frac{d\Phi_N}{dr}(r) = \frac{3}{4a_1}\frac{GM(< r)}{r^2}, \quad (7.233)$$

the internal and external derivatives of the corrective potential terms are more complicated. We do not provide their explicit form here, but note that they are integrals of the form

$$\mathcal{F}(r, r') = \int_{\alpha(r)}^{\beta(r)} dr'\, f(r, r'), \quad (7.234)$$

from which it follows that

$$\frac{d\mathcal{F}(r, r')}{dr} = \int_{\alpha(r)}^{\beta(r)} dr' \frac{df(r, r')}{dr}$$
$$- f(r, \alpha(r))\frac{d\alpha}{dr}(r) + f(r, \beta(r))\frac{d\beta}{dr}(r). \quad (7.235)$$

This expression is evaluated numerically once the integration limits are specified. The Gauss theorem is valid only for the Newtonian part because it scales as $1/r^2$. However, the Gauss theorem no longer holds for the total potential (7.223) because of the non-Newtonian correction. This is not a problem because in $f(R)$ gravity the full conservation laws are determined by the contracted Bianchi identities which guarantee self-consistency [217, 237, 1033].

### 7.3.3 *The cluster mass profiles*

Galaxy clusters are generally considered to be bound gravitational systems with approximate spherical symmetry and in hydrostatic equilibrium if virialized. These assumptions are widely used in spite of the fact that most clusters exhibit more complex morphologies and/or evidence of strong interactions or dynamical activity, especially in their innermost regions [290, 380].

Assuming spherical symmetry and hydrostatic equilibrium, the structure equation can be derived from the collisionless Boltzmann equation

$$\frac{d}{dr}\left[\rho_{gas}(r)\,\sigma_r^2\right] + \frac{2\rho_{gas}(r)}{r}(\sigma_r^2 - \sigma_{\theta,\varphi}^2) = -\rho_{gas}(r)\,\frac{d\Phi(r)}{dr}, \tag{7.236}$$

where $\Phi$ is the gravitational potential of the cluster, $\sigma_r$ and $\sigma_{\theta,\varphi}$ are the mass-weighted velocity dispersions in the radial and tangential directions, respectively, and $\rho$ is the gas mass density. For an isotropic system, it is

$$\sigma_r = \sigma_{\theta,\varphi} \tag{7.237}$$

and the pressure profile can be related to said quantities as

$$P(r) = \sigma_r^2\,\rho_{gas}(r). \tag{7.238}$$

Substituting Eqs. (7.237) and (7.238) into Eq. (7.236) we have, for an isotropic sphere,

$$\frac{dP(r)}{dr} = -\rho_{gas}(r)\,\frac{d\Phi(r)}{dr}. \tag{7.239}$$

For a gas sphere with temperature profile $T(r)$ the velocity dispersion becomes

$$\sigma_r^2 = \frac{K_{BT}(r)}{\mu m_p}, \tag{7.240}$$

where $K_B$ is the Boltzmann constant, $\mu \approx 0.609$ is the mean mass particle, and $m_p$ is the proton mass. The substitution of Eqs. (7.238) and (7.240) into Eq. (7.239) yields

$$\frac{d}{dr}\left[\frac{K_B T(r)}{\mu m_p}\rho_{gas}(r)\right] = -\rho_{gas}(r)\frac{d\Phi}{dr} \tag{7.241}$$

or, equivalently,

$$-\frac{d\Phi}{dr} = \frac{K_B T(r)}{\mu m_p r}\left[\frac{d\ln\rho_{gas}(r)}{d\ln r} + \frac{d\ln T(r)}{d\ln r}\right]. \tag{7.242}$$

Now the total gravitational potential of the cluster is

$$\Phi(r) = \Phi_N(r) + \Phi_C(r), \tag{7.243}$$

where

$$\Phi_C(r) = \Phi_{C,int}(r) + \Phi_{C,ext}(r). \tag{7.244}$$

If we consider only the standard Newtonian potential, then the total mass $M_{cl,N}(r)$ of the cluster is composed by the mass of the gas, plus the mass of galaxies, plus the mass of the cD galaxy and dark matter, and it is given by the expression

$$\begin{aligned} M_{cl,N}(r) &= M_{gas}(r) + M_{gal}(r) + M_{CDgal}(r) + M_{DM}(r) \\ &= -\frac{K_B T(r)}{\mu m_p G} r \left[ \frac{d \ln \rho_{gas}(r)}{d \ln r} + \frac{d \ln T(r)}{d \ln r} \right], \end{aligned} \tag{7.245}$$

where $M_{cl,N}$ denotes the standard estimated Newtonian mass. The contribution from galaxies is usually considered to be negligible in comparison with that of the other two components and

$$\begin{aligned} M_{cl,N}(r) &\approx M_{gas}(r) + M_{DM}(r) \\ &\approx -\frac{K_B T(r)}{\mu m_p} r \left[ \frac{d \ln \rho_{gas}(r)}{d \ln r} + \frac{d \ln T(r)}{d \ln r} \right]. \end{aligned} \tag{7.246}$$

Since estimates of the gas mass are provided by X-ray observations, the equilibrium equation can be used to derive the amount and spatial distribution of dark matter present in a cluster of galaxies. Inserting the previous *extended-corrected* potential (7.243) into Eq. (7.242) yields

$$-\frac{d\Phi_N}{dr} - \frac{d\Phi_C}{dr} = \frac{K_B T(r)}{\mu m_p r} \left[ \frac{d \ln \rho_{gas}(r)}{d \ln r} + \frac{d \ln T(r)}{d \ln r} \right], \tag{7.247}$$

from which the *extended-corrected* mass estimate

$$\begin{aligned} M_{cl,EC}(r) &+ \frac{4a_1}{3G} r^2 \frac{d\Phi_C}{dr}(r) \\ &= \frac{4a_1}{3} \left[ -\frac{K_B T(r)}{\mu m_p G} r \left( \frac{d \ln \rho_{gas}(r)}{d \ln r} + \frac{d \ln T(r)}{d \ln r} \right) \right] \end{aligned} \tag{7.248}$$

follows. Since the use of a corrected potential avoids, in principle, the need for dark matter, the total cluster mass is

$$M_{cl,EC}(r) = M_{gas}(r) + M_{gal}(r) + M_{CDgal}(r) \tag{7.249}$$

and the mass density corresponding to the $\Phi_C$ term is

$$\rho_{cl,EC}(r) = \rho_{gas}(r) + \rho_{gal}(r) + \rho_{CDgal}(r) \tag{7.250}$$

with the density components derived from observations.

We will use Eq. (7.248) to compare the baryonic mass profile $M_{cl,EC}(r)$, estimated from observations with the theoretical deviation from the Newtonian gravitational potential given by the expression $-\frac{4a_1 r^2}{3G} \frac{d\Phi_C}{dr}(r)$. Our goal is to reproduce the observed mass profiles for a sample of galaxy clusters.

### 7.3.4 The galaxy clusters sample

We now apply the formalism of Sect. 7.3.3 to the sample of galaxy clusters of Refs. [1114, 1115] which consists of 13 low redshift clusters spanning a temperature range 0.7–9.0 keV derived from high quality *Chandra* archival data. In all these clusters, the surface brightness and the gas temperature profiles are measured out to large radii, so that mass estimates can be extended up to $r_{500}$ or beyond.

### 7.3.5 The gas density model

The gas density distribution of the clusters in the sample is described by the analytic model proposed in [1114], which modifies the classical $\beta$-model to represent the characteristic properties of the observed X-ray surface brightness profiles, *i.e.*, the power-law type cusps of gas density in the cluster center, instead of a flat core and the steepening of the brightness profiles at large radii. Eventually, a second $\beta$-model with a small core radius is added to improve the model near the cluster cores. The analytical form for particle emission is given by

$$n_p n_e = n_0^2 \frac{(r/r_c)^{-\alpha}}{(1+r^2/r_c^2)^{3\beta-\alpha/2}} \cdot \frac{1}{(1+r^\gamma/r_s^\gamma)^{\varepsilon/\gamma}} + \frac{n_{02}^2}{(1+r^2/r_{c2}^2)^{3\beta_2}}, \tag{7.251}$$

which can be easily converted to a mass density using the relation

$$\rho_{gas} = n_T \mu m_p = 1.1667\, n_e\, m_p, \tag{7.252}$$

where $n_T$ is the total number density of particles in the gas. The resulting model has a large number of parameters, some of which do not have a direct physical interpretation. While this can often be inappropriate and computationally inconvenient, it suits well our situation in which the main requirement is a detailed qualitative description of the cluster profiles.

In [1114], Eq. (7.251) is applied to a restricted range of distances from the cluster centre between an inner cutoff $r_{min}$ chosen to exclude the central temperature bin (10–20 kpc) where the intra-cluster medium is likely to be multi-phase, and $r_{det}$, where the X-ray surface brightness is at least $3\sigma$ significant. We have extrapolated the above function to values outside this restricted range using the following criteria:

- For $r < r_{min}$, we have performed a linear extrapolation of the first three terms out to $r = 0$ kpc;
- For $r > r_{det}$, we have performed a linear extrapolation of the last three terms out to a distance $\bar{r}$ for which $\rho_{gas}(\bar{r}) = \rho_c$, where $\rho_c$ is the critical density of the universe at the cluster redshift: $\rho_c = \rho_{c,0}(1+z)^3$. For radii larger than $\bar{r}$, the gas density is assumed constant at $\rho_{gas}(\bar{r})$.

**Table 7.1** Column 1: cluster name. Column 2: richness. Column 3: total cluster mass. Column 4: gas mass. Column 5: galaxy mass. Column 6: cD galaxy mass (all mass values are estimated at $r = r_{max}$). Column 7: ratio of total galaxy mass to gas mass. Column 8: minimum radius. Column 9: maximum radius.

| name | R | $M_{cl,N}$ $(M_\odot)$ | $M_{gas}$ $(M_\odot)$ | $M_{gal}$ $(M_\odot)$ | $M_{cDgal}$ $(M_\odot)$ | $\frac{gal}{gas}$ | $r_{min}$ (kpc) | $r_{max}$ (kpc) |
|---|---|---|---|---|---|---|---|---|
| A133 | 0 | $4.35874 \cdot 10^{14}$ | $2.73866 \cdot 10^{13}$ | $5.20269 \cdot 10^{12}$ | $1.10568 \cdot 10^{12}$ | 0.23 | 86 | 1060 |
| A262 | 0 | $4.45081 \cdot 10^{13}$ | $2.76659 \cdot 10^{12}$ | $1.71305 \cdot 10^{11}$ | $5.16382 \cdot 10^{12}$ | 0.25 | 61 | 316 |
| A383 | 2 | $2.79785 \cdot 10^{14}$ | $2.82467 \cdot 10^{13}$ | $5.88048 \cdot 10^{12}$ | $1.09217 \cdot 10^{12}$ | 0.25 | 52 | 751 |
| A478 | 2 | $8.51832 \cdot 10^{14}$ | $1.05583 \cdot 10^{14}$ | $2.15567 \cdot 10^{13}$ | $1.67513 \cdot 10^{12}$ | 0.22 | 59 | 1580 |
| A907 | 1 | $4.87657 \cdot 10^{14}$ | $6.38070 \cdot 10^{13}$ | $1.34129 \cdot 10^{13}$ | $1.66533 \cdot 10^{12}$ | 0.24 | 563 | 1226 |
| A1413 | 3 | $1.09598 \cdot 10^{15}$ | $9.32466 \cdot 10^{13}$ | $2.30728 \cdot 10^{13}$ | $1.67345 \cdot 10^{12}$ | 0.26 | 57 | 1506 |
| A1795 | 2 | $5.44761 \cdot 10^{14}$ | $5.56245 \cdot 10^{13}$ | $4.23211 \cdot 10^{12}$ | $1.93957 \cdot 10^{12}$ | 0.11 | 79 | 1151 |
| A1991 | 1 | $1.24313 \cdot 10^{14}$ | $1.00530 \cdot 10^{13}$ | $1.24608 \cdot 10^{12}$ | $1.08241 \cdot 10^{12}$ | 0.23 | 55 | 618 |
| A2029 | 2 | $8.92392 \cdot 10^{14}$ | $1.24129 \cdot 10^{14}$ | $3.21543 \cdot 10^{13}$ | $1.11921 \cdot 10^{12}$ | 0.27 | 62 | 1771 |
| A2390 | 1 | $2.09710 \cdot 10^{15}$ | $2.15726 \cdot 10^{14}$ | $4.91580 \cdot 10^{13}$ | $1.12141 \cdot 10^{12}$ | 0.23 | 83 | 1984 |
| MKW4 | - | $4.69503 \cdot 10^{13}$ | $2.83207 \cdot 10^{12}$ | $1.71153 \cdot 10^{11}$ | $5.29855 \cdot 10^{11}$ | 0.25 | 60 | 434 |
| RXJ1159 | - | $8.97997 \cdot 10^{13}$ | $4.33256 \cdot 10^{12}$ | $7.34414 \cdot 10^{11}$ | $5.38799 \cdot 10^{11}$ | 0.29 | 64 | 568 |

In Table 7.1, the radius limit $r_{min}$ is almost the same as defined above. When the value given by [1114] is less than the cD galaxy radius (defined below), we choose this last one as the lower limit. On the contrary, $r_{max}$ is quite different from $r_{det}$: it is fixed by considering the higher value of the temperature profile and not by imaging methods.

We finally compute the gas mass $M_{gas}(r)$ and the total mass $M_{cl,N}(r)$ for all clusters in our sample, substituting Eq. (7.251) into Eqs. (7.252) and (7.245). The gas temperature profile is described in Sect. 7.3.6. The resulting mass values, estimated at $r = r_{max}$, are listed in Table 7.1.

### *7.3.6 Temperature profiles*

An accurate qualitative description of the radial behavior of the gas properties is needed for our purposes. Standard isothermal or polytropic models, and even the more complex model of [1114], do not provide a good description of the data at all radii and for all clusters in the sample. Therefore, we describe the gas temperature profiles using the X-ray spectral analysis results without introducing any analytic model and the X-ray spectral values of [1116]. A detailed description of the relative spectral analysis is given in [1115].

### *7.3.7 The galaxy distribution model*

The density of galaxies is modelled after [67]; even if the galaxy distribution is a point distribution instead of a continuous function, assuming that galaxies are in equilibrium with the gas, one can use a $\beta$-model $\rho \propto r^{-3}$ for $r < R_c$ from the cluster centre, and a steeper one $\rho \propto r^{-2.6}$ for $r > R_c$, where $R_c$ is the cluster core radius (its value is taken from [1114]). The density of galaxies is given by

$$\rho_{gal}(r) = \begin{cases} \rho_{gal,1}\left[1 + \left(\frac{r}{R_c}\right)^2\right]^{-3/2} & r < R_c, \\ \\ \rho_{gal,2}\left[1 + \left(\frac{r}{R_c}\right)^2\right]^{-1.3} & r > R_c, \end{cases} \tag{7.253}$$

where the constants $\rho_{gal,1}$ and $\rho_{gal,2}$ are chosen as follows:

- Reference [67] provides the central number density of galaxies in rich compact clusters for galaxies located within $1.5h^{-1}$Mpc from the cluster centre and brighter than $m_3 + 2^m$ (where $m_3$ is the magnitude of the third brightest galaxy): $n_{gal,0} \sim 10^3 h^3$ galaxies $\cdot$Mpc$^{-3}$. We then fix $\rho_{gal,1}$ in the range $10^{34} - 10^{36}$ kg/kpc$^3$. For clusters obeying the condition chosen for the galaxy to gas mass ratio, we assume a typical elliptical and cD galaxy mass in the range $10^{12} - 10^{13} M_\odot$.
- The constant $\rho_{gal,2}$ is fixed with the only requirement that the galaxy density function is continuous at $R_c$.

The effect of varying the galaxy density in the range $10^{34} - 10^{36}$ kg/kpc$^3$ is tested on the cluster with the lowest mass, *i.e.*, A262, for which one would expect the largest variations; the result is that the contribution of galaxies and of the cD galaxy gives a variation not larger than 1% to the final estimate of fit parameters.

The cD galaxy density is modelled as in [989], which uses a Jaffe model of the form

$$\rho_{cDgal} = \frac{\rho_{0,J}}{\left(\frac{r}{r_c}\right)^2 \left(1 + \frac{r}{r_c}\right)^2}, \tag{7.254}$$

where $r_c$ is the core radius and the central density is obtained from $M_J = \frac{4}{3}\pi R_c^3 \rho_{0,J}$. The mass of the cD galaxy has been fixed at $1.14 \times 10^{12}\ M_\odot$, with $r_c = R_e/0.76$ and where $R_e = 25$ kpc is the effective radius of the galaxy. The central galaxy for each cluster in the sample is assumed to have approximately this mass in stars.

We assume that the total mass of the galaxy component (*i.e.*, galaxies plus cD galaxy masses) is 20–25% of the gas mass: in [979] the mean fraction of gas versus the total mass (with dark matter) for a cluster is estimated to be 15–20%, while the same quantity for galaxies is 3–5%. This means that the relative mean galaxies-to-gas mass ratio in a cluster is 20–25%. We have varied the parameters $\rho_{gal,1}$, $\rho_{gal,2}$ and $M_J$ in their previous defined ranges to obtain a mass ratio between total galaxy mass and total gas mass which lies in this range. The resulting galaxy mass values and ratios $\frac{\text{gal}}{\text{gas}}$, estimated at $r = r_{max}$, are listed in Table 7.1.

Figures 7.5 and 7.6 show how each component is spatially distributed. The cD galaxy is dominant with respect to the other galaxies only in the inner region (*i.e.*,

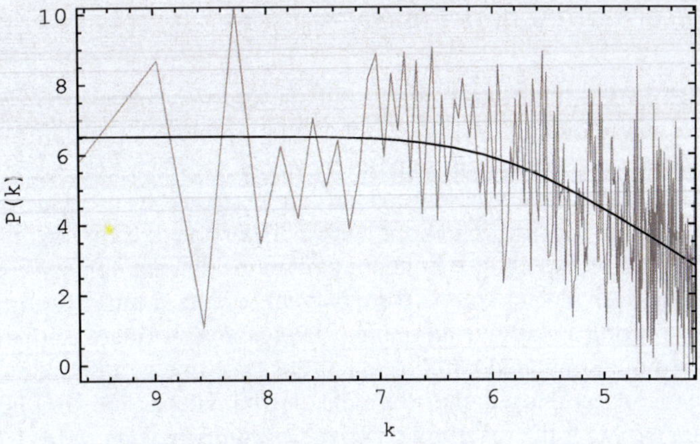


**Fig. 7.5** Power spectrum test on sample chain for the parameter $a_1$ using the method described in Sect. 7.3.9. The black curve is the logarithm of the analytical template Eq. (7.261) for the power spectrum, while the gray curve is the discrete power spectrum obtained using Eqs. (7.259) and (7.260).

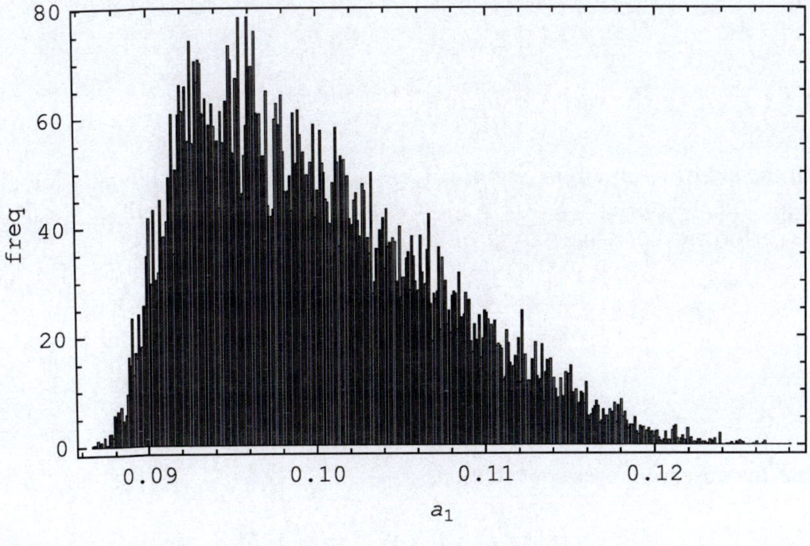


**Fig. 7.6** Histogram of the sample points for the parameter $a_1$ in Abell 383 resulting from the MCMC implementation used to estimate best-fit values and errors for the fitting procedure described in Sect. 7.3.9. Binning (horizontal axis) and relative frequencies (vertical axis) are given by automatic procedure from Mathematica 6.0.

below 100 kpc). The cluster's innermost regions have been excluded from the analysis and therefore the contribution of the cD galaxy is negligible and the gas is the dominant visible component, starting from the innermost regions out to large radii, with the galaxy mass amounting to only 20–25% of the gas mass. Similar features are shown by all the clusters in the sample used.

### 7.3.8 Uncertainties in the mass profiles

Uncertainties in the total mass profiles of the clusters are estimated performing Monte-Carlo simulations [834]. We proceed to simulate temperature profiles and choose random radius-temperature pairs of values for each bin in the temperature data of [1115]. Random temperature values are extracted from a Gaussian distribution centered on the spectral values and with a dispersion fixed to its 68% confidence level. For the radius, we choose a random value inside each bin. We perform 2000 simulations for each cluster and perform two cuts on the simulated profile. First, we exclude those profiles that give an unphysical negative estimate of the mass: this is possible when our simulated pairs of quantities give rise to a temperature gradient that is too large. After this cut, there remain approximately 1500 simulations for any cluster. Then we order the resulting mass values for increasing radii. Extreme mass estimates (outside the 10–90% range) are excluded from the obtained distribution in order to avoid other high mass gradients which originate masses too different from the real data. The resulting limits provide the errors on the total mass. Uncertainties in the electron density profiles have not been included in the simulations because they are negligible in comparison with those of the gas temperature profiles.

### 7.3.9 Fitting the mass profiles

With the aid of X-ray observations it is possible to model theoretically the galaxy distribution and, using Eq. (7.248), to estimate the baryon content of clusters. We have performed a best-fit analysis of the theoretical equation (7.248),

$$M_{bar,th}(r) = \frac{4a_1}{3}\left[-\frac{kT(r)}{\mu m_p G} r \left(\frac{d \ln \rho_{gas}(r)}{d \ln r} + \frac{d \ln T(r)}{d \ln r}\right)\right] - \frac{4a_1}{3G} r^2 \frac{\mathrm{d}\Phi_C}{\mathrm{d}r}(r) \qquad (7.255)$$

versus the observed mass contributions

$$M_{bar,obs}(r) = M_{gas}(r) + M_{gal}(r) + M_{CDgal}(r). \qquad (7.256)$$

Since not all the data involved in the above estimate have measurable errors we cannot perform an exact $\chi^2$ minimization but we can minimize the quantity

$$\chi^2 = \frac{1}{N - n_p - 1} \sum_{i=1}^{N} \frac{(M_{bar,obs} - M_{bar,theo})^2}{M_{bar,theo}}, \qquad (7.257)$$

where $N$ is the number of data and $n_p = 2$ are the free parameters of the model. We minimize the $\chi^2$ using the Markov Chain Monte Carlo Method (MCMC). For each cluster, we run various chains to set the best parameters of the Metropolis-Hastings algorithm used. Beginning with an initial parameter vector $\mathbf{p} = (a_1, a_2)$, we generate a new trial point $\mathbf{p}'$ from a tested proposal density $q(\mathbf{p}', \mathbf{p})$ rcpresenting the conditional probability to get $\mathbf{p}'$, given $\mathbf{p}$. This new point is accepted with probability

$$\alpha(\mathbf{p}, \mathbf{p}') = \min\left\{1, \frac{L(\mathbf{d}|\mathbf{p}')P(\mathbf{p}')q(\mathbf{p}', \mathbf{p})}{L(\mathbf{d}|\mathbf{p})P(\mathbf{p})q(\mathbf{p}, \mathbf{p}')}\right\}, \tag{7.258}$$

where $\mathbf{d}$ are the data, $L(\mathbf{d}|\mathbf{p}') \propto \exp(-\chi^2/2)$ is the likelihood function, and $P(\mathbf{p})$ is the prior on the parameters. The prior on the fit parameters is related to Eq. (7.258); since $L$ is a length, we need to force the ratio $a_1/a_2$ to be positive. The proposal density is Gaussian-symmetric with respect to the vectors $\mathbf{p}$ and $\mathbf{p}'$, *i.e.*, $q(\mathbf{p}, \mathbf{p}') \propto \exp(-\Delta p^2/2\sigma^2)$, with $\Delta\mathbf{p} = \mathbf{p} - \mathbf{p}'$. We decide to fix the dispersion $\sigma$ of any trial distribution of parameters to 20% of the trial $a_1$ and $a_2$ at any step. This means that the parameter $\alpha$ reduces to the ratio between the likelihood functions.

We run one chain of $10^5$ points for each cluster; the convergence of these chains is tested using the power spectrum analysis of [400]. The key idea of this method is simple but powerful: if we take the power spectra of the MCMC samples we have a high correlation on small scales but, when the chain reaches convergence, the spectrum becomes flat (like a white noise spectrum). Hence, by checking the spectrum of just one chain, instead of many parallel chains as in the Gelmann-Rubin test, is sufficient to assess the convergence reached (we refer the reader to [400] for a detailed discussion). The discrete power spectrum of the chains is calculated as

$$P_j = |a_N^j|^2 \tag{7.259}$$

with

$$a_N^j = \frac{1}{\sqrt{N}} \sum_{n=0}^{N-1} x_n \exp\left[i\,\frac{2\pi j}{N} n\right], \tag{7.260}$$

where $N$ and $x_n$ are the length and the element of the sample from the MCMC, respectively, and $j = 1, \ldots, \frac{N}{2} - 1$. The wavenumber $k_j$ of the spectrum is related to the index $j$ by $k_j = 2\pi j/N$. Then this quantity is fitted with the analytical template

$$P(k) = P_0 \frac{(k^*/k)^\alpha}{1 + (k^*/k)^\alpha} \tag{7.261}$$

or, in the equivalent logarithmic form,

$$\ln P_j = \ln P_0 + \ln\left[\frac{(k^*/k_j)^\alpha}{1 + (k^*/k_j)^\alpha}\right] - \gamma + r_j, \tag{7.262}$$

where $\gamma = 0.57216$ is the Euler-Mascheroni constant and $r_j$ are random measurement errors with $< r_j >= 0$ and $< r_i r_j >= \pi^2 \delta_{ij}/6$. From the fit, we estimate the fundamental parameters $P_0$ and $j^*$ (the index corresponding to $k^*$). The first is the value of the power spectrum extrapolated to $k \to 0$ and, from this, we derive the convergence ratio from $r \approx P_0/N$; if $r < 0.01$, convergence is reached. The second parameter is related to the turning point from a power-law to a flat spectrum: it must be larger than 20 in order to be sure that the number of points in the sample coming from the convergence region is larger than the noise points. If these two conditions are verified for all the parameters, then the chain has reached convergence and the statistics derived from MCMC describes well the underlying probability distribution (typical results are shown in Figs. 7.5 and 7.6). Following the prescriptions of [400], we perform the fit over the range $1 \le j \le j_{max}$, with $j_{max} \sim 10 j^*$, where a first estimate of $j^*$ can be obtained from a fit with $j_{max} = 1000$, and then perform a second iteration in order to have a better estimate. Even if convergence is achieved after a few thousand steps in the chain, we run longer chains of $10^5$ points in order to reduce the noise from the histograms and avoid under- or over-estimating the errors in the parameters. The $i\sigma$ confidence levels are easily estimated by deriving from the final sample the 15.87-th and 84.13-th quantiles (which define the 68% confidence interval) for $i = 1$, the 2.28-th and 97.72-th quantiles (which define the 95% confidence interval) for $i = 2$, and the 0.13-th and 99.87-th quantiles (which define the 99% confidence interval) for $i = 3$. Having described the method, let us now comment on the results.

### *7.3.10 Results*

The numerical results of the fitting analysis are summarized in Table 7.2, which gives the best-fit values of the independent fitting parameters $a_1$ and $a_2$ and of the gravitational length $L$, considered as a function of the previous two quantities. Figures 7.5 and 7.6 provide a typical power spectrum and histogram of samples derived by the MCMC in order to assess the convergence reached (flat spectrum at large scales). The baryonic mass profiles versus the radius for some clusters of the sample are plotted in Figs. 7.9–7.13.

The goodness and the properties of the fits are shown in Figs. 7.5 and 7.6. The main feature is the presence of a typical scale for each cluster above which the model works well (typical relative differences are less than 5%), while for lower scales there is a larger difference. Inspection reveals that this turning point is located at a radius ~150 kpc. Except for very large clusters, this value is independent of the cluster, being approximately the same for all members of the sample. There are two main independent explanations that could justify this trend: limitations due to a break in the state of hydrostatic equilibrium, or limitations of the series expansion of the $f(R)$ models.

If the assumption of hydrostatic equilibrium is incorrect, then we are in a regime in which the fundamental relations (7.236)–(7.242) fail. As discussed in [1115],

**Table 7.2** Column 1: cluster name. Column 2: first derivative coefficient $a_1$ of $f(R)$ series. Column 3: $1\sigma$ confidence interval for $a_1$. Column 4: second derivative coefficient $a_2$ of $f(R)$ series. Column 5: $1\sigma$ confidence interval for $a_2$. Column 6: characteristic length $L$ of the modified gravitational potential, derived from $a_1$ and $a_2$. Column 7: $1\sigma$ confidence interval for $L$.

| name | $a_1$ | $[a_1 - 1\sigma, a_1 + 1\sigma]$ | $a_2$ (kpc$^2$) | $[a_2 - 1\sigma, a_2 + 1\sigma]$ (kpc$^2$) | $L$ (kpc) | $[L - 1\sigma, L + 1\sigma]$ (kpc) |
|---|---|---|---|---|---|---|
| A133 | 0.085 | [0.078, 0.091] | $-4.98 \cdot 10^3$ | $[-2.38 \cdot 10^4, -1.38 \cdot 10^3]$ | 591.78 | [323.34, 1259.50] |
| A262 | 0.065 | [0.061, 0.071] | $-10.63$ | $[-57.65, -3.17]$ | 31.40 | [17.28, 71.10] |
| A383 | 0.099 | [0.093, 0.108] | $-9.01 \cdot 10^2$ | $[-4.10 \cdot 10^3, -3.14 \cdot 10^2]$ | 234.13 | [142.10, 478.06] |
| A478 | 0.117 | [0.114, 0.122] | $-4.61 \cdot 10^3$ | $[-1.01 \cdot 10^4, -2.51 \cdot 10^3]$ | 484.83 | [363.29, 707.73] |
| A907 | 0.129 | [0.125, 0.136] | $-5.77 \cdot 10^3$ | $[-1.54 \cdot 10^4, -2.83 \cdot 10^3]$ | 517.30 | [368.84, 825.00] |
| A1413 | 0.115 | [0.110, 0.119] | $-9.45 \cdot 10^4$ | $[-4.26 \cdot 10^5, -3.46 \cdot 10^4]$ | 2224.57 | [1365.40, 4681.21] |
| A1795 | 0.093 | [0.084, 0.103] | $-1.54 \cdot 10^3$ | $[-1.01 \cdot 10^4, -2.49 \cdot 10^2]$ | 315.44 | [133.31, 769.17] |
| A1991 | 0.074 | [0.072, 0.081] | $-50.69$ | $[-3.42 \cdot 10^2, -13]$ | 64.00 | [32.63, 159.40] |
| A2029 | 0.129 | [0.123, 0.134] | $-2.10 \cdot 10^4$ | $[-7.95 \cdot 10^4, -8.44 \cdot 10^3]$ | 988.85 | [637.71, 1890.07] |
| A2390 | 0.149 | [0.146, 0.152] | $-1.40 \cdot 10^6$ | $[-5.71 \cdot 10^6, -4.46 \cdot 10^5]$ | 7490.80 | [4245.74, 15715.60] |
| MKW4 | 0.054 | [0.049, 0.060] | $-23.63$ | $[-1.15 \cdot 10^2, -8.13]$ | 51.31 | [30.44, 110.68] |
| RXJ1159 | 0.048 | [0.047, 0.052] | $-18.33$ | $[-1.35 \cdot 10^2, -4.18]$ | 47.72 | [22.86, 125.96] |

the central (70 kpc) region of every cluster is strongly affected by radiative cooling and thus it cannot be directly related to the depth of the cluster potential well. This means that, in this region, the gas is not in hydrostatic equilibrium but in a multiphase, turbulent state driven by some astrophysical, non-gravitational, interaction. In this case, the gas cannot be used as a good standard tracer.

Another limitation of our modelling must also be taken into account, *i.e.*, the requirement that the function $f(R)$ is analytical. The corrected gravitational potential used is derived in the weak-field limit, which requires that

$$R - R_0 \ll \frac{a_1}{a_2}, \tag{7.263}$$

where $R_0$ is the background curvature. If this condition is not satisfied, the approach does not work [252]. Considering that $a_1/a_2$ has the dimensions of the inverse of a length squared, this condition defines the length scale over which the expansion can work and indicates the limit in which the model can be compared with data.

For the sample employed, the fit of the parameters $a_1$ and $a_2$ spans the length range 19–200 kpc, except for the largest cluster. Every galaxy cluster has its own gravitational length scale. A similar situation, but at completely different scales, has been found for LSB galaxies modelled by $f(R)$ gravity [217].

Considering the data available and the analysis performed, it is not possible to quantify the amount of radiative cooling and the validity of the weak-field limit, which are not mutually exclusive but should be considered in detail in a more refined picture. Other phenomena, including cooling flows, mergers, and asymmetric shapes should also be considered in a more detailed modelling of clusters. Here we are only interested in a proof of principle that extended gravity could be a valid alternative to dark matter in order to explain the cluster dynamics.

Similar issues are present also in [182], in which Metric-Skew-Tensor gravity is used as a generalization of GR to derive the gas mass profile of a sample of clusters with gas as the only baryonic component of the clusters. These authors consider some of the clusters included in the sample used here (in particular, A133, A262, A478, A1413, A1795, A2029, and MKW4), and they find the same different trend for $r \leq 200$ kpc, although with a different behavior. While the model presented here gives lower values than X-ray data on the gas mass, their model gives higher values with respect to the X-ray gas mass data. This fact stresses the need for a more accurate model of the gravitational potential.

As discussed in Chap. 5, in general, the weak-field limit of ETGs provides Yukawa-like corrections to the Newtonian potential [663, 1051]. Specifically, given a theory of gravity of order $(2n + 2)$, there are $n$ Yukawa corrections to the Newtonian potential [920]. This means that, if the effective Lagrangian density of the theory is

$$\mathscr{L} = f(R, \Box R, \ldots, \Box^k R, \ldots, \Box^n R)\sqrt{-g} \tag{7.264}$$

we have

$$\phi(r) = -\frac{GM}{r}\left[1 + \sum_{k=1}^{n} \alpha_k \, e^{-r/L_k}\right]. \tag{7.265}$$

Standard GR with no Yukawa corrections is recovered for $n = 0$ (second order theory), while metric $f(R)$ gravity corresponds to $n = 1$ and any $\Box$ operator introduces derivatives of a further order two in the field equations.

In the series (7.265), $G$ is the value of the gravitational constant at infinity and $L_k$ is the interaction length of the $k$-th component of the non-Newtonian correction. The amplitude $\alpha_k$ of each component is normalized to the standard Newtonian term; the sign of $\alpha_k$ informs us of whether the corrections are attractive or repulsive [1167]. Moreover, the variation of the gravitational coupling is involved. In our case, we take into account only the first, leading term of the series. Let us rewrite (7.223) as

$$\phi(r) = -\frac{GM}{r}\left[1 + \alpha_1 \, e^{-r/L_1}\right]; \tag{7.266}$$

then the effect of non-Newtonian terms can be parametrized by $(\alpha_1, L_1)$ which could be a more useful parameterization than the previous $(a_1, a_2)$ or $(G_{eff}, L)$ with $G_{eff} = 3G/(4a_1)$. For large distances $r \gg L_1$, the exponential term vanishes and the gravitational coupling is $G$. If $r \ll L_1$ the exponential approaches unity and, by differentiating Eq. (7.266) and comparing with the gravitational force measured in the laboratory, one obtains

$$G_{lab} = G\left[1 + \alpha_1\left(1 + \frac{r}{L_1}\right)e^{-r/L_1}\right] \simeq G(1 + \alpha_1), \tag{7.267}$$

where $G_{lab} = 6.67 \times 10^{-8}\ \mathrm{g}^{-1}\,\mathrm{cm}^3\,\mathrm{s}^{-2}$ is the usual Newton constant measured by Cavendish-like experiments. Of course, $G$ and $G_{lab}$ coincide in Newtonian gravity. The inverse square law holds asymptotically but the measured coupling constant differs from $G_{lab}$ by a factor $(1 + \alpha_1)$. In general, any correction introduces a characteristic length that acts at a certain scale for the self-gravitating systems, as in the case of the galaxy clusters studied here. The range $L_k$ of the $k$-th component of the non-Newtonian force can be identified with the mass $m_k$ of a pseudo-particle with effective Compton wavelength

$$L_k = \frac{\hbar}{m_k c}. \tag{7.268}$$

The interpretation is that, in the low-energy limit, fundamental theories attempting to unify gravity with the other forces introduce, in addition to the massless graviton, massive particles which also mediate the gravitational interaction [537] (see [228, 1033] for metric $f(R)$ gravity). These masses are related to effective length scales which can be parametrized as

$$L_k = 2 \times 10^{-5}\left(\frac{1\,\mathrm{eV}}{m_k}\right)\ \mathrm{cm}. \tag{7.269}$$

There have been several attempts to constrain $L_k$ and $\alpha_k$ (and then $m_k$) using experiments on scales in the range $1\,\mathrm{cm} < r < 1000\,\mathrm{km}$ and various techniques [410, 494, 1037]. The expected masses of particles which are the additional carriers of the gravitational force range in the interval $10^{-13}\,\mathrm{eV} < m_k < 10^{-5}\,\mathrm{eV}$. The general outcome of these experiments, even retaining only the term $k = 1$, is that the geophysical window between laboratory and astronomical scales

$$|\alpha_1| \sim 10^{-2}, \qquad L_1 \sim 10^2 - 10^3\ \mathrm{m} \tag{7.270}$$

is not excluded. Fujii [514] suggested that an exponential deviation from the Newtonian potential could arise due to the microscopic interaction coupling nuclear isospin and baryon number.

The astrophysical counterparts of these non-Newtonian corrections seemed ruled out until a few years ago due to the fact that the experimental tests of GR seemed to predict the Newtonian potential in the weak-field limit inside the Solar System. However, it has been shown that several alternative theories can evade the Solar System constraints [228, 712]. There are also indications of an anomalous long-range acceleration in the data of the Pioneer 10/11, Galileo, and Ulysses spacecrafts, which are now almost outside the Solar System. If confirmed, this anomalous acceleration calls into play Yukawa-like corrections again [45]. Besides, it is possible to reproduce phenomenologically the flat rotation curves of spiral galaxies considering the values

$$\alpha_1 = -0.92, \qquad L_1 \sim 40\,\text{kpc}. \tag{7.271}$$

The main assumption of this approach is that the additional gravitational interaction is carried by some ultra-soft boson with mass in the range $m_1 \sim 10^{-27}$–$10^{-28}$ eV. The action of this boson becomes efficient at galactic scales without the need for enormous amounts of dark matter to stabilize these systems [966]. Furthermore, it is possible to use a combination of two exponential corrections and give a detailed explanation of the kinematics of galaxies and galaxy clusters without dark matter [410].

Both spacecrafts anomalies and galactic rotation curves come from outside the Solar System boundaries within which GR has been tested, and they are purely phenomenological. Certain authors (*e.g.*, [786]) interpret also CMB experiments such as *BOOMERANG* and *WMAP* [377, 1038] in the framework of modified Newtonian dynamics, again without invoking dark matter. All these facts point toward the idea that corrections to standard gravity can be taken seriously and should not be excluded dogmatically, especially considering that the direct detection of dark matter still eludes us.

With this philosophy in mind, we plot the trend of $a_1$ as a function of the density in Fig. 7.7. The values of $a_1$ are strongly constrained in a narrow region of the parameter space and $a_1$ can be considered a tracer for the size of gravitational structures. $a_1$ ranges between 0.12–0.8 for larger clusters and 0.4–0.6 for poorer structures (*i.e.*, galaxy groups such as MKW4 and RXJ1159). We expect a particular trend when applying the model to different gravitational structures. Figure 7.7 gives characteristic density values.

Similar considerations hold also for the characteristic gravitational length $L$ directly related to both $a_1$ and $a_2$. The parameter $a_2$ shows a large range of variation (between $-10^6$ and $-10$) in comparison with the density and mass of the clusters. The value of $L$ changes with the size of the gravitational structure (Fig. 7.8) so it can be considered, in addition to the Schwarzschild radius, as a sort of additional gravitational radius. Particular care must be taken when considering Abell 2390, which shows large cavities in the X-ray surface brightness distribution and whose central, highly asymmetric, region is not expected to be in hydrostatic equilibrium. All results at small and medium radii for this cluster could be strongly biased by these effects [1114], and the same can be said for the resulting exceptionally large value of $L$. Figure 7.8 shows how observational properties of the cluster which characterize well its gravitational potential (such as the average temperature and the

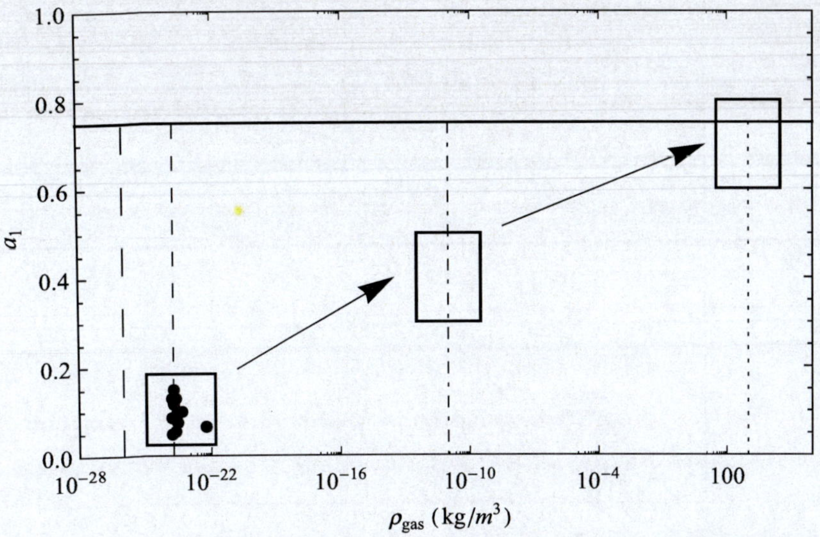


**Fig. 7.7** Density versus $a_1$: predictions on the behavior of $a_1$. The horizontal bold line indicates the Newtonian limit $a_1 \to 3/4$, which we expect to be realized on scales comparable with the size of the Solar System. Vertical lines indicate typical approximate values of the matter density (without dark matter) for various gravitational structures: a universe with critical density $\rho_{crit} \approx 10^{-26}\,\mathrm{kg/m^3}$ (long-dashed); galaxy clusters with $\rho_{cl} \approx 10^{-23}\,\mathrm{kg/m^3}$ (short-dashed); galaxies with $\rho_{gal} \approx 10^{-11}\,\mathrm{kg/m^3}$ (dot-dashed); the Sun with $\rho_{sun} \approx 10^3\,\mathrm{kg/m^3}$ (dot-dashed). Arrows and boxes show the predicted trend for $a_1$.

total cluster mass within $r_{500}$, plotted in the upper and lower panel, respectively), correlate well with the characteristic gravitational length $L$.

For clusters we can define a gas density-weighted mean and a gas mass-weighted mean, both depending on the series parameters $a_{1,2}$ as

$$< L >_\rho = 318\,\mathrm{kpc}, \qquad < a_2 >_\rho = -3.40 \cdot 10^4, \tag{7.272}$$

$$< L >_M = 2738\,\mathrm{kpc}, \qquad < a_2 >_M = -4.15 \cdot 10^5. \tag{7.273}$$

Note the correlation with the sizes of the cluster's cD-dominated central region and the "gravitational" interaction length of the whole cluster. In other words, the parameters $a_{1,2}$ directly related to the first and second derivatives of $f(R)$ determine the characteristic sizes of the self-gravitating structures.

### *7.3.11 Outlooks*

If the previous considerations are correct, the gravitational interaction depends on the scale and its infrared limit is dominated by the expansion coefficient of the gravi-

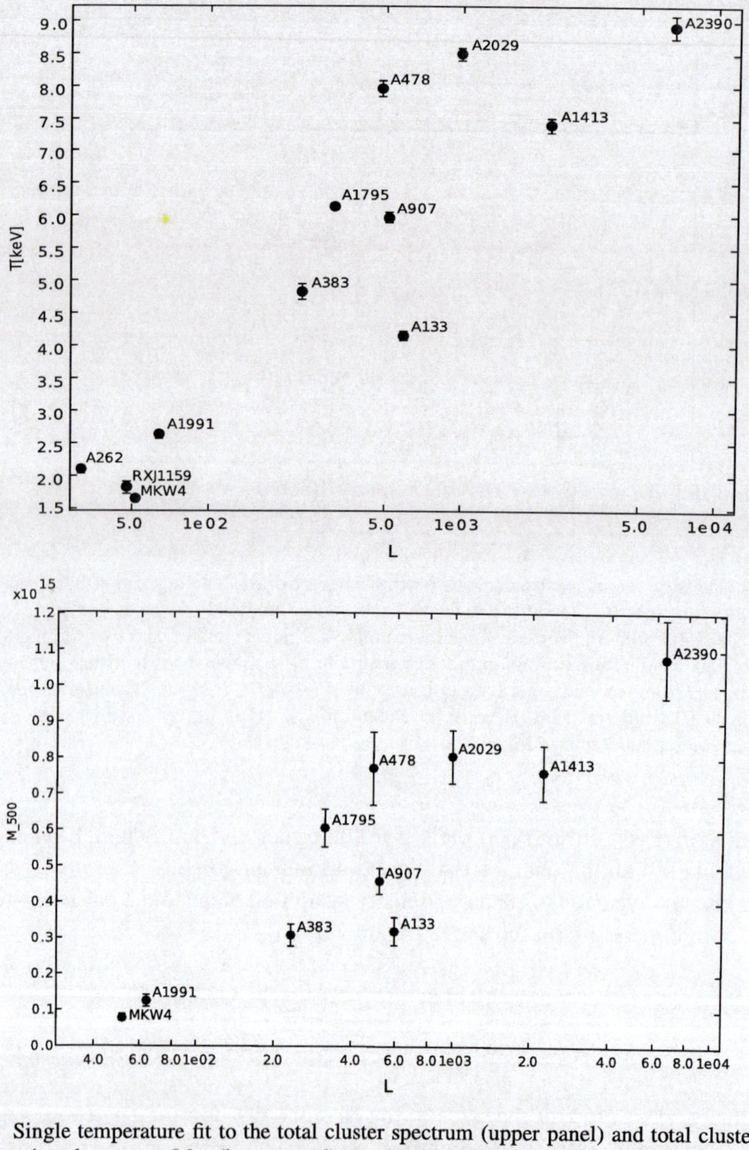


**Fig. 7.8** Single temperature fit to the total cluster spectrum (upper panel) and total cluster mass within $r_{500}$, in solar masses $M_\odot$ (lower panel) are plotted as functions of the characteristic gravitational length $L$. Temperature and mass values are taken from [1114].

tational Lagrangian. Roughly speaking, it is expected that beginning with the cluster scale down to the galactic scale, and then to the smaller Solar System or terrestrial scales, the terms of the series lead the clustering of self-gravitating systems dominating over other non-gravitational phenomena. The Newtonian limit is recovered for $a_1 \to 3/4$ and $L(a_1, a_2) \gg r$ at small scales and for $L(a_1, a_2) \ll r$ at large

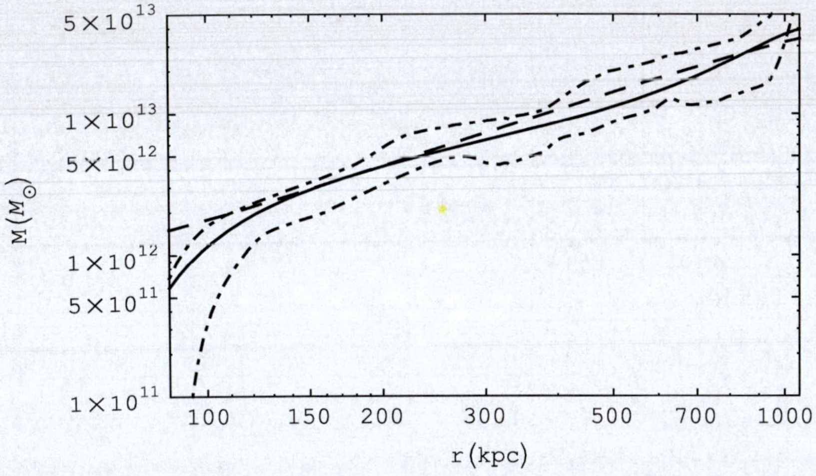


**Fig. 7.9** Baryonic mass versus radius for the cluster A133. The dashed curve is the experimentally observed estimate (7.256) of the baryonic matter component (*i.e.*, gas, galaxies, and cD galaxy); the solid curve is the theoretical estimate (7.255) for the baryonic component. The dotted curves are the $1\sigma$ confidence levels given by errors on fitting the parameters plus statistical errors on mass profiles as discussed in Sect. 7.3.8.

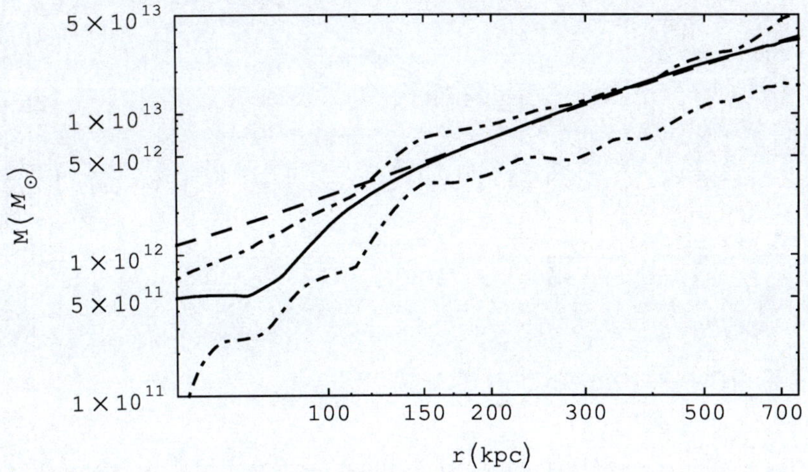


**Fig. 7.10** Baryonic mass versus radius for the cluster A383.

scales. In the first case, the gravitational coupling has to be redefined, in the second $G_\infty \simeq G$. In these limits, the linear Ricci term is dominant in the gravitational Lagrangian and Newtonian gravity is restored [920]. Reversing the argument, this could be the starting point to achieve a theory capable of explaining the strong segregation in masses and sizes of gravitationally bound systems.

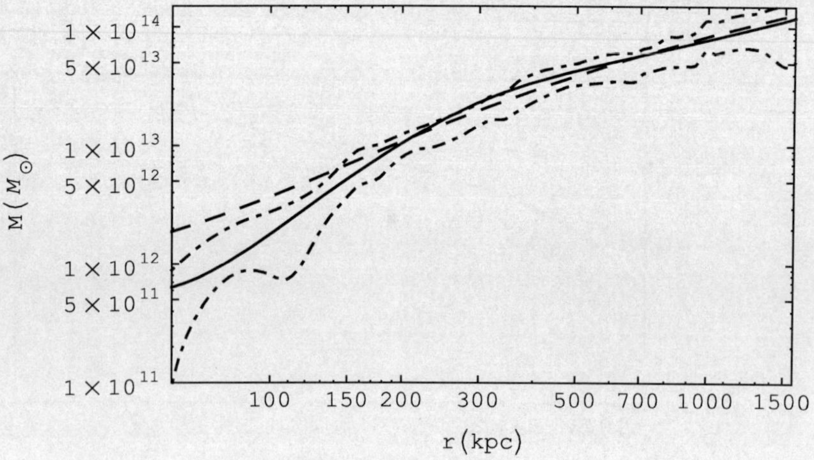


**Fig. 7.11** Baryonic mass versus radius for the cluster A478.

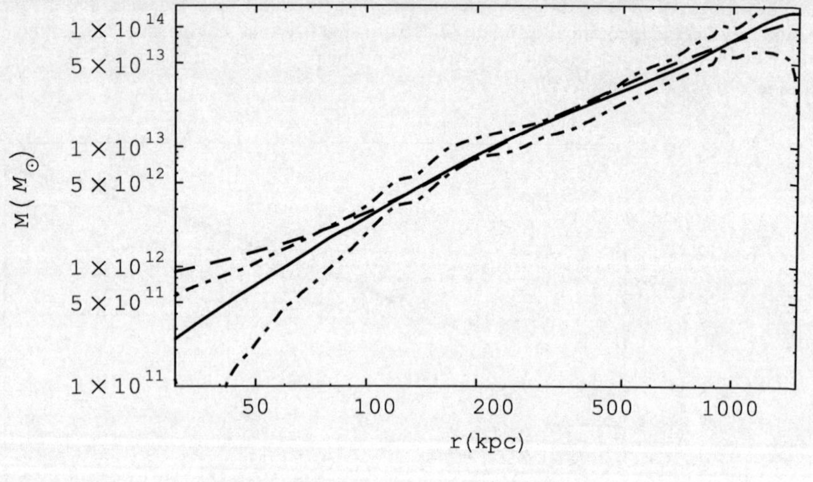


**Fig. 7.12** Baryonic mass versus radius for the cluster A1413.

The present state of cosmology shows that the Standard Cosmological Model based on GR, primordial nucleosynthesis, cosmic abundances and large scale structure, has difficulties, including the lack of a consistent theory for missing matter and cosmic acceleration. These shortcomings originate further difficulties in interpreting the observational data – one can say that we have a book but not the alphabet to read it. There are two main approaches to this problem: many researchers try to solve the problems of the Standard Cosmological Model by assuming that GR is the correct theory of gravity, but this leads to the introduction of exotic and invisible energy and matter components to explain the cosmic dynamics and large scale

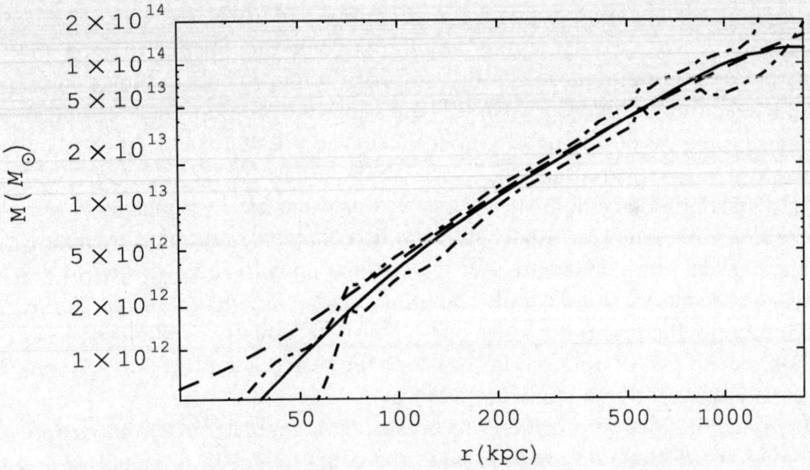


**Fig. 7.13** Baryonic mass versus radius for the cluster A2029.

structures. Other authors purport that GR is not the definitive and comprehensive theory of gravity but it should be revised at ultraviolet (quantum gravity) and infrared (extragalactic and cosmological) scales. In the second approach dark energy and dark matter could be just signals that a more general theory is needed at large scales even if GR works well at Solar System scales. To some extent the situation can be seen as a philosophical debate without solution, but it is possible to pose the question in more physical terms.

$f(R)$ gravity is strictly related to the second approach. It is a fruitful approach to generalize GR even if most models in the literature are purely phenomenological. It is interesting that, as soon as Einstein formulated GR, many authors including himself begun exploring other possibilities [237]. At the beginning these studies were mainly devoted to check the mathematical consistency of GR, but the desire to unify gravity with electromagnetism first and the other interactions later (only electromagnetism was known in the early days of GR), motivated efforts to develop alternative gravity theories. Today, one of the goals of alternative gravity is to understand the effective content and dynamics of the universe. This question has recently assumed dramatic tones because the fact that more than 95% of cosmic matter-energy is unknown at fundamental level is disturbing and alternative gravity could be a way out of this situation. Even if we live in the so-called era of precision cosmology, the present status of the observations does not allow us to discriminate between alternative gravity and dark energy/dark matter. The Large Hadron Collider could resolve many questions by detecting new supersymmetric particles, prime candidates for dark matter.

Cosmography, being by definition a model-independent approach, is a useful tool for discriminating between different cosmological models. Cosmographic parameters can be estimated without assigning an *a priori* cosmological model and cosmography can be used in two ways. First, one can use it to discriminate between GR and alternative gravity. a goal which relies upon the availability of good quality data. Certain minimum sensibility and error requirements in the surveys and the data must be met in order to solve this question: they are not met yet and therefore we are not able to do the necessary job because reliable standard candles are not available at very large redshifts [238]. Second, we can use the cosmographic parameters to constrain cosmological models as was done here for metric $f(R)$ gravity. Since these parameters are model-independent, they are natural priors in any theory. As already said, the accuracy in their determination is a crucial issue. So far, SNeIa have been used as standard candles but other classes of objects and phenomena (the CMB, bright galaxies, gamma ray bursts, baryon acoustic oscillations, weak lensing, *etc.*) should be considered in the quest for better accuracy. Forthcoming space missions will be extremely useful in this sense.

To conclude, it is very difficult to break the degeneracy in the large set of cosmological models attempting to explain the observations in a deductive approach. It would be fruitful to reconstruct the correct cosmological model by an inductive approach without *a priori* assumptions and using the minimum possible number of parameters, according to Occam's razor. This inductive approach would not be fully satisfactory because it would provide only a limited amount of information, but it could nevertheless inspire and guide the theoretical efforts.

## 7.4 Testing cosmological models with observations

The methods developed to test the many candidates proposed to explain the cosmic acceleration are based on measurements of distance and lookback time of astronomical objects identified as standard candles. In this section we discuss the characteristic parameters and constraints for various classes of cosmological models and we emphasize the degeneracy between these models, a signal that more data at low ($0 < z < 1$), medium ($1 < z < 10$) and high ($10 < z < 1000$) redshift are needed to realistically discriminate between models. The large number of viable candidates to explain the accelerated expansion signals that the number of cosmological tests available to discriminate between competing models is too limited and that there is a serious degeneracy problem. Both the SNeIa Hubble diagram and the angular diameter-redshift relation of compact radio sources [294, 908] are distance-based probes of cosmological models in which systematic errors and biases could be iterated. It is interesting, therefore, to look for tests based on time-dependent observables, for example the *lookback time* to distant objects. The lookback time is estimated observationally as the difference between the present-day age of the universe and the age of a given object at redshift $z$. The estimate of its value is possible if the object is a galaxy observed in more than one photometric band, since its color is determined by its age as a consequence of stellar evolution. It is thus possible to estimate the age of the galaxy by measuring its magnitude in different bands and then using stellar evolutionary codes to choose the model that best reproduces the observed colors. A similar approach is pursued by Lima and Alcaniz [740] (see also [640]), who use the age, rather than the lookback time, of old high redshift galaxies to constrain the dark energy equation of state. The same method is applied to braneworld models [21] and to the Chaplygin gas [22]. However, the estimate of the age of a single galaxy may be affected by systematic errors which are difficult to control. This problem can be overcome by considering a sample of galaxies belonging to the same cluster. By averaging the estimates for all galaxies, one obtains an estimate of the cluster age and reduces the systematic errors. This technique was proposed by Dalal *et al.* [355] and used by Ferreras *et al.* [489] to test a class of models in which a quintessence scalar field couples explicitly with matter.

We review various classes of dark energy models discussing methods to constrain them with observational data. Without attempting to be complete, we intend to illustrate the degeneracy problem and the need for further and self-consistent observational surveys at all redshifts to remove the degeneracy.

## 7.4.1 Toward a new cosmological standard model

As a simple classification scheme, candidates to explain the cosmic acceleration may be divided into three wide classes.[9] The first class includes models based on dark energy: the simplest representative is the $\Lambda$CDM scenario and its quintessential generalizations, which will be referred to as QCDM models. The second class comprises UDE models with a single fluid described by an equation of state able to describe all regimes of the cosmic history [214, 259]: they will be referred to as *parametric density models* or *generalized EoS* models. The third class is composed of models describing the accelerated expansion as a manifestation of the breakdown of GR and its Friedmann equations, for example $f(R)$ gravity [211,212,851]. Although not exhaustive, these three classes allow one to explore qualitatively different physics, which is done in the following.

### 7.4.1.1 The $\Lambda$CDM model and its generalizations

A cosmological constant $\Lambda$ is certainly capable of driving a period of accelerated expansion of the universe and the $\Lambda$CDM model is the best-fit to a combined analysis of completely different astrophysical data ranging from SNeIa to the CMB

[9] A conceptually different explanation proposed is that, in the context of GR, the cosmic acceleration is due to the backreaction of inhomogeneities on the dynamics of an (averaged) background [193–195, 685, 686, 724, 725, 775, 776, 890, 922, 1163, 1164]. We do not discuss this proposal here because it has not yet been proved that the backreaction (which is undeniably present) has the correct magnitude to explain the cosmic acceleration, nor that the backreaction term in the averaged acceleration equation even gives a positive contribution to $\ddot{a}$. Extending this idea to scalar-tensor of $f(R)$ gravity in the hope of improving the situation does not bring definitive answers, either [1128].

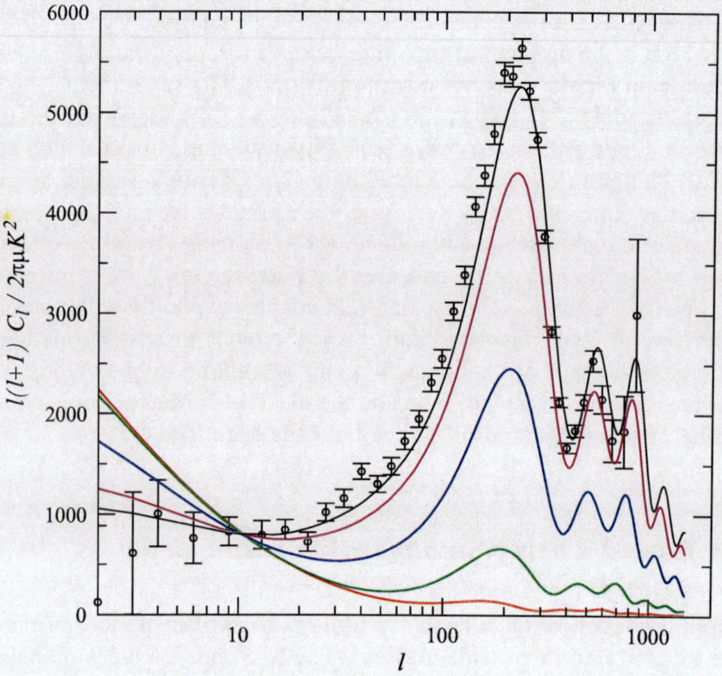


**Fig. 7.14** CMB anisotropy spectrum for various values of $w$. Data points are the *WMAP* measurements and the best-fit is obtained for $w \simeq -1$.

anisotropy spectrum (see Fig. 7.14) and galaxy clustering [926, 1038, 1070]. A simple generalization is the QCDM scenario in which the effective equation of state parameter $w \equiv p/\rho$ becomes negative at a certain epoch, with $w = -1$ corresponding to the standard cosmological constant. One of the goals of observational cosmology is testing whether $w$ deviates or not from $-1$. How such a negative pressure fluid can drive the cosmic acceleration is easily understood by inspection of the Friedmann equations

$$H^2 \equiv \frac{\kappa}{3}\left(\rho_M + \rho_Q\right), \tag{7.274}$$

$$2\frac{\ddot{a}}{a} + H^2 = -\kappa\, P_Q = -8\pi G w \rho_Q, \tag{7.275}$$

where the universe is assumed to be spatially flat as suggested by the position of the first peak in the CMB anisotropy spectrum ([377, 1038, 1055], see also Fig. 7.14).

The continuity equation $\dot{\rho} + 3H\,(P + \rho) = 0$ for the $i$-th fluid component with $P_i = w_i \rho_i$ yields

$$\Omega_i = \frac{\Omega_{i,0}}{a^{3(1+w_i)}} = \Omega_{i,0}(1+z)^{3(1+w_i)}, \tag{7.276}$$

where $\Omega_i \equiv \rho_i/\rho_{crit}$ is the density parameter for the $i$-th fluid. Equation (7.274) then gives

$$H(z) = H_0\sqrt{\Omega_{M,0}(1+z)^3 + \Omega_{Q,0}(1+z)^{3(1+w)}}. \tag{7.277}$$

Using Eqs. (7.274) and (7.275), the deceleration parameter $q \equiv -a\ddot{a}/\dot{a}^2$ becomes

$$q_0 = \frac{1}{2} + \frac{3w}{2}(1-\Omega_{M,0}). \tag{7.278}$$

The Hubble diagram of SNeIa, the large scale galaxy clustering and the CMB anisotropy spectrum can all be fitted by the $\Lambda$CDM model with $(\Omega_{M,0}, \Omega_Q) \simeq (0.3, 0.7)$, giving $q_0 \simeq -0.55$ describing an accelerated universe today. The simplicity of the model and its ability to fit most data explain why the $\Lambda$CDM scenario is the leading candidate from a purely observational point of view. Nonetheless its generalization, the QCDM models incorporating mechanisms for the evolution of $\Lambda$, are invoked to solve the coincidence problem.

#### 7.4.1.2 Generalizing the EoS: parametric density models

A phenomenological class of UDE models [214, 259] introduces a single fluid[10] with energy density

$$\rho(a) = A_{norm}\left(1+\frac{s}{a}\right)^{\beta-\alpha}\left[1+\left(\frac{b}{a}\right)^{\alpha}\right] \tag{7.279}$$

with $0 < \alpha < \beta$, $s$ and $b$ (with $s < b$) two scaling factors, and $A_{norm}$ a normalization constant. It is convenient to rewrite the energy density as a function of redshift. Replacing $a = (1+z)^{-1}$ in Eq. (7.279), one obtains

$$\rho(z) = A_{norm}\left(1+\frac{1+z}{1+z_s}\right)^{\beta-\alpha}\left[1+\left(\frac{1+z}{1+z_b}\right)^{\alpha}\right], \tag{7.280}$$

having defined $z_s = 1/s - 1$ and $z_b = 1/b - 1$. It is $\rho \propto a^{-\beta}$ for $a \ll s$, $\rho \propto a^{-\alpha}$ for $s \ll a \ll b$, and $\rho \propto$ const. for $a \gg b$. By setting $(\alpha, \beta) = (3, 4)$ the energy density smoothly interpolates between a radiation-dominated phase and a matter-dominated period, finally approaching a de Sitter phase. $A_{norm}$ may be estimated by inserting Eq. (7.279) into Eq. (7.274) and evaluating the result at the present time,

[10] This model may be interpreted not only as comprising a single fluid with an exotic equation of state, but also as composed of dark matter and scalar field dark energy, or in the framework of modified Friedmann equations.

$$A_{norm} = \frac{\rho_{crit,0}}{(1+s)^{\beta-\alpha}\,(1+b^\alpha)}. \tag{7.281}$$

The continuity equation may be recast in a form more convenient for computing the pressure and the parameter $w \equiv p/\rho$, obtaining [214, 259]

$$w = \frac{[(\alpha-3)a + (\beta-3)s]b^\alpha - [3(a+s) + (\alpha-\beta)s]a^\alpha}{3(a+s)(a^\alpha + b^\alpha)}, \tag{7.282}$$

which shows that $w$ depends strongly on the scale factor and hence on the redshift. Combining the Friedmann equations, the deceleration parameter $q = (1+3w)/2$ is

$$q = \frac{[(\alpha-2)a + (\beta-2)s]b^\alpha - [2(a+s) + (\alpha-\beta)s]a^\alpha}{2(a+s)(a^\alpha + b^\alpha)}. \tag{7.283}$$

The present day acceleration parameter is obtained by setting $a = 1$,

$$q_0 = \frac{(y-1)\alpha + z_s[\alpha\, y - 2(1+y)] + (\beta-4)(1+y)}{2(2+z_s)(1+y)} \tag{7.284}$$

with $y = (1+z_b)^{-\alpha}$. It is straightforward to derive the constraints on $q_0$ [214, 259]

$$\frac{1}{2}\left(\frac{\beta-\alpha}{2+z_s} - 2\right) \le q_0 \le \frac{1}{2}\left[\frac{\alpha z_s + 2\beta}{2(2+z_s)} - 2\right]. \tag{7.285}$$

It is convenient to solve Eq. (7.284) with respect to $z_b$ in order to express this as the function of $q_0$ and $z_s$

$$z_b = \left[\frac{\alpha(1+z_s) + \beta - (2+z_s)(2q_0+2)}{\alpha - \beta + (2+z_s)(2q_0+2)}\right]^{1/\alpha} - 1. \tag{7.286}$$

This parametric density model is fully characterized by five parameters chosen as the asymptotic slopes $(\alpha, \beta)$, the present-day deceleration and Hubble parameters $(q_0, H_0)$, and the scaling redshift $z_s$. As in [214, 259], we set $(\alpha, \beta) = (3, 4)$ and $z_s = 3454$ so that $(q_0, H_0)$ are the parameters to be constrained by the data. Any generalized EoS approach can be reduced to this scheme when fitting the data; the phenomenological parameters are usually given also a physical meaning.

#### 7.4.1.3 Curvature quintessence

As already discussed, metric $f(R)$ gravity can be seen as a curvature fluid which, added to the matter fluid, produces the effective energy density and pressure

$$\rho_{tot} = \rho_M + \rho_{curv}\,, \qquad P_{tot} = P_M + P_{curv}, \tag{7.287}$$

where [24, 211, 275, 851]

$$\rho_{curv} = \frac{1}{f'(R)} \left\{ \frac{1}{2} \left[ f(R) - Rf'(R) \right] - 3H\dot{R}f''(R) \right\} , \tag{7.288}$$

$$P_{curv} = \frac{1}{f'(R)} \left\{ \left( 2H\dot{R} + \ddot{R} \right) f''(R) + \dot{R}^2 f'''(R) + \frac{1}{2} \left[ f(R) - Rf'(R) \right] \right\} . \tag{7.289}$$

and

$$w_{curv} = -1 + \frac{f''(R)\ddot{R} + \left[ f'''(R)\dot{R} - Hf''(R) \right]}{\left[ f(R) - Rf'(R) \right]/2 - 3Hf''(R)} . \tag{7.290}$$

As an example [24, 211, 275, 851], consider the choice $f(R) = f_0 R^n$ (with $f_0$ a constant). Then, there exist power-law solutions $a(t) = (t/t_0)^\alpha$ with

$$\alpha = \frac{2n^2 - 3n + 1}{2 - n} \tag{7.291}$$

and constant deceleration parameter

$$q(t) = q_0 = \frac{1 - \alpha}{\alpha} = -\frac{2n^2 - 2n - 1}{2n^2 - 3n + 1} . \tag{7.292}$$

For acceleration ($\alpha > 0$, $q_0 < 0$) the parameter $n$ must satisfy

$$n \in \left( -\infty, \frac{1 - \sqrt{3}}{2} \right) \cup \left( \frac{1 + \sqrt{3}}{2}, \infty \right) . \tag{7.293}$$

Models with $n$ in the first (second) interval of this range will be referred to as *CurvDown* (*CurvUp*) models, respectively. The method described in the following subsection can be used to constrain the parameters $(n, H_0)$ of this model. A simple power-law $f(R)$ model is not sufficient to realistically reproduce the transient matter era needed for large scale structure formation and the transition to acceleration [245]. However, this simple example constitutes a proof of principle that accelerated behavior can be recovered in a simple way by extending GR without exotic dark energy.

### 7.4.2 *Methods to constrain models*

We now discuss how cosmological models can be constrained, in principle, using suitable distance and/or time indicators. Theoretical cosmological models must be matched with observations by using the redshift $z$ as the natural time variable for the

Hubble parameter, *i.e.*,

$$H(z) = -\frac{\dot{z}}{z+1}. \tag{7.294}$$

Interesting redshift ranges are $100 < z < 1000$ for the early universe probed by CMB experiments, $10 < z < 100$ probed by large scale structure surveys, and $0 < z < 10$ probed by SNeIa and radiogalaxies. The method consists of building a reasonable patchwork of data from different epochs and then matching them with the same cosmological solution spanning, in principle, the cosmic history from inflation to the present. In order to constrain the parameters contained in the cosmological solution, a reasonable approach is to maximize the likelihood function

$$\mathcal{L} \propto \exp\left[-\frac{\chi^2(\mathbf{p})}{2}\right], \tag{7.295}$$

where $\mathbf{p}$ are the parameters characterizing the cosmological solution. The $\chi^2$ merit function is

$$\chi^2(\mathbf{p}) \equiv \sum_{i=1}^{N}\left[\frac{y^{th}(z_i,\mathbf{p}) - y_i^{obs}}{\sigma_i}\right]^2 + \left[\frac{\mathcal{R}(\mathbf{p}) - 1.716}{0.062}\right]^2 + \left[\frac{\mathcal{A}(\mathbf{p}) - 0.469}{0.017}\right]^2. \tag{7.296}$$

The terms in Eq. (7.296) can be characterized as follows: the dimensionless coordinate distances $y$ to objects at redshifts $z$ in the first term are defined as

$$y(z) \equiv \int_0^z \frac{dz'}{E(z')}, \tag{7.297}$$

where $E(z) \equiv H(z)/H_0$ is the normalized Hubble parameter, the main quantity which allows to compare the theoretical results with data. The function $y$ is related to the luminosity distance $D_L = (1+z)r(z)$. A sample of data on $y(z)$ for 157 SNeIa is discussed in the Gold dataset of [940] and in the 20 radiogalaxies sample of [357]. These works fit with good accuracy the linear Hubble law at low redshift $z < 0.1$, obtaining the dimensionless Hubble parameter $h = 0.664 \pm 0.008$. This value is in agreement with $H_0 = 72 \pm 8\,\mathrm{km\,s^{-1}\,Mpc^{-1}}$ provided by the Hubble Space Telescope Key project [506] and based on the local distance ladder, estimates from time delays in multiply-imaged quasars [258], and the Sunyaev-Zel'dovich effect in X-ray clusters [973, 990].

The second term in Eq. (7.296) allows one to extend the redshift range to probe $y(z)$ up to the last scattering surface, $z \geq 1000$. The *shift parameter* $\mathcal{R} \equiv \sqrt{\Omega_M}\, y(z_{lss})$ [1149, 1150] can be determined from the CMB anisotropy spectrum, where $z_{lss}$ is the redshift of the last scattering surface which can be approximated as $z_{lss} = 1048\left(1 + 0.00124\omega_b^{-0.738}\right)\left(1 + g_1\omega_M^{g_2}\right)$ with $\omega_i = \Omega_i h^2$ ($i = b, M$ for baryons and total matter, respectively), and $(g_1, g_2)$ given in [607]. The parameter $\omega_b$ is constrained by baryogenesis calculations compared with the observed abundances of primordial elements. Using this method, the value $\omega_b = 0.0214 \pm 0.0020$

is found [672]. In any case, the exact value of $z_{lss}$ has a negligible impact on the results, and setting $z_{lss} = 1100$ does not change constraints and priors on the other parameters of a given model.

The third term in $\chi^2$ takes into account the acoustic peak of the large scale correlation function at $100\,h^{-1}$ Mpc separation, detected by using 46748 luminous red galaxies (LRG) selected from the Sloan Digital Sky Survey (SDSS) Main Sample [417, 1056]. The quantity

$$\mathscr{A} = \frac{\sqrt{\Omega_M}}{z_{LRG}} \left[ \frac{z_{LRG}}{E(z_{LRG})} y^2(z_{LRG}) \right]^{1/3}, \tag{7.298}$$

where $z_{LRG} = 0.35$ is the effective redshift of the sample, is related to the position of the acoustic peak. $\mathscr{A}$ depends on the dimensionless coordinate distance (and thus on the integrated expansion rate) and on $\Omega_M$ and $E(z)$. This dependence removes some of the degeneracies intrinsic in the distance fitting methods. Therefore, it is particularly interesting to include $\mathscr{A}$ as a further constraint on the model parameters using its measured value $\mathscr{A} = 0.469 \pm 0.017$ [417]. Although similar to the usual $\chi^2$ introduced in statistics, the reduced $\chi^2$, *i.e.*, the ratio between $\chi^2$ and the number of degrees of freedom, is not forced to be unity for the best-fit model because of the presence of priors on $\mathscr{R}$ and $\mathscr{A}$ and because the uncertainties $\sigma_i$ are not Gaussianly distributed but take care of both statistical errors and systematic uncertainties. With the definition (7.295) of the likelihood function, the best-fit model parameters are those that maximize $\mathscr{L}(\mathbf{p})$.

Using the method sketched above, the classes of models of interest here can be constrained by observations. However, most of the tests recently used to constrain cosmological parameters (such as the SNeIa Hubble diagram and the angular size-redshift test) are essentially distance-based methods. The proposal of Dalal *et al.* [355] of using the lookback time to high redshift objects is thus particularly interesting since it relies on a completely different observable. The lookback time is defined as the difference between the present-day age of the universe and its age at redshift $z$ and may be computed as

$$t_L(z, \mathbf{p}) = t_H \int_0^z \frac{dz'}{(1+z')E(z', \mathbf{p})}, \tag{7.299}$$

where $t_H = H_0^{-1} = 9.78\mathrm{h}^{-1}$ Gyr is the Hubble time (with $h$ the Hubble constant in units of $100\,\mathrm{km\,s^{-1}\,Mpc^{-1}}$). By definition, the lookback time is not sensitive to the present-day age of the universe $t_0$ so that it is, at least in principle, possible that a model fits well the data on the lookback time but nonetheless predicts a completely wrong value for $t_0$. This parameter can be evaluated from Eq. (7.299) by simply changing the upper integration limit from $z$ to infinity. This shows that it is indeed a different quantity since it depends on the full history of the universe and not only on how the universe evolves from redshift $z$ to now. This is the reason why this quantity can be explicitly introduced as an additional constraint. As an example, let us discuss how to use the lookback time and the age of the universe to test a given cosmological model. Consider an object $i$ at redshift $z$ and denote by $t_i(z)$ its age defined as the difference between the age of the universe when the object was born (at the formation redshift $z_F$) and the one at $z$. It is

$$t_i(z) = \int_z^\infty \frac{dz'}{(1+z')E(z',\mathbf{p})} - \int_{z_F}^\infty \frac{dz'}{(1+z')E(z',\mathbf{p})}$$
$$= \int_z^{z_F} \frac{dz'}{(1+z')E(z',\mathbf{p})} = t_L(z_F) - t_L(z), \tag{7.300}$$

where the definition (7.299) has been used. Assume that we have $N$ objects and that we are able to estimate the age $t_i$ of the $i$-th object at redshift $z_i$ $(i = 1, 2, \ldots, N)$. We can estimate the lookback time as

$$t_L^{obs}(z_i) = t_L(z_F) - t_i(z)$$
$$= [t_0^{obs} - t_i(z)] - [t_0^{obs} - t_L(z_F)]$$
$$= t_0^{obs} - t_i(z) - df, \tag{7.301}$$

where $t_0^{obs}$ is the present age of the universe and

$$df \equiv t_0^{obs} - t_L(z_F) \tag{7.302}$$

is a *delay factor* which is introduced to take into account our ignorance about the formation redshift $z_F$ of the object. Actually, what can be measured is the age $t_i(z)$ of the object at redshift $z$. In order to estimate $z_F$, one should use Eq. (7.300) assuming a background cosmological model. Since our goal is to determine the background cosmological model, we cannot infer $z_F$ from the measured age so that this quantity is completely undetermined. In principle, $df$ should be different for each object in the sample unless there is a theoretical reason to assume the same redshift at the formation of all the objects. If this is indeed the case (as will be assumed later), then it is computationally convenient to consider $df$ rather than $z_F$ as the unknown parameter to be determined from the data. Again, a likelihood function can be defined as

$$\mathcal{L}_{lt}(\mathbf{p}, df) \propto \exp\left[-\frac{\chi^2_{lt}(\mathbf{p}, df)}{2}\right] \tag{7.303}$$

with

$$\chi^2_{lt} = \frac{1}{N - N_p + 1}\left\{\left[\frac{t_0^{theor}(\mathbf{p}) - t_0^{obs}}{\sigma_{t_0^{obs}}}\right]^2 + \sum_{i=1}^{N}\left[\frac{t_L^{theor}(z_i,\mathbf{p}) - t_L^{obs}(z_i)}{\sqrt{\sigma_i^2 + \sigma_t^2}}\right]^2\right\}, \tag{7.304}$$

where $N_p$ is the number of parameters of the model, $\sigma_t$ is the uncertainty on $t_0^{obs}$, $\sigma_i$ the one on $t_L^{obs}(z_i)$, and the superscript *theor* denotes the predicted value of a given quantity. The delay factor enters the definition of $\chi^2_{lt}$ since it determines $t_L^{obs}(z_i)$ from $t_l(z)$ through Eq. (7.301), but the theoretical lookback time does not depend on $df$. In principle, this method should discriminate efficiently between the various dark energy models, but this is not the case due to the scarcity of data available, which leads to large uncertainties on the estimated parameters. In order to alleviate this problem it is convenient to add further constraints on the models by using Gaussian priors[11] on the Hubble constant, *i.e.* redefining the likelihood function as

$$\mathscr{L}(\mathbf{p}) \propto \mathscr{L}_{lt}(\mathbf{p}) \exp\left[-\frac{1}{2}\left(\frac{h-h^{obs}}{\sigma_h}\right)^2\right] \propto \exp\left[-\frac{\chi^2(\mathbf{p})}{2}\right], \tag{7.305}$$

where $df$ is absorbed in the set of parameters $\mathbf{p}$ and

$$\chi^2 \equiv \chi^2_{lt} + \left(\frac{h-h^{obs}}{\sigma_h}\right)^2 \tag{7.306}$$

with $h^{obs}$ the estimated value of $h$ and $\sigma_h$ its uncertainty. The Hubble Space Telescope Key project results [506] set $(h, \sigma_h) = (0.72, 0.08)$. This estimate, obtained by local distance ladder methods, is independent of the cosmological model. The best-fit model parameters $\mathbf{p}$ may be obtained by maximizing $\mathscr{L}(\mathbf{p})$, which is equivalent to minimize the $\chi^2$ (7.306). A qualitative comparison of different models may be obtained by comparing the values of this pseudo-$\chi^2$, although this should not be considered as a definitive evidence against a given model. Having more than one parameter, one obtains the best-fit value of each single parameter $p_i$ as the value which maximizes the marginalized likelihood for that parameter

$$\mathscr{L}_{p_i} \propto \int dp_1 \ldots \int dp_{i-1} \int dp_{i+1} \ldots \int dp_n \, \mathscr{L}(\mathbf{p}). \tag{7.307}$$

After normalizing the marginalized likelihood to unity at the maximum, one computes the $1\sigma$ and $2\sigma$ confidence limits (CL) on that parameter by solving $\mathscr{L}_{p_i} = \exp(-0.5)$ and $\mathscr{L}_{p_i} = \exp(-2)$, respectively. To summarize, taking into account the procedures described for distance and time measurements, one can reasonably constrain a given cosmological model. In any case, the main problem is the access to sufficiently large and sufficiently high quality datasets.

---

[11] The need for priors to reduce the parameter uncertainties is often advocated for cosmological tests. For instance, a strong prior on $\Omega_M$ is introduced in [740] to constrain the dark energy equation of state. It is likely that extending the dataset to higher redshifts and reducing the uncertainties on the age estimate will avoid the need for priors.

### 7.4.3 Data samples for constraining models: large scale structure

In order to apply the method outlined above, a set of distant objects is needed whose age can be somehow estimated. Galaxy clusters seem ideal candidates in this sense because they can be detected up to high redshift and their redshift at formation[12] is almost the same for all clusters. Furthermore, it is relatively easy to estimate their age using only photometric data. The color of their component galaxies, in particular the reddest ones, is needed for this purpose. The stellar populations of the reddest galaxies become redder and redder as they evolve. Then, it is just a matter of assuming a stellar population synthesis model and looking at how old the latest episode of star formation should have happened in the galaxy's past to produce colors as red as the observed ones (this is referred to as the *color age*). The main limitations of this method are the stellar population synthesis model and a few unknown ingredients, including the metallicity and the star formation rate. The choice of the evolutionary model is a crucial step in estimating the color age and is also the main source of uncertainty [1165]. An alternative and more robust route to cluster age is to consider the color scatter (see [162] for an early application of this approach). The argument, qualitatively, goes this way: if galaxies have an extreme similarity in their color and nothing is conspiring to make the color scatter surreptitiously small, then the latest episode of star formation should happen in the galaxy's far past, otherwise the observed color scatter would be larger. Quantitatively, the scatter in color should thus be equal to the derivative of the color with time, multiplied by the scatter of star formation times. The first quantity may be predicted using population synthesis models and turns out to be almost the same for all evolutionary models, reducing significantly the systematic uncertainty. We refer to the age estimated by this method as *scatter age*. The dataset needed to apply the method described earlier may now be obtained using the following procedure. First, for a given redshift $z_i$, we collect the colors of the reddest galaxies in a cluster at that redshift, and then we use one of the two methods outlined above to determine the color or the scatter age of the cluster. If more than one cluster is available at that redshift, we average the results from different clusters in order to reduce the systematic error. Having obtained $t_i(z_i)$, we then use Eq. (7.301) to estimate the lookback time at that redshift. What we measure is $t_L^{obs}(z_i) + df$, that is, the quantity that enters the definition (7.304) of $\chi^2_{lt}$ and then the likelihood function. To estimate the color age, following [49], we choose the Kodama and Arimoto model [678] among the various available stellar population synthesis models. Unlike other models, this one allows for chemical evolution which is otherwise neglected. This choice provides three points on the $z\left(t_L^{obs}\right)$

[12] In the literature, the cluster formation redshift is defined as the redshift at which the last episode of star formation occurred. In this sense, the definition of *df* given here should be modified by adding a constant term which accounts for the duration of the star formation process and the time elapsed from the beginning of the universe to the birth of the first galaxy cluster. For this reason, it is still possible to consider the delay factor to be the same for all clusters, but it is not possible to infer $z_F$ from the fitted value of *df* because the details of star formation are unknown. This approach is particularly useful since it avoids considering lower limits to the age of the universe at redshift $z$ rather than the actual values.

**Table 7.3** Main properties of the cluster sample used in the analysis. The data on the left refer to clusters whose age has been estimated from the color of the reddest galaxies (color age), while those on the right have been obtained using the color scatter (scatter age). The redshift $z$, number $N$ of clusters used, age estimate, and the relevant reference are reported for each data point.

| Color age | | | | Scatter age | | | |
|---|---|---|---|---|---|---|---|
| $z$ | N | Age (Gyr) | Ref. | $z$ | N | Age (Gyr) | Ref. |
| 0.60 | 1 | 4.53 | [49] | 0.10 | 55 | 10.65 | [48] |
| 0.70 | 3 | 3.93 | [49] | 0.25 | 103 | 8.89 | [48] |
| 0.80 | 2 | 3.41 | [49] | 1.27 | 1 | 1.60 | [145] |

diagram obtained by applying the method to a set of six clusters at three different redshifts as detailed in Table 7.3. Using a large sample of low redshift SDSS clusters it is possible to evaluate the scatter age for clusters at $z = 0.10$ and $z = 0.25$ [48]. Blakeslee *et al.* [145] apply the same method to a single high redshift ($z = 1.27$) cluster. Collecting data using both the color and the scatter ages, we end up with a sample of 159 clusters at six redshifts (listed in Table 7.3) which probe the redshift range $(0.10, 1.27)$. This sample overlaps nicely with the one probed by the Hubble diagram of SNeIa, making it possible to compare our results with those from SNeIa. We assume $\sigma = 1$ Gyr as uncertainty on the cluster age, no matter what method is used to obtain that estimate. This is a conservative choice: if the error on the age were so large, the color-magnitude relation for the reddest cluster galaxies should have a large scatter that is not observed. We have, however, chosen such a large error to account qualitatively for the systematic uncertainties related to the choice of the evolutionary model.

Finally, we need an estimate of $t_0^{obs}$ to apply this method. Following Rebolo *et al.* [926], one can choose $(t_0^{obs}, \sigma_t) = (14.4, 1.4)$ Gyr obtained by a combined analysis of the *WMAP* and *VSA* data on the CMB anisotropy spectrum and SDSS galaxy clustering. This estimate is model-dependent since Rebolo *et al.* [926] implicitly assume that the $\Lambda$CDM model is the correct one. However, this value is in good agreement with $t_0^{obs} = 12.6^{+3.4}_{-2.4}$ Gyr determined from the age of globular clusters [696] and $t_0^{obs} > 12.5 \pm 3.5$ Gyr from radioisotopic studies [286]. For this reason, one can be reasonably confident that systematic errors are not introduced in the method adopted by the estimate of Rebolo *et al.* for $t_0^{obs}$, even when testing cosmological models different from the $\Lambda$CDM one.

### 7.4.4 *Testing cosmological models: an example*

The method described can be applied to dark energy models in order to constrain their parameters and determine whether they can be phenomenologically viable candidates to explain the cosmic acceleration. All the models presented are roughly described by few parameters which are[13] $(\Omega_M, h, w)$ for the QCDM model, $(q_0, h)$ for the parametric density model, and $(n, h)$ for curvature quintessence models. For the three classes of models, another parameter is needed, *i.e.*, the delay factor $df$ which will be marginalized over since it is not interesting for our purposes.

Let us consider the QCDM model first: the results are shown in Figs. 7.15 and 7.16. In the first plot the estimated cluster age is compared with

$$\tau(z) = t_L(z) + df \tag{7.308}$$

using the best-fit values for the model parameters and the delay factor, which turn out to be

$$(\Omega_M, h, w) = (0.25, 0.70, -0.81), \qquad df = 4.5\,\text{Gyr}, \tag{7.309}$$

giving $\chi^2 \simeq 0.04$. The $\chi^2$ value for the best-fit parameters (both for the QCDM model and the other dark energy models considered) is quite small, suggesting that errors have been seriously overestimated. This is not surprising given the arbitrary way in which the uncertainties on the estimated age of the clusters have been fixed. That this is likely to be the case is also suggested by a qualitative argument: one could rescale the errors on $t_L^{obs}(z_i)$ in such a way that $\chi^2 = 1$ for the best-fit model. Since for the best-fit QCDM model $\chi^2 \simeq \chi^2_{lt}$, this rescaling leads to multiply by almost $1/5$ the uncertainties on $t_L^{obs}(z_i)$. If the error on $t_0$ was negligible, we should reduce the uncertainty on the cluster age from 1 Gyr to 0.2 Gyr, which is a more realistic value. The presence of an error in $t_0^{obs}$ slightly complicates this argument

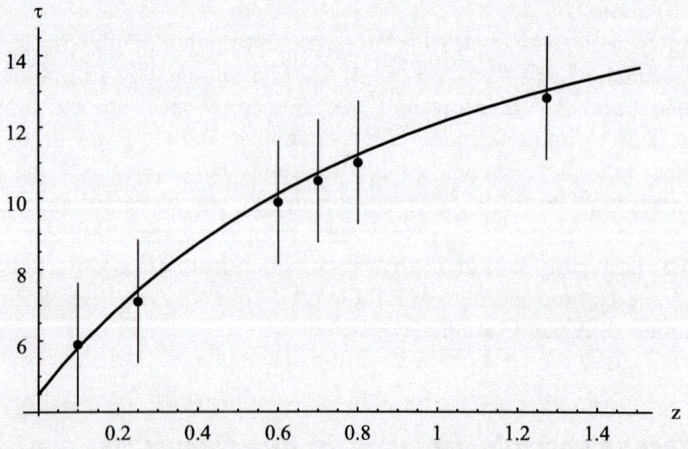


**Fig. 7.15** Comparison between predicted and observed values of $\tau = t_L(z) + df$ for the best-fit QCDM model.

---

[13] From now on we drop the subscript 0 from $\Omega_{M,0}$ without the risk of confusion and we use the dimensionless parameter $h$ instead of the Hubble constant $H_0$.

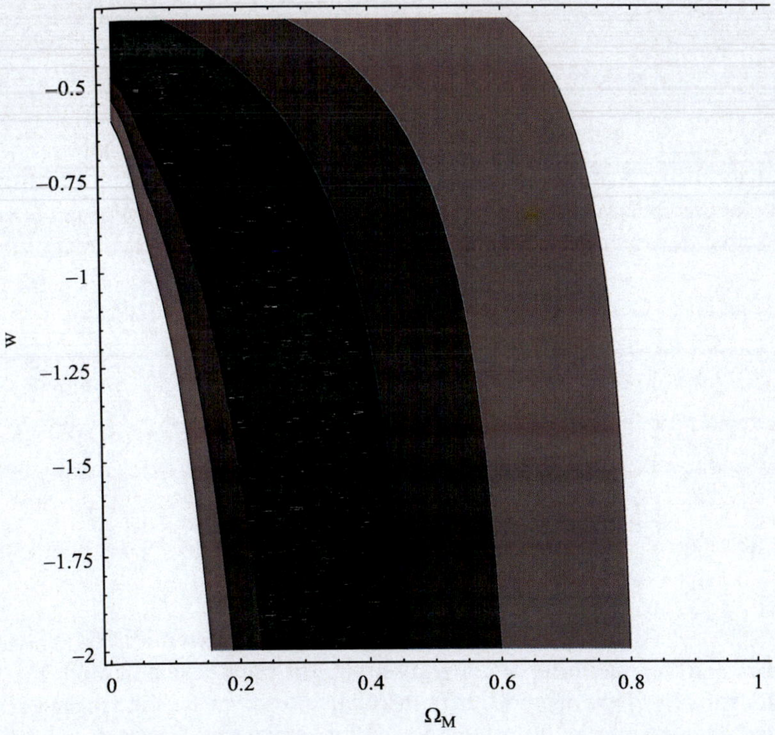


**Fig. 7.16** $1\sigma$ and $2\sigma$ confidence regions in the $(\Omega_M, w)$ plane for the QCDM model after marginalizing over the Hubble constant $h$ and the delay factor $df$

but does not change the main conclusion. We are thus confident that the very low value of the $\chi^2$ obtained for the best-fit model is due only to overestimating the uncertainties on the clusters ages. However, we do not perform any rescaling of the uncertainties since, to this end, we should select *a priori* a model as fiducial, which is contrary to the philosophy adopted here. Such a rescaling would not affect the main results anyway.

Figure 7.16 shows the $1\sigma$ and $2\sigma$ confidence limits in the $(\Omega_M, w)$ plane obtained by marginalizing over the Hubble constant and the delay factor. Two interesting considerations may be drawn: first, phantom models are allowed by the data, which agrees with fitting the QCDM model to the SNeIa Hubble diagram and the CMB anisotropy spectrum [571]. Unfortunately, a direct comparison is not possible since the marginalized likelihood is too flat to provide constraints on $w$ and all the values in the range $-2 \leq w \leq 1/3$ tested are well within the $1\sigma$ CL. Due to the scarcity of the data, this result is also a consequence of not having used priors on $\Omega_M$, as is usually done. By using the procedure described, we obtain the estimates on the other QCDM parameters

$$\Omega_M \in (0.13, 0.39)\,, \qquad h \in (0.63, 0.77) \quad (1\sigma \text{ CL}), \tag{7.310}$$

$$\Omega_M \in (0.01, 0.62), \qquad h \in (0.56, 0.84) \quad (2\sigma \text{ CL}). \tag{7.311}$$

Given that we are able to produce only weak constraints on the QCDM model, from now on we focus on the case $w = -1$ of the $\Lambda$CDM model and no longer discuss the results for the QCDM model. The best-fit parameters for the cosmological constant model are

$$(\Omega_M, h) = (0.22, 0.71)\,, \qquad df = 4.05\,\text{Gyr} \quad (\chi^2 \simeq 0.09), \tag{7.312}$$

originating the curve $\tau(z)$ of Fig. 7.17, while Fig. 7.18 reports the confidence regions in the $(\Omega_M, h)$ plane after marginalizing over the delay factor. From the marginalized likelihood functions one obtains

$$\Omega_M \in (0.10, 0.35), \qquad h \in (0.63, 0.78) \quad (1\sigma \text{ CL}), \tag{7.313}$$

$$\Omega_M \in (0.06, 0.59)\,, \qquad h \in (0.56, 0.85) \quad (2\sigma \text{ CL}). \tag{7.314}$$

The $\Lambda$CDM model has been widely tested against a large set of different astrophysical data and this fact offers us the possibility of cross-checking both the model and the validity of the method. It is instructive to compare the results presented here with those from the fit to the SNeIa Hubble diagram. Barris *et al.* [87] use a set of 120 SNeIa up to $z = 1.03$, finding $\Omega_M = 0.33$ as the best-fit value with a large uncertainty (not quoted explicitly, but noticeable in their Fig. 7.12) in good agreement with our result. Another result by Riess *et al.* [940] uses a SNeIa Hubble

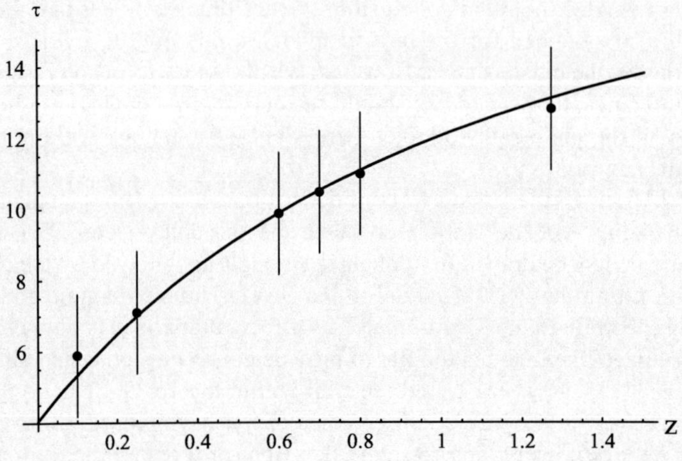


**Fig. 7.17** Comparison between predicted and observed values of $\tau = t_L(z) + df$ for the best-fit $\Lambda$CDM model.

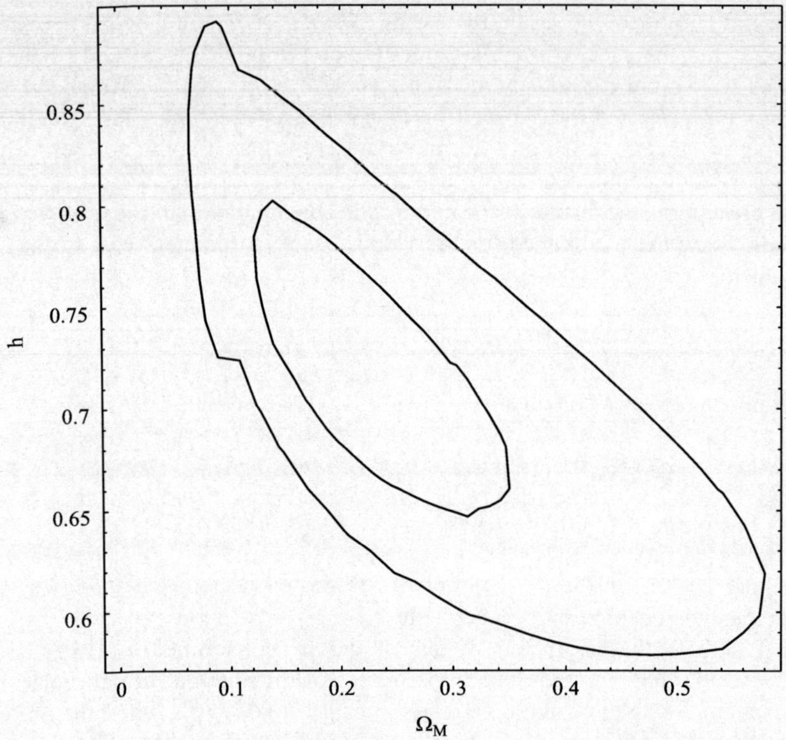


**Fig. 7.18** $1\sigma$ and $2\sigma$ confidence regions in the $(\Omega_M, h)$ plane for the $\Lambda$CDM model.

diagram extending to $z = 1.55$, finding $\Omega_M = 0.29^{+0.05}_{-0.03}$, still in agreement with our result. The $\Lambda$CDM model has also been tested by means of the angular size-redshift relation. Using a catalogue of ultracompact radio sources and taking into account carefully the systematic uncertainties and selection effects, Jackson [636] finds $\Omega_M = 0.24^{+0.09}_{-0.07}$, again in agreement with the estimate above.

Neither Barris *et al.* [87] nor Riess *et al.* [940] quote a best-fit value for $h$ since this parameter is infinitely degenerate with the supernovae absolute magnitude $M$ when fitting the SNeIa Hubble diagram. Nonetheless, SNeIa may still be used to determine $h$ by fitting the linear Hubble law to low redshifts ($z < 0.1$) SNeIa. Using this method, Daly and Djorgovski [357] find $h = 0.664 \pm 0.08$. Our estimate for $h$ agrees also with estimates using different methods such as various local standard candles [506], the Sunyaev-Zel'dovich effect in galaxy clusters [973], and time delays in multiply-imaged quasars [258]. Finally, we quote the results of Tegmark *et al.* [1070], who perform a combined fit of the $\Lambda$CDM model to both the *WMAP* data on the CMB anisotropy spectrum and the galaxy power spectrum measured by more than 200,000 galaxies surveyed by the SDSS collaboration. They find $\Omega_M = 0.30 \pm 0.04$ and $h = 0.70^{+0.04}_{-0.03}$.

The most interesting result of testing the $\Lambda$CDM model with the lookback method is not the success of this model (since the latter has already been tested in many ways), but the substantial agreement between these estimates of the parameters $(\Omega_M, h)$ and those based on distance measurements. This agreement is encouraging since it constitutes an important cross-check applicable, in principle, to any cosmological model.

Let us examine the results of the parametric density model in Figs. 7.19 and 7.20. The best-fit model is obtained for the values of the parameters $(q_0, h)$ and of the delay factor

$$(q_0, h) = (-0.68, 0.71), \qquad df = 4.20\,\text{Gyr} \qquad (\chi^2 \simeq 0.07). \tag{7.315}$$

Marginalizing over $df$ one obtains

$$q_0 \in (-0.81, -0.47), \qquad h \in (0.64, 0.78) \qquad (1\sigma), \tag{7.316}$$

$$q_0 \in (-0.89, -0.24), \qquad h \in (0.58, 0.85) \qquad (2\sigma). \tag{7.317}$$

The upper $2\sigma$ CL on the $q_0$ parameter is truncated because it extends to values higher than what is physically acceptable.

In [214, 259] the parameters of this model are constrained by using both the SNeIa Hubble diagram and the angular size-redshift relation. In particular, fitting the model to the SNeIa Hubble diagram gives $h = 0.64^{+0.08}_{-0.05}$, while the physically acceptable range for $q_0$ turns out to be in agreement with the data for $q_0 = -0.42$

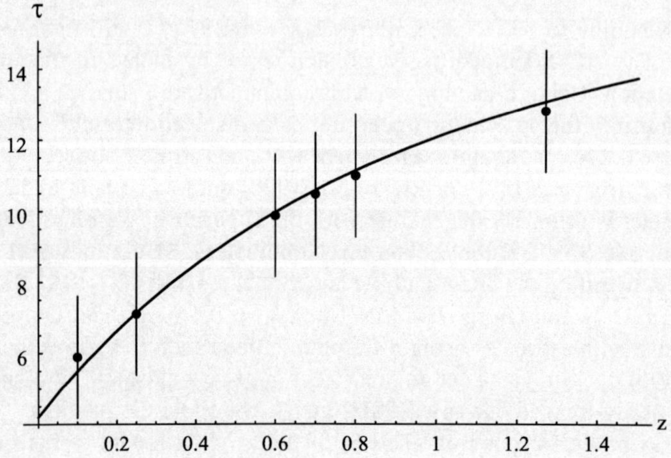


**Fig. 7.19** Comparison between predicted and observed values of $\tau = t_L(z) + df$ for the best-fit parametric density model.

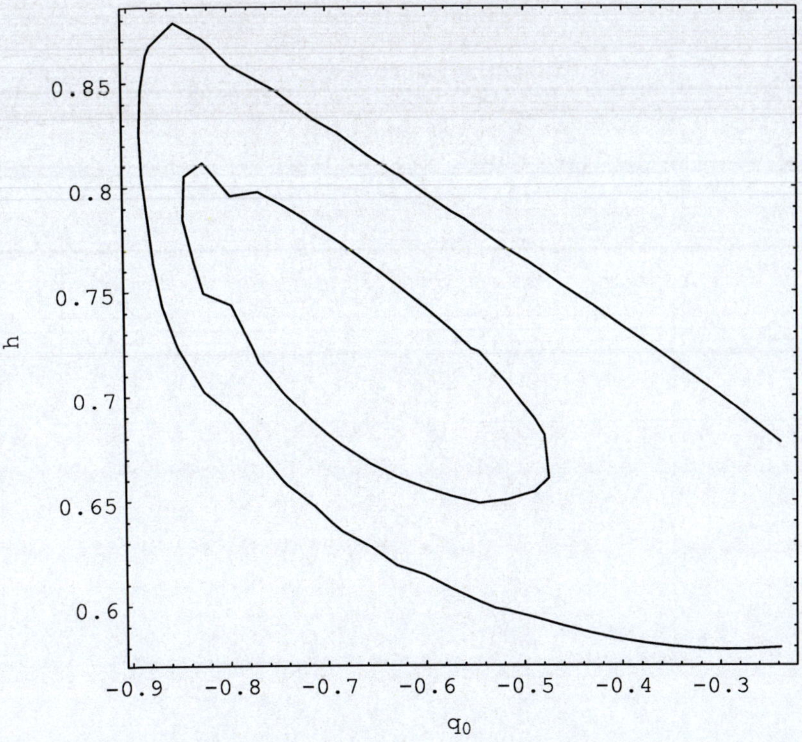


**Fig. 7.20** $1\sigma$ and $2\sigma$ confidence regions in the $(q_0, h)$ plane for the parametric density model.

as best-fit value. The present-day deceleration parameter $q_0$ is better constrained using the data of [636] to perform the angular size-redshift test, obtaining $q_0 = -0.64^{+0.10}_{-0.12}$ [214, 259]. Both these results are in good agreement with the present estimates and it is concluded that the parametric density model is a viable candidate alternative to the $\Lambda$CDM model which is, of course, a case of degeneracy.

Finally, let us discuss the results for curvature quintessence. Figures 7.21 and 7.22 report the confidence regions in the $(n, h)$ plane for the CurvUp and CurvDown regimes, respectively, after marginalizing over the delay factor. A striking feature is that the contour plots are not closed so that the marginalized likelihood function gives only an upper (lower) limit to the parameter $n$ in the CurvUp (CurvDown) regime. Formally, the estimates for the best-fit values in the CurvUp and CurvDown regime are

$$(n, h) = (1.367, 0.71)\,, \qquad df = 4.80\,\text{Gyr} \quad (\chi^2 \simeq 0.23), \tag{7.318}$$

$$(n, h) = (-0.367, 0.74), \qquad df = 4.80\,\text{Gyr} \quad (\chi^2 \simeq 0.21), \tag{7.319}$$

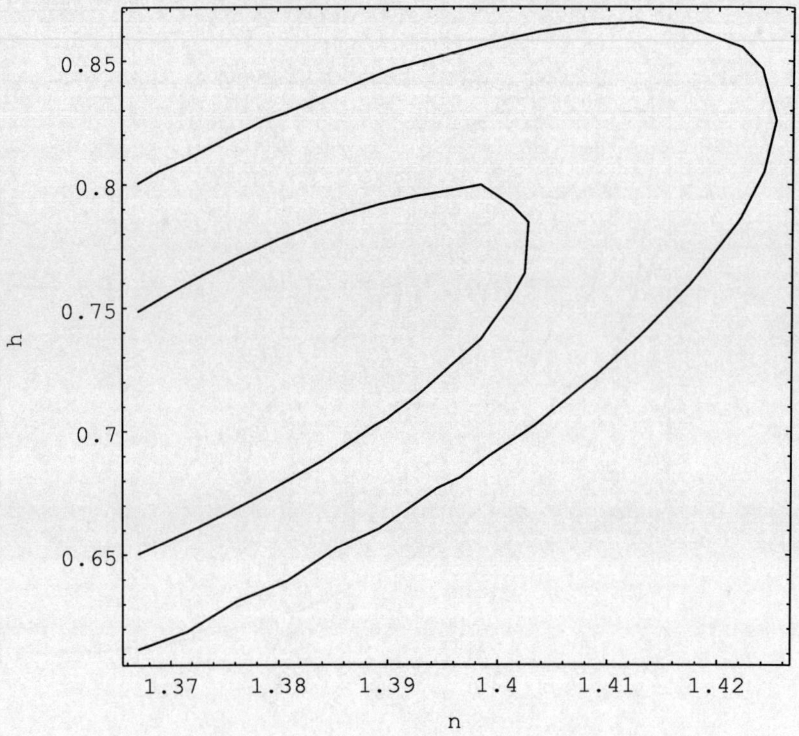


**Fig. 7.21** $1\sigma$ and $2\sigma$ confidence regions in the $(n, h)$ plane for curvature quintessence in the CurvUp regime.

but the best-fit value for $n$ actually lies outside the range investigated. Since the confidence regions are open, it is meaningless to give constraints on $h$, but it is nonetheless possible to infer the limits

$$n \leq 1.402 \quad (1\sigma), \quad n \leq 1.424 \quad (2\sigma), \tag{7.320}$$

$$n \geq -0.508 \quad (1\sigma), \quad n \geq -0.606 \quad (2\sigma), \tag{7.321}$$

for the CurvUp and CurvDown regimes, respectively. These limits do not contradict the ranges determined by fitting the SNeIa Hubble diagram [212] but they seem unrealistic. The deceleration parameter corresponding to these values of $n$ is too small ($q_0 \sim -0.01$), contradicting the evidence for an accelerating universe. Moreover, the results in [212] have been obtained by using an old sample of SNeIa, including certain SNeIa that have now been discarded from the Gold dataset of [940]. On the other hand, fitting a power-law scale factor to the angular size-redshift relation for compact radio sources gives $\alpha \simeq 1$ [635] which, using Eq. (7.291), translates in an estimate for $n$ in agreement with the result presented.

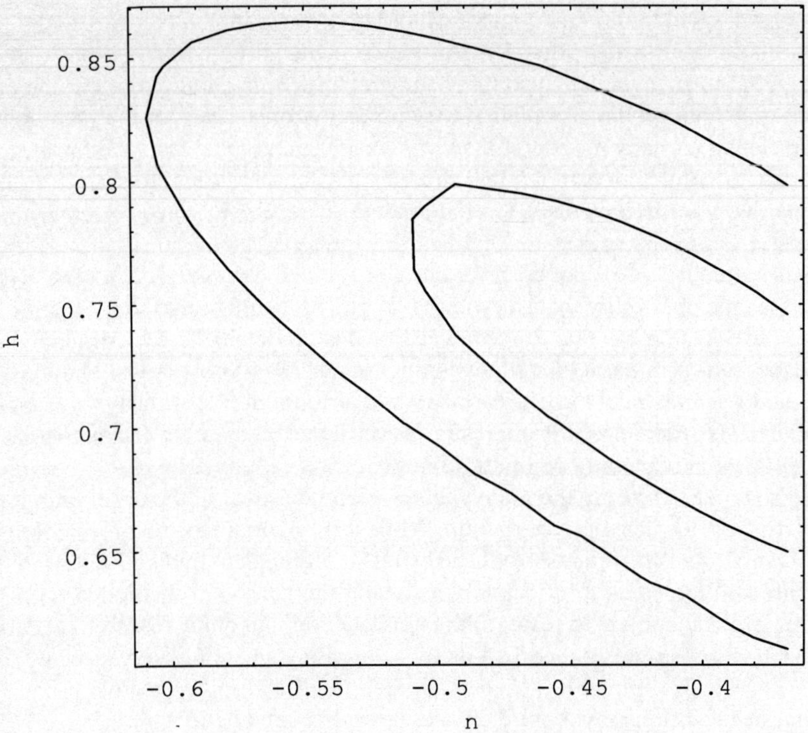


**Fig. 7.22** $1\sigma$ and $2\sigma$ confidence regions in the $(n, h)$ plane for curvature quintessence in the CurvDown regime.

## 7.5 Conclusions

The 1998 discovery of the cosmic acceleration came as a shock to most cosmologists and it is likely that its explanation requires new physics. The cosmological constant is problematic because of the cosmological constant problem and the coincidence problem. Dark energy scenarios fit the data but dark energy seems an *ad hoc* fix. The backreaction of inhomogeneities in GR has been invoked, but it is not clear whether it contributes to accelerating or decelerating the expansion of the universe and, at present, there is no consensus on the magnitude of its effect. Modifications of Einstein gravity at large scales could be an explanation. While astronomers, less exposed to the fancy theories and speculative world of high energy physics, are naturally more conservative and not inclined to modify GR, theoretical physicists have plenty of motivation to modify it, as we have seen in Chap. 1 (although, usually, modifications are contemplated in the ultraviolet and not in the infrared sector).

We are confronting a plethora of possible theoretical models to explain the cosmic acceleration. These must be confronted with the observational data. For asymptotically flat spacetimes, the Parametrized Post-Newtonian (PPN) formalism has been devised in order to compare alternative gravitational theories with GR and to constrain them using Solar System experiments [1167]. In cosmology, the role of the PPN formalism is played by cosmography. While certain models can be definitely ruled out on the ground that they predict wrong values of the cosmographic parameters, too many models still survive the cosmographic test. We have outlined methods to compare cosmological theories and observations, focusing on metric $f(R)$ gravity for definiteness. Similar analyses can be performed for other theories as well.

Since gravity is still poorly tested at galactic and extragalactic scales, we have discussed the possibility that (metric) $f(R)$ gravity could replace dark matter at the cluster scale using an $f(R)$ toy model for illustration. If the experimental search for dark matter (in particular supersymmetric particles) turns out to elude us, then modified gravities may have to be taken into account more seriously. While such a conclusion is premature, it is interesting to speculate on possible alternatives to dark matter to avoid becoming dogmatic about it.

To conclude, there are too many theoretical models that fit the observational data and "explain" the cosmic acceleration. While it is important to enlarge and refine the observations, the constraints produced by these, although unthinkable a few decades ago, are still too weak to effectively weed out the garden of theoretical scenarios. We believe that one has to resort to theoretical criteria to guide research on dark energy and its substitutes rather than merely compiling a list of observationally viable models. It is also important to keep an open mind and to be creative because the cosmic acceleration may require physics never explored before.

# Chapter 8
# From the early to the present universe

*There are more things in heaven and earth, Horatio, than are dreamt of in your philosophy.*
*– William Shakespeare*

In this final chapter we bring together elements developed in the rest of the book and we further discuss cosmology in ETGs. We begin with the early universe and carry on to the present universe. Inflation in the early universe is now widely accepted as a paradigm of modern cosmology, although it is still largely a speculation. While the 1992 detection of temperature anisotropies in the CMB by the *COBE* satellite [1020] and their further studies culminating in the detailed maps provided by the *WMAP* experiment [1038] (soon to be enhanced by the future *PLANCK* results [907]) are regarded as providing at least partial support for inflation, it is healthy to keep in mind that inflation is not proven beyond doubt, and alternative scenarios should not be discarded *a priori* but should be given due consideration.

The state of the universe before inflation is unknown, but the idea that inflation begins in a quantum gravity regime setting initial conditions for the inflationary expansion has become quite popular. In the past, the quantum regime of the early universe was addressed within the speculative formalism of quantum cosmology based on canonical quantization of a Hamiltonian for gravity and matter [668]. More recently, loop quantum cosmology has provided an alternative approach to the fundamental questions previously addressed by quantum cosmology [55, 150–154]. Here we limit ourselves to examine the main aspects of more "traditional" quantum cosmology in ETGs. We then continue by discussing inflation in scalar-tensor and quadratic gravities, including cosmological perturbations, the constraints on these theories coming from primordial nucleosynthesis, and we finish with a discussion of the present accelerated universe.

## 8.1 Quantum cosmology

Several different points of view can be adopted to introduce and motivate quantum cosmology. The latter is not a well-defined and finished theory: on the contrary, it has both mathematical problems in its foundations and problems of physical

S. Capozziello and V. Faraoni, *Beyond Einstein Gravity: A Survey of Gravitational Theories for Cosmology and Astrophysics*, Fundamental Theories of Physics 170,
DOI 10.1007/978-94-007-0165-6_8, © Springer Science+Business Media B.V. 2011

interpretation. However, quantum cosmology can be considered as the first step toward the construction of a non-perturbative quantum theory of the universe, a system with strong gravity. A primary motivation for studying quantum cosmology is finding initial conditions from which our classical universe evolved. A peculiar feature is that, contrary to familiar physical theories such as electromagnetism, GR, or non-relativistic quantum mechanics, initial and boundary conditions for the system "universe" cannot be imposed from the outside. Familiar physical theories contain a fundamental dynamical law (*e.g.*, the Maxwell, Einstein, or Schrödinger equations) and the specification, from the outside, of initial conditions. In cosmology, by definition, there is no "outside the universe", hence boundary conditions must be part of the fundamental laws of physics. Moreover, time disappears from quantum cosmology because its fundamental equation, the Wheeler-DeWitt (WDW) equation, is analogous to a time-independent Schrödinger equation with zero energy eigenvalue. Quantum cosmology can be considered as an autonomous branch of physics due to this issue of initial conditions and time [577].

It is considerably difficult for quantum cosmology to achieve the status of a consistent theory due not only to these conceptual difficulties, but also to mathematical ones. First, the superspace of geometrodynamics [799–803, 1158] has infinitely many degrees of freedom and, in practice, it is impossible to integrate the full WDW equation. Second, the Hilbert space of states describing the quantum universe is ill-defined [388]. Finally, the probabilistic interpretation of ordinary quantum mechanics fails for the WDW equation. In spite of these shortcomings, several results have been obtained and quantum cosmology has become a tool with which to address important questions in theoretical physics, albeit there is no doubt that these results will have to be replaced by rigorous ones when better computational and interpretational schemes are available. For example, the full, infinite-dimensional, superspace can be restricted by *fiat* to a finite-dimensional configuration space (the *minisuperspace*) and most of the mathematical difficulties can be avoided, making it possible to integrate the WDW equation.

The Hartle-Hawking *no-boundary condition* [578, 579] and Vilenkin's *tunneling from nothing* condition [1117, 1118, 1120, 1121] provide attempts to guess initial conditions from which the classical universe could originate. The *Hartle criterion* [576] is an interpretive scheme for solutions of the WDW equation and consists of searching for peaks of the wavefunction of the universe. If the latter is strongly peaked, then there are correlations between the geometrical and matter degrees of freedom. If this wavefunction does not peak, correlations are lost. In the first case, the emergence of classical trajectories (*i.e.*, universes) is expected. The analogy with non-relativistic quantum mechanics is immediate: in the presence of a potential barrier, a wavefunction which solves the Schrödinger equation exhibits oscillations on and outside the barrier, and a decreasing exponential behavior through the barrier. A similar situation occurs in quantum cosmology, in which the potential barrier is replaced by the superpotential $U\left(h^{ij}, \varphi\right)$, where $h^{ij}$ are the components of the three-metric of geometrodynamics and $\varphi$ is a scalar field adopted as a simple description of the matter content of the universe. The wavefunction of the universe is written as

$$\Psi\left[h_{ij}(x),\varphi(x)\right] \sim \mathrm{e}^{i\,m_{Pl}^2\,S}\,, \tag{8.1}$$

where $m_{Pl}$ is the Planck mass and

$$S \equiv S_0 + m_{Pl}^{-2}S_1 + \mathrm{O}\left(m_{Pl}^{-4}\right)\,, \tag{8.2}$$

is an action expanded in inverse powers of $m_{Pl}$. There is no normalization factor because there is no Born probability interpretation of the wavefunction of the universe.

Inserting the action $S$ into the WDW equation and equating term to term equal powers of $m_{Pl}$, one obtains the Hamilton-Jacobi equation for $S_0$. Similarly, equations for $S_1, S_2, \ldots$ are obtained, which can then be solved perturbatively. With $S_0$ only, the semiclassical limit of quantum cosmology is recovered [566–568]. If $S_0$ is a real number, one obtains oscillating WKB modes and the Hartle criterion is recovered because $\Psi$ is peaked on a region of the phase space defined by

$$\pi_{ij} = m_{Pl}^2\,\frac{\delta S_0}{\delta h^{ij}}\,, \qquad \pi_\varphi = m_{Pl}^2\,\frac{\delta S_0}{\delta\varphi}\,, \tag{8.3}$$

where $\pi_{ij}$ and $\pi_\varphi$ are classical momenta conjugated to $h^{ij}$ and $\varphi$. The semiclassical region of superspace in which $\Psi$ has an oscillating structure is either a Lorentzian or an Euclidean one. If it is Euclidean, then $S = iI$ and

$$\Psi \sim e^{-m_{Pl}^2\,I}\,, \tag{8.4}$$

where $I$ is the action for the Euclidean solutions of the classical field equations (*instantons*). This scheme, solves, at least at the semiclassical level, the problem of initial conditions. Given an action $S_0$, Eqs. (8.3) imply $n$ free parameters (one for each dimension of the configuration space $\mathcal{Q} \equiv \left(h^{ij},\varphi\right)$) and $n$ first integrals of motion. However, the general solution of the field equations involves $(2n-1)$ parameters, one for each Hamilton equation of motion except for the energy (Hamiltonian) constraint. Consequently, the wavefunction is peaked on a subset of the general solution. In this sense, the boundary conditions on the wavefunction (Hartle-Hawking, Vilenkin, or others) imply initial conditions for the classical solutions. The issue is now whether there exist methods to select these constants of motion. In other words, can the Hartle criterion and the emergence of classical trajectories be implemented by some general approach without arbitrarily choosing regions of the phase space where the momenta (8.3) are constant? This question is addressed in the following. We will show that in ETGs the existence of Noether symmetries determines at least a subset of the general solution of the WDW equation corresponding to oscillatory behavior. *Vice-versa*, the Hartle criterion is always connected to the presence of a Noether symmetry and then to the emergence of *classical trajectories* defined as solutions of the classical field equations of the theory. The discussion is restricted to minisuperspace models but the method could be extended to the full theory.

### 8.1.1 Noether symmetries in quantum cosmology

Minisuperspaces are restrictions of the superspace of geometrodynamics. They are finite-dimensional configuration spaces on which point-like Lagrangians can be defined. Physically relevant cosmological models (*e.g.*, Bianchi models) can be defined on such minisuperspaces. From the point of view of quantum cosmology, any symmetry selects a constant conjugate momentum since, according to the Euler-Lagrange equations it is

$$\frac{\partial \tilde{L}}{\partial Q^i} = 0 \Longleftrightarrow \frac{\partial \tilde{L}}{\partial \dot{Q}^i} = \Sigma_i \,. \tag{8.5}$$

*Vice-versa*, the existence of a constant conjugate momentum means that a cyclical variable and a Noether symmetry exists. This means that the minisuperspace $\left(q^i, \dot{q}^i\right)$ is mapped by the Noether symmetry into the minisuperspace $\left(Q^i, \dot{Q}^i\right)$, where the cyclical variable is evident.

We will examine time-independent, non-degenerate Lagrangians $L(q^i, \dot{q}^j)$ with

$$\frac{\partial L}{\partial t} = 0\,, \qquad \det\left(H_{ij}\right) \equiv \det||\frac{\partial^2 L}{\partial \dot{q}^i \partial \dot{q}^j}|| \neq 0\,, \tag{8.6}$$

where $H_{ij}$ is the Hessian. As usual in analytical mechanics, $L$ can be set in the form

$$L = T(q^i, \dot{q}^i) - V(q^i)\,, \tag{8.7}$$

where $T$ is a quadratic form in $\dot{q}^j$ and $V(q^i)$ is a potential term. The energy function associated with $L$ is

$$E_L \equiv \frac{\partial L}{\partial \dot{q}^i}\dot{q}^i - L\left(q^j, \dot{q}^j\right) \tag{8.8}$$

and, using the Legendre transformation

$$H = \pi_j \dot{q}^j - L(q^j, \dot{q}^j)\,, \qquad \pi_j = \frac{\partial L}{\partial \dot{q}^j}\,, \tag{8.9}$$

one obtains the Hamiltonian and the conjugate momenta.

Considering again the symmetry, the condition $\mathscr{L}_X L = 0$ and the vector field $X$ give a homogeneous polynomial of second degree in the velocities plus an inhomogeneous term in $q^j$. Thanks to the condition $\mathscr{L}_X L = 0$, this polynomial vanishes identically and then all its coefficient vanish. If $n$ is the dimension of the configuration space (*i.e.*, the dimension of the minisuperspace), there are $[1 + n(n+1)/2]$ PDEs whose solutions specify the symmetry. Such a system is overdetermined and, if a solution exists, it is expressed in terms of integration constants instead of boundary conditions.

In the Hamiltonian formalism, we have

$$\left[\Sigma_j, H\right] = 0 \qquad (1 \leq j \leq m)\,, \tag{8.10}$$

as it must be for conserved momenta in quantum mechanics and the Hamiltonian must satisfy

$$\mathscr{L}_\Gamma H = 0 \tag{8.11}$$

in order to obtain a Noether symmetry, where the vector $\Gamma$ is defined by [774]

$$\Gamma = \dot{q}^i \frac{\partial}{\partial q^i} + \ddot{q}^i \frac{\partial}{\partial \dot{q}^i}\,. \tag{8.12}$$

Let us now consider the minisuperspace of quantum cosmology and the semiclassical interpretation of the wavefunction of the universe. Straightforward canonical quantization yields (in units $\hbar = c = 1$)

$$\pi_j \longrightarrow \hat{\pi}_j = -i\,\partial_j\,, \tag{8.13}$$

$$H \longrightarrow \hat{H}\left(q^j, -i\,\partial_{q^j}\right). \tag{8.14}$$

It is well known that the Hamiltonian constraint gives the WDW equation so that, if $|\Psi>$ is a state of the system (*i.e.*, the wavefunction of the universe), the dynamics are given by

$$H|\Psi>= 0\,. \tag{8.15}$$

If a Noether symmetry exists, the reduction procedure outlined above can be applied and then Eqs. (8.5) and (8.9) imply that

$$\pi_1 \equiv \frac{\partial L}{\partial \dot{Q}^1} = i_{X_1}\theta_L = \Sigma_1\,, \tag{8.16}$$

$$\pi_2 \equiv \frac{\partial L}{\partial \dot{Q}^2} = i_{X_2}\theta_L = \Sigma_2\,, \tag{8.17}$$

$$\ldots\,,$$

depending on the number of Noether symmetries. Here $i_{X_j}\theta_L = \Sigma_j$ is the contraction of $X_j$ with the Cartan one-form defined in Chap. 2. After quantization, one gets

$$-\,i\,\partial_1|\Psi> = \Sigma_1|\Psi>\,, \tag{8.18}$$

$$-i\,\partial_2|\Psi> = \Sigma_2|\Psi>\,, \tag{8.19}$$

$$\ldots\,,$$

which are translations along the $Q^j$ axis singled out by the corresponding symmetry. Equations (8.18) and (8.19) can be immediately integrated and, since the $\Sigma_j$ are real constants, oscillatory behavior occurs for $|\Psi>$ in the directions of the symmetries, *i.e.*,

$$|\Psi>=\sum_{j=1}^{m} e^{i\Sigma_j Q^j}|\chi(Q^l)> \qquad (m<l\leq n)\,, \tag{8.20}$$

where $m$ is the number of symmetries, $l$ are the directions along which symmetries do not exist, and $n$ is the total dimension of the minisuperspace. *Vice-versa*, the dynamics given by Eq. (8.15) can be reduced using Eqs. (8.18) and (8.19) if and only if it is possible to define constant conjugate momenta as in (8.16) and (8.17), that is, oscillatory behavior for a subset of solutions $|\Psi>$ arises only if a Noether symmetry exists [239].

The $m$ symmetries provide first integrals of motion and the possibility to select classical trajectories. In one- and two-dimensional minisuperspaces, the existence of a Noether symmetry allows one to obtain the complete solution of the problem and the full semiclassical limit of quantum cosmology, as summarized in the

**Theorem**: *In the semiclassical limit of quantum cosmology, the reduction procedure of the dynamics related to the existence of Noether symmetries allows one to select a subset of solutions of the WDW equation with oscillatory behavior. According to the Hartle criterion on the wavefunction of the universe, this fact provides conserved momenta and trajectories which can be interpreted as classical cosmological solutions. Vice-versa, if a subset of solutions of the WDW equation has an oscillatory behavior, due to Eq. (8.19), conserved momenta and Noether symmetries exist or, Noether symmetries select classical universes.*

Examples of minisuperspace cosmological models derived from ETGs are given in the following.

### 8.1.2 Scalar-tensor quantum cosmology

To illustrate the theorem, consider a scalar-tensor theory described by the action

$$S_{ST}=\int d^4x\sqrt{-g}\left[f(\varphi)\,R-\frac{1}{2}g^{\mu\nu}\nabla_\mu\varphi\nabla_\nu\varphi_\nu-V(\varphi)\right]. \tag{8.21}$$

Restricting, for simplicity, to a FLRW cosmology with scale factor $a(t)$, the point-like Lagrangian is

$$L=6a\dot{a}^2f+6a^2\dot{a}\dot{f}-6Kaf+a^3\left(V-\frac{\dot{\varphi}^2}{2}\right), \tag{8.22}$$

with configuration space $\mathscr{Q} \equiv (a, \varphi)$ (a two-dimensional minisuperspace). A Noether symmetry exists if the condition $\mathscr{L}_X L = 0$ is satisfied, in which case

$$X = \alpha \frac{\partial}{\partial a} + \beta \frac{\partial}{\partial \varphi} + \dot{\alpha} \frac{\partial}{\partial \dot{a}} + \dot{\beta} \frac{\partial}{\partial \dot{\varphi}}, \tag{8.23}$$

where $\alpha = \alpha(a, \varphi), \beta = \beta(a, \varphi)$. The PDE system given by $\mathscr{L}_X L = 0$ is

$$f(\varphi)\left(\alpha + 2a \frac{\partial \alpha}{\partial a}\right) + a f'(\varphi)\left(\beta + a \frac{\partial \beta}{\partial a}\right) = 0, \tag{8.24}$$

$$3\alpha + 12 f'(\varphi) \frac{\partial \alpha}{\partial \varphi} + 2a \frac{\partial \beta}{\partial \varphi} = 0, \tag{8.25}$$

$$a\beta f''(\varphi) + \left(2\alpha + a \frac{\partial \alpha}{\partial a} + \frac{\partial \beta}{\partial \varphi}\right) f'(\varphi) + 2 \frac{\partial \alpha}{\partial \varphi} f(\varphi) + \frac{a^2}{6} \frac{\partial \beta}{\partial a} = 0, \tag{8.26}$$

$$\left[3\alpha V(\varphi) + a\beta V'(\varphi)\right] a^2 + 6K[\alpha f(\varphi) + a\beta f'(\varphi)] = 0, \tag{8.27}$$

where a prime now denotes differentiation with respect to $\varphi$ (there are four equations since $n = 2$). Several solutions exist for this system [230,231,235], determining also the form of the model since the system (8.24)–(8.27) gives $\alpha, \beta, f(\varphi)$, and $V(\varphi)$. For example, with spatially flat sections ($K = 0$), a solution is

$$\alpha = -\frac{2}{3} p(s) \beta_0 a^{s+1} \varphi^{m(s)-1}, \qquad \beta = \beta_0 a^s \varphi^{m(s)}, \tag{8.28}$$

$$f(\varphi) = D(s)\varphi^2, \qquad V(\varphi) = \lambda \varphi^{2p(s)}, \tag{8.29}$$

where

$$D(s) = \frac{(2s+3)^2}{48(s+1)(s+2)}, \qquad p(s) = \frac{3(s+1)}{2s+3}, \qquad m(s) = \frac{2s^2 + 6s + 3}{2s+3}, \tag{8.30}$$

and $(s, \lambda)$ are free parameters. The variable change yields

$$w = \sigma_0 a^3 \varphi^{2p(s)}, \qquad z = \frac{3}{\beta_0 \chi(s)} a^{-s} \varphi^{1-m(s)}, \tag{8.31}$$

where $\sigma_0$ is an integration constant and

$$\chi(s) = -\frac{6s}{2s+3}. \tag{8.32}$$

The Lagrangian (8.22) becomes, for $K = 0$,

$$L = \gamma(s) w^{s/3} \dot{z} \dot{w} - \lambda w \tag{8.33}$$

where $z$ is cyclical and

$$\gamma(s) = \frac{2s+3}{12\sigma_0^2(s+2)(s+1)}\,. \tag{8.34}$$

The conjugate momenta are

$$\pi_z = \frac{\partial L}{\partial \dot{z}} = \gamma(s)\, w^{s/3}\dot{w}\,, \qquad \pi_w = \frac{\partial L}{\partial \dot{w}} = \gamma(s)\, w^{s/3}\,\dot{z}\,, \tag{8.35}$$

the Hamiltonian is

$$\tilde{H} = \frac{\pi_z \pi_w}{\gamma(s) w^{s/3}} + \lambda w\,, \tag{8.36}$$

and the Noether symmetry is given by

$$\pi_z = \Sigma_0\,. \tag{8.37}$$

Quantizing Eqs. (8.35), one obtains

$$\pi \longrightarrow -i\,\partial_z\,, \quad \pi_w \longrightarrow -i\,\partial_w\,, \tag{8.38}$$

and the WDW equation

$$\left[(i\,\partial_z)(i\,\partial_w) + \tilde{\lambda} w^{1+s/3}\right]|\Psi>= 0\,, \tag{8.39}$$

where $\tilde{\lambda} = \gamma(s)\lambda$. The quantum version of the constraint (8.37) is

$$-i\partial_z|\Psi>= \Sigma_0|\Psi>\,, \tag{8.40}$$

and the dynamics are reduced. A straightforward integration of Eqs. (8.39) and (8.40) yields

$$|\Psi>= |\Omega(w)>|\chi(z)> \propto \mathrm{e}^{i\,\Sigma_0 z}\,\mathrm{e}^{-i\,\tilde{\lambda} w^{2+s/3}}\,, \tag{8.41}$$

which is an oscillating wavefunction and the Hartle criterion is recovered. In the semiclassical limit, we have two first integrals of motion: $\Sigma_0$ (*i.e.*, the equation for $\pi_z$) and $E_{\mathscr{L}} = 0$, *i.e.*, the Hamiltonian (8.36) which becomes the equation for $\pi_w$. Classical trajectories in the configuration space $\tilde{\mathscr{Q}} \equiv (w, z)$ are immediately recovered,

$$w(t) = (k_1 t + k_2)^{\frac{3}{s+3}}\,, \tag{8.42}$$

$$z(t) = (k_1 t + k_2)^{\frac{s+6}{s+3}} + z_0\,. \tag{8.43}$$

Going back to $\mathcal{Q} \equiv (a, \varphi)$, the classical cosmological behavior

$$a(t) = a_0(t - t_0)^{l(s)}, \tag{8.44}$$

$$\varphi(t) = \varphi_0(t - t_0)^{q(s)}, \tag{8.45}$$

is recovered, where

$$l(s) = \frac{2s^2 + 9s + 6}{s(s+3)}, \qquad q(s) = -\frac{2s+3}{s}. \tag{8.46}$$

Depending on the value of $s$, the classical universe exhibits Friedmann, power-law, or pole-like behavior.

If one considers instead generic Bianchi models with distinct scale factors $a_1, a_2, a_3$, the configuration space is $\mathcal{Q} \equiv (a_1, a_2, a_3, \varphi)$ and more than one symmetry can exist [241]. The considerations on the oscillatory regime of the wavefunction of the universe and the recovery of the classical regime are repeated without change.

### *8.1.3 The quantum cosmology of fourth order gravity*

Arguments similar to those exposed above work for higher order gravity. In particular, consider the fourth order theory with purely gravitational sector

$$S = \int d^4x \sqrt{-g}\, f(R). \tag{8.47}$$

As seen in Chap. 6, the corresponding point-like action for FLRW cosmology is

$$S = \int dt\, L\left(a, \dot{a}, R, \dot{R}\right), \tag{8.48}$$

where an overdot denotes again differentiation with respect to time and the scale factor $a$ and the Ricci scalar $R$ are the canonical variables, as usual in canonical quantization [233, 985, 988, 1119]. The expression of the Ricci scalar in terms of $(a, \dot{a}, \ddot{a})$ introduces a Lagrangian constraint which eliminates second and higher order derivatives in the action (8.48) and yields a system of second order differential equations for $a$ and $R$. The action (8.48) can be written as

$$S = 2\pi^2 \int dt \left\{ a^3 f(R) - \lambda \left[ R - 6\left( \frac{\ddot{a}}{a} + \frac{\dot{a}^2}{a^2} + \frac{K}{a^2} \right) \right] \right\}, \tag{8.49}$$

where the Lagrange multiplier $\lambda$ is derived by varying with respect to $R$:

$$\lambda = a^3 f'(R) \tag{8.50}$$

(a prime now denotes differentiation with respect to $R$). We know that $f(R)$ gravity is a special scalar-tensor theory, which we highlight introducing the scalar degree of freedom

$$p \equiv f'(R)\,, \tag{8.51}$$

so that the Lagrangian (8.49) becomes

$$L = 6a\dot{a}^2 p + 6a^2\dot{a}\dot{p} - 6Kap - a^3 W(p) \tag{8.52}$$

of the same form of (8.22) apart from the kinetic term. This is an Helmholtz-like Lagrangian [769] with $(a, p)$ as independent variables and with potential

$$W(p) = h(p)p - r(p)\,, \tag{8.53}$$

where

$$r(p) = \int f'(R)dR = \int pdR = f(R)\,, \qquad h(p) = R\,, \tag{8.54}$$

with $h = (f')^{-1}$ the inverse of $f'$. The configuration space is now $\mathcal{Q} \equiv (a, p)$ and $p$ is equivalent to $\varphi$. The condition $\mathcal{L}_X L = 0$ is realized by the vector field

$$X = \alpha\,(a, p)\,\frac{\partial}{\partial a} + \beta\,(a, p)\,\frac{\partial}{\partial p} + \dot{\alpha}\,\frac{\partial}{\partial \dot{a}} + \dot{\beta}\,\frac{\partial}{\partial \dot{p}} \tag{8.55}$$

and gives

$$p\left(\alpha + 2a\,\frac{\partial\alpha}{\partial a}\right)p + a\left(\beta + a\frac{\partial\beta}{\partial a}\right) = 0\,, \tag{8.56}$$

$$a^2\,\frac{\partial\alpha}{\partial p} = 0\,, \tag{8.57}$$

$$2\alpha + a\,\frac{\partial\alpha}{\partial a} + 2p\frac{\partial\alpha}{\partial p} + a\,\frac{\partial\beta}{\partial p} = 0\,, \tag{8.58}$$

$$6K\,(\alpha p + \beta a) + a^2\left(3\alpha W + a\beta\frac{\partial W}{\partial p}\right) = 0\,. \tag{8.59}$$

Solving this system, *i.e.*, finding the Noether symmetry, provides $\alpha$, $\beta$ and $W(p)$. The system is satisfied by

$$\alpha = \alpha(a)\,, \qquad \beta\,(a, p) = \beta_0 a^s p\,, \tag{8.60}$$

where $s$ is a parameter and $\beta_0$ is an integration constant. In particular,

$$s = 0 \longrightarrow \alpha(a) = -\frac{\beta_0}{3}a\,, \qquad \beta(p) = \beta_0\, p\,, \qquad W(p) = W_0\, p\,, \qquad K = 0\,, \tag{8.61}$$

$$s = -2 \longrightarrow \alpha(a) = -\frac{\beta_0}{a}\,, \qquad \beta(a,p) = \beta_0\,\frac{p}{a^2}\,, \qquad W(p) = W_1 p^3\,, \quad \forall\, K\,, \tag{8.62}$$

where $W_0$ and $W_1$ are constants. As before, the new set of variables $Q^j(q^i)$ is adapted to the foliation induced by $X$. Let us discuss separately the solutions (8.61) and (8.62).

#### 8.1.3.1 The case $s = 0$

The induced change of variables $\mathcal{Q} \equiv (a, p) \longrightarrow \tilde{Q} \equiv (w, z)$ can be chosen as

$$w\,(a, p) = a^3 p\,, \qquad z(p) = \ln p \tag{8.63}$$

and the Lagrangian (8.52) becomes

$$\tilde{L}\,(w, \dot{w}, \dot{z}) = \dot{z}\,\dot{w} - 2w\,\dot{z}^2 + \frac{\dot{w}^2}{w} - 3W_0 w\,, \tag{8.64}$$

with $z$ the cyclical variable. The conjugate momenta are

$$\pi_z \equiv \frac{\partial \tilde{L}}{\partial \dot{z}} = \dot{w} - 4\dot{z} = \Sigma_0\,, \tag{8.65}$$

$$\pi_w \equiv \frac{\partial \tilde{L}}{\partial \dot{w}} = \dot{z} + 2\frac{\dot{w}}{w}\,, \tag{8.66}$$

and the Hamiltonian is

$$H\,(w, \pi_w, \pi_z) = \pi_w \pi_z - \frac{\pi_z^2}{w} + 2w\pi_w^2 + 6W_0 w\,. \tag{8.67}$$

By quantizing canonically, the reduced dynamics are given by

$$\left(\partial_z^2 - 2w^2\partial_w^2 - w\partial_w\partial_z + 6W_0 w^2\right)|\Psi >= 0\,, \tag{8.68}$$

$$-i\,\partial_z|\Psi >= \Sigma_0\,|\Psi >\,. \tag{8.69}$$

We have taken simple factor ordering in the WDW equation (8.68). The wavefunction has an oscillatory factor, with

$$|\Psi >\sim e^{i\,\Sigma_0 z}\,|\chi(w) >\,, \tag{8.70}$$

where the function $|\chi>$ satisfies the Bessel equation

$$\left[w^2\partial_w^2 + i\,\frac{\Sigma_0}{2}\,w\,\partial_w + \left(\frac{\Sigma_0^2}{2} - 3W_0 w^2\right)\right]\chi(w) = 0\,. \tag{8.71}$$

The solutions are linear combinations of Bessel functions $Z_\nu(w)$,

$$\chi(w) = w^{1/2 - i\Sigma_0/4}\, Z_\nu(\lambda w)\,, \tag{8.72}$$

where

$$\nu = \pm\frac{1}{4}\sqrt{4 - 9\Sigma_0^2 - i4\Sigma_0}\,, \qquad \lambda = \pm 9\sqrt{\frac{W_0}{2}}\,. \tag{8.73}$$

The oscillatory regime for this component depends on $\nu$ and $\lambda$ being real or not. From the Noether symmetry (8.61), the wavefunction of the universe is then

$$\Psi(z, w) \sim \mathrm{e}^{i\Sigma_0[z - (1/4)\ln w]}\, w^{1/2} Z_\nu(\lambda w)\,. \tag{8.74}$$

For large values of $w$, the Bessel functions have an exponential behavior [10] and the wavefunction (8.74) is written as

$$\Psi \sim \mathrm{e}^{i[\Sigma_0 z - (\Sigma_0/4)\ln w \pm \lambda w]}\,. \tag{8.75}$$

Due to the oscillatory behavior of $\Psi$, Hartle's criterion is satisfied. By identifying the exponential factor of (8.75) with $S_0$, we recover the conserved momenta $\pi_z$, $\pi_w$ and select classical trajectories. In terms of the original variables, we obtain the cosmological solutions

$$a(t) = a_0 \mathrm{e}^{\frac{\lambda t}{6}} \exp\left(-\frac{z_1}{3}\,\mathrm{e}^{-\frac{2\lambda t}{3}}\right), \tag{8.76}$$

$$p(t) = p_0 \mathrm{e}^{(\lambda/6)t} \exp\left[z_1\,\mathrm{e}^{-(2\lambda/3)t}\right], \tag{8.77}$$

where $a_0$, $p_0$, and $z_1$ are integration constants. It is clear that $\lambda$ plays the role of a cosmological constant and then inflationary behavior is recovered asymptotically.

#### 8.1.3.2 The case $s = -2$

The new variables adapted to the foliation for the solution (8.62) are now

$$w(a, p) = ap\,, \qquad z(a) = a^2\,, \tag{8.78}$$

and the Lagrangian (8.52) assumes the form

$$\tilde{L}(w, \dot{w}, \dot{z}) = 3\dot{z}\dot{w} - 6Kw - W_1 w^3 . \tag{8.79}$$

The conjugate momenta are

$$\pi_z = \frac{\partial \tilde{L}}{\partial \dot{z}} = 3\dot{w} = \Sigma_1 , \tag{8.80}$$

$$\pi_w = \frac{\partial \tilde{L}}{\partial \dot{w}} = 3\dot{z} , \tag{8.81}$$

and the Hamiltonian is

$$H(w, \pi_w, \pi_z) = \frac{1}{3}\pi_z \pi_w + 6Kw + W_1 w^3 . \tag{8.82}$$

Repeating the discussion given above, the wavefunction of the universe is found to be

$$\Psi(z, w) \sim e^{i[\Sigma_1 z + 9Kw^2 + (3W_1/4)w^4]} \tag{8.83}$$

and the classical cosmological solutions are

$$a(t) = \pm\sqrt{h(t)} , \qquad p(t) = \pm\frac{c_1 + (\Sigma_1/3)\,t}{\sqrt{h(t)}} , \tag{8.84}$$

where

$$h(t) = \left(\frac{W_1 \Sigma_1^3}{36}\right) t^4 + \left(\frac{W_1 w_1 \Sigma_1}{6}\right) t^3 + \left(K\Sigma_1 + \frac{W_1 w_1^2 \Sigma_1}{2}\right) t^2$$
$$+ w_1 \left(6K + W_1 w_1^2\right) t + z_2 \tag{8.85}$$

and where $w_1$, $z_1$ and $z_2$ are integration constants. At late times one has the power-law expansion

$$a(t) \sim t^2 , \qquad p(t) \sim \frac{1}{t} . \tag{8.86}$$

### 8.1.4 *Quantum cosmology with gravity of order higher than fourth*

Minisuperspaces suitable for the previous analysis can be found for theories with gravitational action

$$S = \int d^4x \sqrt{-g}\, f(R, \Box R) , \tag{8.87}$$

with configuration space $\mathcal{Q} \equiv (a, R, \Box R)$ [129, 233, 551, 985, 988]. The FLRW point-like Lagrangian is $L\left(a, \dot{a}, R, \dot{R}, \Box R, (\dot{\Box R})\right)$ with the constraints

$$R = 6\left[\frac{\ddot{a}}{a} + \left(\frac{\dot{a}}{a}\right)^2 + \frac{K}{a^2}\right], \tag{8.88}$$

$$\Box R = -\left(\ddot{R} + 3\frac{\dot{a}}{a}\dot{R}\right). \tag{8.89}$$

Introducing Lagrange multipliers as usual, the point-like Lagrangian is

$$L = 6a\dot{a}^2 p + 6a^2\dot{a}\dot{p} - 6Kap - a^3\dot{h}q - a^3 W(p,q), \tag{8.90}$$

where

$$p \equiv \frac{\partial f}{\partial R}, \qquad q \equiv \frac{\partial f}{\partial \Box R}, \tag{8.91}$$

$$W(p,q) = h(p)p + g(q)q - f, \tag{8.92}$$

and

$$h(p) = R, \qquad g(q) = \Box R, \qquad f = f(R, \Box R). \tag{8.93}$$

The minisuperspace is now three-dimensional but, again, Noether symmetries can be found. Cases of physical interest include [129, 551]

$$f(R, \Box R) = f_0 R + f_1 R^2 + f_2 R\Box R, \tag{8.94}$$

$$f(R, \Box R) = f_0 R + f_1 \sqrt{R\Box R}, \tag{8.95}$$

discussed in detail in [239]. Once symmetries (if they exist) are identified, suitable changes of variables

$$\mathcal{Q} \equiv (a, R, \Box R) \longrightarrow \tilde{\mathcal{Q}} \equiv (z, u, w) \tag{8.96}$$

can be found, where one or two variables are cyclical in the Lagrangian (8.90). For example, for (8.95) one obtains

$$\tilde{L} = 3\left(w\dot{w}^2 - Kw\right) - f_1\left(3w\dot{w}^2 u + 3w^2\dot{w}\dot{u} + \frac{w^3\dot{z}\dot{u}}{2u^2} - 3Kwu\right), \tag{8.97}$$

where we assumed $f_0 = 1/2$, the standard Einstein coupling, $z$ is the cyclical variable, and

$$z = R, \qquad u = \sqrt{\frac{\Box R}{R}}, \qquad w = a. \tag{8.98}$$

The conserved quantity is

$$\Sigma_0 = \frac{w^3 \dot{u}}{2u^2} \,. \tag{8.99}$$

Using canonical quantization and deriving the WDW equation from Eq. (8.97), the wavefunction of the universe is

$$|\Psi> \sim e^{i\Sigma_0 z}|\chi(u)> |\Theta(w)>\,, \tag{8.100}$$

where $\chi(u)$ and $\Theta(w)$ are combinations of Bessel functions. The oscillatory part of the solution is evident and the Hartle criterion is satisfied. In the semiclassical limit, using the conserved momentum (8.99), the cosmological behaviors

$$a(t) = a_0\, t\,, \qquad a(t) = a_0\, t^{1/2}\,, \qquad a(t) = a_0\, e^{k_0 t}\,, \tag{8.101}$$

are obtained, depending on the choice of boundary conditions.

These results conclude our discussion of the connection between Noether symmetries for minisuperspace cosmological models and classical solutions. If the wavefunction of the universe is related to the probability of a given classical cosmology, the existence of symmetries tells us the conditions under which the Hartle criterion works. The wavefunction is related only to the probability of a certain behavior but is not the probability amplitude since quantum cosmology is not unitary. Furthermore, the Hartle criterion is meaningful in the context of Everett's many-worlds interpretation of quantum mechanics [432,492,575], in which the universe branches into a large number of copies whenever a measurement is made. For ordinary quantum mechanics, the many-world interpretation is just one of the possible consistent formulations designed to deal with correlations internal to isolated systems. The Hartle criterion provides an operative interpretation of such correlations. In particular, if the wavefunction is strongly peaked around some region of the configuration space, we predict that we will observe the correlations characteristic of that region. If instead the wavefunction is rather flat in some region, correlations characteristic of that region are precluded to the observations. If the wavefunction is neither peaked nor flat, no observable prediction is possible; the correlation of some region of minisuperspace can be seen as a causal connection.

As noted, the analogy with non-relativistic quantum mechanics is straightforward. By considering situations in which an individual system consists of a large number of identical subsystems, one can derive from the above interpretation the usual probabilistic interpretation of quantum mechanics for the subsystems [566–568,577]. Hartle's criterion is recovered without arbitrariness if one or more Noether symmetries are present in a given minisuperspace model, when strongly peaked subsets of the wavefunction of the universe are found. *Vice-versa*, oscillatory segments of the wavefunction can be always connected to conserved momenta and then to Noether symmetries.

## 8.2 Inflation in ETGs

Inflation, a very short period of accelerated, superluminal expansion of the universe preceding the radiation-dominated era, has the advantages of solving the horizon, flatness, and monopole problems of the standard Big Bang model, while providing a mechanism to generate primordial density fluctuations that later can grow into the structures observed today in the universe, *i.e.*, galaxies, clusters, and superclusters. It is possible that (as was believed in the 1980s) inflation follows an earlier radiation-dominated era, or (an idea that is more popular today) that inflation begins in a quantum gravity regime [688, 728, 741]. Since GR is modified by quantum corrections at high energies, it is likely that inflation is correctly described not by GR, but by a theory incorporating corrections to the Hilbert-Einstein action, such as scalar-tensor or quadratic gravity [1044]. In this section we discuss inflation in ETGs.

### *8.2.1 Scalar-tensor gravity: extended and hyperextended inflation*

Guth's scenario known as *old inflation* [564], based on the idea of the universe going through a spontaneous first order phase transition from a metastable vacuum, and later abandoned because of its difficulties, was resurrected in *extended inflation* [701]. In Guth's old inflation the cosmic expansion is de Sitter-like because the inflaton $\psi$ is trapped in a false vacuum state (*supercooling* of the universe). While $\psi$ is fixed to the value $\psi = 0$ in this state, its constant potential $V(0) > 0$ acts as a cosmological constant $\Lambda = 8\pi G V(0)$ causing exponential expansion $a(t) = a_0\, e^{Ht}$. Old inflation is supposed to end because the field $\psi$ tunnels from the false to the true vacuum spontaneously nucleating bubbles of true vacuum. The nucleation rate of true vacuum bubbles per Hubble time and per Hubble volume is $\eta_V = \Gamma/H^4$, where $\Gamma$ is the nucleation rate per unit time. In most field theories, $\Gamma$ is a constant determined by the shape of the potential barrier separating the true and the false vacua. In GR-based old inflation $H$ is constant and the nucleation rate $\eta_V$ is also constant. Then, in order to complete the phase transition ending inflation, a sufficient number of bubbles must nucleate per Hubble time and volume, or $\eta_V \sim 1$. But if inflation lasts long enough to solve the problems of the standard Big Bang model, the bound $\eta_V \ll 1$ must hold. This limit arises because, if $\eta_V$ is too large, a large number of true vacuum bubbles stops inflation too early.

The old inflationary scenario was abandoned because of these conflicting constraints and replaced by other GR-based scenarios in which the universe expands only approximately with an exponential law, *i.e.*, $a(t) = a_0 \exp(H(t))$ where $H(t) = H_0 + H_1\, t + \cdots$ with the constant $H_0$ dominating the expansion in the slow-roll approximation. Slow-roll inflation occurs because the potential $V(\psi)$ has a flat section mimicking a cosmological constant. Inflation is then terminated by a second order phase transition, assuming that the potential becomes a steep well and that the inflaton starts rolling fast on it, breaking the slow-roll approximation. The inflaton travels down the steep potential, reaches the minimum $V = 0$ of its potential, overshoots and oscillates around it, dissipating potential energy in its oscillations around the minimum. These oscillations create particles during the *reheating* regime due the coupling of $\psi$ to some other field, and rising the temperature of the universe, which by then is very cold because of the inflationary expansion.

The first order phase transition terminating old inflation works better in Brans-Dicke theory than in GR. This happens because in Brans-Dicke gravity the solution of the field equations with a cosmological constant as the only matter source is not de Sitter space, but the power-law solution[1] (6.86) and (6.87) [779]

$$a(t) \propto t^{\omega+1/2}, \qquad \phi(t) \propto t^2 . \tag{8.102}$$

This solution describes power-law inflation[2] if $\omega > 1/2$. The Hubble parameter $H = (\omega + 1/2)/t$ is not constant, then the nucleation rate of true vacuum bubbles $\eta_V = \dfrac{\Gamma}{H^4} \propto t^4$ is time-dependent, and this is the remedy for the problem of old inflation. Now $\eta_V$ is small at early times, when few true vacuum bubbles nucleate and inflation can proceed for a sufficient number of e-folds, after which the nucleation of true vacuum bubbles becomes more efficient, and the cosmic expansion slows down. The phase transition is then completed and the false vacuum energy $V(0)$ disappears as tunneling continues. The latent heat of the transition is then dissipated via collisions of the rapidly moving bubbles, producing a hot thermal bath of particles and gravitational waves. This spontaneous mechanism of exit from inflation looks in principle more appealing than a second order phase transition obtained by changing by hand the shape of the potential $V(\psi)$ (hence the equation of motion of the inflaton). However, a serious shortcoming called the *big bubble problem* was found [702, 1152]: early on, the nucleation rate $\eta_V$ is small but not exactly zero and a few true vacuum bubbles nucleate early. They are much smaller than the Hubble radius and, therefore, they are not affected appreciably by the spacetime curvature, evolving as if they were in flat spacetime and expanding at the speed of light. When they become comparable in size with the Hubble radius, they begin expanding at the inflationary cosmic rate and eventually reach a cosmological size. Since these early-generated bubbles are few and they end up being so large, they do not thermalize like the much more numerous small bubbles nucleated later, and they leave a significant imprint in the CMB. They generate inhomogeneities that dominate those generated by quantum fluctuations of the inflaton and the metric tensor. These bubble-induced fluctuations would be at a level detectable with present technology [684, 702, 733, 1152] and, since they are not observed, extended inflation is severely constrained. Detailed calculations of the bubble spectrum [733] set the

---

[1] This solution is a special case of the Nariai solution (6.70)–(6.73).

[2] Power-law inflation was originally introduced using GR and an exponential potential $V(\phi)$ [4, 92, 727, 754]. The perturbation spectrum for power-law inflation can be calculated analytically [433, 760], while for most other scenarios the solution can only be obtained numerically.

constraint $\omega \leq 20$ on the Brans-Dicke parameter, in gross violation of the limit $\omega > 40,000$ coming from the Cassini experiment[3] [133].

Various alternative scenarios of extended inflation were devised to solve the big bubble problem [12, 596, 597, 702, 708, 785]. As usual, the simplest way to circumvent the Solar System bound $\omega > 40,000$ consists of giving the scalar $\phi$ a short range by introducing a potential $U(\phi)$ [702]. This potential is supposed to begin dominating the dynamics of $\phi$ after the end of inflation, to have a minimum (at $\phi_0 > 0$), in which $\phi$ can sit, with $\phi_0 = G^{-1}$ as the inverse of the present-day value of the gravitational coupling. Keeping $\phi$ constant ensures that GR is recovered, provided that $V(\phi_0) \simeq 0$ (otherwise a cosmological constant is introduced in the field equations). With this modification, bubble nucleation proceeds slowly with a low value of $\omega$, but the Solar System limits on $\omega$ are circumvented, avoiding the big bubble problem. Another problem raises its head, however: at the level of perturbations, this remedy for extended inflation alters the spectrum of scalar perturbations [729], introducing a significant tilt from a scale-invariant Harrison-Zeldovich spectrum and violating the observational constraints on the spectral index $n_s$. The new spectral index $n_s(\omega)$ is a monotonically increasing function of the Brans-Dicke parameter and the constraint $\omega \leq 20$ implies $n \leq 0.8$, violating already the old $1\sigma$ *COBE* limit $n_S = 1.2 \pm 0.3$ [71, 126, 549, 591]. The only way to reconcile the predicted value of the spectral index with observations is by admitting a higher value of $\omega$, which spoils extended inflation.

Other proposals to solve the big bubble problem which appeared in the literature include the possibility of allowing extended inflation to terminate via a first order phase transition, followed by a second phase of inflation in the slow-roll regime and terminating the latter by a second order phase transition to erase the bubble perturbations in the CMB left behind by the first order phase transition ending the previous stage of extended inflation (*plausible double inflation*). However, the main motivation of the original extended inflation model, *i.e.*, ending inflation via a spontaneous first order phase transition, is lost in double inflationary models.

Extended inflation was also considered in the context of stochastic inflation, a semiclassical scenario in which short wavelength quantum fluctuations of both the inflaton and the Brans-Dicke field act as noise for long wavelength classical modes and induce Brownian motion described by a Fokker-Planck diffusion equation [521–524, 1060, 1062, 1063]. The trajectory of the scalar field peaks around classical values and the problems associated with classical scenarios persist.

The big bubble problem of extended inflation can be avoided by setting the inflationary scenario in the context of a scalar-tensor theory instead of Brans-Dicke gravity, which is done in *hyperextended inflation*. Allowing the coupling function $\omega(\phi)$ to be time-dependent, this quantity can go from relatively small values at the beginning of inflation to large values after the end of inflation. In this way, there is no need to introduce a potential giving a short range to the Brans-Dicke-like

[3] At the time in which the big bubble problem was pointed out, the experimental constraint on the Brans-Dicke parameter was $\omega > 500$ [1167].

scalar field. While this is a possibility, the value of $\omega$ changes by a large amount in a time that is very short in comparison with the age of the universe. Early versions of hyperextended inflation [100] were criticized because of fine-tuning in the mechanism achieving such a large change [734].

A version of hyperextended inflation proposed in [734,1049] has $\omega$ initially large, which entails the production of an acceptable spectrum of density perturbations. Later on, $\omega$ decreases, with only a few true vacuum bubbles nucleated at this time which has the advantage of leaving the CMB unaltered. Inflation stops because of the dynamics of the scalar, not because of a first order phase transition with nucleation of bubbles. A significant nucleation of true vacuum bubbles due to tunneling occurs only after inflation is finished, removing the effective vacuum energy of the potential $V(\psi = 0)$. This scenario relies on a scalar-tensor theory described by the action

$$S = \int d^4x \sqrt{-g} \left[ \frac{1}{16\pi} \left( \phi R - \frac{\omega(\phi)}{\phi} g^{\mu\nu} \nabla_\mu \phi \nabla_\nu \phi - U(\phi) \right) \right.$$
$$\left. - \frac{1}{2} g^{\alpha\beta} \nabla_\alpha \psi \nabla_\beta \psi - V(\psi) \right] , \tag{8.103}$$

where the inflaton potential $V(\psi)$ has a minimum $V(0) > 0$ corresponding to the metastable state. The dynamics of the inflaton $\psi$ can again be described in the slow-roll approximation [95, 523, 1081].

In *intermediate inflation* [93,105,819] the cosmic expansion interpolates between exponential and power-law and yields a Harrison-Zeldovich ($n_s = 1$) or a blue ($n_s > 1$) spectrum of density perturbations. Since no large bubbles are created, the big bubble problem is not an issue. Various models that seem to work have been found [552, 708]. However, as for the alternatives to extended inflation, the original motivation for extended inflation, which was ending inflation via a first order phase transition, is lost in hyperextended models.

### 8.2.2 *Inflation with quadratic corrections*

Inflation can be implemented not only with scalar fields but also in the context of higher derivative theories which contain in the action a term proportional to $R^2$, a special case of metric $f(R)$ gravity corresponding to $f(R) = R + \alpha R^2$. Indeed, the first model of inflation [1044] was of this kind, with extra terms quadratic in the Ricci tensor, Ricci scalar, and their derivatives added to $f(R)$. However, the simple addition of an $R^2$ term to the Hilbert-Einstein Lagrangian density works and is not to be regarded as a toy model: it was derived from supergravity in [260] and has by now been the subject of many studies [103, 401, 679, 796, 1046, 1058, 1160]. In the literature, $f(R) = R + \alpha R^2$ inflation is usually reduced to scalar field inflation by using the equivalence between metric $f(R)$ and $\omega = 0$ Brans-Dicke gravity discussed earlier. The effective scalar field degree of freedom is $\phi = f'(R) = 1 + 2\alpha R$ and its potential is

$$V(\phi) = Rf'(R) - f(R)\big|_{R=R(\phi)} = \frac{(\phi - 1)^2}{4\alpha} . \tag{8.104}$$

In the Einstein frame (which seems to be more common than the Jordan frame in the literature) the action is

$$S = \int d^4x \sqrt{-\tilde{g}} \left[ \frac{\tilde{R}}{2\kappa} - \frac{1}{2} \tilde{g}^{\mu\nu} \tilde{\nabla}_\mu \tilde{\phi} \tilde{\nabla}_\nu \tilde{\phi} - U\left(\tilde{\phi}\right) \right] , \tag{8.105}$$

where

$$\tilde{\phi} = \sqrt{\frac{3}{2\kappa}} \ln \phi \tag{8.106}$$

and

$$U\left(\tilde{\phi}\right) = \frac{\mathrm{e}^{-2\sqrt{\frac{2\kappa}{3}}\tilde{\phi}}}{8\kappa\alpha} \left( \mathrm{e}^{\sqrt{\frac{2\kappa}{3}}\tilde{\phi}} - 1 \right)^2 . \tag{8.107}$$

The computed spectral index of scalar perturbations is $n_S = 0.96$ and the gravitational wave contribution to anisotropies in the CMB is negligible [730].

Inflation and quantum cosmology in a mixed Brans-Dicke-$R^2$ theory of gravity were studied in [337, 586].

## 8.3 Cosmological perturbations

It is difficult to overstate the importance of perturbations of a FLRW universe: in addition to growing and forming superclusters, clusters, galaxies, and stars and making life possible, primordial perturbations are important for the theoretical physicist because they generate a permanent imprint as temperature fluctuations in the CMB. These temperature fluctuations were detected in 1992 by the *COBE* satellite [1020] and have been the subject of intense scrutiny ever since. Different theories and models of the early universe usually predict different spectra of fluctuations, hence density, gravitational wave, and possibly vorticity perturbations around a FLRW background provide a way to test theoretical predictions about the early universe.

In addition to solving the horizon, flatness, and monopole problems of standard Big Bang cosmology, inflation provides a natural mechanism to create density perturbations, generated as quantum fluctuations of the inflaton field. Such a mechanism is missing in the Big Bang model. There, an initial spectrum of perturbations must be assumed without a cause. The physical wavelength $\lambda_{phys} = a\lambda$ of these perturbations (where $\lambda$ is the *comoving* wavelength) redshifts as the scale factor increases, while the size of the horizon $H^{-1}$ remains practically constant during inflation. As a consequence, perturbations cross outside the horizon during inflation and the spectrum of density perturbations is usually specified by their amplitude at this horizon crossing. While they stay outside the horizon, regions separated by a distance larger than $H^{-1}$ are not in causal contact with each other and the perturbations do not evolve. After the end of inflation, $H^{-1}$ grows as a power of the scale factor and the perturbations eventually re-enter the horizon during the radiation- or the matter-dominated era. These perturbations act as seeds and can then grow non-linearly during the matter era and begin to form cosmic structures.

This mechanism of generation of structure works in GR as well as in ETGs; however, the growth history of these perturbations after they go non-linear depends on the theory of gravity and, therefore, large scale structure surveys can potentially discriminate between GR and its competitors. Here we focus on the primordial generation of scalar and tensor perturbations during inflation, which leave an imprint in the CMB.

Many studies of primordial perturbations of a FLRW universe can be found in the literature, especially in the context of scalar-tensor gravity [76, 102, 132, 183, 296, 307, 372, 434, 436–439, 525, 531, 565, 653, 654, 687, 691, 692, 729, 755, 760, 770, 771, 896, 963, 964, 1048, 1061, 1094, 1095]. A formalism applicable to generalized gravitational theories incorporating both scalar-tensor and modified $f(R)$ gravity was formulated by Hwang and his collaborators [611, 612, 616–618, 841]. This treatment is probably the most convenient for our purposes and we review it in this section, referring the reader to the original references for a more detailed discussion. The formalism is valid in regions of the phase space in which the field equations do not have singular points (see [8, 9, 457, 471, 515, 518, 600, 1045] for discussions of singular points in scalar-tensor theories).

### 8.3.1 *Scalar perturbations*

The perturbed FLRW line element is written as

$$ds^2 = -(1+2\alpha)\, dt^2 - \chi_{,i}\, dt\, dx^i + a^2(t)\left[(1+2\varphi)\,\delta_{ij} + 2H_T Y_{ij}\right] dx^i\, dx^j \,, \tag{8.108}$$

where $\alpha$, $\chi$, and $\varphi$ can be expressed in terms of scalar ($Y$), vector ($Y_i$), and tensor ($Y_{ij}$) spherical harmonics. The usual scalar spherical harmonics $Y$ satisfy the equation

$$\nabla^2 Y = -k^2\, Y \tag{8.109}$$

(where $k$ is an eigenvalue), while the vector and tensor spherical harmonics $Y_i$ and $Y_{ij}$ are related to $Y$ by

$$Y_i = \frac{1}{k^2}\,\partial_i Y \,, \tag{8.110}$$

$$Y_{ij} = \frac{1}{k^2}\,\partial_i\partial_j Y + \frac{1}{3}\,\delta_{ij}\, Y \,. \tag{8.111}$$

The scalar field is decomposed as

$$\phi(t, \mathbf{x}) = \phi_0(t) + \delta\phi(t, \mathbf{x}) , \tag{8.112}$$

with the background field $\phi_0$ depending only on the comoving time, while the perturbation $\delta\phi$ depends on both space and time. With the observational results in mind, we focus on a spatially flat FLRW universe setting to zero the curvature index $K$. The background quantities are classical while the perturbations are quantum-mechanical and are quantized by a canonical procedure associating quantum operators to these quantities.

Since the perturbations are inhomogeneous, they suffer from the notorious gauge-dependence problems of cosmology and one needs to study gauge-invariant quantities. We use the covariant and gauge-invariant formalism of Bardeen, Ellis, Bruni, Hwang and Vishniac [425–427, 621] in the version adapted by Hwang [612, 616–618] to generalized gravity described by the action

$$S = \int d^4x \sqrt{-g} \left[ \frac{1}{2} f(\phi, R) - \frac{\omega(\phi)}{2} \nabla^\alpha\phi \nabla_\alpha\phi - V(\phi) \right] \tag{8.113}$$

(since the inflaton dominates other forms of matter during inflation, we omit the matter part of the action). A convenient procedure consists of writing the field equations in the form of effective Einstein equations,

$$G_{\mu\nu} = T_{\mu\nu}[\phi] , \tag{8.114}$$

where

$$\begin{aligned} T_{\mu\nu}[\phi] = \frac{1}{F} \bigg[ & \omega \left( \nabla_\mu\phi \nabla_\nu\phi - \frac{1}{2} g_{\mu\nu} \nabla^\alpha\phi \nabla_\alpha\phi \right) - \frac{1}{2} (RF - f + 2V) g_{\mu\nu} \\ & + \nabla_\mu \nabla_\nu F - g_{\mu\nu} \Box F \bigg] \end{aligned} \tag{8.115}$$

is an effective stress-energy tensor for the scalar $\phi$ and we adopt the notation

$$F(\phi, R) \equiv \frac{\partial f}{\partial R} . \tag{8.116}$$

The gauge-invariant treatment for cosmological perturbations in GR [737, 812] can then be applied. The effective stress-energy tensor has the form of an imperfect fluid tensor

$$T_{\mu\nu} = (P + \rho) u_\mu u_\nu + P g_{\mu\nu} + 2q_{(\mu} u_{\nu)} + \pi_{\mu\nu} , \tag{8.117}$$

where $q^\mu$ and $\pi_{\mu\nu} = \pi_{\nu\mu}$ are the heat current density and the anisotropic stress tensor, respectively. They are purely spatial quantities, *i.e.*,

$$q_\beta u^\beta = 0 , \qquad \pi_{\mu\nu} u^\nu = \pi_{\mu\nu} u^\mu = 0 . \tag{8.118}$$

The evolution of the scalar field $\phi$ is ruled by

$$\Box\phi + \frac{1}{2\omega}\left(\frac{d\omega}{d\phi}\nabla^{\beta}\phi\,\nabla_{\beta}\phi + F - 2\frac{dV}{d\phi}\right) = 0\,. \tag{8.119}$$

Following the gauge-invariant formalism for perturbations in GR, one considers the Bardeen gauge-invariant variables [84] and the gauge-invariant variable

$$\delta\phi_{\varphi} \equiv \delta\phi - \frac{\dot{\phi}}{H}\varphi \equiv -\frac{\dot{\phi}}{H}\varphi_{\delta\phi}\,, \tag{8.120}$$

where an overdot denotes differentiation along the direction parallel to the fluid four-velocity $u^{\mu}$, *i.e.*, $\dot{f} \equiv u^{\beta}\,\nabla_{\beta} f$. To first order, this operation is nothing but the differentiation with respect to the comoving time of the FLRW background. Scalar perturbations are described by the second order action [613–615, 757, 758, 811]

$$\begin{aligned} S^{(pert)} &= \int dt\, d^3\mathbf{x}\, \mathscr{L}^{(pert)} = \frac{1}{2}\int dt\, d^3\mathbf{x}\, a^3 Z \\ &\times \left\{\delta\dot{\phi}_{\varphi}^2 - \frac{1}{a^2}\,\delta\phi_{\varphi}{}^{,i}\,\delta\phi_{\varphi,i} + \frac{1}{a^3 Z}\frac{H}{\dot{\phi}}\frac{d}{dt}\left[a^3 Z\frac{d}{dt}\left(\frac{\dot{\phi}}{H}\right)\right]\delta\phi_{\varphi}^2\right\} \end{aligned} \tag{8.121}$$

generalizing the action for inflationary perturbations in GR [737], where

$$Z(t) = \frac{2\omega + \frac{3(\dot{F})^2}{F(\dot{\phi})^2}}{2\left(\frac{\dot{F}}{2HF}+1\right)^2}\,. \tag{8.122}$$

In the limit to Einstein theory, $F = \kappa^{-1}$, $\omega = 1$ and $Z = 1$. The variation of the action (8.121) yields the evolution equation for the perturbations $\delta\phi_{\varphi}$

$$\delta\ddot{\phi}_{\varphi} + \frac{(a^3 Z)^{\cdot}}{a^3 Z}\delta\dot{\phi}_{\varphi} - \left\{\frac{\nabla^2}{a^2} + \frac{1}{a^3 Z}\frac{H}{\dot{\phi}}\frac{d}{dt}\left[a^3 Z\frac{d}{dt}\left(\frac{\dot{\phi}}{H}\right)\right]\right\}\delta\phi_{\varphi} = 0\,. \tag{8.123}$$

By introducing the auxiliary variables

$$z(t) \equiv \frac{a\,\dot{\phi}}{H}\sqrt{Z}\,, \tag{8.124}$$

$$v(t,\mathbf{x}) \equiv z\,\frac{H}{\dot{\phi}}\,\delta\phi_{\varphi} = a\sqrt{Z}\,\delta\phi_{\varphi}\,, \tag{8.125}$$

Equation (8.123) is written as

$$v_{\eta\eta} - \left(\nabla^2 + \frac{z_{\eta\eta}}{z}\right) v = 0\,, \tag{8.126}$$

where $\eta$ is the conformal time defined by $dt = a d\eta$. The Heisenberg picture is used in the quantization of the perturbations in order to keep the vacuum state time-independent. The quantum operators $\delta\hat{\phi}(t, \mathbf{x})$ and $\hat{\varphi}$ are associated with the classical variables $\delta\phi(t, \mathbf{x})$ and $\varphi$, with (a hat denotes a quantum operator)

$$\delta\hat{\phi}_\varphi = \delta\hat{\phi} - \frac{\dot{\phi}}{H}\,\hat{\varphi}\,. \tag{8.127}$$

The operator $\delta\hat{\phi}_\varphi$ is then Fourier-decomposed in the usual way

$$\delta\hat{\phi}_\varphi = \frac{1}{(2\pi)^{3/2}} \int d^3\mathbf{k} \left[\hat{a}_k\, \delta\phi_{\varphi k}(t)\, \mathrm{e}^{i\,\mathbf{k}\cdot\mathbf{x}} + \hat{a}_k^\dagger\, \delta\phi_{\varphi_k}^*(t)\, \mathrm{e}^{-i\,\mathbf{k}\cdot\mathbf{x}}\right]. \tag{8.128}$$

The creation and annihilation operators $\hat{a}_k^\dagger$ and $\hat{a}_k$ obey the canonical commutation relations

$$\left[\hat{a}_k, \hat{a}_{k'}^\dagger\right] = \delta^{(3)}\left(\mathbf{k} - \mathbf{k}'\right), \tag{8.129}$$

$$\left[\hat{a}_k, \hat{a}_{k'}\right] = \left[\hat{a}_k^\dagger, \hat{a}_{k'}^\dagger\right] = 0\,, \tag{8.130}$$

and the complex Fourier coefficients evolve in time according to

$$\delta\ddot{\phi}_{\varphi k} + \frac{(a^3 Z)^{\cdot}}{a^3 Z}\,\delta\dot{\phi}_{\varphi k} + \left\{\frac{k^2}{a^2} - \frac{1}{a^3 Z}\frac{H}{\dot{\phi}}\left[a^3 Z\,\frac{d}{dt}\left(\frac{\dot{\phi}}{H}\right)\right]\right\}\delta\phi_{\varphi k} = 0\,. \tag{8.131}$$

Now $v$ is expanded as the Fourier integral

$$v(t, \mathbf{x}) = \frac{1}{(2\pi)^{3/2}} \int d^3\mathbf{k} \left[v_k(t)\, \mathrm{e}^{i\,\mathbf{k}\cdot\mathbf{x}} + v_k^*(t)\, \mathrm{e}^{-i\,\mathbf{k}\cdot\mathbf{x}}\right], \tag{8.132}$$

and its quantum counterpart is

$$\hat{v} = \frac{zH}{\dot{\phi}}\,\delta\hat{\phi}_\varphi = a\sqrt{Z}\,\delta\hat{\phi}_\varphi\,, \tag{8.133}$$

where the components $v_k(t)$ are the solutions of

$$(v_k)_{\eta\eta} + \left(k^2 - \frac{z_{\eta\eta}}{z}\right) v_k = 0\,. \tag{8.134}$$

The classical momentum conjugated to $\delta\phi_\varphi$ is

$$\delta\pi_\varphi(t,\mathbf{x}) = \frac{\partial \mathcal{L}^{(pert)}}{\partial(\delta\dot{\phi}_\varphi)} = a^3 Z\, \delta\dot{\phi}_\varphi(t,\mathbf{x})\,, \tag{8.135}$$

with associated quantum operator $\delta\hat{\pi}_\varphi$. The operators $\delta\hat{\phi}_\varphi$ and $\delta\hat{\pi}_\varphi$ satisfy the equal time commutation relations

$$\left[\delta\hat{\phi}_\varphi(t,\mathbf{x}), \delta\hat{\phi}_\varphi(t,\mathbf{x}')\right] = \left[\delta\hat{\pi}_\varphi(t,\mathbf{x}), \delta\hat{\pi}_\varphi(t,\mathbf{x}')\right] = 0\,, \tag{8.136}$$

$$\left[\delta\hat{\phi}_\varphi(t,\mathbf{x}), \delta\hat{\pi}_\varphi(t,\mathbf{x}')\right] = \frac{i}{a^3 Z}\,\delta^{(3)}\left(\mathbf{x}-\mathbf{x}'\right)\,, \tag{8.137}$$

with the $\delta\phi_{\varphi k}(t)$ obeying

$$\delta\phi_{\varphi k}\,\delta\dot{\phi}^*_{\varphi k} - \delta\phi^*_{\varphi k}\,\delta\dot{\phi}_{\varphi k} = \frac{i}{a^3 Z}\,. \tag{8.138}$$

The relation

$$\frac{z_{\eta\eta}}{z} = \frac{m}{\eta^2}\,, \tag{8.139}$$

where $m$ is a constant, is used at this point in GR-based inflation. This relation can be generalized to the slow-roll inflationary regime of theories of the form (8.113), which also have de Sitter attractors in phase space. The use of Eq. (8.139) reduces Eq. (8.134) for $v_k$ to

$$(v_k)_{\eta\eta} + \left[k^2 - \frac{(\nu^2 - 1/4)}{\eta^2}\right] v_k = 0\,, \tag{8.140}$$

where

$$\nu = \sqrt{m + \frac{1}{4}}\,. \tag{8.141}$$

In order to solve Eq. (8.140) we introduce the variable $s = k\eta$, which turns $v_k$ into $v_k = \sqrt{s}\,J(s)$ and Eq. (8.140) into the Bessel equation

$$\frac{d^2 J}{ds^2} + \frac{1}{s}\frac{dJ}{ds} + \left(1 - \frac{\nu^2}{s^2}\right) J = 0\,, \tag{8.142}$$

which has as solutions the Bessel functions of order $\nu$ and

$$v_k(\eta) = \sqrt{k\eta}\, J_\nu(k\eta)\,. \tag{8.143}$$

The solutions of Eq. (8.125) are the Fourier coefficients

$$\delta\phi_{\varphi k}(\eta) = \frac{\dot{\phi}}{zH} v_k(\eta) = \frac{1}{a\sqrt{Z}} v_k(\eta) \,. \tag{8.144}$$

The quantities $v_k(\eta)$ can be rewritten by expressing the Bessel functions $J_\nu$ in terms of Hankel functions $H_\nu^{(1,2)}$, yielding

$$v_k(\eta) = \frac{\sqrt{\pi|\eta|}}{2}\left[c_1(\mathbf{k})\; H_\nu^{(1)}(k|\eta|) + c_2(\mathbf{k})\; H_\nu^{(2)}(k|\eta|)\right] \tag{8.145}$$

and Eq. (8.144) gives

$$\delta\phi_{\varphi k}(\eta) = \frac{\sqrt{\pi|\eta|}}{2a\sqrt{Z}}\left[c_1(\mathbf{k})H_\nu^{(1)}(k|\eta|) + c_2(\mathbf{k})H_\nu^{(2)}(k|\eta|)\right] . \tag{8.146}$$

The coefficients $c_1$ and $c_2$ are normalized according to

$$|c_2(\mathbf{k})|^2 - |c_1(\mathbf{k})|^2 = 1 \,, \tag{8.147}$$

to preserve Eq. (8.138).

In the limit of small wavelengths $\frac{z_{\eta\eta}}{z} \ll k^2$, the vacuum state must correspond to positive frequency solutions and Eq. (8.134) assumes the form

$$(v_k)_{\eta\eta} - k^2 v_k = 0 \,, \tag{8.148}$$

the solutions of which are $v_k \propto \mathrm{e}^{\pm i\,k\eta}$, hence

$$\delta\phi_{\varphi k} = \frac{1}{a\sqrt{Z}\sqrt{2k}}\left[c_1(\mathbf{k})\;\mathrm{e}^{i\,k|\eta|} + c_2(\mathbf{k})\;\mathrm{e}^{-i\,k|\eta|}\right] \tag{8.149}$$

(the same result is obtained by expanding Eq. (8.146) for $k|\eta| \gg 1$). The positive frequency solution in the small wavelength limit is obtained for

$$c_1(\mathbf{k}) = 0 \,, \qquad c_2(\mathbf{k}) = 1 \,, \tag{8.150}$$

which is equivalent to selecting the Bunch-Davies vacuum for de Sitter space and yields

$$\delta\phi\,(\eta,\mathbf{x}) = \frac{1}{(2\pi)^{3/2}}\int d^3\mathbf{x}\left[c_2(\mathbf{k})\;\mathrm{e}^{i\,(\mathbf{k}\cdot\mathbf{x}-k\eta)} + c_2^*(\mathbf{k})\;\mathrm{e}^{i\,(-\mathbf{k}\cdot\mathbf{x}+k|\eta|)}\right] . \tag{8.151}$$

The power spectrum of a quantity $g(t,\mathbf{x})$ is defined as

$$\mathscr{P}(k,t) \equiv \frac{k^3}{2\pi^2}\int d^3\mathbf{r}\,\langle g\,(\mathbf{x}+\mathbf{r},t)\,g\,(\mathbf{x},t)\rangle_{\mathbf{x}}\;\mathrm{e}^{-i\,\mathbf{k}\cdot\mathbf{r}} = \frac{k^3}{2\pi^2}\,|g_k(t)|^2 \,, \tag{8.152}$$

where $\langle\ \rangle_{\mathbf{x}}$ denotes an average over the spatial coordinates $\mathbf{x}$ and $g_k(t)$ are the Fourier coefficients of the expansion

$$g\,(t,\mathbf{x}) = \frac{1}{(2\pi)^{3/2}} \int d^3\mathbf{k}\, \left[g_k(t)\, \mathrm{e}^{i\,\mathbf{k}\cdot\mathbf{x}} + g_k^*(t)\, \mathrm{e}^{-i\,\mathbf{k}\cdot\mathbf{x}}\right] . \tag{8.153}$$

The power spectrum of the gauge-invariant operator $\delta\hat{\phi}_{\varphi k}$ is obtained with the choice $g\,(t,\mathbf{x}) = \langle 0|\delta\hat{\phi}_{\varphi k}|0\rangle$, which yields

$$\mathscr{P}_{\delta\hat{\phi}_\varphi}(k,t) = \frac{k^3}{2\pi^2} \int d^3\mathbf{r}\, \langle 0\,|\delta\hat{\phi}_{\varphi k}\,(\mathbf{x}+\mathbf{r},t)\; \delta\hat{\phi}_{\varphi k}(\mathbf{x},t)|0\,\rangle_{\mathbf{x}}\; \mathrm{e}^{-i\,\mathbf{k}\cdot\mathbf{r}} . \tag{8.154}$$

Long wavelength perturbations cross outside the horizon during inflation and are described by the large scale limit of Eq. (8.131), which admits the solution

$$\delta\phi_\varphi(t,\mathbf{x}) = -\frac{\dot{\phi}}{H}\left[C(\mathbf{x}) - D(\mathbf{x}) \int_0^t dt'\, \frac{1}{a^3 Z}\, \frac{H^2}{\dot{\phi}}\right] , \tag{8.155}$$

where $C(\mathbf{x})$ and $D(\mathbf{x})$ are the coefficients of a growing and a decaying component, respectively (from now on we omit the latter). In the long wavelength limit $k|\eta| \ll 1$, the solution $\delta\phi_{\varphi k}(\eta)$ for $\nu \neq 0$ becomes

$$\delta\phi_{\varphi k}(\eta) = \frac{i\,\sqrt{|\eta|}\,\Gamma(\nu)}{2a\sqrt{\pi Z}} \left(\frac{k|\eta|}{2}\right)^{-\nu} [c_2(\eta) - c_1(\eta)] , \tag{8.156}$$

where $\Gamma$ is the Gamma function. The power spectrum of $\delta\phi_{\varphi k}$ then turns out to be

$$\mathscr{P}^{1/2}_{\delta\hat{\phi}_\varphi}(k,\eta) = \frac{\Gamma(\nu)}{\pi^{3/2} a\,|\eta|\sqrt{Z}} \left(\frac{k|\eta|}{2}\right)^{\frac{3}{2}-\nu} |c_2(\mathbf{k}) - c_1(\mathbf{k})| \tag{8.157}$$

for $\nu \neq 0$. Equation (8.155) without the decaying mode yields

$$C(\mathbf{x}) = -\frac{H}{\dot{\phi}}\,\delta\phi_\varphi(t,\mathbf{x}) \tag{8.158}$$

and the definition (8.152) of power spectrum gives

$$\mathscr{P}^{1/2}_{C}(k,t) = \left|\frac{H}{\dot{\phi}}\right| \mathscr{P}^{1/2}_{\delta\phi_\varphi}(k,t) . \tag{8.159}$$

Equations (8.120) and (8.158) then yield

$$\varphi_{\delta\phi} = -\frac{H}{\dot{\phi}}\,\delta\phi_\varphi = C . \tag{8.160}$$

The quantity $C(\mathbf{x})$ is proportional to the CMB percent temperature fluctuation $\delta T/T$ [617],

$$\frac{\delta T}{T} = \frac{C}{5} . \tag{8.161}$$

As a consequence, the temperature anisotropy spectrum is

$$\sqrt{\mathscr{P}_{\delta T/T}(k,t)} = \frac{1}{5}\sqrt{\mathscr{P}_C(k,t)} \tag{8.162}$$

and Eq. (8.159) yields

$$\mathscr{P}^{1/2}_{\delta T/T}(k,t) = \frac{1}{5}\left|\frac{H}{\dot{\phi}}\right| \mathscr{P}^{1/2}_{\delta\phi_\varphi}(k,t) \tag{8.163}$$

where Eq. (8.157) provides $\mathscr{P}^{1/2}_{\delta\phi_\varphi}$.

Equation (8.162) now yields the spectral index of scalar perturbations

$$n_S \equiv 1 + \frac{d\left(\ln \mathscr{P}_{\delta\hat{\varphi}\delta\phi}\right)}{d\,(\ln k)} , \tag{8.164}$$

which is computed as

$$n_S = 1 + \frac{d\,(\ln \mathscr{P}_C)}{d\,(\ln k)} . \tag{8.165}$$

Equations (8.159) and (8.157) yield

$$n_S = 4 - 2\nu \qquad (\nu \neq 0) . \tag{8.166}$$

While in slow-roll inflation in GR it is sufficient to introduce two slow-roll parameters, in the slow-roll regime of generalized gravity *four* such parameters are needed [618, 841]:

$$\varepsilon_1 = \frac{\dot{H}}{H^2} = -\varepsilon_H , \tag{8.167}$$

$$\varepsilon_2 = \frac{\ddot{\phi}}{H\dot{\phi}} = -\eta_H , \tag{8.168}$$

$$\varepsilon_3 = \frac{\dot{F}}{2\,H\,F} , \tag{8.169}$$

$$\varepsilon_4 = \frac{\dot{\alpha}}{2\,H\alpha} , \tag{8.170}$$

where

$$\alpha = F \left[ \omega + \frac{3\left(\dot{F}\right)^2}{2F\left(\dot{\phi}\right)^2} \right]. \tag{8.171}$$

The first two parameters $\varepsilon_1$ and $\varepsilon_2$ coincide, apart from the sign, with the familiar slow-roll parameters $\varepsilon_H$ and $\eta_H$ introduced in the Hubble slow-roll approximation to inflation in GR. The remaining two parameters $\varepsilon_{3,4}$ are characteristic of generalized inflation. The four slow-roll parameters and their time derivatives $\dot{\varepsilon}_i$ remain small during slow-roll inflation. By neglecting $\dot{\varepsilon}_i$ one has

$$\frac{z_{\eta\eta}}{z} = a^2 H^2 \left(2 - 2\varepsilon_1 + 3\varepsilon_2 - 3\varepsilon_3 + 3\varepsilon_4\right) \tag{8.172}$$

to first order.

To proceed, we use the standard relation

$$\eta \simeq -\frac{1}{aH}\frac{1}{1+\varepsilon_1}, \tag{8.173}$$

which is proved as follows. In de Sitter space it is $a = a_0 \exp(H_0 t)$ and integrating the definition of conformal time $dt = a\, d\eta$ yields

$$\eta = -\frac{1}{aH} = -\frac{e^{-H_0 t}}{a_0 H_0}. \tag{8.174}$$

For expanding ($H_0 > 0$) de Sitter spaces, $t \to +\infty$ corresponds to $\eta \to 0$, while for contracting ($H_0 < 0$) de Sitter spaces, $t \to +\infty$ corresponds to $\eta \to +\infty$. In the approximation in which the derivatives $\dot{\varepsilon}_i$ are negligible, Eq. (8.174) reduces to Eq. (8.173) [613–615].

Equation (8.173) then yields

$$m = 2 + 3\left(-2\varepsilon_1 + \varepsilon_2 - \varepsilon_3 + \varepsilon_4\right), \tag{8.175}$$

$$\nu = \frac{3}{2} - 2\varepsilon_1 + \varepsilon_2 - \varepsilon_3 + \varepsilon_4, \tag{8.176}$$

and Eqs. (8.166) and (8.176) give the spectral index of scalar perturbations

$$n_S = 1 + 2\left(2\varepsilon_1 - \varepsilon_2 + \varepsilon_3 - \varepsilon_4\right), \tag{8.177}$$

The right hand side of Eq. (8.177) is evaluated when the perturbations cross outside the horizon. We obtain a nearly Harrison-Zeldovich spectrum with $n_S$ not very different from unity. In the GR limit $f(\phi, R) = R/\kappa$, $\omega = 1$, $\dot{F} = 0$, $\varepsilon_3$ and $\varepsilon_4$ vanish and the spectral index reduces to the usual one for GR inflation with a minimally coupled scalar $n_S = 1 - 4\varepsilon_H + 2\eta_H$ [688, 728, 737].

### 8.3.2 Gravitational wave perturbations

Gravitational waves (tensor) perturbations are generated during inflation as quantum fluctuations of the metric $g_{\mu\nu}$. Tensor perturbations in generalized gravity inflation have also been computed in [615,841] following the now canonical method for GR-based inflation, and we report here also this calculation.

Tensor modes are described by the gauge-invariant perturbation $H_T$ in Eq. (8.108). The latter is now rewritten using

$$c_{ij}(t,\mathbf{x}) \equiv H_T(t)\,Y_{ij}(\mathbf{x})\,, \tag{8.178}$$

where the $Y_{ij}$ are tensor spherical harmonics. The $c_{ij}$ are transverse and traceless,

$$\nabla^i c_{ij} = 0\,, \qquad c^i_i = 0\,. \tag{8.179}$$

The action for tensor perturbations is [37,610]

$$S^{(gw)} = \int dt \int d^3\mathbf{x}\,\mathscr{L}^{(gw)} = \int dt \int d^3\mathbf{x}\,\frac{a^3F}{2}\left[\dot{c_{ij}}\dot{c^{ji}} - \frac{1}{a^2}\,c_{ij,k}\nabla^k c^{ij}\right]. \tag{8.180}$$

The variation of this action yields the classical evolution equation

$$\ddot{c_{ij}} + \left(3H + \frac{\dot{F}}{F}\right)\dot{c_{ij}} - \frac{\nabla^2 c_{ij}}{a^2} = 0\,. \tag{8.181}$$

It is convenient to use the quantities

$$z_g \equiv a\sqrt{F}\,, \tag{8.182}$$

$$v_g(t,\mathbf{x}) \equiv z_g\,c_{ij}(t,\mathbf{x})\,, \tag{8.183}$$

to reduce the evolution equation (8.181) to

$$(v_g)_{\eta\eta} - \left[\frac{(z_g)_{\eta\eta}}{z} + \nabla^2\right] v_g = 0\,. \tag{8.184}$$

In the long wavelength limit $(z_g)_{\eta\eta}/z_g \gg k^2$, the solution is

$$c_{ij}(t,\mathbf{x}) = C_{ij}(\mathbf{x}) - D_{ij}(\mathbf{x})\int_0^t \frac{dt}{a^3F}\,, \tag{8.185}$$

where the second term on the right hand side decays and will be omitted in the following.

In the small wavelength limit, the asymptotic solution for the Fourier coefficients of $c_{ij}$ as functions of conformal time is

$$c_{ij}(\eta, \mathbf{k}) = \frac{1}{a\sqrt{F}}\left[c_{ij}^{(1)}(\mathbf{k})\, e^{i\,k\eta} + c_{ij}^{(2)}(\mathbf{k})\, e^{-i\,k\eta}\right]. \tag{8.186}$$

The classical perturbations $c_{ij}$ are decomposed into the two possible polarizations "+" and "×" as

$$\begin{aligned} c_{ij}(t, \mathbf{x}) &= \frac{L^{3/2}}{(2\pi)^3}\int d^3\mathbf{k} \sum_{l=1}^{2} h_{l\mathbf{k}}(t)\, e_{ij}^{(l)}(\mathbf{k})\, e^{i\mathbf{k}\cdot\mathbf{x}} \\ &\equiv \frac{L^{3/2}}{(2\pi)^3}\int d^3\mathbf{k}\, \tilde{C}_{ij}(t, \mathbf{x}, \mathbf{k}), \end{aligned} \tag{8.187}$$

where $L^3$ is an irrelevant normalization volume which disappears in the final results and $l = +$ or $\times$. The $e_{ij}^{(l)}$ are polarization tensors describing the two polarization states and obey

$$e_{ij}^{(l)}(\mathbf{k})\, e^{(l')\,ij}(\mathbf{k}) = 2\,\delta_{ll'}. \tag{8.188}$$

Now we introduce

$$\begin{aligned} h_l(t, \mathbf{x}) &\equiv \frac{L^{3/2}}{2\,(2\pi)^3}\int d^3\mathbf{k}\, \tilde{C}_{ij}(t, \mathbf{x}, \mathbf{k})\, e^{(l)\,ij}(\mathbf{k}) \\ &= \frac{L^{3/2}}{(2\pi)^3}\int d^3\mathbf{k}\, h_{l\mathbf{k}}(t)\, e^{i\,\mathbf{k}\cdot\mathbf{x}}, \end{aligned} \tag{8.189}$$

and the classical power spectrum

$$\mathscr{P}_{c_{ij}}(t, \mathbf{k}) = \frac{k^3}{2\pi^2}\int d^3\mathbf{r}\, \langle c_{ij}(t, \mathbf{x}+\mathbf{r})\, c^{ij}(t, \mathbf{x})\rangle_{\mathbf{x}}\, e^{-i\,\mathbf{k}\cdot\mathbf{r}}, \tag{8.190}$$

where $\langle\ \ \rangle_{\mathbf{x}}$ denotes a spatial average, and it is

$$\mathscr{P}_{c_{ij}}(t, \mathbf{k}) = 2\sum_{l=1}^{2} \mathscr{P}_{h_l}(t, \mathbf{k}) = 2\sum_{l=1}^{2} \frac{k^3}{2\pi^2}\, |h_{l\mathbf{k}}(t)|^2 \tag{8.191}$$

with $h_{+\mathbf{k}} = h_{\times\mathbf{k}} \equiv h_{\mathbf{k}}$.

The tensor modes are quantized by associating quantum operators $\hat{c}_{ij}$ to the classical variables $c_{ij}$. These operators are Fourier-expanded as

$$\widehat{c}_{ij}\,(t,\mathbf{x}) = \frac{1}{(2\pi)^{3/2}} \int d^3\mathbf{k} \sum_{l=1}^{2} \Big[\widetilde{h}_{l\mathbf{k}}(t)\,\widehat{a}_{l\mathbf{k}}\; e_{ij}^{(l)}\,(\mathbf{k})\; \mathrm{e}^{i\,\mathbf{k}\cdot\mathbf{x}}$$

$$+\, \widetilde{h}^*_{l\mathbf{k}}(t)\,\widehat{a}^\dagger_{l\mathbf{k}}\; e_{ij}^{(l)}\,(\mathbf{k})\; \mathrm{e}^{-i\,\mathbf{k}\cdot\mathbf{x}}\Big]$$

$$\equiv \frac{1}{2\,(2\pi)^{3/2}} \int d^3\mathbf{k}\; \hat{C}_{ij}\,(t,\mathbf{x},\mathbf{k})\;, \tag{8.192}$$

where the $\tilde{h}_{l\mathbf{k}}(t)$ are mode functions. The creation and annihilation operators $\hat{a}^\dagger_{l\mathbf{k}}$ and $\hat{a}_{l\mathbf{k}}$ satisfy the canonical commutation relations

$$\left[\hat{a}_{l\mathbf{k}}, \hat{a}^\dagger_{l'\mathbf{k}'}\right] = \delta_{ll'}\;\delta^{(3)}\left(\mathbf{k}-\mathbf{k}'\right)\;, \tag{8.193}$$

$$\left[\hat{a}_{l\mathbf{k}}, \hat{a}_{l'\mathbf{k}'}\right] = \left[\hat{a}^\dagger_{l\mathbf{k}}, \hat{a}^\dagger_{l'\mathbf{k}'}\right] = 0\,. \tag{8.194}$$

The quantum operators

$$\hat{h}_l\,(t,\mathbf{x}) \equiv \frac{1}{2\,(2\pi)^{3/2}} \int d^3\mathbf{k}\; \hat{C}_{ij}\,(t,\mathbf{x},\mathbf{k})\; e^{(l)\,ij}\,(\mathbf{k})$$

$$= \frac{1}{(2\pi)^{3/2}} \int d^3\mathbf{k}\left[\tilde{h}_{l\mathbf{k}}\,(t)\;\hat{a}_{l\mathbf{k}}\,\mathrm{e}^{i\,\mathbf{k}\cdot\mathbf{x}} + \tilde{h}^*_{l\mathbf{k}}\,(t)\;\hat{a}^\dagger_{l\mathbf{k}}\,\mathrm{e}^{-i\,\mathbf{k}\cdot\mathbf{x}}\right] \tag{8.195}$$

are associated to the classical variables and the equation of motion for the $\hat{c}_{ij}$ gives

$$\ddot{\hat{h}}_l + \left(3H + \frac{\dot{F}}{F}\right)\dot{\hat{h}}_l - \frac{\nabla^2 \hat{h}_l}{a^2} = 0\,. \tag{8.196}$$

The momenta canonically conjugated to $\hat{h}_l$ are

$$\delta\,\hat{\pi}_{h_l}\,(t,\mathbf{x}) = \frac{\partial \mathscr{L}^{(gw)}}{\partial\left(\dot{\hat{h}}_l\right)} = 2\,a^3\;F\;\hat{h}_l \tag{8.197}$$

and satisfy the equal time commutation relation

$$\left[\hat{h}_l\,(t,\mathbf{x})\,,\delta\,\hat{\pi}_{h_l}\left(t,\mathbf{x}'\right)\right] = i\;\delta^{(3)}\left(\mathbf{x}-\mathbf{x}'\right)\,, \tag{8.198}$$

while the mode functions $\tilde{h}_{l\mathbf{k}}\,(t,\mathbf{k})$ satisfy the Wronskian condition

$$\widetilde{h_{l\mathbf{k}}}\;\dot{\widetilde{h_{l\mathbf{k}}}}^* - \widetilde{h}^*_{l\mathbf{k}}\;\dot{\widetilde{h_{l\mathbf{k}}}} = \frac{i}{a^3\,F}\,. \tag{8.199}$$

If the assumption

$$\frac{(z_g)_{\eta\eta}}{z_g} = \frac{m_g}{\eta^2} \tag{8.200}$$

(where $m_g$ is a constant) is valid, then the equation for the mode functions admits the analytical solution in terms of conformal time

$$\widetilde{h_{l\mathbf{k}}}(\eta) = \frac{\sqrt{\pi|\eta|}}{2a\sqrt{2F}}\left[c_{l1}(\mathbf{k})\, H^{(1)}_{\nu_g}(k|\eta|) + c_{l2}(\mathbf{k})\, H^{(2)}_{\nu_g}(k|\eta|)\right], \tag{8.201}$$

where

$$\nu_g = \sqrt{m_g + \frac{1}{4}}\,. \tag{8.202}$$

The normalization

$$|c_{l2}(k)|^2 - |c_{l1}(k)|^2 = 1 \tag{8.203}$$

must be satisfied by each polarization state in order to satisfy Eq. (8.199). The power spectrum of tensor modes

$$\mathscr{P}_{\hat{c}_{ij}}(\eta, k) = \frac{k^3}{2\pi^2}\int d^3\mathbf{r}\, \langle 0\left|\hat{c}_{ij}(t, \mathbf{x}+\mathbf{r})\, \hat{c}^{ij}(t, \mathbf{x})\right| 0\rangle\, \mathrm{e}^{-i\mathbf{k}\cdot\mathbf{r}}, \tag{8.204}$$

then is computed as

$$\mathscr{P}_{\hat{c}_{ij}}(\eta, k) = 2\sum_{l=1}^{2}\mathscr{P}_{\hat{h}_l}(\eta, k) = 2\sum_{l=1}^{2}\frac{k^3}{2\pi^2}\left|\tilde{h}_{l\mathbf{k}}(\eta)\right|^2. \tag{8.205}$$

The vacuum state is identified by imposing the condition that quantum field theory in Minkowski space is recovered in the limit of short wavelengths, which is equivalent to setting $c_1 = 0$ and $c_2 = 1$.

For $\nu_g \neq 0$ one obtains

$$\sqrt{\mathscr{P}_{\hat{c}_{ij}}(\eta, k)} = \frac{H}{2\pi\sqrt{2F}}\,\frac{1}{aH|\eta|}\,\frac{\Gamma(\nu_g)}{\Gamma(3/2)}\left(\frac{k|\eta|}{2}\right)^{\frac{3}{2}-\nu_g}. \tag{8.206}$$

The spectral index of tensor perturbations is defined as

$$n_T \equiv \frac{d\left(\ln\mathscr{P}_{\hat{c}_{ij}}\right)}{d(\ln k)}, \tag{8.207}$$

and therefore turns out to be

$$n_T = 3 - 2\,\nu_g\,. \tag{8.208}$$

In the slow-roll regime, the derivatives $\dot{\varepsilon}_i$ of the four slow-roll parameters (8.167)–(8.170) are small and can be neglected obtaining

$$\frac{(z_g)_{\eta\eta}}{z_g} = a^2 H^2 (2 + \varepsilon_1 + 3\varepsilon_3) \tag{8.209}$$

to first order. The relation $aH = -(1 - \varepsilon_1)/\eta + \mathrm{O}(2)$ then yields

$$m_g = 2 - 3\,\varepsilon_1 + 3\,\varepsilon_3\,, \tag{8.210}$$

$$\nu_g = \frac{3}{2} - (\varepsilon_1 - \varepsilon_3)\,, \tag{8.211}$$

and the spectral index of tensor perturbations is simply

$$n_T = 2\,(\varepsilon_1 - \varepsilon_3)\,. \tag{8.212}$$

If inflation enters a *superacceleration* regime, defined by increasing Hubble parameter $\dot{H} > 0$, then $\varepsilon_1 = \dot{H}/H^2 > 0$ and the spectral index of tensor modes Eq. (8.212) can be positive, giving a *blue spectrum* with more power at small wavelengths than the usual inflationary spectrum [754]. A blue spectrum of gravitational waves is physically interesting because the possibility of detecting cosmological gravitational waves (*e.g.*, [266, 745, 904]) is thus enhanced. Blue spectra of gravitational waves cannot be obtained with minimally coupled scalar fields in GR (for which $n_T = 4\dot{H}/H^2$ is always non-positive).

The temperature anisotropies observed in the CMB sky are decomposed in spherical harmonics[4] $Y^{lm}$ and then the monopole (average temperature) and the dipole (caused by the peculiar motion of the Solar System) terms are removed, leaving

$$\frac{\delta T}{T} = \sum_{l=2}^{+\infty} \sum_{m=-l}^{+l} a_{lm}\, Y^{lm}(\theta, \varphi)\,, \tag{8.213}$$

with contributions from both scalar and tensor modes,

$$\frac{\delta T}{T} = \left(\frac{\delta T}{T}\right)_S + \left(\frac{\delta T}{T}\right)_T. \tag{8.214}$$

Using the multipole moments

$$C_l^{(S,T)} \equiv \sum_{m=-l}^{+l} |a_{lm}|^2 \tag{8.215}$$

[4] We now change the notation for the scalar spherical harmonics to the more familiar $Y^{lm}$ instead of $Y$.

for both scalar and tensor modes, the relative importance of scalar and tensor modes in producing the temperature fluctuations is expressed by the ratio

$$\mathscr{R} \equiv \frac{C_l^{(T)}}{C_l^{(S)}} . \tag{8.216}$$

This ratio for quadrupole modes is computed by Hwang [841] obtaining

$$\mathscr{R} = 4\pi \left( -\frac{\omega}{8\pi GF} \,\varepsilon_1 + 3\,\varepsilon_3^2 \right) . \tag{8.217}$$

In GR it is $\frac{\omega}{8\pi GF} = 1$, $\varepsilon_3 = 0$, and Eq. (8.217) can be rewritten as

$$\mathscr{R} = -2\pi\, n_T ; \tag{8.218}$$

this relation is independent of the inflationary model and potential $V(\phi)$ considered, hence it is regarded as a consistency relation for the paradigm of inflation [728]. In general, this reduction is not possible for slow-roll inflation in ETGs.

## 8.4 Constraints on ETGs from primordial nucleosynthesis

The standard Big Bang model is supported by three major pieces of observational evidence: the redshift of galaxies, the existence, temperature and spectrum of the CMB, and primordial nucleosynthesis. The prediction of the relative abundances of light elements processed in the radiation-dominated universe ([28, 29, 603, 894, 1134], see also [1143]) matches the observed abundances. It is only during the radiation era that, when the decreasing temperature of the expanding universe drops to $\sim 10^9$ K (corresponding to an age of three minutes for the universe), primordial nucleosynthesis begins with the production of $\sim$25% by mass of $^4$He and traces of $^2$H, $^3$He and $^7$Li [196, 277]. There is almost no production of heavier elements (see [884, 992] for reviews).

The abundance of $^4$He and other light elements depends crucially on the expansion rate of the universe; this sensitive dependence provides a way to constrain ETGs by using observational limits on the relative abundances of light elements [13, 283, 390, 397, 554, 555, 790, 967, 1007, 1009, 1082]. During nucleosynthesis, practically all the neutrons that are present are used to manufacture helium nuclei and the helium mass fraction $X\left(^4\text{He}\right)$ depends on the ratio of the neutron and proton number densities ($n_n$ and $n_p$, respectively) at that time,

$$X\left(^4\text{He}\right) = 2 \left. \frac{\frac{n_n}{n_p}}{1 + \frac{n_n}{n_p}} \right|_{nucleosynthesis} . \tag{8.219}$$

Before the time $t_F$ known as *freeze-out*, the weak interaction maintains neutrons and protons in chemical equilibrium. After this time, this is no longer possible and the ratio $n_p/n_n$ remains frozen to its value at $t_F$, which is

$$\left.\frac{n_n}{n_p}\right|_{t_F} = \exp\left(\frac{m_n - m_p}{K_B T_F}\right), \tag{8.220}$$

where $m_n$ and $m_p$ are the neutron and proton masses, respectively, $K_B$ is the Boltzmann constant, and $T_F$ is the freeze-out temperature. Neutrons decay freely due to $\beta$-decay during the time between freeze-out and nucleosynthesis and the length of this time interval determines the final abundance of neutrons at nucleosynthesis and the mass fraction of $^4$He produced. If the cosmic expansion proceeds faster than in GR, freeze-out occurs earlier and at a higher temperature $T_F$, hence the ratio $n_n/n_p$ is closer to unity. In addition, the time between freeze-out and nucleosynthesis is shorter, fewer neutrons decay) assuming that the lifetime of the free neutron is the same as in GR (which is not to be taken for granted if the fundamental constants of physics are allowed to vary), and more neutrons are available. These coincidences determine a higher mass fraction $X\left(^4\text{He}\right)$. The expression $2x\,(1+x)^{-1}$ on the right hand side of Eq. (8.219) is an increasing function of its argument $x \equiv n_n/n_p$, hence a slower cosmic expansion translates into an underproduction of $^4$He.

The deviation of the cosmic expansion rate from that of standard Einstein gravity is measured by the Hubble parameter $H_{ST}$ in the ETG considered divided by the Hubble parameter $H$ appropriate for GR with the same forms and amounts of matter in a spatially flat FLRW universe,

$$\xi_n \equiv \frac{H_{ST}}{H} = H_{ST}\sqrt{\frac{8\pi G}{3}\,\rho}\,. \tag{8.221}$$

The ratio (8.221) is called *speedup factor* and is equal to unity in Einstein gravity. If $\xi_n$ deviates from this value, there is over- or under-production of $^4$He during nucleosynthesis [90, 363, 1132].

In scalar-tensor gravity, the speedup factor is calculated as [361, 362]

$$\xi_n = \frac{1}{\Omega(\phi_n)\sqrt{1+\left[\frac{1}{G\Omega}\frac{d\Omega}{d\phi}\,|_{\phi_0}\right]^2}}, \tag{8.222}$$

where $\Omega(\phi) = \sqrt{G\phi}$ is the scale factor of the conformal transformation mapping the Jordan frame variables to the Einstein frame, and $\phi_n$ and $\phi_0$ are the values of the scalar field at the time of nucleosynthesis and now, respectively. While there is some consensus on the estimate of the value of $\xi_n$ in the range $0.8 \leq \xi_n \leq 1.2$, more conservative interpretations of the observed abundances yield the tighter limits $0.95 \leq \xi_n \leq 1.03$ [283, 1007]. The mass fraction of $^4$He produced during nucleosynthesis is [283, 867, 993]

$$X\left(^4\mathrm{He}\right) = 0.228 + 0.010\ln\left(10^{10}\,\frac{n_b}{n_\gamma}\right) + 0.012\,(N_\nu - 3)$$

$$+0.185\left(\frac{\tau_\nu - 889.8\,\mathrm{s}}{889.8\,\mathrm{s}}\right) + 0.327\ln\xi_n\,, \tag{8.223}$$

where $n_b$ and $n_\gamma$ are the number densities of baryons and photons, respectively, and $N_\nu$ is the number of light (*i.e.*, mass $< 1\,\mathrm{MeV}$) neutrino species, and $\tau_\nu$ is the lifetime of the neutrino. On the basis on the experimental limits the reasonable choice $N_\nu = 3$ and $\tau_\nu < 885$ s provides the limit [967] $X\left(^4\mathrm{He}\right) < 0.250$, which implies the bound $\ln\xi_n \leq 0.0797$ on the speedup factor.

In scalar-tensor gravity, the bounds on $\xi_n$ translate into constraints on the coupling function $\omega(\phi)$. Unfortunately, these constraints are not universal but depend on the particular scalar-tensor model. For example, the lower bound on the present value of $\omega$ is $\omega(\phi_0) \geq 10^7$ in the scalar-tensor models studied in [363, 967], while the much tighter constraint $\omega(\phi_0) \geq 10^{20}$ holds for a different class of scalar-tensor gravities [397, 1007]. This large range of lower bounds arises because, in the absence of self-interaction of the Brans-Dicke-like scalar field, two mechanisms acting in different directions rule the convergence of scalar-tensor to Einstein gravity, and therefore the magnitude of $\xi_n$ at any given time. One mechanism attracts the theory toward GR and dominates in certain scalar-tensor theories satisfying particular boundary conditions, which leads to a monotonic behavior of the speedup factor. This rapid convergence produces nucleosynthesis bounds that are much more stringent than those obtained in scalar-tensor models in which the attractor mechanism is less efficient [1008]. A repulsive mechanism counteracts the attractor one and shows its effects in theories with non-monotonic or oscillating speedup factor $\xi_n$, in which the nucleosynthesis bounds turn out to be much less stringent.

When a self-interaction potential $V(\phi)$ is included in the picture, the dynamics of $\phi$ change. For example, if the potential has a minimum, $\phi$ is attracted toward a fixed value corresponding to this minimum, turning the scalar-tensor theory into GR. However, the situation is more complicated because the dynamics of $\phi$ are regulated not only by $V(\phi)$ but also by the source terms $\left(-8\pi\,T^{(m)} - \dot{\omega}\,\dot{\phi}\right)/\,(2\omega + 3)$ in the equation of motion for $\phi$. The theory may converge to GR but, in certain situations, only asymptotically (a behavior reported for a massive Brans-Dicke-like scalar in a spatially flat FLRW universes, or for a massless scalar in a $K = -1$ universe [968]).

To conclude, the nucleosynthesis bounds are model-dependent but, near the present epoch, it is meaningful to consider an expansion about the present-day value $\phi_0$ of $\phi$. In this approximation, valid up to the epoch of primordial nucleosynthesis, a scalar-tensor model is described only by the first two coefficients of the expansion of the conformal factor $\Omega$. Introducing $a_{1,2}$ as in

$$\ln\Omega = a_1\,(\phi - \phi_0) + \frac{a_2}{2}\,(\phi - \phi_0)^2 + \cdots\,, \tag{8.224}$$

$$\frac{d\,(\ln\Omega)}{d\phi} = a_1 + a_2\,(\phi - \phi_0) + \cdots\,, \tag{8.225}$$

it is

$$\xi_n = \left[ G\phi_n \left(1 + \frac{a_1^2}{G^2}\right)\right]^{-1/2} . \tag{8.226}$$

The coefficients $a_{1,2}$ are related to the post-Newtonian parameters of the scalar-tensor theory used in the analysis of Solar System experiments [1167] by the equations [360, 855]

$$\gamma - 1 \simeq \frac{-2\,a_1^2}{1 + a_1^2} , \tag{8.227}$$

$$\beta - 1 \simeq \frac{a_1^2\, a_2}{2\left(1 + a_1^2\right)^2} . \tag{8.228}$$

## 8.5 The present universe: $f(R)$ gravity as an alternative to dark energy

In this section we summarize various elements introduced earlier in this book and try to provide a unifying picture. The varied astronomical and cosmological observations of the last decade, including CMB studies and supernovae and large scale structure surveys, concur to provide the picture of a universe with a present expansion that is accelerating and with an energy content comprising 4% of ordinary baryonic matter, 20% of dark matter, and 76% of a mysterious dark energy with exotic properties [60, 417, 940, 1039]. A cosmological constant $\Lambda$ would be the most natural explanation for dark energy but it suffers from the well known cosmological constant problems [271, 1154]: the observed value of the cosmological constant is enormously smaller than the vacuum energy predicted by quantum mechanics; and the energy densities of $\Lambda$ and of matter are comparable only during a short time in the history of the universe, so why is this coincidence happening today when humans are present to observe it? (This is the *coincidence problem.*)

Anthropic arguments can be advanced in order to explain the smallness of $\Lambda$ [106, 280]. More recently, anthropic reasoning has been revived in the context of the *landscape* of string theory, assuming that our universe corresponds to only one of an enormous number of vacuum states [1064]. However, advocating anthropic arguments is perceived by most physicists as a last ditch attempt showing the lack of better physical arguments.

Rejecting the cosmological constant, in order to remain within the context of GR, one must postulate the existence of dark energy, an unknown form of energy which escapes direct detection and does not cluster as ordinary matter. In order to accelerate the universe (*i.e.*, $\ddot{a} > 0$), this dark energy must posses a very exotic negative pressure, as follows from the Friedmann equation for the scale factor

$$\frac{\ddot{a}}{a} = -\frac{4\pi G}{3}(\rho + 3P) , \tag{8.229}$$

and the strong energy condition $\rho + 3P \geq 0$ [1139] must be violated by dark energy. The most popular scenarios for explaining a dynamical dark energy are known as quintessence and they are usually based on scalar fields acting the same way as the inflaton does in the early universe [68, 204, 272, 873, 895, 923, 1148, 1156]. Quintessence scenarios share the common feature of not being theoretically motivated and not providing a satisfactory solution to one of the biggest puzzles of current theoretical physics. In fact, the mass of the proposed quintessence scalar field is typically many orders of magnitude smaller than the natural mass scales of particle physics, and it is not clear why such fields do not couple explicitly to matter (no mechanism or symmetry is enforced to prevent this coupling to occur [269]).

The present acceleration of the universe ends the matter-dominated era when dark energy (or modified gravity, or whatever else causes the acceleration) becomes dominant over dust. During the matter era, the density fluctuations generated during inflation, which could not grow during the radiation era, go nonlinear and form the present-day cosmic structures that we observe. Therefore, it is essential for a theoretical model to incorporate a matter era that lasts for a sufficiently long time.

The "standard" picture of the present universe is summarized in the *concordance model* or *$\Lambda$CDM model* supplemented by some inflationary scenario. While regarded as highly successful, this model is plagued by the cosmological constant problems and lacks a convincing explanation of the nature of dark energy and seems to constitute more an empirical fit to the data with poor theoretical motivation than a complete model, or a consistent theory of the universe. It is natural therefore, that alternatives to dark energy have been sought for. None of these proposals are free of problems.

As already mentioned, a possible explanation of the cosmic acceleration could be that gravity is not described by GR, but by a theory that is "close" to GR at small scales but deviates from it in the infrared sector, causing effects only at large scales. Can such a theory explain the present cosmic acceleration? As long as the dark energy problem is not solved in a plausible and satisfactory way, it is certainly worth pursuing alternatives. Moreover, questioning the gravitational theory itself definitely provides a deeper understanding of gravity and highlights how things could be different.

### 8.5.1 Background universe

Assuming the spatially flat FLRW line element

$$ds^2 = -dt^2 + a^2(t)\left[dr^2 + r^2\left(d\theta^2 + \sin^2\theta d\phi^2\right)\right] \tag{8.230}$$

and a universe filled with a perfect fluid with energy-momentum tensor $T^{\mu\nu} = (\rho + P)\, u^\mu u^\nu + P\, g^{\mu\nu}$, the field equations of metric $f(R)$ gravity yield

$$H^2 = \frac{1}{3f'}\left[\kappa\,\rho + \frac{Rf' - f}{2} - 3H\dot{R}f''\right], \tag{8.231}$$

$$2\dot{H} + 3H^2 = -\frac{1}{f'}\Big[\kappa\, P + (\dot{R})^2 f''' + 2H\dot{R}f'' + \ddot{R}f'' + \frac{1}{2}(f - Rf')\Big], \tag{8.232}$$

where it is assumed that $f' > 0$ to keep the effective gravitational coupling positive, and that $f'' > 0$ to avoid local instabilities [396, 460, 851]. The effective energy density and pressure of the $f(R)$ fluid are

$$\rho_{eff} = \frac{Rf' - f}{2f'} - \frac{3H\dot{R}f''}{f'}, \tag{8.233}$$

$$P_{eff} = \frac{\dot{R}^2 f''' + 2H\dot{R}f'' + \ddot{R}f'' + \frac{1}{2}(f - Rf')}{f'}. \tag{8.234}$$

The effective density $\rho_{eff}$ is non-negative in a spatially flat FLRW universe, as follows from Eq. (8.231) in the limit $\rho \to 0$. $f(R)$ gravity can produce accelerated expansion without the need for dark energy or an inflaton. *In vacuo*, Eqs. (8.231) and (8.232) assume the form [211, 219]

$$H^2 = \frac{\kappa}{3}\rho_{eff}, \tag{8.235}$$

$$\frac{\ddot{a}}{a} = -\frac{\kappa}{6}\left(\rho_{eff} - 3P_{eff}\right) \tag{8.236}$$

but, if a cosmological fluid if present, it couples to gravity with the effective strength $\kappa/f'$. One can define the effective EoS parameter

$$w_{eff} \equiv \frac{P_{eff}}{\rho_{eff}} = \frac{\dot{R}^2 f''' + 2H\dot{R}f'' + \ddot{R}f'' + \frac{1}{2}(f - Rf')}{\frac{Rf'-f}{2} - 3H\dot{R}f''}. \tag{8.237}$$

For example, a metric $f(R)$ model mimics the de Sitter equation of state $w_{eff} = -1$ when

$$\frac{f'''}{f''} = \frac{\dot{R}H - \ddot{R}}{(\dot{R})^2}. \tag{8.238}$$

By introducing explicitly the scalar degree of freedom of metric $f(R)$ gravity $\phi \equiv f'(R)$, the effective EoS parameter becomes

$$w_{eff} = -1 + 2\,\frac{(\ddot{\phi} - H\dot{\phi})}{R\phi - f - 6H\dot{\phi}} = -1 + \frac{\kappa\,(\ddot{\phi} - H\dot{\phi})}{3\phi H^2}, \tag{8.239}$$

while

$$\rho_{eff} + P_{eff} = \frac{\ddot{\phi} - H\dot{\phi}}{\phi} = \frac{\dot{\phi}}{\phi}\frac{d}{dt}\left[\ln\left(\frac{\dot{\phi}}{a}\right)\right]. \tag{8.240}$$

and a de Sitter solution corresponds to $\dot{\phi} = f''(R)\dot{R} = 0$.

The ODEs describing spatially homogeneous and isotropic cosmologies are of fourth order in the scale factor $a(t)$. When matter is absent (a situation of interest in early time inflation or in a late universe completely dominated by $f(R)$ corrections), $a(t)$ appears only through the Hubble parameter $H \equiv \dot{a}/a$. In this situation it is convenient to adopt $H$, instead of $a$, as the dynamical variable. First, $H$ is a cosmological observable; second, the field equations (8.231) and (8.232) are of third order in $H$. This elimination of $a$ is not possible when the spatial sections are not flat or when a fluid with density $\rho(a)$ is present.

The dynamical fields of the theory are the metric $g_{\mu\nu}$ and the massive scalar degree of freedom $\phi \equiv f'(R)$. That quadratic corrections to the Hilbert-Einstein action introduce a massive scalar field was noted early on in [190, 1051, 1052, 1057, 1100, 1122], and this conclusion is relevant for any metric $f(R)$ theory [484, 590, 869]. The metric tensor contains, in principle, various degrees of freedom: spin two modes, vector and scalar modes, and all of these can be massless or massive. GR contains only a massless graviton but when nonlinear corrections depending on $R$, $R_{\mu\nu}R^{\mu\nu}$, $R_{\mu\nu\rho\sigma}R^{\mu\nu\rho\sigma}$ are included in the action, other modes raise their head. In $f(R)$ gravity these include only a massive scalar mode, which is dynamical in the metric formalism but not in the Palatini formalism.

$f(R)$ gravity can achieve cosmic acceleration through the effective equation of state parameter $w_{eff} \simeq -1$, as is well known from $R^2$-inflation. This is possible also in the late universe, and it has even been attempted to unify early inflation and late time acceleration in the context of modified gravity [70, 842, 848, 850, 852, 853]. However, modelling the late-time cosmic acceleration should not spoil the successes of the standard cosmological model which requires early inflation, a radiation era allowing Big Bang nucleosynthesis, a matter era during which matter overdensities can grow and form structures, and the present accelerated epoch leading to an uncertain future era the prediction of which is model-dependent (a de Sitter attractor solution or a Big Rip singularity are common predictions). The transitions between consecutive eras must be smooth. Smoothness may not be guaranteed in all $f(R)$ models and the radiation-matter transition, in particular, was claimed to originate problems in specific but representative $f(R)$ models, including $f = R - \mu^{2(n+1)}/R^n$, $n > 0$ [36, 40, 181, 245, 852]. However, the prototypical toy model $f(R) = R - \mu^4/R$ which reportedly could not terminate the radiation era was analyzed in detail with singular perturbation methods [431] and a sufficiently long matter era was found. In general, although a *caveat* exists about terminating the radiation era and allowing a sufficiently long matter era, one can always find choices of the function $f(R)$ which achieve the correct cosmological dynamics (or any prescribed evolutionary history) by first assigning the desired form of the scale factor $a(t)$ and then by integrating a differential equation for $f(R)$ that produces the desired scale factor (*designer* $f(R)$ *gravity*) [218, 245, 381, 478, 481, 482, 605, 606, 815, 844, 845, 852, 1029]. The result is a rather contrived form of the function $f(R)$. More important, the solution $f(R)$ is not unique [815, 1024, 1025, 1042], which shows that observational data providing information on (a segment of) the cosmic expansion history $a(t)$ cannot suffice for a reconstruction of the function $f(R)$ specifying the theory of gravity. Additional information is necessary, and it may come from the growth history of cosmological density perturbations, which depends on the theory of gravity.

As usual, analytical solutions of the equations of FLRW cosmology are rare and phase space analyses (a powerful tool in cosmology [333, 1136]) are necessary, and common in the literature, originating with pre-1998 studies of $R^2$-inflation (not limited to spatially flat FLRW spaces) [35, 246, 813, 1044]. The possibility of chaos in metric $f(R)$ gravity was discussed in [98, 99], and many dynamical system studies appeared due to the recent interest in $f(R)$ models of the cosmic acceleration [6, 36, 39–41, 261–263, 265, 274, 322, 323, 327, 409, 481, 482, 545, 546, 709, 710, 719, 846, 965].

## 8.5.2 *Perturbations*

As already remarked, the FLRW metric is an analytical solution of the field equations of most gravitational theories, hence $f(R)$ gravity, dark energy models, or other theories cannot be discriminated if only the unperturbed FLRW cosmological model is probed. From the observational point of view, this means that using only probes sensitive to the expansion history of the universe will not identify the correct theory. But there is hope, for the growth of density inhomogeneities depends on the theory of gravity and potentially provides the means to discriminate between dark energy and modified gravity. The effects of modifying gravity on the growth of structures should be visible in the CMB and in galaxy surveys [677, 681, 694, 719, 722, 1000, 1012, 1013, 1019, 1029, 1041, 1088, 1159, 1173]. Most efforts to constrain $f(R)$ gravity with CMB data [41, 52, 264, 605, 719, 720, 909, 1042, 1089, 1092, 1151] rely on specific choices of the function $f(R)$, but a few general results have also been obtained. The growth and evolution of local scalar perturbations in metric $f(R)$ gravity theories which reproduce GR at high curvatures were studied in [278, 382, 1029] by assuming a scale factor evolution typical of a $\Lambda$CDM model. Vector and tensor modes are unaffected by $f(R)$ corrections. The well known result that $f''(R) > 0$ is required for the stability of scalar perturbations is also recovered [1029]. The qualitative effects of correcting the Hilbert-Einstein action include lowering the large angle anisotropy of the CMB (perhaps helping to obtain the desired low quadrupole reported by the observations), and different correlations between the CMB and galaxy surveys than in GR [1029]. Currently, we are at a stage in which the viability of $f(R)$ gravity in comparison with the $\Lambda$CDM model begins to be challenged, but more precise observations are needed [18, 114, 116, 707, 1173], while previous approximations (*e.g.*, [1172]) are criticized and refined. For example, in [382] matter inhomogeneities are studied in the longitudinal gauge using a full fourth order equation for the density contrast $\delta\rho/\rho$ (this reduces to a second order one only for sub-horizon modes). The quasi-static approximation invalid for general forms of the function $f(R)$ is found to hold for physically motivated choices of this function. The relation between the gravitational potentials appearing in the metric and responsible for gravitational lensing and the matter overdensities depends on the theory of gravity [1173].

Density inhomogeneities generated in the Palatini formalism have been studied in [30,278,681–683,714,715,721,1097]: two different formalisms developed in [619, 683] and [756] were compared for the specific model $f(R) = R - \mu^{2(n+1)}/R^n$ and it is was found that the two models agree for scenarios that are "close" in parameter space to the standard concordance model, but give different results for models that differ significantly from the $\Lambda$CDM model. However, Palatini $f(R)$ gravity in its present form is not viable and these studies do not seem applicable to "cured" versions of the Palatini formalism which will necessarily include extra degrees of freedom and extra modes in the solutions of the evolution equations for density perturbations.

## 8.6 Conclusions

We begun our discussion by reviewing the main pieces of evidence pointing to the need for theories of gravity alternative to GR and incorporating new features. The latter include extra degrees of freedom, such as scalar fields, and novel couplings of these fields to gravity (in the Jordan frame) or to matter (in the Einstein frame). Higher order corrections to the Hilbert-Einstein action also appear naturally.

As soon as one leaves GR, many new possibilities appear. For example, the metric and Palatini variations originate different field equations, the extra fields introduced bring with them the freedom of choosing coupling functions and potentials, *etc.* The weak-field limit of these alternative gravities provides too little information on how to choose these functions. As we have seen, cosmology provides an important arena for the study of alternative theories and the discovery of the present acceleration of the universe has renewed the interest in modifications of GR as a possible way to avoid the introduction of an *ad hoc* dark energy. Again, the modifications of gravity lead to an enormous variety of possible models. Being introduced to explain some modern observations (the luminosity distance versus redshift relation of type Ia supernovae, the CMB data, *etc.*), these models must fit the observational data, however too many models satisfy this requirement, which is not sufficient to weed out competing candidates. It is, therefore, necessary that theory provide guidelines to direct the exploration of new models for gravity. It has been proved very useful to examine closely various criteria for the viability of a given theory. They include possessing Newtonian and post-Newtonian limits compatible with Solar System and terrestrial experiments; being ghost-free; admitting a well-posed Cauchy problem; producing all the cosmological eras needed for a successful cosmological model from the early universe to the present accelerated era, with smooth transitions between these different epochs; local and global stability; compatibility with the Standard Model of particle physics and with stellar evolution, *etc*. Over the years, the study of these criteria has led to the formulation of the PPN formalism, cosmography, to an understanding of the Palatini formalism, of attractor mechanisms toward GR in cosmology, of the chameleon mechanism for gravity, and of other interesting issues.

Ultimately, neither theoretical criteria not fitting the data can select a unique theory or class of theories and we are always presented with a degeneracy of models. Researchers often argue in favor of this or that theory on the basis of a particular criterion being more important than others, but this kind of argument invariably reveals itself to be a matter of taste and not compelling.

To conclude our excursion into the territory that lies beyond Einstein's GR, it is fair to say that we do not have a complete view of this territory but, rather, we have just taken a glimpse and charted only a small part of it. As soon as new gravitational degrees of freedom are excited or new couplings introduced, the phenomenology becomes richer and the situation unfamiliar and more complex. Given our very limited perspective on this new territory, it would be presumptuous to claim that this or that theory is validated by cosmology, theoretical arguments, or data. The theories studied thus far, most notably scalar-tensor and $f(R)$ gravity, can work as toy models to satisfy many pressing theoretical and observational needs, but they should be seen more as temporary models, proofs of principle that modifying gravity satisfies these needs and avoids dark energy and dark matter, rather than final theories. The possibility that gravity is not fundamental but emerges in the same way that fluid mechanics and thermodynamics describe collective averaged properties of microscopic constituents of matter, should also be kept in mind at all times.

Progress in gravitational physics is slow due to the scarcity of experimental data and recent developments have largely been stimulated by observational progress in cosmology. We hope to have made clear that our current knowledge of gravity is extremely limited and that we cannot rest satisfied with Einstein's theory. Although it is a milestone of twentieth century physics and a major intellectual achievement of mankind, GR is almost a century old and was created before quantum mechanics was fully developed. The fact that it is not yet fully understood does not diminish the need to go beyond it and to understand how things could be different (and, most likely, how they *are* different) at energy and spatial scales not fully explored. The recognition of this state of affairs in our present knowledge is the first step needed to make progress.

# Appendix A
# Physical constants and astrophysical and cosmological parameters

Digits in parenthesis denote the 1-$\sigma$ uncertainty in the previous two digits. The values are taken from Ref. [1169] and the Particle Data Group tables [42].

## A.1 Physical constants

Speed of light *in vacuo* $c = 2.99792458 \cdot 10^{10}$ cm/s

gravitational constant $G = 6.67259(85) \cdot 10^{-8}\,\mathrm{cm}^3 \cdot \mathrm{g}^{-1} \cdot \mathrm{s}^{-2}$

Planck constant $h = 6.6260755(40) \cdot 10^{-27}\,\mathrm{erg} \cdot \mathrm{s}$

reduced Planck constant $\hbar \equiv \frac{h}{2\pi} = 1.05457266(63) \cdot 10^{-27}\,\mathrm{erg} \cdot \mathrm{s}$

Boltzmann constant $K_B = 1.380658(12) \cdot 10^{-16}\,\mathrm{erg/K}$

Stefan-Boltzmann constant $\sigma = 5.67051(19) \cdot 10^{-5}\,\mathrm{erg} \cdot \mathrm{cm}^{-2} \cdot \mathrm{s}^{-1} \cdot \mathrm{K}^{-4}$

electron mass $m_e = 9.1093897(54) \cdot 10^{-28}\,\mathrm{g} \simeq 511.0\,\mathrm{keV}$

proton mass $m_p = 1.6726231(10) \cdot 10^{-24}\,\mathrm{g} \simeq 938.3\,\mathrm{MeV}$

atomic mass unit $1\,\mathrm{a.m.u.} = 1.6605402(10) \cdot 10^{-24}\,\mathrm{g} \simeq 931.5\,\mathrm{MeV}$

fine structure constant $\alpha = 7.29735308(33) \cdot 10^{-3} \approx \frac{1}{137}$

Compton wavelength of the electron $\lambda_c = 2.426 \cdot 10^{-10}$ cm.

## A.2 Conversion factors

Armstrong: $1\,\text{Å} = 10^{-8}$ cm $= 10^{-10}$ m

Fermi: $1$ fm $= 10^{-13}$ cm $= 10^{-15}$ m

$1$ erg $= 10^{-7}$ J

$1$ eV $= 1.602177 \cdot 10^{-19}$ J $= 1.602177 \cdot 10^{-12}$ erg

Astronomical Unit $1$ A.U. $= 1.496 \cdot 10^{13}$ cm

light year: $1$ ly $= 9.46073 \cdot 10^{17}$ cm

parsec: $1$ pc $= 30.85678 \cdot 10^{17}$ cm$=$ $3.2616$ ly

## A.3 Astrophysical and cosmological quantities

Standard acceleration of gravity $g = 9.806 \cdot 10^{2}\,\text{cm/s}^2$

mass of the Sun $M_\odot = 1.989 \cdot 10^{33}$ g

mass of the Earth $M_e = 5.978 \cdot 10^{27}$ g

average radius of the Earth $R_e = 6.370 \cdot 10^{8}$ cm

radius of the Sun $R_\odot = 6.96 \cdot 10^{10}$ cm

Chandrasekhar mass (upper limit to the mass of a white dwarf) $M_{Ch} \simeq 1.46 M_\odot$

Schwarzschild radius $R_s = 2GM/c^2 = 3\,(M/M_\odot)$ km

Schwarzschild black hole temperature $T = \dfrac{\hbar c^3}{8\pi G K_B M} \simeq 10^{-7}\,(M_\odot/M)\,\text{K}$

Schwarzschild black hole evaporation time $\tau = 10^{66}\,(M/M_\odot)^3$ yr

Hubble parameter $H_0 = 72 \pm 3\,\text{km} \cdot \text{s}^{-1} \cdot \text{Mpc}^{-1}$

reduced Hubble parameter $h = H_0/\left(100\,\text{km} \cdot \text{s}^{-1} \cdot \text{Mpc}^{-1}\right)$

total matter density parameter $\Omega_m$: $\Omega_m h^2 = 0.133 \pm 0.006$

baryon density parameter $\Omega_b$: $\Omega_b h^2 = 0.0227 \pm 0.0006$

radiation density parameter $\Omega_r$: $\Omega_r h^2 = 2.47 \cdot 10^{-5}$

cosmological constant density parameter $\Omega_\Lambda$: $\Omega_\Lambda = 0.74 \pm 0.03$

spectral index of density perturbations $n_S = 0.963^{+0.014}_{-0.015}$

tensor to scalar ratio for perturbations $r < 0.43$ (95% confidence level).

## A.4 Planck scale quantities

Planck length $l_{Pl} = \sqrt{\frac{\hbar G}{c^3}} = 1.6 \cdot 10^{-33}$ cm

Planck mass $m_{Pl} = \sqrt{\frac{\hbar c}{G}} = 2.2 \cdot 10^{-5}$ g

Planck time $t_{Pl} = \frac{l_{Pl}}{c} = 5.4 \cdot 10^{-44}$ s

Planck energy $E_{Pl} = m_{Pl} c^2 = 2.0 \cdot 10^{16}$ erg $= 1.3 \cdot 10^{19}$ GeV

Planck temperature $T_{Pl} = \frac{E_{Pl}}{K_B} = 1.4 \cdot 10^{32}$ K

Planck mass density $\rho_{Pl} = \frac{c^2}{G\, l_{Pl}^2} = 5.2 \cdot 10^{93}\ \mathrm{g/cm^3}$.

# Appendix B
# The Noether symmetry approach to $f(R)$ gravity

The field equations used in the Noether symmetry approach to spherical symmetry and to FLRW cosmology in metric $f(R)$ gravity are reported below.

## B.1 The field equations and the Noether vector for spherically symmetric $f(R)$ gravity

The field equations of metric $f(R)$ gravity with spherical symmetry are

$$\begin{aligned} H_{00} &= 2A^2B^2Mf + \left\{BMA'^2 - A\left[2BA'M' + M\left(2BA'' - A'B'\right)\right]\right\} f_R \\ &\quad + \left(-2A^2MB'R' + 4A^2BM'R' + 4A^2BMR''\right) f_{RR} \\ &\quad + 4A^2BMR'^2 f_{RRR} = 0\,, \end{aligned} \tag{B.1}$$

$$\begin{aligned} H_{rr} &= 2\,A^2B^2M^2f + \left(BM^2A'^2 + AM^2A'B' + 2A^2MB'M' + 2A^2BM'^2\right. \\ &\quad \left. - 2ABM^2A'' - 4A^2BMM''\right) f_R + \left(2ABM^2A'R'\right. \\ &\quad \left. + 4A^2BMM'R'\right) f_{RR} = 0\,, \end{aligned} \tag{B.2}$$

$$\begin{aligned} H_{\theta\theta} &= 2AB^2Mf + \left(4AB^2 - BA'M' + AB'M' - 2ABM''\right) f_R \\ &\quad + \left(2BMA'R' - 2AMB'R' + 2ABM'R' + 4ABMR''\right) f_{RR} \\ &\quad + 4ABMR'^2 f_{RRR} = 0\,, \\ H_{\varphi\varphi} &= \sin^2\theta\, H_{\theta\theta} = 0\,. \end{aligned} \tag{B.3}$$

The trace of the field equations is

$$\begin{aligned} H = g^{\mu\nu}H_{\mu\nu} &= 4AB^2Mf - 2AB^2MRf_R + 3\left(BMA'R' - AMB'R'\right. \\ &\quad \left. + 2ABM'R' + 2ABMR''\right) f_{RR} + 6ABMR'^2 f_{RRR} = 0\,. \end{aligned} \tag{B.4}$$

The system (4.77) is derived from the condition for the existence of a Noether symmetry $\mathscr{L}_{\mathbf{X}}L = 0$. Considering the configuration space $\underline{q} = (A\,, M\,, R)$ and defining the Noether vector $\underline{\alpha} = (\alpha_1\,, \alpha_2\,, \alpha_3)$, the system (4.77) assumes the explicit form

$$\xi\left(\frac{\partial\alpha_2}{\partial A} f_R + M\,\frac{\partial\alpha_3}{\partial A} f_{RR}\right) = 0\,, \tag{B.5}$$

$$\frac{A}{M}\left[(2+MR)\,\alpha_3 f_{RR} - \frac{2\alpha_2}{M} f_R\right] f_R$$
$$+\xi\left[\left(\frac{\alpha_1}{M} + 2\,\frac{\partial\alpha_1}{\partial M} + \frac{2A}{M}\frac{\partial\alpha_2}{\partial M}\right) f_R + A\left(\frac{\alpha_3}{M} + 4\,\frac{\partial\alpha_3}{\partial M}\right) f_{RR}\right] = 0\,, \tag{B.6}$$

$$\xi\left(M\,\frac{\partial\alpha_1}{\partial R} + 2A\frac{\partial\alpha_2}{\partial R}\right) f_{RR} = 0\,, \tag{B.7}$$

$$\alpha_2\,(f - Rf_R)\,f_R - \xi\left[\left(\alpha_3 + M\,\frac{\partial\alpha_3}{\partial M} + 2A\,\frac{\partial\alpha_3}{\partial A}\right) f_{RR}\right.$$
$$\left.+\left(\frac{\partial\alpha_2}{\partial M} + \frac{\partial\alpha_1}{\partial A} + \frac{A}{M}\frac{\partial\alpha_2}{\partial A}\right) f_R\right] = 0\,, \tag{B.8}$$

$$[M\,(2+MR)\,\alpha_3 f_{RR} - 2\alpha_2 f_R]\,f_{RR} + \xi\left[f_R\,\frac{\partial\alpha_2}{\partial R}\right.$$
$$\left.+\left(2\alpha_2 + M\,\frac{\partial\alpha_1}{\partial A} + 2A\,\frac{\partial\alpha_2}{\partial A} + M\,\frac{\partial\alpha_3}{\partial R}\right)\alpha_3 f_{RR} + M f_{RRR}\right] = 0\,, \tag{B.9}$$

$$2A\,[(2+MR)\,\alpha_3 f_{RR} - (f - Rf_R)\,\alpha_2]\,f_{RR} + \xi\left[\left(\frac{\partial\alpha_1}{\partial R} + \frac{A}{M}\frac{\partial\alpha_2}{\partial R}\right) f_R\right.$$
$$\left.+\left(2\alpha_1 + 2A\,\frac{\partial\alpha_3}{\partial R} + M\,\frac{\partial\alpha_1}{\partial M} + 2A\,\frac{\partial\alpha_2}{\partial M}\right) f_{RR} + 2A\,\alpha_3 f_{RRR}\right] = 0\,, \tag{B.10}$$

with the condition $\xi = (2+MR)\,f_R - Mf \neq 0$ to guarantee the non-vanishing of the Hessian of the Lagrangian (4.69).

## B.2 Noether symmetries in metric $f(R)$ cosmology

Here it is shown that Eq. (6.208) is indeed a Noether solution with $K = 0$ and $\mu_0 = 0$. First, from $H(a)$ given by

$$H^2 = -\frac{4d_1 d_2(-c_3)^{9/2}}{a^4} + \frac{24 d_1 d_2(-c_3)^{7/2}}{a^2} + \frac{\rho_{0m} d_2(-c_3)^{5/2}}{a^4}$$
$$-36 d_1 d_2(-c_3)^{5/2}$$

$$+\frac{2\sqrt{3}\rho_{r0}\tanh^{-1}\left(\frac{\sqrt{3}a}{\sqrt{-c_3}}\right)d_2c_3^2}{a^4}+\frac{10\rho_{r0}d_2(-c_3)^{3/2}}{a^3}$$

$$+\frac{12\sqrt{3}\rho_{r0}\tanh^{-1}\left(\frac{\sqrt{3}a}{\sqrt{-c_3}}\right)d_2c_3}{a^2}-\frac{18\rho_{r0}d_2\sqrt{-c_3}}{a}$$

$$+18\sqrt{3}\rho_{r0}\tanh^{-1}\left(\frac{\sqrt{3}a}{\sqrt{-c_3}}\right)d_2\,, \tag{B.11}$$

we calculate the expression of $R(a)$

$$R = 6\left(2H^2 + a\,H\,H'\right)$$

$$= -\frac{144d_1d_2(-c_3)^{7/2}}{a^2}+432d_1d_2(-c_3)^{5/2}-\frac{48d_2\rho_{r0}(-c_3)^{3/2}}{a^3}$$

$$-\frac{72\sqrt{3}d_2\rho_{r0}\tanh^{-1}\left(\frac{\sqrt{3}a}{\sqrt{-c_3}}\right)c_3}{a^2}+\frac{216d_2\rho_{r0}\sqrt{-c_3}}{a}$$

$$-216\sqrt{3}d_2\rho_{r0}\tanh^{-1}\left(\frac{\sqrt{3}a}{\sqrt{-c_3}}\right). \tag{B.12}$$

Since both $H$ and $R$ are known one finds, using Eq. (6.194),

$$f(a) = -\frac{8d_1c_3^2}{a^3}-\frac{24d_1c_3}{a}-\frac{3\rho_{r0}}{a^4}+\frac{4\sqrt{3}\rho_{r0}\tanh^{-1}\left(\frac{\sqrt{3}a}{\sqrt{-c_3}}\right)}{a^3\sqrt{-c_3}}-\frac{12\rho_{r0}}{a^2\,c_3}$$

$$-\frac{12\sqrt{3}\rho_{r0}\tanh^{-1}\left(\frac{\sqrt{3}a}{\sqrt{-c_3}}\right)}{a(-c_3)^{3/2}}. \tag{B.13}$$

The expressions of $H$, $R$, and $f$ satisfy Eq. (6.129). The system admits the constant of motion $\mu_0 = 0$, which is the Noether charge given by Eq. (6.155).

Let us consider now the equivalent scalar-tensor picture. The action

$$S = \int d^4x\sqrt{-g}\,f(R) + S^{(m)} \tag{B.14}$$

can be rewritten as

$$S = \int d^4x\sqrt{-g}\,[\phi\,R - V(\phi)] + S^{(m)}\,, \tag{B.15}$$

where $\phi = f'(R)$ and $V = Rf'(R) - f(R)\,|_{R=R(\phi)}$. Using Eqs. (B.11)–(B.13), one can write explicitly the potential

$$
\begin{aligned}
V(\phi) &= 3456 d_1 d_2^3 \phi^3 (-c_3)^{13/2} - 10368 d_2^4 \rho_{r0} \phi^4 c_3^6 \\
&-1728\sqrt{3} d_2^3 \rho_{r0} \phi^3 \tanh^{-1}\left[\frac{\sqrt{3}\left(6 d_2 \phi (-c_3)^{5/2} + \sqrt{-36 d_2^2 \phi^2 c_3^5 - c_3}\right)}{\sqrt{-c_3}}\right] c_3^4 \\
&+1728 d_2^3 \rho_{r0} \phi^3 \sqrt{-36 d_2^2 \phi^2 c_3^5 - c_3} (-c_3)^{7/2} - 288 d_1 d_2 \phi (-c_3)^{5/2} \\
&+432 d_2^2 \rho_{r0} \phi^2 c_3^2 + 288\sqrt{3} d_2^2 \rho_{r0} \phi^2 \sqrt{-36 d_2^2 \phi^2 c_3^5 - c_3} \\
&\cdot \tanh^{-1}\left[\frac{\sqrt{3}\left(6 d_2 \phi (-c_3)^{5/2} + \sqrt{-36 d_2^2 \phi^2 c_3^5 - c_3}\right)}{\sqrt{-c_3}}\right] (-c_3)^{3/2} \\
&+144\sqrt{3} d_2 \rho_{r0} \phi \tanh^{-1}\left[\frac{\sqrt{3}\left(6 d_2 \phi (-c_3)^{5/2} + \sqrt{-36 d_2^2 \phi^2 c_3^5 - c_3}\right)}{\sqrt{-c_3}}\right] \\
&-16 d_1 \sqrt{-36 d_2^2 \phi^2 c_3^5 - c_3} - \frac{96 d_2 \rho_{r0} \phi \sqrt{-36 d_2^2 \phi^2 c_3^5 - c_3}}{\sqrt{-c_3}} - \frac{9\rho_{r0}}{c_3^2} \\
&-576 d_1 d_2^2 \phi^2 \sqrt{-36 d_2^2 \phi^2 c_3^5 - c_3}\, c_3^4 + 8\sqrt{3}\rho_{r0} (-c_3)^{-5/2} \sqrt{-36 d_2^2 \phi^2 c_3^5 - c_3} \\
&\cdot \tanh^{-1}\left[\frac{\sqrt{3}\left(6 d_2 \phi (-c_3)^{5/2} + \sqrt{-36 d_2^2 \phi^2 c_3^5 - c_3}\right)}{\sqrt{-c_3}}\right]. \qquad \text{(B.16)}
\end{aligned}
$$

In order to assess whether the evolution of the background universe constitutes a viable cosmic history, it is sufficient to check whether the Hubble parameter given by Eq. (B.11) fits the data from primordial nucleosynthesis to the current accelerated era.

# Appendix C
# The weak-field limit of metric $f(R)$ gravity

Here it is shown that the harmonic gauge can be reduced to the form (5.27)–(5.30). This gauge is usually characterized by the condition $g^{\sigma\tau}\Gamma^{\mu}_{\sigma\tau} = 0$. For $\mu = 0$, one has

$$2g^{\sigma\tau}\Gamma^{0}_{\sigma\tau} \approx g^{(2)}{}^{0,0}_{\ 0} - 2g^{(3)}{}^{0,m}_{\ m} + g^{(2)}{}^{m,0}_{\ m} = 0 \tag{C.1}$$

and, for $\mu = i$,

$$2g^{\sigma\tau}\Gamma^{i}_{\sigma\tau} \approx g^{(2)}{}^{0,i}_{\ 0} + 2g^{(2)}{}^{mi}_{\ ,m} - g^{(2)}{}^{m,i}_{\ m} = 0\,. \tag{C.2}$$

By differentiating Eq. (C.1) with respect to $x^0$, $x^j$ and Eq. (C.2) with respect to $x^0$, one obtains

$$g^{(2)}{}^{0}_{\ 0,00} - 2\,g^{(3)}{}^{m}_{\ 0,0m} + g^{(2)}{}^{m}_{\ m,00} = 0\,, \tag{C.3}$$

$$g^{(2)}{}^{0}_{\ 0,0j} - 2\,g^{(3)}{}^{m}_{\ 0,jm} + g^{(2)}{}^{m}_{\ m,0j} = 0\,, \tag{C.4}$$

$$g^{(2)}{}^{0}_{\ 0,0i} + 2\,g^{(2)}{}^{m}_{\ i,0m} - g^{(2)}{}^{m}_{\ m,0i} = 0\,. \tag{C.5}$$

Combining Eqs. (C.4) and (C.5) yields

$$g^{(2)}{}^{m}_{\ m,0i} - g^{(2)}{}^{m}_{\ i,0m} - g^{(3)}{}^{m}_{\ 0,mi} = 0\,. \tag{C.6}$$

Finally, differentiating Eq. (C.2) with respect to $x^j$, one obtains

$$g^{(2)}{}^{0}_{\ 0,ij} + 2g^{(2)}{}^{m}_{\ i,jm} - g^{(2)}{}^{m}_{\ m,ij} = 0 \tag{C.7}$$

and redefining the indices as $(i, j) \to (j, i)$, we get

$$g^{(2)}{}^{0}_{\ 0,ij} + 2g^{(2)}{}^{m}_{\ j,im} - g^{(2)}{}^{m}_{\ m,ij} = 0\,. \tag{C.8}$$

Combining Eqs. (C.7) and (C.8) now yields

$$g^{(2)0}{}_{0,ij} + g^{(2)m}{}_{i,jm} + g^{(2)m}{}_{j,im} - g^{(2)m}{}_{m,ij} = 0\,. \tag{C.9}$$

The relations (C.3), (C.6), and (C.9) guarantee the viability of (5.27)–(5.30).

# References


1. B. Abbott *et al.*, *Phys. Rev. D* **72**, 042002 (2005).
2. B. Abbott *et al.*, *Phys. Rev. D* **76**, 082003 (2007).
3. L.F. Abbott, *Nucl. Phys. B* **185**, 233 (1981).
4. L.F. Abbott, M.B. Wise, *Nucl. Phys. B* **244**, 541 (1984).
5. T. Abbott *et al.* [Dark Energy Survey Collaboration], arXiv:astro-ph/0510346.
6. M. Abdelwahab, S. Carloni, P.K.S. Dunsby, *Class. Quantum Grav.* **25**, 135002 (2008).
7. L.R. Abramo, R.P. Woodard, *Phys. Rev. D* **65**, 063516 (2002).
8. L.R. Abramo, L. Brenig, E. Gunzig, A. Saa, *Phys. Rev. D* **67**, 027301 (2003).
9. L.R. Abramo, L. Brenig, E. Gunzig, A. Saa, *Int. J. Theor. Phys.* **42**, 1145 (2003).
10. M. Abramowitz, I.A. Stegun, *Handbook of Mathematical Functions* (Dover, New York 1972).
11. J.P. Abreu, P. Crawford, J.P. Mimoso, *Class. Quantum Grav.* **11**, 1919 (1994).
12. F.S. Accetta, J.J. Trester, *Phys. Rev. D* **39**, 2854 (1989).
13. F.S. Accetta, L.M. Krauss, P. Romanelli, *Phys. Lett. B* **248**, 146 (1990).
14. A.J. Accioly, S. Ragusa, E.C. de Rey Neto, H. Mukai, *Nuovo Cimento B* **114**, 595 (1999).
15. A.J. Accioly, A.N. Vaidya, M.M. Som, *Phys. Lett. A* **138**, 95 (1983).
16. A.J. Accioly, A.N. Vaidya, M.M. Som, *Phys. Rev. D* **27**, 2282 (1983).
17. F. Acernese *et al.*, *Class. Quantum Grav.* **24**, S381 (2007).
18. N. Agarwal, R. Bean, *Class. Quantum Grav.* **25**, 165001 (2008).
19. M. Akbar, R.-G. Cai, *Phys. Lett. B* **635**, 7 (2006).
20. A. Albrecht *et al.*, *Dark Energy Task Force Final Report*, FERMILAB-FN-0793-A, arXiv:astro-ph/0609591.
21. J.S. Alcaniz, D. Jain, A. Dev, *Phys. Rev. D* **66**, 067301 (2002).
22. J.S. Alcaniz, D. Jain, A. Dev, *Phys. Rev. D* **67**, 043514 (2003).
23. G. Aldering *et al.* [SNAP collaboration], SLAC-PUB-11420, arXiv:astro-ph/0405232; see also `snap.lbl.gov`.
24. G. Allemandi, A. Borowiec, M. Francaviglia, *Phys. Rev. D* **70**, 103503 (2004).
25. G. Allemandi, M. Francaviglia, M.L. Ruggiero, A. Tartaglia, *Gen. Relat. Gravit.* **37**, 1891 (2005).
26. G. Allemandi, M. Capone, S. Capozziello, M. Francaviglia, *Gen. Relat. Gravit.* **38**, 33 (2006).
27. B. Allen, *Nucl. Phys. B* **226**, 232 (1983).
28. R.A. Alpher, R. Herman, *Rev. Mod. Phys.* **22**, 153 (1950).
29. R.A. Alpher, H.A. Bethe, G. Gamow, *Phys. Rev.* **73**, 803 (1948).
30. M. Amarzguioui, O. Elgaroy, D.F. Mota, T. Multamaki, *Astron. Astrophys.* **454**, 707 (2006).
31. L. Amendola, *Mon. Not. R. Astron. Soc.* **312**, 521 (2000).
32. L. Amendola, *Phys. Rev. D* **62**, 043511 (2000).
33. L. Amendola, *Phys. Rev. Lett.* **86**, 196 (2001).
34. L. Amendola, A. Battaglia-Mayer, S. Capozziello, S. Gottlöber, V. Müller, F. Occhionero, H.-J. Schmidt, *Class. Quantum Grav.* **10**, L43 (1993).

35. L. Amendola, S. Capozziello, M. Litterio, F. Occhionero, *Phys. Rev. D* **45**, 417 (1992).
36. L. Amendola, R. Gannouji, D. Polarski, S. Tsujikawa, *Phys. Rev. D* **75**, 083504 (2007).
37. L. Amendola, M. Litterio, F. Occhionero, *Phys. Lett. B* **231**, 43 (1989).
38. L. Amendola, M. Litterio, F. Occhionero, *Int. J. Mod. Phys. A* **5** 3861 (1990).
39. L. Amendola, D. Polarski, S. Tsujikawa, *Int. J. Mod. Phys. D* **16**, 1555 (2007).
40. L. Amendola, D. Polarski, S. Tsujikawa, *Phys. Rev. Lett.* **98**, 131302 (2007).
41. L. Amendola, S. Tsujikawa, *Phys. Lett. B* **660**, 125 (2008).
42. C. Amsler *et al.* (Particle Data Group), *Phys. Lett. B* **667**, 1 (2008).
43. L.A. Anchordoqui, S.E. Perez–Bergliaffa, M.L. Trobo, G.S. Birman, *Phys. Rev. D* **57**, 829 (1998).
44. J.D. Anderson *et al.*, *Phys. Rev. D* **65**, 082004 (2002).
45. J.D. Anderson *et al.*, *Phys. Rev. Lett.* **81**, 2858 (1998).
46. J.L. Anderson, *Phys. Rev. D* **3**, 1689 (1971).
47. M. Ando *et al.*, *Class. Quantum Grav.* **19**, 1615 (2002).
48. S. Andreon, *Astron. Astrophys.* **409**, 37 (2003).
49. S. Andreon, C. Lobo, A. Iovino, *Mont. Not. R. Astron. Soc.* **349**, 889 (2004).
50. I. Antoniadis, C. Bachas, J. Ellis, D.V. Nanopoulos, *Phys. Lett. B* **211**, 393 (1988).
51. T. Appelquist, A. Chodos, P.G.O. Freund, *Modern Kaluza-Klein Theories* (Addison-Wesley, Reading, Massachusetts, 1987).
52. S.A. Appleby, RA. Battye, *J. Cosmol. Astropart. Phys.* **0805**, 019 (2008).
53. V.I. Arnold, *Mathematical Methods of Classical Mechanics* (Springer, Berlin, 1978).
54. R. Arnowitt, S. Deser, C.W. Misner, in *Gravitation: An Introduction to Current Research*, ed. by L. Witten (Wiley, New York, 1962).
55. A. Ashtekar, M. Bojowald, J. Lewandowski, *Adv. Theor. Math. Phys.* **7**, 233 (2003).
56. A. Ashtekar, A. Corichi, *Class. Quantum Grav.* **20**, 4473 (2003).
57. A. Ashtekar, B. Krishnan, *Living Rev. Relativity* **7**, 10 (2004).
58. A. Ashtekar, B. Krishnan, *Phys. Rev. D* **68**, 104030 (2003).
59. A. Ashtekar, B. Krishnan, *Phys. Rev. Lett.* **89**, 261101 (2002).
60. P. Astier *et al.*, *Astron. Astrophys.* **447**, 31 (2006).
61. E. Ayon-Beato, C. Martinez, R. Troncoso, J. Zanelli, *Phys. Rev. D* **71**, 104037 (2005).
62. E. Ayon-Beato, C. Martinez, J. Zanelli, *Gen. Relat. Gravit.* **38**, 145 (2006).
63. D. Babusci, L. Baciotti, F. Fucito, A. Nagar, *Phys. Rev. D* **64**, 062001 (2001).
64. C. Baccigalupi, A. Balbi, S. Matarrese, F. Perrotta, N. Vittorio, *Nucl. Phys. Proc. Suppl.* **124**, 68 (2003).
65. C. Baccigalupi, S. Matarrese, F. Perrotta, *Phys. Rev. D* **62**, 123510 (2000).
66. S. Baghram, M. Farhang, S. Rahvar, *Phys. Rev. D* **75**, 044024 (2007).
67. N.A. Bahcall, in *Formation of Structure in the Universe*, 13th Jerusalem Winter School in Theoretical Physics, 1995, arXiv:astro-ph/9611148.
68. N.A. Bahcall, J.P. Ostriker, S. Perlmutter, P.J. Steinhardt, *Science* **284**, 1481 (1999).
69. D. Bailin, A. Love, *Rep. Prog. Phys.* **50**, 1087 (1987).
70. K. Bamba, S.D. Odintsov, *J. Cosmol. Astropart. Phys.* **0804**, 024 (2008).
71. A.J. Banday *et al.*, *Astrophys. J.* **475**, 393 (1997).
72. A. Banerjee, S.B. Dutta Choudhury, N. Banerjee, *J. Math. Phys.* **25**, 3041 (1984).
73. A. Banerjee, N.O. Santos, *J. Math. Phys.* **22**, 1075 (1981).
74. N. Banerjee, R.K. Jain, D.P. Jaktar, *Gen. Relat. Gravit.* **40**, 93 (2008).
75. N. Banerjee, S. Sen, *Phys. Rev. D* **56**, 1334 (1997).
76. A.B. Baptista, J.C. Fabris, S.V.B. Gonçalves, *Astr. Sp. Sci.* **246**, 315 (1997).
77. E. Barausse, T.P. Sotiriou, *Phys. Rev. Lett.* **101**, 099001 (2008).
78. E. Barausse, T.P. Sotiriou, J. Miller, *Class. Quantum Grav.* **25**, 062001 (2008).
79. E. Barausse, T.P. Sotiriou, J. Miller, *Class. Quantum Grav.* **25**, 105008 (2008).
80. E. Barausse, T.P. Sotiriou, J. Miller, *EAS Publ. Ser.* **30**, 189 (2008), arXiv:0801.4852.
81. G.A. Barber, arXiv:gr-qc/0302088.
82. C. Barcelo, S. Liberati, M. Visser, *Living Rev. Relativity* **8**, 12 (2005).
83. C. Barcelo, M. Visser, *Class. Quantum Grav.* **17**, 3843 (2000).
84. J.M. Bardeen, *Phys. Rev. D* **22**, 1882 (1980).

85. J.M. Bardeen, B. Carter, S.W. Hawking, *Comm. Math. Phys.* **31**, 161 (1973).
86. D.E. Barraco, V.H. Hamity, *Phys. Rev. D* **62**, 044027 (2000).
87. B.J. Barris *et al.*, *Astrophys. J.* **602**, 571 (2004).
88. A. Barros, C. Romero, *Phys. Lett. A* **245**, 31 (1998).
89. A. Barroso, J. Casasayas, P. Crawford, P. Moniz, A. Nunes, *Phys. Lett. B* **275**, 264 (1992).
90. J.D. Barrow, *Nature* **272**, 211 (1978).
91. J.D. Barrow, *Nucl. Phys. B* **310**, 743 (1988).
92. J.D. Barrow, *Phys. Lett. B* **187**, 12 (1987).
93. J.D. Barrow, *Phys. Lett. B* **249**, 406 (1990).
94. J.D. Barrow, *Phys. Rev. D* **47**, 5329 (1993).
95. J.D. Barrow, *Phys. Rev. D* **51**, 2729 (1994).
96. J.D. Barrow, T. Clifton, *Class. Quantum Grav.* **23**, L1 (2006).
97. J.D. Barrow, S. Cotsakis, *Phys. Lett. B* **214**, 515 (1988).
98. J.D. Barrow, S. Cotsakis, *Phys. Lett. B* **232**, 172 (1989).
99. J.D. Barrow, S. Cotsakis, *Phys. Lett. B* **258**, 299 (1991).
100. J.D. Barrow, K. Maeda, *Nucl. Phys. B* **341**, 294 (1990).
101. J.D. Barrow, J.P. Mimoso, *Phys. Rev. D* **50**, 3746 (1994).
102. J.D. Barrow, J.P. Mimoso, M.R. de Garcia Maia, *Phys. Rev. D* **48**, 3630 (1993).
103. J.D. Barrow, A.C. Ottewill, *J. Phys. A* **16**, 2757 (1983).
104. J.D. Barrow, P. Parsons, *Phys. Rev. D* **55**, 1906 (1997).
105. J.D. Barrow, P. Saich, *Phys. Lett. B* **249**, 406 (1990).
106. J.D. Barrow, F.J. Tipler, *The Anthropic Cosmological Principle* (Clarendon Press, Oxford, 1986).
107. N.H. Barth, S.M. Christensen, *Phys. Rev. D* **28**, 1876 (1983).
108. D.F. Bartlett, *Am. J. Phys.* **62**, 8, (1994).
109. D.F. Bartlett, *Int. J. Mod. Phys. A* **16S1B**, 680 (2001).
110. B.A. Bassett, M. Kunz, D. Parkinson, C. Ungarelli, *Phys. Rev. D* **68**, 043504 (2003).
111. B.A. Bassett, S. Liberati, *Phys. Rev. D* **58**, 021302 (1998).
112. M. Bateman, *Proc. Lon. Math. Soc.* **8**, 223 (1910).
113. A. Battaglia-Mayer, H.-J. Schmidt, *Class. Quantum Grav.* **10**, 2441 (1993).
114. R. Bean, D. Bernat, L. Pogosian, A. Silvestri, M. Trodden, *Phys. Rev. D* **75**, 064020 (2007).
115. R. Bean, J. Magueijo, *Phys. Lett. B* **517**, 177 (2001).
116. R. Bean, M. Tangmatithan, *Phys. Rev. D* **81**, 083534 (2010).
117. A. Beesham, *Mod. Phys. Lett. A* **13**, 805 (1998).
118. J.D. Bekenstein, *Ann. Phys.* **82**, 535 (1974).
119. J.D. Bekenstein, *Lett. Nuovo Cimento* **4**, 737 (1972).
120. J.D. Bekenstein, *Phys. Rev. D* **7**, 2333 (1973).
121. J.D. Bekenstein, *Phys. Rev. D* **9**, 3292 (1974).
122. V.A. Belinsky, L.P. Grishchuk, I.M. Khalatnikov, Ya.B. Zel'dovich, *Phys. Lett. B* **155**, 232 (1985).
123. V.A. Belinsky, I.M. Khalatnikov, *Sov. Phys. JETP* **36**, 591 (1973).
124. S. Bellucci, S. Capozziello, M. De Laurentis, V. Faraoni, *Phys. Rev. D* **79**, 104004 (2009).
125. K. Benabed, F. Bernardeau, *Phys. Rev. D* **64**, 083501 (2001).
126. C.L. Bennett *et al.*, *Astrophys. J. Lett.* **464**, L1 (1996).
127. M.C. Bento, O. Bertolami, N.C. Santos, *Phys. Rev. D* **65**, 067301 (2002).
128. P.G. Bergmann, *Int. J. Theor. Phys.* **1**, 25 (1968).
129. A.L. Berkin, K. Maeda, *Phys. Lett. B* **245**, 348 (1990).
130. O. Bertolami *et al.*, arXiv:0704.1733.
131. O. Bertolami, C.G. Böhmer, F.S.N. Lobo, *Phys. Rev. D* **75**, 104016 (2007).
132. O. Bertolami, J. Martins, *Phys. Rev. D* **61**, 064007 (2000).
133. B. Bertotti, L. Iess, P. Tortora, *Nature* **425**, 374 (2003).
134. A. Bhadra, K. Sarkar, *Gen. Relat. Gravit.* **37**, 2189 (2005).
135. K.S. Bhamra, *Austral. J. Phys.* **27**, 541 (1974).
136. G.V. Bicknell, A.H. Klotz, *J. Phys. A* **9**, 1647 (1976).
137. N. Bilić, B. Guberina, R. Horvat, H. Nikolić, H. Stefancić, *Phys. Lett. B* **657**, 232 (2007).

138. A. Billyard, A.A. Coley, *Mod. Phys. Lett. A* **12**, 2121 (1997).
139. A. Billyard, A.A. Coley, J. Ibanez, *Phys. Rev. D* **59** 023507 (1999).
140. P. Binètruy, *Int. J. Theor. Phys.* **39**, 1859 (2000).
141. J. Binney, S. Tremaine, *Galactic Dynamics* (Princeton University Press, Princeton, 1987).
142. G.D. Birkhoff, *Relativity and Modern Physics* (Harvard University Press, Cambridge, USA, 1923), p. 253.
143. N.D. Birrell, P.C.W. Davies, *Phys. Rev. D* **22**, 322 (1980).
144. N.D. Birrell, P.C.W. Davies, *Quantum Fields in Curved Space* (Cambridge University Press, Cambridge, 1982).
145. J.P. Blakeslee *et al.*, *Astrophys. J.* **596**, L143 (2003).
146. S. Blanco, G. Domenech, C. El Hasi, O.A. Rosso, *Gen. Relat. Gravit.* **26**, 1131 (1994).
147. A. Blasi, N. Maggiore, S.P. Sorella, L.C.Q. Vilar, *Phys. Rev. D* **59**, 121701 (1999).
148. C.G. Boehmer, T. Harko, F.S.N. Lobo, *Astropart. Phys.* **29**, 386 (2008).
149. B. Boisseau, G. Esposito-Farèse, D. Polarski, A.A. Starobinsky, *Phys. Rev. Lett.* **85**, 2236 (2000).
150. M. Bojowald, *Living Rev. Relativity* **8**, 11 (2005).
151. M. Bojowald, *Living Rev. Relativity* **11**, 4 (2008).
152. M. Bojowald, *Nature Phys.* **3N8**, 523 (2007).
153. M. Bojowald, *Sci. Am.* **299N4**, 28 (2008).
154. M. Bojowald, H. Hernandez, M. Kagan, P. Singh, A. Skirzewski, *Phys. Rev. Lett.* **98**, 031301 (2007).
155. L. Bombelli, F. Lombardo, M. Castagnino, *J. Math. Phys.* **39**, 6040 (1998).
156. A. Bonanno, *Phys. Rev. D* **52**, 969 (1995).
157. A. Bonanno, D. Zappalá, *Phys. Rev. D* **55**, 6135 (1997).
158. N. Bonasia, M. Gasperini, *Phys. Rev. D* **71**, 104020 (2005).
159. H. Bondi, *Cosmology* (Cambridge University Press, Cambridge, 1952).
160. H. Bondi, *Mon. Not. R. Astron. Soc.* **107**, 410 (1947).
161. I. Booth, *Can. J. Phys.* **83**, 1073 (2005).
162. R.G. Bower, T. Kodama, A. Terlevich, *Mont. Not. R. Astron. Soc.* 299, 1193 (1998).
163. D. Branch, G.A. Tammann, *Annu. Rev. Astron. Astrophys.* **30**, 359 (1992).
164. C.H. Brans, *Phys. Rev.* **125**, 2194 (1962).
165. C.H. Brans, R.H. Dicke, *Phys. Rev.* **124**, 925 (1961).
166. P. Brax, J. Martin, arXiv:astro-ph/0210533.
167. P. Brax, J. Martin, *Phys. Lett. B* **468**, 40 (1999).
168. P. Brax, C. van de Bruck, A.C. Davis, *Rep. Prog. Phys.* **67**, 2183 (2004).
169. I. Brevik, S. Nojiri, S.D. Odintsov, L. Vanzo, *Phys. Rev. D* **70**, 043520 (2004).
170. F. Briscese, E. Elizalde, *Phys. Rev. D* **77**, 044009 (2008).
171. K.A. Bronnikov, *J. Math. Phys.* **43**, 6096 (2002).
172. K.A. Bronnikov, *Phys. Rev. D* **64**, 064013 (2001).
173. K.A. Bronnikov, M.S. Chernakova, arXiv:gr-qc/0503025.
174. K.A. Bronnikov, M.S. Chernakova, *Gravit. Cosmol.* **11**, 305 (2005).
175. K.A. Bronnikov, M.S. Chernakova, J.C. Fabris, N. Pinto-Neto, M.E. Rodrigues, *Int. J. Mod. Phys. D* **17**, 25 (2008).
176. K.A. Bronnikov, G. Clement, C.P. Constantinidis, J.C. Fabris, *Gravit. Cosmol.* **4**, 128 (1998).
177. K.A. Bronnikov, G. Clement, C.P. Constantinidis, J.C. Fabris, *Phys. Lett. A* **243**, 121 (1998).
178. K.A. Bronnikov, C.P. Constantinidis, R.L. Evangelista, J.C. Fabris, *Int. J. Mod. Phys. D* **8**, 481 (1999).
179. K.A. Bronnikov, J.C. Fabris, N. Pinto-Neto, M.E. Rodrigues, arXiv:gr-qc/0604055.
180. K.A. Bronnikov, G.N. Shikin, *Gravit. Cosmol.* **8**, 107 (2002).
181. A.W. Brookfield, C. van de Bruck, L.M.H. Hall, *Phys. Rev. D* **74**, 064028 (2006).
182. J.R. Brownstein, J.W. Moffat, *Mont. Not. R. Astron. Soc.* **367**, 527 (2006).
183. R. Brustein, M. Gasperini, M. Giovannini, V.F. Mukhanov, G. Veneziano, *Phys. Rev. D* **51**, 6744 (1995).
184. R. Brustein, D. Gorbonos, M. Hadad, *Phys. Rev. D* **79**, 044025 (2009).
185. I.L. Buchbinder, *Fortschr. Phys.* **34**, 605 (1986).

186. I.L. Buchbinder, S.D. Odintsov, *Lett. Nuovo Cimento* **42**, 379 (1985).
187. I.L. Buchbinder, S.D. Odintsov, *Sov. J. Nucl. Phys.* **40**, 848 (1983).
188. I.L. Buchbinder, S.D. Odintsov, I. Lichzier, *Class. Quantum Grav.* **6**, 605 (1989).
189. I.L. Buchbinder, S.D. Odintsov, I. Shapiro, in *Group-Theoretical Methods in Physics*, ed. by M. Markov (Nauka, Moscow, 1986).
190. I.L. Buchbinder, S.D. Odintsov, I. Shapiro, *Effective Action in Quantum Gravity* (IOP Publishing, Bristol, 1992).
191. H. Buchdahl, *Acta Math.* **85**, 63 (1951).
192. A. Buchdahl, *Nuovo Cimento B* **12**, 269 (1972).
193. T. Buchert, *Gen. Relat. Gravit.* **32**, 105 (2000).
194. T. Buchert, M. Carfora, *Class. Quantum Grav.* **19**, 6109 (2002).
195. T. Buchert, M. Carfora, *Class. Quantum Grav.* **25** 195001 (2008).
196. S. Burles, K.M. Nollett, M.S. Turner, *Astrophys. J. Lett.* **552**, L1 (2001).
197. A.J. Bustelo, D.E. Barraco, *Class. Quantum Grav.* **24**, 2333 (2007).
198. R.-G. Cai, *Phys. Lett. B* **582**, 237 (2004).
199. R.-G. Cai, L.-M. Cao, *Nucl. Phys. B* **785**, 135 (2007).
200. R.-G. Cai, L.-M. Cao, *Phys. Rev. D* **75**, 064008 (2007).
201. R.-G. Cai, L.-M. Cao, Y.-P. Hu, *J. High Energy Phys.* **0808**, 090 (2008).
202. R.R. Caldwell, *Phys. Lett. B* **545**, 23 (2002).
203. R.R. Caldwell, M. Kamionkowski, N.N. Weinberg, *Phys. Rev. Lett.* **91**, 071301 (2003).
204. R.R. Caldwell, D. Rahul, P.J. Steinhardt, *Phys. Rev. Lett.* **80**, 1582 (1998).
205. C.G. Callan Jr., S. Coleman, R. Jackiw, *Ann. Phys. (NY)* **59**, 42 (1970).
206. C.G. Callan, R.C. Myers, M. Perry, *Nucl. Phys. B* **311**, 673 (1988).
207. F. Calogero, A. Degasperis, *Am. J. Phys.* **72**, 1202 (2004).
208. E. Calzetta, C. El Hasi, *Class. Quantum Grav.* **10**, 1825 (1993).
209. E. Calzetta, C. El Hasi, *Phys. Rev. D* **51**, 2713 (1995).
210. M. Campanelli, C.O. Lousto, *Int. J. Mod. Phys. D* **2**, 451 (1993).
211. S. Capozziello, *Int. J. Mod. Phys. D* **11**, 483 (2002).
212. S. Capozziello, V.F. Cardone, S. Carloni, A. Troisi, *Int. J. Mod. Phys. D* **12**, 1969 (2003).
213. S. Capozziello, V.F. Cardone, S. Carloni, A. Troisi, *Phys. Lett. A* **326**, 292 (2004).
214. S. Capozziello, V.F. Cardone, E. Elizalde, S. Nojiri, S.D. Odintsov, *Phys. Rev. D* **73**, 043512, (2006).
215. S. Capozziello, V.F. Cardone, E. Piedipalumbo, M. Sereno, A. Troisi, *Int. J. Mod. Phys. D* **12**, 381 (2003).
216. S. Capozziello, V.F. Cardone, A. Troisi, *J. Cosmol. Astropart. Phys.* **0608**, 001 (2006).
217. S. Capozziello, V.F. Cardone, A. Troisi, *Mont. Not. R. Astr. Soc.* **375**, 1423 (2007).
218. S. Capozziello, V.F. Cardone, A. Troisi, *Phys. Rev. D* **71**, 043503 (2005).
219. S. Capozziello, S. Carloni, A. Troisi, *Rec. Res. Dev. Astron. Astrophys.* **1**, 625 (2003).
220. S. Capozziello, R. Cianci, C. Stornaiolo, S. Vignolo, *Class. Quantum Grav.* **24**, 6417 (2007).
221. S. Capozziello, R. Cianci, C. Stornaiolo, S. Vignolo, *Int. J. Geom. Meth. Mod. Phys.* **5**, 765 (2008).
222. S. Capozziello, R. Cianci, C. Stornaiolo, S. Vignolo, *Physica Scripta* **78**, 065010 (2008).
223. S. Capozziello, C. Corda, *Int. J. Mod. Phys. D* **15**, 1119 (2006).
224. S. Capozziello, C. Corda, M. De Laurentis, *Mod. Phys. Lett. A* **22**, 1097 (2007).
225. S. Capozziello, C. Corda, M. De Laurentis, *Phys. Lett. B* **699**, 255 (2008).
226. S. Capozziello, A. De Felice, *J. Cosmol. Astropart. Phys.* **0808**, 016 (2008).
227. S. Capozziello, M. De Laurentis, M. Francaviglia, S. Mercadante, *Found. Phys.* **39** 1161 (2009).
228. S. Capozziello, M. De Laurentis, S. Nojiri, S.D. Odintsov, *Gen. Relat. Gravit.* **41**, 2313 (2009).
229. S. Capozziello, M. De Laurentis, A. Stabile, *Class. Quantum Grav.* **27**, 165008 (2010).
230. S. Capozziello, R. de Ritis, *Class. Quantum Grav.* **11**, 107 (1994).
231. S. Capozziello, R. de Ritis, *Phys. Lett. A* **177**, 1 (1993).
232. S. Capozziello, R. de Ritis, A.A. Marino, *Class. Quantum Grav.* **14**, 3243 (1997).
233. S. Capozziello, R. de Ritis, A.A. Marino *Gen. Relat. Gravit.* **30**, 1247 (1998).

234. S. Capozziello, R. de Ritis, C. Rubano, P. Scudellaro, *Riv. Nuovo Cimento* **4**, 1 (1996).
235. S. Capozziello, M. Demianski, R. de Ritis, C. Rubano, *Phys. Rev. D* **52**, 3288 (1995).
236. S. Capozziello, E. Elizalde, S. Nojiri, S.D. Odintsov, *Phys. Lett. B* **671**, 193 (2009).
237. S. Capozziello, M. Francaviglia, *Gen. Relat. Gravit.* **40**, 357 (2008).
238. S. Capozziello, L. Izzo, *Astron. Astrophys.* **490**, 31 (2008).
239. S. Capozziello, G. Lambiase, *Gen. Relat. Gravit.* **32**, 295 (2000).
240. S. Capozziello, G. Lambiase, C. Stornaiolo, *Ann. Phys. (Leipzig)*, **10**, 713 (2001).
241. S. Capozziello, G. Marmo, C. Rubano, P. Scudellaro, *Int. J. Mod. Phys. D* **6**, 491 (1997).
242. S. Capozziello, P. Martin-Moruno, C. Rubano, *Phys. Lett. B* **689**, 117 (2010).
243. S. Capozziello, S. Nesseris, L. Perivolaropoulos, *J. Cosmol. Astropart. Phys.* **0712**, 009 (2007).
244. S. Capozziello, S. Nojiri, S.D. Odintsov, *Phys. Lett. B* **634**, 93 (2006).
245. S. Capozziello, S. Nojiri, S.D. Odintsov, A. Troisi, *Phys. Lett. B* **639**, 135 (2006).
246. S. Capozziello, F. Occhionero, L. Amendola, *Int. J. Mod. Phys. D* **1**, 615 (1993).
247. S. Capozziello, A. Stabile, *Class. Quantum Grav.* **26**, 085019 (2009).
248. S. Capozziello, A. Stabile, A. Troisi, *Class. Quantum Grav.* **24**, 2153 (2007).
249. S. Capozziello, A. Stabile, A. Troisi, *Class. Quantum Grav.* **25**, 085004 (2008).
250. S. Capozziello, A. Stabile, A. Troisi, *Mod. Phys. Lett. A* **21**, 2291 (2006).
251. S. Capozziello, A. Stabile, A. Troisi, *Mod. Phys. Lett. A* **24**, 659 (2009).
252. S. Capozziello, A. Stabile, A. Troisi, *Phys. Rev. D* **76**, 104019 (2008).
253. S. Capozziello, A. Troisi, *Phys. Rev. D*, **72**, 044022 (2005).
254. S. Capozziello, S. Tsujikawa, *Phys. Rev. D* **77**, 107501 (2008).
255. S. Capozziello, S. Vignolo, *Class. Quantum Grav.* **26**, 168001 (2009).
256. S. Capozziello, S. Vignolo, *Class. Quantum Grav.* **26**, 175013 (2009).
257. S. Capozziello, S. Vignolo, *Int. J. Geom. Meth. Mod. Phys.* **6**, 985 (2009).
258. V.F. Cardone, S. Capozziello, V. Re, E. Piedipalumbo, *Astron. Astrophys.* **382**, 792 (2002).
259. V.F. Cardone, A. Troisi, S. Capozziello, *Phys. Rev. D* **69**, 083517 (2004).
260. G.L. Cardoso, B.A. Ovrut, *Int. J. Mod. Phys. D* **3**, 215 (1994).
261. S. Carloni, S. Capozziello, J.A. Leach, P.K.S. Dunsby, *Class. Quantum Grav.* **25**, 035008 (2008).
262. S. Carloni, P.K.S. Dunsby, *J. Phys. A* **40**, 6919 (2007).
263. S. Carloni, P.K.S. Dunsby, S. Capozziello, A. Troisi, *Class. Quantum Grav.* **22**, 4839 (2005).
264. S. Carloni, P.K.S. Dunsby, A. Troisi, *Phys. Rev. D* **77**, 024024 (2008).
265. S. Carloni, A. Troisi, P.K.S. Dunsby, *Gen. Relat. Gravit.* **41**, 2313 (2009).
266. B. Caron *et al.*, *Class. Quantum Grav.* **14**, 1461 (1997).
267. M. Carrera, D. Giulini, *Phys. Rev. D* **81**, 043521 (2010).
268. M. Carrera, D. Giulini, *Rev. Mod. Phys.* **82**, 169 (2010).
269. S.M. Carroll, arXiv:astro-ph/0107571.
270. S.M. Carroll, arXiv:gr-qc/9712019.
271. S.M. Carroll, *Living Rev. Relativity* **4**, 1 (2001).
272. S.M. Carroll, *Phys. Rev. Lett.* **81**, 3067 (1998).
273. S.M. Carroll, *Spacetime and Geometry: An Introduction to General Relativity* (Addison-Wesley, San Francisco, 2004).
274. S.M. Carroll, A. De Felice, V. Duvvuri, D.A. Easson, M. Trodden, M.S. Turner, *Phys. Rev. D* **71**, 063513 (2005).
275. S.M. Carroll, V. Duvvuri, M. Trodden, M.S. Turner, *Phys. Rev. D* **70**, 043528 (2004).
276. S.M. Carroll, M. Hoffman, M. Trodden, *Phys. Rev. D* **68**, 023509 (2004).
277. S.M. Carroll, M. Kaplinghat, *Phys. Rev. D* **65**, 063507 (2002).
278. S.M. Carroll, I. Sawicki, A. Silvestri, M. Trodden, *New J. Phys.* **8**, 323 (2006).
279. E. Cartan, *Ann. Ec. Norm.* **42**, 17 (1925).
280. B. Carter, in *IAU Symposium 63: Confrontation of Cosmological Theories with Observational Data*, ed. by M.S. Longair (Reidel, Dordrecht, 1974).
281. F.C. Carvalho, A. Saa, *Phys. Rev. D* **70**, 087302 (2004).
282. R. Casadio, A. Gruppuso, *Int. J. Mod. Phys. D* **11**, 703 (2002).
283. J.A. Casas, J. Garcia-Bellido, M. Quiros, *Mod. Phys. Lett. A* **7**, 447 (1995).

284. R. Catena *et al.*, arXiv:astro-ph/0406152.
285. C. Cattoen, M. Visser, *Class. Quantum Grav.* **24**, 5985 (2007).
286. R. Cayrel *et al.*, *Nature* **409**, 691 (2001).
287. J.M. Cline, S. Jeon, G.D. Moore, *Phys. Rev. D* **70**, 043543 (2004).
288. J.A.R. Cembranos, *Phys. Rev. D* **73**, 064029 (2006).
289. J.L. Cervantes-Cota, *Class. Quantum Grav.* **16**, 3903 (1999).
290. D. Chakrabarty, E. de Filippis, H. Russell, *Astron. Astrophys.* **487**, 75 (2008).
291. P. Chauvet, J.L. Cervantes-Cota, *Phys. Rev. D* **52**, 3416 (1995).
292. B. Chauvineau, *Class. Quantum Grav.* **20**, 2617 (2003).
293. B. Chauvineau, *Gen. Relat. Gravit.* **39**, 297 (2007).
294. G. Chen, B. Ratra, *Astrophys. J.* **582**, 586 (2003).
295. X. Chen, in *Proceedings COSMO 2000, 4th International Workshop on Particle Physics and the Early Universe*, Cheju Island, Korea, 2000, ed. by J.E. Kim, P. Ko, K. Lee (World Scientific, Singapore, 2001).
296. X. Chen, M. Kamionkowski, *Phys. Rev. D* **60**, 104036 (1999).
297. X. Chen, R.J. Scherrer, G. Steigman, *Phys. Rev. D* **63**, 123504 (2001).
298. N.A. Chernikov, E.A. Tagirov, *Ann. Inst. H. Poincarè A* **9**, 109 (1968).
299. M. Chevallier, D. Polarski, *Int. J. Mod. Phys. D* **10**, 213 (2001).
300. T. Chiba, *J. Cosmol. Astropart. Phys.* **0503**, 008 (2005).
301. T. Chiba, *Phys. Lett. B* **575**, 1 (2003).
302. T. Chiba, *Phys. Rev. D* **60**, 083508 (1999).
303. T. Chiba, *Phys. Rev. D* **64**, 103503 (2001).
304. T. Chiba, T. Nakamura, *Prog. Theor. Phys.* **100**, 1077 (1998).
305. T. Chiba, T. Okabe, M. Yamaguchi, *Phys. Rev. D* **62**, 023511 (2000).
306. T. Chiba, T.L. Smith, A. Erickcek, *Phys. Rev. D* **75**, 124014 (2007).
307. T. Chiba, N. Sugiyama, J. Yokoyama, *Nucl. Phys. B* **530**, 304 (1998).
308. L.P. Chimento, A. Jakubi, D. Pavón, *Phys. Rev. D* **62**, 063508 (2000).
309. L.P. Chimento, A. Jakubi, N.A. Zuccalá, *Phys. Rev. D* **63**, 103508 (2001).
310. G. Chirco, S. Liberati, *Phys. Rev. D* **81**, 024016 (2010).
311. J.S. Chisolm, *Nucl. Phys.* **26**, 469 (1961)
312. Y.M. Cho, *Class. Quantum Grav.* **14**, 2963 (1997).
313. Y.M. Cho, *Phys. Rev. Lett.* **68**, 3133 (1992).
314. K. Choi, in *Proceedings, International Symposium on Supersymmetry, Supergravity and Superstring*, Seoul, Korea, 1999, ed. by J.E. Kim, C. Lee (World Scientific, Singapore, 2000).
315. K. Choi, *Phys. Rev. D* **62**, 043509 (2000).
316. D.J.H. Chung, L.L. Everett, A. Riotto, *Phys. Lett. B* **556**, 61 (2003).
317. C.A. Clarkson, PhD thesis, University of Glasgow, UK, arXiv:astro-ph/0008089.
318. C.A. Clarkson, R.K. Barrett, *Class. Quantum Grav.* **16**, 3781 (1999).
319. C.A. Clarkson, A.A. Coley, E.S.D. O'Neill, *Phys. Rev. D* **64**, 063510 (2001).
320. C.A. Clarkson, A.A. Coley, E.S.D. O'Neill, R.A. Sussman, R.K. Barrett, *Gen. Relat. Gravit.* **35**, 969 (2003).
321. T. Clifton, *Class. Quantum Grav.* **23**, 7445 (2006).
322. T. Clifton, *Class. Quantum Grav.* **24**, 5073 (2007).
323. T. Clifton, PhD thesis, University of Cambridge, UK, 2006, arXiv:gr-qc/0610071.
324. T. Clifton, *Phys. Rev. D* **77**, 024041 (2008).
325. T. Clifton, J.D. Barrow, *Class. Quantum Grav.* **23**, 2951 (2006).
326. T. Clifton, J.D. Barrow, *Phys. Rev. D* **72**, 103005 (2005).
327. T. Clifton, J.D. Barrow, *Phys. Rev. D* **72**, 123003 (2005).
328. T. Clifton, D.F. Mota, J.D. Barrow, *Mon. Not. R. Astron. Soc.* **358**, 601 (2004).
329. J. Cocke, J.M. Cohen, *J. Math. Phys.* **9**, 971 (1968).
330. G. Cognola, E. Elizalde, S. Nojiri, S.D. Odintsov, L. Sebastiani, S. Zerbini, *Phys. Rev. D* **77**, 046009 (2008).
331. G. Cognola, E. Elizalde, S. Nojiri, S.D. Odintsov, S. Zerbini, *J. Cosmol. Astropart. Phys.* **0502**, 010 (2005).
332. S. Coleman, J. Wess, B. Zumino, *Phys. Rev.* **177**, 2239 (1969).

333. A.A. Coley, *Dynamical Systems and Cosmology* (Kluwer Academic, Dordrecht, 2003).
334. A.A. Coley, J. Ibanez, R.J. van der Hoogen, *Int. J. Mod. Phys. D* **38**, 5256 (1997).
335. P.D.B. Collins, A.D. Martin, E.J. Squires, *Particle Physics and Cosmology* (Wiley, New York, 1989).
336. C.P. Constantinidis, J.C. Fabris, R.G. Furtado, M. Picco, *Phys. Rev. D* **61**, 043503 (2000).
337. C. Contreras, R. Herrera, S. del Campo, *Phys. Rev. D* **52**, 4349 (1995).
338. A. Cooray, D. Huterer, D.E. Holz, *Phys. Rev. Lett.* **96**, 021301 (2006).
339. E.J. Copeland, N.J. Nunes, F. Rosati, *Phys. Rev. D* **62**, 123503 (2000).
340. E.J. Copeland, M. Sami, S. Tsujikawa, *Int. J. Mod. Phys. D* **15**, 1753 (2006).
341. P.S. Corasaniti *et al.*, *Mont. Not. Astron. Soc.* **369**, 798 (2006).
342. V. Corbin, N.J. Cornish, *Class. Quantum Grav.* **23**, 2435 (2006).
343. C. Corda, *J. Cosmol. Astropart. Phys.* **0704**, 009 (2007).
344. N.J. Cornish, E.P.S. Shellard, *Phys. Rev. Lett.* **81**, 3571 (1998).
345. L.A. Correa-Borbonet, *Braz. J. Phys.* **35**, 1145 (2005).
346. S. Cotsakis, P.J. Saich, *Class. Quantum Grav.* **11**, 383 (1994).
347. D.H. Coule, *Class. Quantum Grav.* **9**, 2353 (1992).
348. D.H. Coule, *Phys. Lett. B* **450**, 48 (1999).
349. D.H. Coule, K. Maeda, *Class. Quantum Grav.* **7**, 955 (1990).
350. R. Courant, D. Hilbert, *Methods of Mathematical Physics* (Interscience, New York, 1962).
351. J.D.E. Creighton, R.B. Mann, *Phys. Rev. D* **52**, 4569 (1995).
352. J. Crowder, N.J. Cornish, *Phys. Rev. D* **72**, 083005 ( 2005).
353. E. Cunningham, *Proc. Lon. Math. Soc.* **8**, 77 (1909).
354. M.P. Dabrowski, T. Denkiewicz, D. Blaschke, *Ann. Phys. (Leipzig)* **16**, 237 (2007).
355. N. Dalal, K. Abazajian, E. Jenkins, A.V. Manohar, *Phys. Rev. Lett.* **87**, 141302 (2001).
356. R.A. Daly, arXiv:astro-ph/0212107.
357. R.A. Daly, S.G. Djorgovski, *Astrophys. J.* **612**, 652 (2004).
358. R.A. Daly, E.J. Guerra, *Astron. J.* **124**, 1831 (2002).
359. R.A. Daly, M.P. Mory, E.J. Guerra, arXiv:astro-ph/0203113.
360. T. Damour, G. Esposito-Farèse, *Class. Quantum Grav.* **9**, 2093 (1992).
361. T. Damour, K. Nordtvedt, *Phys. Rev. D* **48**, 3436 (1993).
362. T. Damour, K. Nordtvedt, *Phys. Rev. Lett.* **70**, 2217 (1993).
363. T. Damour, B. Pichon, *Phys. Rev. D* **59**, 123502 (1999).
364. T. Damour, A.M. Polyakov, *Nucl. Phys. B* **423**, 532 (1994).
365. A. Das, *Prog. Theor. Phys.* **24**, 915 (1960).
366. A. Davidson, *Class. Quantum Grav.* **22**, 1119 (2005).
367. T. Davis *et al.*, *Astrophys. J.* **666**, 716 (2007).
368. A. Degasperis, S.N.M. Ruijsenaars, *Ann. Phys. (N.Y.)* **293**, 92 (2001).
369. H. Dehnen, O. Obregon, *Astr. Sp. Sci.* **14**, 454 (1971).
370. M. Demianski *et al.*, *Phys. Rev. D* **44**, 3136 (1991).
371. D.A. Demir, B. Pulice, *Phys. Lett. B* **638**, 1 (2006).
372. N. Deruelle, C. Gundlach, D. Langlois, *Phys. Rev. D* **46**, 5337 (1992).
373. S. Deser, *Ann. Phys. (NY)* **59**, 248 (1970).
374. S. Deser, *Gen. Relat. Gravit.* **37**, 2251 (2005).
375. S. Deser, J. Franklin, *Am. J. Phys.* **73**, 261 (2005).
376. S. Deser, R.I. Nepomechie, *Ann. Phys. (NY)* **154**, 396 (1984).
377. P. de Bernardis *et al.*, *Nature* **404**, 955 (2000).
378. W.J.G. de Blok, *Astrophys. J.* **634**, 227 (2005).
379. W.J.G. de Blok, A. Bosma, *Astron. Astrophys.* **385**, 816 (2002).
380. E. De Filippis, M. Sereno, G. Bautz, M.W. Longo, *Astrophys. J.* **625**, 108 (2005).
381. A. de la Cruz-Dombriz, A. Dobado, *Phys. Rev. D* **74**, 087501 (2006).
382. A. de la Cruz-Dombriz, A. Dobado, A.L. Maroto, *Phys. Rev. D* **77**, 123515 (2008).
383. A. de la Macorra, H. Vucetich, *J. Cosmol. Astropart. Phys.* **012**, 1475 (2004).
384. R. de Ritis, A.A. Marino, P. Scudellaro, *Phys. Rev. D* **62**, 043506 (2000).
385. V. De Sabbata, C. Sivaram, *Astr. Sp. Sci.* **176**, 141 (1991).
386. J.C.C. de Souza, V. Faraoni, *Class. Quantum Grav.* **24**, 3637 (2007).

387. B.S. DeWitt, in *Battelle Rencontres*, ed. by C. DeWitt, J.A. Wheeler (Benjamin, New York, 1968).
388. B.S. DeWitt, *Phys. Rev.* **160**, 1113 (1967).
389. R. Dick, *Gen. Relativ. Gravit.* **36**, 217 (2004).
390. R.H. Dicke, *Astrophys. J.* **152**, 1 (1968).
391. R.H. Dicke, in *Gravitation and Relativity*, ed. by H.-Y. Chiu and W.F. Hoffmann (Benjamin, New York, 1964).
392. R.H. Dicke, *Phys. Rev.* **125**, 2163 (1962).
393. R.H. Dicke, P.J.E. Peebles, *Phys. Rev. Lett.* **12**, 435 (1964).
394. J.O. Dickey *et al.*, *Science* **265**, July 1994.
395. P.A.M. Dirac, *Proc. R. Soc. Lon.* **A 333**, 403 (1973).
396. A.D. Dolgov, M. Kawasaki *Phys. Lett. B* **573**, 1 (2003).
397. R. Dominguez-Tenreiro, G. Yepes, *Astron. Astrophys.* **177**, 5 (1987).
398. S.L. Dubovsky, S.M. Sibiryakov, *Phys. Lett. B* **638**, 509 (2006).
399. S.L. Dubovsky, P.G. Tiyakov, I.I. Tkachev, *Phys. Rev. Lett.* **94**, 181102 (2005).
400. J. Dunkley, M. Bucher, P.G. Ferreira, K. Moodley, C. Skordis, *Mont. Not. R. Astron. Soc.* **356**, 925 (2005).
401. J.P. Duruisseau, R. Kerner, *Class. Quantum Grav.* **3**, 187 (1986).
402. J.P. Duruisseau, R. Kerner, *Gen. Relat. Gravit.* **15**, 797 (1983).
403. S.B. Dutta Choudhury, D. Battacharya, *J. Math. Phys.* **23**, 1366 (1982).
404. S.B. Dutta Choudhury, D. Battacharyya, *J. Phys. A* **13**, 3643 (1980).
405. G.R. Dvali, G. Gabadadze, M. Porrati, *Phys. Lett. B* **485**, 208 (2000).
406. F.J. Dyson, *Phys. Rev.* **73**, 929 (1948).
407. D.M. Eardley, D.L. Lee, A.P. Lightman, *Phys. Rev. D* **10**, 3308 (1973).
408. D.M. Eardley, D.L. Lee, A.P. Lightman, R.V. Wagoner, C.M. Will, *Phys. Rev. Lett.* **30**, 884 (1973).
409. D.A. Easson, *Int. J. Mod. Phys. A* **19**, 5343 (2004).
410. D.H. Eckhardt, *Phys. Rev. D* **48**, 3762 (1993).
411. G. Efstathiou, S.L. Bridle, A.N. Lasenby, M.P. Hobson, R.S. Ellis, arXiv:astro-ph/9812226.
412. J. Ehlers, P. Geren, R.K. Sachs, *J. Math. Phys.* **9**, 1344 (1968).
413. J. Ehlers, A. Krasinski, *Gen. Relat. Gravit.* **38**, 1329 (2006).
414. A. Einstein, *Ann. der Physik* **49**, 769 (1916).
415. A. Einstein, *Sitzung-ber. Preuss. Akad. Wiss.*, 414 (1925).
416. L.P. Eisenhart, *Riemannian Geometry* (Princeton University Press, Princeton, 1955).
417. D. Eisenstein *et al.*, *Astrophys. J.* **633**, 560 (2005).
418. C. Eling, B.Z. Foster, T. Jacobson, A.C. Wall, *Phys. Rev. D* **75**, 101503 (2007).
419. C. Eling, R. Guedens, T. Jacobson, *Phys. Rev. Lett.* **96**, 121301 (2006).
420. E. Elizalde *et al.*, *Phys. Lett. B* **574**, 1 (2003).
421. E. Elizalde, S. Naftulin, S.D. Odintsov, *Int. J. Mod. Phys. A* **9**, 933 (1994).
422. E. Elizalde, S. Nojiri, S.D. Odintsov, *Phys. Rev. D* **70**, 043539 (2004).
423. E. Elizalde, S.D. Odintsov, *Phys. Lett. B* **333**, 331 (1994).
424. E. Elizalde, P.J. Silva, *Phys. Rev. D* **78**, 061501(R) (2008).
425. G.F.R. Ellis, M. Bruni, *Phys. Rev. D* **40**, 1804 (1989).
426. G.F.R. Ellis, M. Bruni, J.-C. Hwang, *Phys. Rev. D* **42**, 1035 (1990).
427. G.F.R. Ellis, J.-C. Hwang, M. Bruni, *Phys. Rev. D* **40**, 1819 (1989).
428. G.F.R. Ellis, D.R. Matravers, R. Treciokas, *Ann. Phys. (NY)* **150**, 455 (1983).
429. R. Emparan, J. Garriga, *J. High Energy Phys.* **0603**, 028 (2006).
430. A.L. Erickcek, T.L. Smith, M. Kamionkowski, *Phys. Rev. D* **74**, 121501 (2006).
431. J.D. Evans, L.M.H. Hall, P. Caillol, *Phys. Rev. D* **77**, 083514 (2008).
432. H. Everett, *Rev. Mod. Phys.* **29**, 454 (1957).
433. R. Fabbri, F. Lucchin, S. Matarrese, *Phys. Lett. B* **166**, 49 (1986).
434. J.C. Fabris, A.B. Batista, J.P. Baptista, *C. Rend. Acad. Sci. Paris* **309**, 791 (1989).
435. J.C. Fabris, R.G. Furtado, P. Peter, N. Pinto-Neto, *Phys. Rev. D* **67**, 124003 (2002).
436. J.C. Fabris, R. Kerner, J. Tossa, *Int. J. Mod. Phys. D* **9**, 111 (2000).
437. R. Fakir, S. Habib, *Mod. Phys. Lett. A* **8**, 2827 (1993).

438. R. Fakir, S. Habib, W.G. Unruh, *Astrophys. J.* **394**, 396 (1992).
439. R. Fakir, W.G. Unruh, *Phys. Rev. D* **41**, 1792 (1990).
440. R. Fakir, W.G. Unruh, *Phys. Rev. D* **41**, 3012 (1990).
441. V. Faraoni, *Ann. Phys. (NY)* **317**, 366 (2005).
442. V. Faraoni, arXiv:gr-qc/9807066.
443. V. Faraoni, *Class. Quantum Grav.* **26**, 145014 (2009).
444. V. Faraoni, *Class. Quantum Grav.* **26**, 168002 (2009).
445. V. Faraoni, *Class. Quantum Grav.* **26**, 195013 (2009).
446. V. Faraoni, *Cosmology in Scalar-Tensor Gravity* (Kluwer Academic, Dordrecht, 2004).
447. V. Faraoni, in *Entropy in Quantum Gravity*, special issue of *Open Astron. J.* **12**, 1246 (2010), ed. by R. Garattini.
448. V. Faraoni, in *The Problems of Modern Cosmology, volume in honour of Professor S.D. Odintsov on the occasion of his 50th birthday*, ed. by P.M. Lavrov (Tomsk State Pedagogical University Press, Tomsk, 2009), p. 185, arXiv:0811.1870.
449. V. Faraoni, *Int. J. Theor. Phys.* **40**, 2259 (2001).
450. V. Faraoni, *Int. J. Mod. Phys. D* **11**, 471 (2002).
451. V. Faraoni, *Phys. Lett. A* **245**, 26 (1998).
452. V. Faraoni, *Phys. Lett. B* **665**, 135 (2008).
453. V. Faraoni, *Phys. Rev. D* **53**, 6813 (1996).
454. V. Faraoni, *Phys. Rev. D* **59**, 084021 (1999).
455. V. Faraoni, *Phys. Rev. D* **62**, 023504 (2000).
456. V. Faraoni, *Phys. Rev. D* **68**, 063508 (2003).
457. V. Faraoni, *Phys. Rev. D* **70**, 047301 (2004).
458. V. Faraoni, *Phys. Rev. D* **70**, 081501(R) (2004).
459. V. Faraoni, *Phys. Rev. D* **74**, 023529 (2006).
460. V. Faraoni, *Phys. Rev. D* **74**, 104017 (2006).
461. V. Faraoni, *Phys. Rev. D* **75**, 067302 (2007).
462. V. Faraoni, *Phys. Rev. D* **80**, 044013 (2009).
463. V. Faraoni, *Phys. Rev. D* **81**, 044002 (2010).
464. V. Faraoni, F.I. Cooperstock, *Eur. J. Phys.* **19**, 419 (1998).
465. V. Faraoni, C. Gao, X. Chen, Y.-G. Shen, *Phys. Lett. B* **671**, 7 (2009).
466. V. Faraoni, E. Gunzig, *Int. J. Mod. Phys. D* **8**, 177 (1998).
467. V. Faraoni, E. Gunzig, P. Nardone, *Fund. Cosm. Phys.* **20**, 121 (1998).
468. V. Faraoni, A. Jacques, *Phys. Rev. D* **76**, 063510 (2007).
469. V. Faraoni, M.N. Jensen, S.A. Theuerkauf, *Class. Quantum Grav.* **23**, 4215 (2006).
470. V. Faraoni, N. Lanahan-Tremblay, *Phys. Rev. D* **77**, 108501 (2008).
471. V. Faraoni, N. Lanahan-Tremblay, *Phys. Rev. D* **78**, 064017 (2008).
472. V. Faraoni, S. Nadeau, *Phys. Rev. D* **72**, 124005 (2005).
473. V. Faraoni, S. Nadeau, *Phys. Rev. D* **75**, 023501 (2007).
474. V. Faraoni, S. Sonego, in *Proceedings of the 5th Canadian Conference on General Relativity and Relativistic Astrophysics*, Waterloo, Canada 1993, ed. by R.B. Mann, R.G. McLenaghan (World Scientific, Singapore, 1994), p. 386.
475. V. Faraoni, A.F. Zambrano Moreno, *Phys. Rev. D* **81**, 12405 (2010).
476. G. Esposito-Farèse, in *Proceedings of the 9th Marcel Grossmann Meeting*, Rome, 2000 (World Scientific, Singapore, 2002), p. 1749, arXiv:gr-qc/0011115.
477. G. Esposito-Farèse, D. Polarski, *Phys. Rev. D* **63**, 063504 (2001).
478. T. Faulkner, M. Tegmark, E.F. Bunn, Y. Mao, *Phys. Rev. D* **76**, 063505 (2007).
479. S. Fay, *Class. Quantum Grav.* **17**, 891 (2000).
480. S. Fay, *Class. Quantum Grav.* **18**, 45 (2001).
481. S. Fay, S. Nesseris, L. Perivolaropoulos, *Phys. Rev. D* **76**, 063504 (2007).
482. S. Fay, R. Tavakol, S. Tsujikawa, *Phys. Rev. D* **75**, 063509 (2007).
483. J.J. Ferrando, J.A. Morales, M. Portilla, *Phys. Rev. D* **46**, 578 (1992).
484. M. Ferraris, M. Francaviglia, G. Magnano, *Class. Quantum Grav.* **5** L95 (1988).
485. M. Ferraris, M. Francaviglia, C. Reina, *Gen. Relat. Gravit.* **14**, 243 (1982).
486. M. Ferraris, M. Francaviglia, I. Volovich, *Class. Quantum Grav.* **11**, 1505 (1994).

487. P.G. Ferreira, M. Joyce, *Phys. Rev. D* **58**, 023503 (1998).
488. P.G. Ferreira, M. Joyce, *Phys. Rev. Lett.* **79**, 4740 (1998).
489. I. Ferreras, A. Melchiorri, D. Tocchini Valentini, *Mont. Not. R. Astron. Soc.* **344**, 257 (2003).
490. M. Fierz, *Helv. Phys. Acta* **29**, 128 (1956).
491. A.V. Filippenko, A.G. Riess, *Phys. Rep.* **307**, 31 (1998).
492. D. Finkelstein, *Trans. N.Y. Acad. Sci.* **25**, 621 (1963).
493. L.S. Finn, P.J. Sutton, *Phys. Rev. D* **65**, 044022, (2002).
494. E. Fischbach, D. Sudarsky, A. Szafer, C. Talmadge, S.H. Aronson, *Phys. Rev. Lett.* **56**, 3 (1986).
495. E.E. Flanagan, *Class. Quantum Grav.* **21**, 417 (2004).
496. E.E. Flanagan, *Class. Quantum Grav.* **21**, 3817 (2004).
497. E.E. Flanagan, *Phys. Rev. Lett.* **92**, 071101 (2004).
498. O.A. Fonarev, *Class. Quantum Grav.* **12**, 1739 (1995).
499. L.H. Ford, *Phys. Rev. D* **32**, 1409 (1987).
500. L.H. Ford, D.J. Toms, *Phys. Rev. D* **25**, 1510 (1982).
501. S. Foster, *Class. Quantum Grav.* **15**, 3485 (1998).
502. P.H. Frampton, arXiv:hep-th/0302007.
503. P.H. Frampton, in *Proceedings, 30th Conference on High Energy Physics and Cosmology*, Coral Gables, Florida, 2001, ed. by B.N. Kursunoglu, S.L. Mintz, S.L. Perlmutter (American Institute of Physics, Melville, 2002).
504. P.H. Frampton, *Phys. Lett. B* **555**, 139 (2003).
505. P.H. Frampton, T. Takahashi, *Phys. Lett. B* **557**, 135 (2003).
506. W.L. Freedman *et al.*, *Astrophys. J.* **553**, 47 (2001).
507. K. Freese, M. Lewis, *Phys. Lett. B* **540**, 1 (2002).
508. J.A. Frieman, *Comments Astrophys.* **18**, 323 (1997).
509. J.A. Frieman, C. Hill, A. Stebbins, J. Waga, *Phys. Rev. D* **50**, 4895 (1995).
510. C. Frigerio Martins, P. Salucci, *Mon. Not. R. Astron. Soc.* **381**, 1103 (2007).
511. V.P. Frolov, I.D. Novikov, *Black Hole Physics* (Kluwer Academic, Dordrecht, 1998).
512. J. Froyland, *Phys. Rev. D* **25**, 1470 (1982).
513. Y. Fujii, *Gravit. Cosmol.* **6**, 107 (2000).
514. Y. Fujii, *Phys. Lett. B* **202**, 246 (1988).
515. T. Futamase, K. Maeda, *Phys. Rev. D* **39**, 399 (1989).
516. Y. Fujii, K. Maeda, *The Scalar-Tensor Theory of Gravity* (Cambridge University Press, Cambridge, 2003).
517. Y. Fujii, T. Nishioka, *Phys. Rev. D* **62**, 044011 (2000).
518. T. Futamase, T. Rothman, R. Matzner, *Phys. Rev. D* **39**, 405 (1989).
519. T. Futamase, M. Tanaka, *Phys. Rev. D* **60**, 063511 (1999).
520. C. Gao, X. Chen, V. Faraoni, Y-G. Shen, *Phys. Rev. D* **78**, 024008 (2008).
521. J. Garcia-Bellido, *Nucl. Phys. B* **423**, 221 (1994).
522. J. Garcia-Bellido, A. Linde, *Phys. Rev. D* **52**, 6730 (1995).
523. J. Garcia-Bellido, A. Linde, D. Linde, *Phys. Rev. D* **50**, 730 (1994).
524. J. Garcia-Bellido, D. Wands, *Phys. Rev. D* **52**, 5636 (1995).
525. J. Garcia-Bellido, D. Wands, *Phys. Rev. D* **52**, 6739 (1995).
526. J. Garriga, T. Tanaka, *Phys. Rev. Lett.* **84**, 2778 (2000).
527. M. Gasperini, *Phys. Lett. B* **470**, 67 (1999).
528. M. Gasperini, *Phys. Rev. D* **64**, 043510 (2001).
529. M. Gasperini, F. Piazza, G. Veneziano, *Phys. Rev. D* **65**, 023508 (2002).
530. M. Gasperini, G. Veneziano, *Phys. Lett. B* **277**, 256 (1992).
531. E. Gaztanaga, J.A. Lobo, *Astrophys. J.* **548**, 47 (2001).
532. M. Gell-Mann, *Suppl. Nuovo Cimento* **4**, 10 (1956).
533. J.-M. Gerard, I. Mahara, *Phys. Lett. B* **346**, 35 (1995).
534. G. Germán, *Phys. Rev. D* **32**, 3307 (1985).
535. R. Geroch, *Comm. Math. Phys.* **13**, 180 (1969).
536. G.W. Gibbons, S.W. Hawking, *Phys. Rev. D* **15**, 2752 (1977).
537. G.W. Gibbons, B.F. Whiting, *Nature* **291**, 636 (1981).

538. P.B. Gilkey, *Compositio Math.* **38**, 201 (1979).
539. P.B. Gilkey, *J. Diff. Geom.* **10**, 601 (1975).
540. R.L. Gilliland, P.E. Nugent, M.M. Phillips, *Astrophys. J.* **521**, 30 (1999).
541. F. Girelli, S. Liberati, L. Sindoni, *Phys. Rev. D* **78**, 084013 (2008).
542. A. Giveon, M. Porrati, E. Rabinovici, *Phys. Rep.* **244**, 443 (1994).
543. E.N. Glass, J.J.G. Scanio, *Am. J. Phys.* **45**, 344 (1977).
544. K. Gödel, *Rev. Mod. Phys.* **21**, 447 (1949).
545. N. Goheer, J.A. Leach, P.K.S. Dunsby, *Class. Quantum Grav.* **24**, 5689 (2007).
546. N. Goheer, J.A. Leach, P.K.S. Dunsby, *Class. Quantum Grav.* **25**, 035013 (2008).
547. Y. Gong, A. Wang, *Phys. Rev. Lett.* **99**, 211301 (2007).
548. A. Goobar, S. Perlmutter, *Astrophys. J.* **450**, 14 (1995).
549. K.M. Górski *et al.*, *Astrophys. J. Lett.* **464**, L11 (1996).
550. S. Gott, H.-J. Schmidt, A.A. Starobinsky, *Class. Quantum Grav.* **7**, 893 (1990).
551. S. Gottlöber, H.-J. Schmidt, A.A. Starobinsky, *Class. Quantum Grav.* **7**, 893 (1990).
552. A.M. Green, A.R. Liddle, *Phys. Rev. D* **54**, 2557 (1996).
553. M.B. Green, J.H. Schwarz, E. Witten, *Superstring Theory* (Cambridge University Press, Cambridge, 1987).
554. G.S. Greenstein, *Astr. Sp. Sci.* **2**, 155 (1968).
555. G.S. Greenstein, *Astrophys. Lett.* **1**, 139 (1968).
556. A.A. Grib, E.A. Poberii, *Helv. Phys. Acta* **68**, 380 (1995).
557. A.A. Grib, W.A. Rodrigues, *Gravit. Cosmol.* **1**, 273 (1995).
558. D. Grumiller, W. Kummer, D.V. Vassilevich, *J. High Energy Phys.* **7**, 009 (2003).
559. C. Gunnarsson, T. Dahlen, A. Goobar, J. Jönsson, E. Mörtsell, *Astrophys. J.* **640**, 417 (2006).
560. E. Gunzig *et al.*, *Class. Quantum Grav.* **17**, 1783 (2000).
561. E. Gunzig *et al.*, *Mod. Phys. Lett. A* **15**, 1363 (2000).
562. E. Gunzig *et al.*, *Phys. Rev. D* **63**, 067301 (2001).
563. L.E. Gurevich, A.M. Finkelstein, V.A. Ruban, *Astr. Sp. Sci.* **22**, 231 (1973).
564. A.H. Guth, *Phys. Rev. D* **23**, 347 (1981).
565. A.H. Guth, B. Jain, *Phys. Rev. D* **45**, 426 (1992).
566. J.J. Halliwell, in *Quantum Cosmology and Baby Universes*, ed. by S. Coleman, J.B. Hartle, T. Piran, S. Weinberg (World Scientific, Singapore, 1991).
567. J.J. Halliwell, *Nucl. Phys. B* **266**, 228 (1986).
568. J.J. Halliwell, *Phys. Rev. D* **36**, 3626 (1987).
569. J.J. Halliwell, R. Laflamme, *Class. Quantum Grav.* **6**, 1839 (1989).
570. S. Hanany *et al.*, *Astrophys. J. Lett.* **545**, L5 (2000).
571. S. Hannestad, E. Mortsell, *Phys. Rev. D* **66**, 063508 (2002).
572. T. Harada, T. Chiba, K. Nakao, T. Nakamura, *Phys. Rev. D* **55**, 20246 (1997).
573. T. Harada, C. Goymer, B.J. Carr, *Phys. Rev. D* **66**, 104023 (2002).
574. E.R. Harrison, *Phys. Rev. D* **6**, 2077 (1972).
575. J.B. Hartle, *Am. J. Phys.* **36**, 704 (1968).
576. J.B. Hartle, in *Gravitation in Astrophysics*, Proceedings, Cargese 1986, ed. by B. Carter, J.B. Hartle (Plenum, New York, 1986).
577. J.B. Hartle, *Spacetime Quantum Mechanics and the Quantum Mechanics of Spacetime in Gravitation and Quantization*, Proceeding of the 1992 Les Houches Summer School, vol. LVII, ed. by B. Julia, J. Zinn-Justin (North Holland, Amsterdam, 1995).
578. J.B. Hartle, S.W. Hawking, *Phys. Rev. D* **28**, 2960 (1983).
579. S.W. Hawking, in *Relativity, Group and Topology II*, Les Houches Session XL, ed. by B. DeWitt, R. Stora (North Holland, Amsterdam, 1984).
580. S.W. Hawking, *Comm. Math. Phys.* **25**, 152 (1972).
581. S.W. Hawking, *Comm. Math. Phys.* **25**, 167 (1972).
582. S.W. Hawking, G.F.R. Ellis, *The Large Scale Structure of Space-Time* (Cambridge University Press, Cambridge, 1973).
583. F.W. Hehl, J.D. Mc Crea, E.W. Milke, Y. Ne'eman, *Phys. Rep.* **258**, 1 (1995).
584. F.W. Hehl, P. von der Heyde, G.D. Kerlick, J.M. Nester, *Rev. Mod. Phys.* **48**, 393.
585. S. Hellermann, N. Kaloper, L. Susskind, *J. High Energy Phys.* **0106**, 003 (2001).

586. R. Herrera, C. Contreras, S. del Campo, *Class. Quantum Grav.* **12**, 1937 (1995).
587. P.W. Higgs, *Nuovo Cimento* **11**, 816 (1959).
588. C.T. Hill, A.K. Leibovitch, *Phys. Rev. D* **66**, 075010 (2002).
589. C.T. Hill, D.S. Salopek, *Ann. Phys. (NY)* **213**, 21 (1992).
590. A. Hindawi, B.A. Ovrut, D. Waldram, *Phys. Rev. D* **53**, 5597 (1996).
591. G. Hinshaw *et al.*, *Astrophys. J. Lett.* **464**, L17 (1996).
592. G. Hinshaw *et al.*, *Astrophys. J. Suppl.* **148**, 135 (2003).
593. W.A. Hiscock, *Class. Quantum Grav.* **7**, L35 (1990).
594. D.J. Holden, D. Wands, *Class. Quantum Grav.* **15**, 3271 (1998).
595. D.J. Holden, D. Wands, *Phys. Rev. D* **61**, 043506 (2000).
596. R. Holman, E.W. Kolb, S. Vadas, Y. Wang, *Phys. Lett. B* **269**, 252 (1991).
597. R. Holman, E.W. Kolb, Y. Wang, *Phys. Rev. Lett.* **65**, 17 (1990).
598. D.E. Holz, E.V. Linder, *Astrophys. J.* **631**, 678 (2005).
599. D.E. Holz, R.M. Wald, *Phys. Rev. D* **58**, 063501 (1998).
600. Y. Hosotani, *Phys. Rev. D* **32**, 1949 (1985).
601. P.S. Howe, G. Papadopoulos, *Phys. Lett. B* **379**, 80 (1996).
602. C.D. Hoyle *et al.*, *Phys. Rev. Lett.* **86**, 1418 (2001).
603. F. Hoyle, R.J. Tayler, *Nature* **203**, 1108 (1964).
604. B.-L. Hu, *Phys. Lett. A* **90**, 375 (1982).
605. W. Hu, I. Sawicki, *Phys. Rev. D* **76**, 064004 (2007).
606. W. Hu, I. Sawicki, *Phys. Rev. D* **76**, 104043 (2007).
607. W. Hu, N. Sugiyama, *Astrophys. J.* **471**, 542 (1996).
608. L. Hui, P.B. Greene, *Phys. Rev. D* **73**, 123526 (2006).
609. D. Huterer, M.S. Turner, *Phys. Rev. D* **60**, 081301 (1999).
610. J.-C. Hwang, *Class. Quantum Grav.* **7**, 1613 (1990).
611. J.-C. Hwang, *Class. Quantum Grav.* **8**, 195 (1991).
612. J.-C. Hwang, *Class. Quantum Grav.* **14**, 1981 (1997).
613. J.-C. Hwang, *Class. Quantum Grav.* **14**, 3327 (1997).
614. J.-C. Hwang, *Class. Quantum Grav.* **15**, 1387 (1998).
615. J.-C. Hwang, *Class. Quantum Grav.* **15**, 1401 (1998).
616. J.-C. Hwang, *Phys. Rev. D* **42**, 2601 (1990).
617. J.-C. Hwang, *Phys. Rev. D* **53**, 762 (1996).
618. J.-C. Hwang, H. Noh, *Phys. Rev. D* **54**, 1460 (1996).
619. J.-C. Hwang, H. Noh, *Phys. Rev. D* **65**, 023512 (2002).
620. J.-C. Hwang, H. Noh, *Phys. Rev. Lett.* **81**, 5274 (1998).
621. J.-C. Hwang, E.T. Vishniac, *Astrophys. J.* **353**, 1 (1990).
622. A. Iglesias, N. Kaloper, A. Padilla, M. Park, *Phys. Rev.* **76**, 104001 (2007).
623. A. Iorio, L. O'Raifeartaigh, I. Sachs, C. Wiesendanger, *Nucl. Phys. B* **495**, 433 (1997).
624. L. Iorio, *Adv. High Energy Phys.* **2009**, 90731 (2009).
625. R.A. Isaacson, *Phys. Rev.* **166**, 1272 (1968).
626. J. Ishikawa, *Phys. Rev. D* **28**, 2445 (1983).
627. W. Israel, *Nuovo Cimento* **44B**, 1 (1966); *Erratum*, **49B**, 463 (1967).
628. C. Itzykson, J.-C. Zuber, *Quantum Field Theory* (McGraw-Hill, New York, 1980).
629. V. Iyer, R.M. Wald, *Phys. Rev. D* **50**, 846 (1994).
630. V. Iyer, R.M. Wald, *Phys. Rev. D* **52**, 4430 (1995).
631. T. Jacobson, *Phys. Rev. Lett.* **75**, 1260 (1995).
632. T. Jacobson, G. Kang, R.C. Myers, *Phys. Rev. D* **49**, 6587 (1994).
633. T. Jacobson, R.C. Myers, *Phys. Rev. Lett.* **70**, 3684 (1993).
634. T. Jacobson, A.C. Wall, *Found. Phys.* **40**, 1076 (2010).
635. B. Jain, A. Dev, J.S. Alcaniz, *Class. Quantum Grav.* **20**, 4163 (2003).
636. J.C. Jackson, *J. Cosmol. Astropart. Phys.* **0411**, 007 (2004).
637. J.S. Jeavons, C.G.B. McIntosh, D.K. Sen, *J. Math. Phys.* **16**, 320 (1975).
638. J.T. Jebsen, *Ark. Mat. Ast. Fys. (Stockholm)* **15**, nr. 18 (1921), reprinted in *Gen. Relat. Gravit.* **37**, 2253 (2005).
639. P. Jetzer, *Phys. Rep.* **220**, 163 (1993).

640. R. Jimenez, L. Verde, T. Treu, D. Stern, *Astrophys. J.* **593**, 622 (2003).
641. X.H. Jin, D.J. Liu, X.Z. Li, arXiv:astro-ph/0610854.
642. N.V. Johansen, F. Ravndal, *Gen. Relat. Gravit.* **38**, 537 (2006).
643. M.V. John, *Astrophys. J.* **614**, 1 (2004).
644. M.V. John, *Astrophys. J.* **630**, 667 (2005).
645. M. Johnson, *Nuovo Cimento* **4**, 323 (1972).
646. V.B. Johri, *Class. Quantum Grav.* **19**, 5959 (2002).
647. J. Jönsson, T. Dahlen, A. Goobar, C. Gunnarsson, E. Mörtsell, K. Lee, *Astrophys. J.* **639**, 991 (2006).
648. P. Jordan, *Naturwiss.* **26**, 417 (1938).
649. P. Jordan, *Schwerkraft und Weltfall, Grundlagen der Theoretische Kosmologie* (Vieweg & Sohns, Braunschweig, 1952).
650. K. Kainulainen, J. Piilonen, V. Reijonen, D. Sunhede, *Phys. Rev. D* **76**, 024020 (2007).
651. K. Kainulainen, V. Reijonen, D. Sunhede, *Phys. Rev. D* **76**, 043503 (2007).
652. K. Kainulainen, D. Sunhede, *Phys. Rev. D* **78**, 063511 (2008).
653. D.I. Kaiser, arXiv:astro-ph/9507048.
654. D.I. Kaiser, *Phys. Rev. D* **52**, 4295 (1995).
655. N. Kaiser and the PanSTARRS Team, in *Bull. Am. Astron. Soc.*, p. 1049 (2005).
656. M. Kaku, *Quantum Field Theory* (Oxford University Press, Oxford, 1993).
657. T. Kaluza, *Sitzungsber. Preuss. Akad. Wiss. Berlin (Math. Phys.)* **1921**, 966 (1921).
658. S. Kalyana Rama, *Phys. Rev. D* **56**, 6130 (1997).
659. S. Kalyana Rama, *Phys. Rev. Lett.* **78**, 1620 (1997).
660. S. Kamefuchi, L. O'Raifeartaigh, A. Salam, *Nucl. Phys.* **28**, 529 (1961).
661. A. Kamenshchik, U. Moschella, V. Pasquier, *Phys. Lett. B* **511**, 265 (2001).
662. G. Kang, *Phys. Rev. D* **54**, 7483 (1996).
663. M. Kenmoku, Y. Okamoto, K. Shigemoto, *Phys. Rev. D* **48**, 578 (1993).
664. J. Kerimo, *Phys. Rev. D* **62**, 104005 (2000).
665. J. Kerimo, D. Kalligas, *Phys. Rev. D* **58**, 104002 (1992).
666. J. Khoury, A. Weltman, *Phys. Rev. D* **69**, 044026 (2004).
667. J. Khoury, A. Weltman, *Phys. Rev. Lett.* **93**, 171104 (2004).
668. K. Kiefer, B. Sandhofer, in *Beyond the Big Bang*, ed. by R. Vaas (Springer, New York, 2008), arXiv:0804.0672.
669. J.E. Kim, *J. High Energy Phys.* **9905**, 022 (1999).
670. H. Kim, *Nuovo Cimento B* **112**, 329 (1997).
671. A.G. Kim, E.V. Linder, R. Miquel, N. Mostek, *Mont. Not. R. Astron. Soc.* **347**, 909 (2004).
672. D. Kirkman, D. Tyler, N. Suzuki, N., J.M. O'Meara, D. Lubin, *Astrophys. J. (Suppl.)* **149**, 1 (2003).
673. O. Klein, *New Theories in Physics* 77 (Intern. Inst. of Intellectual Co-operation, League of Nations, 1938).
674. O. Klein, *Z. Phys.* **37**, 895 (1926).
675. S. Kluske, H.-J. Schmidt, *Astron. Nachr.* **37**, 337 (1996).
676. R.A. Knop *et al.*, *Astrophys. J.* **598**, 102 (2003).
677. L. Knox, Y.-S. Song, A.J. Tyson, *Phys. Rev. D* **74**, 023512 (2006).
678. T. Kodama, N. Arimoto, *Astron. Astrophys.* **320**, 41 (1997).
679. L. Kofman, V.F. Mukhanov, D.Y. Pogosyan, *Sov. Phys. JETP* **66**, 441 (1987).
680. J. Koga, K. Maeda, *Phys. Rev. D* **58**, 064020 (1998).
681. T. Koivisto, *Phys. Rev. D* **73**, 083517 (2006).
682. T. Koivisto, *Phys. Rev. D* **76**, 043527 (2007).
683. T. Koivisto, H. Kurki-Suonio, *Class. Quantum Grav.* **23**, 2355 (2006).
684. E.W. Kolb, *Physica Scripta* **T36**, 376 (1991).
685. E.W. Kolb, V. Marra, S. Matarrese, *Gen. Relat. Gravit.* **42**, 1399 (2010).
686. E.W. Kolb, S. Matarrese, A. Riotto, *New J. Phys.* **8**, 322 (2006).
687. E.W. Kolb, D. Salopek, M.S. Turner, *Phys. Rev. D* **42**, 3925 (1990).
688. E.W. Kolb, M.S. Turner, *The Early Universe* (Addison-Wesley, Reading, Massachusetts, 1994).

689. S.J. Kolitch, *Ann. Phys. (NY)* **246**, 121 (1996).
690. S.J. Kolitch, D.M. Eardley, *Ann. Phys. (NY)* **241**, 128 (1995).
691. E. Komatsu, T. Futamase, *Phys. Rev. D* **58**, 023004 (1998).
692. E. Komatsu, T. Futamase, *Phys. Rev. D* **59**, 064029 (1999).
693. D. Kothawala, S. Sarkar, T. Padmanabhan, *Phys. Lett. B* **652**, 338 (2007).
694. K. Koyama, R. Maartens, *J. Cosmol. Astropart. Phys.* **0601**, 016 (2006).
695. A. Krasinski, *Inhomogeneous Cosmological Models* (Cambridge University Press, Cambridge, 1997).
696. L.M. Krauss, *Astrophys. J. Lett.* **596**, L1 (2003).
697. L.M. Krauss, M. White, *Astrophys. J.* **397**, 357 (1992).
698. K.D. Krori, D.R. Bhattacharjee, *J. Math. Phys.* **23**, 637 (1982).
699. K.D. Krori, D. Nandy, *J. Phys. A* **10**, 993 (1977).
700. Y.A. Kubyshin, *J. Math. Phys.* **35**, 310 (1994).
701. D. La, P.J. Steinhardt, *Phys. Rev. Lett.* **62**, 376 (1989).
702. D. La, P.J. Steinhardt, E. Bertschinger, *Phys. Lett. B* **231**, 231 (1989).
703. Y.-M. Lam, *Phys. Rev. D* **7**, 2943 (1973).
704. N. Lanahan-Tremblay, V. Faraoni, *Class. Quantum Grav.* **24**, 5667 (2007).
705. L.D. Landau, E.M. Lifschitz, *The Classical Theory of Fields* (Pergamon Press, Oxford, 1989).
706. A.E. Lange *et al.*, *Phys. Rev. D* **63**, 042001 (2001).
707. I. Laszlo, R. Bean, *Phys. Rev. D* **77**, 024048 (2008).
708. A.M. Laycock, A.R. Liddle, *Phys. Rev. D* **49**, 1827 (1994).
709. J.A. Leach, S. Carloni, P.K.S. Dunsby, *Class. Quantum Grav.* **23**, 4915 (2006).
710. J.A. Leach, P.K.S. Dunsby, S. Carloni, in *Proceedings of the 11th Marcel Grossmann Meeting on General Relativity*, Berlin 2006, p. 1213, arXiv:gr-qc/0702122.
711. P. Leaci *et al.*, *Class. Quantum Grav.* **25**, 195018 (2008).
712. O.M. Lecian, G. Montani, *Class. Quantum Grav.* **26**, 045014 (2009).
713. J. Lee *et al.*, *Phys. Rev. D* **61**, 027301 (1999).
714. S. Lee, arXiv:0710.2395.
715. S. Lee, *Mod. Phys. Lett. A* **23**, 1388 (2008).
716. G. Lemaître, *Ann. Soc. Sci. Brussels A* **53**, 51 (1933), reprinted in *Gen. Relat. Gravit.* **29**, 641 (1997).
717. T. Levi-Civita, *The Absolute Differential Calculus* (Blackie & Son, London, 1929).
718. J.J. Levin, K. Freese, *Nucl. Phys. B* **421**, 635 (1994).
719. B. Li, J.D. Barrow, *Phys. Rev. D* **75**, 084010 (2007).
720. B. Li, J.D. Barrow, D.F. Mota, *Phys. Rev. D* **76**, 104047 (2007).
721. B. Li, K.-C. Chan, M.-C. Chu, *Phys. Rev. D* **76**, 024002 (2007).
722. B. Li, M.-C. Chu, *Phys. Rev. D* **74**, 104010 (2006).
723. X.-Z. Li, J.-G. Hao, D.-J. Liu, *Class. Quantum Grav.* **19**, 6049 (2002).
724. N. Li, D.J. Schwarz, *Phys. Rev. D* **76**, 083011 (2007).
725. N. Li, D.J. Schwarz, *Phys. Rev. D* **78**, 083531 (2008).
726. X.-Z. Li, J.-G. Hao, *Phys. Rev. D* **69**, 107303 (2004).
727. A.R. Liddle, *Phys. Lett. B* **220**, 502 (1989).
728. A.R. Liddle, D.H. Lyth, *Cosmological Inflation and Large Scale Structure* (Cambridge University Press, Cambridge, 2000).
729. A.R. Liddle, D.H. Lyth, *Phys. Lett. B* **291**, 391 (1992).
730. A.R. Liddle, D.H. Lyth, *Phys. Rep.* **231**, 1 (1993).
731. A.R. Liddle, M.S. Madsen, *Int. J. Mod. Phys. D* **1**, 101 (1992).
732. A.R. Liddle, R.J. Scherrer, *Phys. Rev. D* **59**, 023509 (1999).
733. A.R. Liddle, D. Wands, *Mon. Not. R. Astron. Soc.* **253**, 637 (1991).
734. A.R. Liddle, D. Wands, *Phys. Rev. D* **45**, 2665 (1992).
735. J.E. Lidsey, *Class. Quantum Grav.* **13**, 2449 (1996).
736. J.E. Lidsey, *Phys. Rev. D* **52**, 5407 (1995).
737. J.E. Lidsey *et al.*, *Rev. Mod. Phys.* **69**, 373 (1997).
738. J.E. Lidsey, D. Wands, E.J. Copeland, *Phys. Rep.* **337**, 343 (2000).

739. *LIGO* homepage http://www.ligo.caltech.edu/
740. J.A.S. Lima, J.S. Alcaniz, *Mont. Not. R. Astron. Soc.* **317**, 893 (2000).
741. A.D. Linde, *Particle Physics and Inflationary Cosmology* (Hardwood Academic, Chur, Switzerland, 1990).
742. A.D. Linde, *Sov. Phys. JETP Lett.* **30**, 447 (1980).
743. E.V. Linder, *Am. J. Phys.* **76**, 197 (2008).
744. E.V. Linder, *Phys. Rev. Lett.* **90**, 091301 (2003).
745. *LISA* homepage http://lisa.nasa.gov/
746. www.lisa.esa.int
747. F.S.N. Lobo, arXiv: 0807.1640.
748. M. Lockwood, *The Labyrinth of Time*, 2nd edn. (Oxford University Press, Oxford, 2007).
749. M.A. Lorentz, *Collected Papers*, vol. 5 (Nijhoff, The Hague, 1937), p. 363.
750. D. Lorentz-Petzold, *Astr. Sp. Sci.* **98**, 249 (1984).
751. D. Lorentz-Petzold, in *Lecture Notes in Physics* vol. 105, ed. by C. Hoenselaers, W. Dietz (Springer, Berlin, 1984).
752. C. Lousto, M. Campanelli, in *The Origin of Structure in the Universe*, Proceedings, Pont d'Oye, Belgium 1992, ed. by E. Gunzig, P. Nardone (Kluwer Academic, Dordrecht, 1993), p. 123.
753. M. Lu, M. Wise, *Phys. Rev. D* **47**, 3095 (1993).
754. F. Lucchin, S. Matarrese, *Phys. Rev. D* **32**, 1316 (1985).
755. F. Lucchin, S. Matarrese, M.D. Pollock, *Phys. Lett. B* **167**, 163 (1986).
756. A. Lue, R. Scoccimarro, G. Starkman, *Phys. Rev. D* **69**, 044005 (2004).
757. V.N. Lukash, *Sov. Phys. JETP* **52**, 807 (1980).
758. V.N. Lukash, *Sov. Phys. JETP Lett.* **31**, 596 (1980).
759. G. Lyra, *Math. Z.* **54**, 2 (1951).
760. D.H. Lyth, E.D. Stewart, *Phys. Lett. B* **274**, 168 (1992).
761. R. Maartens, D.R. Taylor, *Gen. Relat. Gravit.* **26**, 599 (1994).
762. M.S. Madsen, *Gen. Relat. Gravit.* **25**, 855 (1993).
763. H. Maeda, arXiv:0704.2731.
764. K. Maeda, *Phys. Rev. D* **39**, 3159 (1989).
765. H. Maeda, T. Harada, B.J. Carr, *Phys. Rev. D* **77**, 024023 (2008).
766. M. Maggiore, *Phys. Rep.* **331**, 283 (2000).
767. G. Magnano, M. Ferraris, M. Francaviglia, *Class. Quantum Grav.* **7**, 557, 1990.
768. G. Magnano, M. Ferraris, M. Francaviglia, *Gen. Relat. Gravit.* **19**, 465 (1987).
769. G. Magnano, L.M. Sokolowski, *Phys. Rev. D* **50**, 5039 (1994).
770. N. Makino, M. Sasaki, *Prog. Theor. Phys.* **86**, 103 (1991).
771. S. Mallik, D. Rai Chaudhuri, *Phys. Rev. D* **56**, 625 (1997).
772. P.D. Mannheim, D. Kazanas, *Astrophys. J.* **342**, 635 (1989).
773. I. Maor, R. Brustein, J. McMahon, P.J. Steinhardt, *Phys. Rev. D* **65**, 123003 (2002).
774. G. Marmo, E.J. Saletan, A. Simoni, B. Vitale, *Dynamical Systems: A Differential Geometric Approach to Symmetry and Reduction* (Wiley, New York, 1985).
775. V. Marra, E.W. Kolb, S. Matarrese, *Phys. Rev D* **77**, 023003 (2008).
776. V. Marra, E.W. Kolb, S. Matarrese, A. Riotto, *Phys. Rev. D* **76**, 123004 (2007).
777. B. Mashhoon, *Phys. Lett. A* **163**, 7 (1992).
778. S. Matarrese, C. Baccigalupi, F. Perrotta, *Phys. Rev. D* **70**, 061301 (2004).
779. C. Mathiazhagan, V.B. Johri, *Class. Quantum Grav.* **1**, L29 (1984).
780. T. Matsuda, *Prog. Theor. Phys.* **47**, 738 (1962).
781. T. Matsuda, H. Nariai, *Prog. Theor. Phys.* **49**, 1195 (1973).
782. A.E. Mayo, J.D. Bekenstein, *Phys. Rev. D* **54**, 5059 (1996).
783. M.L. McClure, C.C. Dyer, *Class. Quantum Grav.* **23**, 1971 (2006).
784. M.L. McClure, C.C. Dyer, *Gen. Relat. Grav.* **38**, 1347 (2006).
785. J. McDonald, *Phys. Rev. D* **48**, 2462 (1993).
786. S.S. McGaugh, *Astrophys. J. Lett.* **541**, L33 (2000).
787. B. McInnes, *J. High Energy Phys.* **0208**, 029 (2002).

788. B. McInnes, in *On the Nature of Dark Energy*, Proceedings of the XVIIIth Colloquium of the Institut d'Astrophysique de Paris, Paris 2002, ed. by P. Brax, J. Martin, J.-P. Uzan, p. 265, arXiv:astro-ph/0210321.
789. G.C. McVittie, *Mon. Not. R. Astron. Soc.* **93**, 325 (1933).
790. A. Meisels, *Astrophys. J.* **252**, 403 (1982).
791. A. Melchiorri *et al.*, *Astrophys. J. Lett.* **536**, L63 (2000).
792. A. Melchiorri, L. Mersini, C.J. Ödman, M. Trodden, *Phys. Rev. D* **68**, 043509 (2003).
793. S. Mendoza, Y.M. Rosas-Guevara, *Astron. Astrophys.* **472**, 367 (2007).
794. X. Meng, P. Wang, *Gen. Relat. Gravit.* **36**, 1947 (2004).
795. S. Mignemi, D.L. Wiltshire, *Phys. Rev. D* **D46**, 1475 (1992).
796. M.B. Mijic, M.S. Morris, W.-M. Suen, *Phys. Rev. D* **34**, 2934 (1986).
797. A.D. Miller *et al.*, *Astrophys. J. Lett.* **524**, L1 (1999).
798. J.P. Mimoso, D. Wands, *Phys. Rev. D* **51**, 477 (1995).
799. C.W. Misner, *Phys. Rev.* **186**, 1319 (1969).
800. C.W. Misner, *Phys. Rev. D* **8**, 3271 (1973).
801. C.W. Misner, *Phys. Rev. Lett.* **22**, 1071 (1969).
802. C.W. Misner, in *Magic Without Magic*, ed. by J. Klauder (Freeman, San Francisco, 1972).
803. C.W. Misner, in *Relativity*, ed. by M. Carmeli, L. Fickler, L. Witten (Plenum, San Francisco, 1970).
804. C.W. Misner, K.S. Thorne, J.A. Wheeler, *Gravitation* (Freeman, San Francisco, 1973).
805. B. Modak, S. Kamilya, *Int. J. Mod. Phys. A* **22**, 3915 (1998).
806. H. Mosheni Sadjadi, *Phys. Rev. D* **76**, 10424 (2007).
807. H. Mohseni Sadjadi, *Phys. Rev. D* **77**, 103501 (2008).
808. G. Morandi, C. Ferrario, G. Lo Vecchio, G. Marmo, C. Rubano, *Phys. Rep.* **188**, 149 (1990).
809. R.E. Morganstern, *Phys. Rev. D* **4**, 282 (1971).
810. R.E. Morganstern, *Phys. Rev. D* **4**, 946 (1971).
811. V.F. Mukhanov, *Sov. Phys. JETP* **68**, 1297 (1988).
812. V.F. Mukhanov, H.A. Feldman, R.H. Brandenberger, *Phys. Rep.* **215**, 203 (1992).
813. V. Muller, H.-J. Schmidt, A.A. Starobinsky, *Class. Quantum Grav.* **7**, 1163 (1990).
814. T. Multamaki, A. Putaja, E.C. Vagenas, I. Vilja, *Class. Quantum Grav.* **25**, 075017 (2008).
815. T. Multamaki, I. Vilja, *Phys. Rev. D* **73**, 024018 (2006).
816. T. Multamaki, I. Vilja, *Phys. Rev. D* **74**, 064022 (2006).
817. T. Multamaki, I. Vilja, *Phys. Rev. D* **76**, 064021 (2007).
818. T. Multamaki, I. Vilja, *Phys. Lett. B* **659**, 843 (2008).
819. A.G. Muslimov, *Class. Quantum Grav.* **7**, 231 (1990).
820. T.S. Muta, S.D. Odintsov, *Mod. Phys. Lett. A* **6**, 3641 (1991).
821. R.C. Myers, *Phys. Lett. B* **199**, 371 (1987).
822. R.C. Myers, in *Black Holes, Gravitational Radiation and the Universe: Essays in Honor of C.V. Vishveshwara*, ed. by C.V. Vishveshwara, B.R. Iyer, B. Bhawal (Kluwer Academic, Dordrecht, 1999), p. 121, arXiv:gr-qc/9811042.
823. R.C. Myers, J.Z. Simon, *Gen. Relat. Gravit.* **21**, 761 (1989).
824. R.C. Myers, J.Z. Simon, *Phys. Rev. D* **38**, 2434 (1988).
825. T. Nakamura, T. Chiba, *Mon. Not. R. Astron. Soc.* **306**, 696 (1999).
826. K.K. Nandi, A. Bhadra, P.M. Alsing, T.B. Nayak, *Int. J. Mod. Phys. D* **10**, 529 (2001).
827. H. Nariai, *Prog. Theor. Phys.* **40**, 49 (1968).
828. H. Nariai, *Prog. Theor. Phys.* **47**, 1824 (1971).
829. H. Nariai, *Prog. Theor. Phys.* **48**, 763 (1972).
830. I. Navarro, K. van Acoleyen, *Phys. Lett. B* **622**, 1 (2005).
831. I. Navarro, K. Van Acoleyen, *J. Cosmol. Astropart. Phys.* **0603**, 008 (2006).
832. I. Navarro, K. Van Acoleyen, *J. Cosmol. Astropart. Phys.* **0702**, 022 (2007).
833. B. Nelson, P. Panangaden, *Phys. Rev. D* **25**, 1019 (1982).
834. D.M. Neumann, H. Böhringer, *Astron. Astrophys.* **301**, 865 (1995).
835. E.T. Newman, E. Couch, K. Chinnapared, A. Exton, A. Prakash, R. Torrence, *J. Math. Phys.* **6**, 918 (1965).
836. E.T. Newman, A.I. Janis, *J. Math. Phys.* **6**, 915 (1965).

837. S.C.C. Ng, *Phys. Lett. B* **485**, 1 (2000).
838. A.B. Nielsen, *Gen. Relat. Gravit.* **41**, 1539 (2009).
839. A. Nishizawa, A. Taruya, K. Hayama, S. Kawamura, M. Sakagami, *Phys. Rev. D* **79**, 082002 (2009).
840. D.R. Noakes, *J. Math. Phys.* **24**, 1846 (1983).
841. H. Noh, J.-C. Hwang, *Phys. Lett. B* **515**, 231 (2001).
842. S. Nojiri, S.D. Odintsov, arXiv:0801.4843.
843. S. Nojiri, S.D. Odintsov, *Int. J. Geom. Meth. Mod. Phys.* **4**, 115 (2007).
844. S. Nojiri, S.D. Odintsov, *J. Phys. A* **40**, 6725 (2007).
845. S. Nojiri, S.D. Odintsov, *J. Phys. Conf. Ser.* **66**, 012005 (2007).
846. S. Nojiri, S.D. Odintsov, *Gen. Relat. Gravit.* **36**, 1765 (2004).
847. S. Nojiri, S.D. Odintsov, *Phys. Lett. B* **595**, 1 (2004).
848. S. Nojiri, S.D. Odintsov, *Phys. Lett. B* **657**, 238 (2007).
849. S. Nojiri, S.D. Odintsov, *Phys. Lett. B* **657**, 245 (2007).
850. S. Nojiri, S.D. Odintsov, *Phys. Lett. B* **659**, 821 (2008).
851. S. Nojiri, S.D. Odintsov, *Phys. Rev. D* **68**, 123512 (2003).
852. S. Nojiri, S.D. Odintsov, *Phys. Rev. D* **77**, 026007 (2008).
853. S. Nojiri, S.D. Odintsov, *Phys. Rev. D* **78**, 046006 (2008).
854. J. Nordin, A. Goobar, J. Jönsson, *J. Cosmol. Astropart. Phys.* **02**, 008 (2008).
855. K. Nordvedt, *Astrophys. J.* **161**, 1059 (1970).
856. K. Nordtvedt, *Phys. Rev.* **169**, 1017 (1968).
857. J. Novak, *Phys. Rev. D* **57**, 4789 (1998).
858. M. Novello, M. Visser, G. Volovik eds., *Artificial Black Holes* (World Scientific, River Edge, USA, 2002).
859. M. Nozawa, H. Maeda, *Class. Quantum Grav.* **25**, 055009 (2008).
860. A. Nunez, S. Solganik, arXiv:hep-th/0403159.
861. A. Nunez, S. Solganik, *Phys. Lett. B* **608**, 189 (2005).
862. A.-M. Nzioki, MSc thesis, University of Cape Town (2009).
863. A.-M. Nzioki, S. Carloni, R. Goswami, P.K.S. Dunsby, *Phys. Rev. D* **81**, 084028 (2010).
864. S.D. Odintsov, *Fortschr. Phys.* **39**, 621 (1991).
865. S.D. Odintsov, S. Nojiri, *Phys. Lett. B* **576**, 5 (2003).
866. T. Okamura, in *Proceedings of the Asia-Pacific Conference on Gravitation and Cosmology*, Seoul, Korea 1996, ed. by Y.M. Cho, C.H. Lee, S.-W. Kim (World Scientific, Singapore, 1998).
867. K.A. Olive *et al.*, *Phys. Lett. B* **36**, 454 (1990).
868. G.J. Olmo, *Phys. Rev. D* **72**, 083505 (2005).
869. G.J. Olmo, *Phys. Rev. D* **75**, 023511 (2007).
870. G.J. Olmo, *Phys. Rev. D* **77**, 084021 (2008).
871. G.J. Olmo, *Phys. Rev. Lett.* **95**, 261102 (2005).
872. V.K. Onemli, R.P. Woodard, *Class. Quantum Grav.* **19**, 4607 (2002).
873. J.P. Ostriker, P.J. Steinhardt, *Nature* **377**, 600 (1995).
874. A. Oukuiss, *Nucl. Phys. B* **486**, 413 (1997).
875. J.M. Overduin, F.I. Cooperstock, *Phys. Rev. D* **58**, 043506 (1998).
876. J.M. Overduin, P.S. Wesson, *Phys. Rep.* **283**, 303 (1997).
877. I. Ozsvath, E. Schücking, in *Recent Developments in General Relativity*, (Pergamon, New York, 1962), p. 339.
878. J. O'Hanlon, *J. Phys. A* **5**, 803 (1972).
879. J. O'Hanlon, B. Tupper, *Nuovo Cimento B* **7**, 305 (1972).
880. T. Padmanabhan, *Class. Quantum Grav.* **19**, 5387 (2002).
881. T. Padmanabhan, *Phys. Rev. D* **66**, 021301 (2002).
882. T. Padmanabhan, *Phys. Rep.* **380**, 235 (2003).
883. T. Padmanabhan, *Phys. Rep.* **406**, 49 (2005).
884. B.E.J. Pagel, *Nucleosynthesis and Chemical Evolution of Galaxies* (Cambridge University Press, Cambridge, 1997).
885. F.M. Paiva, M. Reboucas, M. MacCallum, *Class. Quantum Grav.* **10**, 1165 (1993).

886. F.M. Paiva, C. Romero, *Gen. Relat. Gravit.* **25**, 1305 (1993).
887. A. Palatini, *Rend. Circ. Mat. Palermo* **43**, 203 (1919).
888. G. Papadopoulos, P.K. Townsend, *Nucl. Phys. B* **444**, 245 (1995).
889. A. Paranjape, S. Sarkar, T. Padmanabhan, *Phys. Rev. D* **74**, 104015 (2006).
890. A. Paranjape, T.P. Singh, *Phys. Rev. D* **76**, 044006 (2007).
891. L. Parker, A. Raval, *Phys. Rev. D* **62**, 083503 (2000).
892. L. Parker, D.J. Toms, *Phys. Rev. D* **29**, 1584 (1985).
893. L. Parker, D.J. Toms, *Phys. Rev. D* **32**, 1409 (1985).
894. P.J.E. Peebles, *Astrophys. J.* **146**, 542 (1996).
895. P.J.E. Peebles, B. Ratra, *Astrophys. J. Lett.* **325**, L17 (1988).
896. P.J.E. Peebles, J.T. Yu, *Astrophys. J.* **162**, 815 (1970).
897. R. Penrose, *Proc. R. Soc. Lon.* **A 284**, 159 (1965).
898. R. Penrose, in *Relativity, Groups and Topology*, Les Houches Lectures, ed. by C. Morette, B.S. DeWitt, B.S (Gordon and Breach, New York, 1964).
899. R. Penrose, unpublished, quoted in [976].
900. S.E. Perez Bergliaffa, *Phys. Lett. B* **642**, 311 (2006).
901. S. Perlmutter *et al.*, *Astrophys. J.* **517**, 565 (1999).
902. S. Perlmutter *et al.*, *Nature* **391**, 51 (1998).
903. F. Perrotta, C. Baccigalupi, S. Matarrese, *Phys. Rev. D* **61**, 023507 (2000).
904. S. Phinney *et al.*, "The Big Bang Observer: direct detection of gravitational waves from the birth of the Universe to the present", NASA Mission Concept Study 2003.
905. F. Piazza, C. Marinoni, *Phys. Rev. Lett.* **91**, 141301 (2003).
906. L.O. Pimentel, J. Stein-Schabes, *Phys. Lett. B* **216**, 27 (1989).
907. *PLANCK* homepage http://www.rssd.esa.int/index.php?project=Planck
908. S. Podariu, R.A. Daly, M.P. Mory, B. Ratra, *Astrophys. J.* **584**, 577 (2003).
909. L. Pogosian, A. Silvestri, *Phys. Rev. D* **77**, 023503 (2008).
910. M.D. Pollock, *Phys. Lett. B* **215**, 635 (1988).
911. M.D. Pollock, D. Sadhev, *Phys. Lett. B* **222**, 1 (1989).
912. N. Poplawski, *Class. Quantum Grav.* **24**, 3013 (2007).
913. N. Poplawski, *Phys. Lett. B* **640**, 135 (2006).
914. A. Pradhan, I., Iotemshi, G.P. Singh, *Astrophys. Sp. Sci.* **288**, 315 (2003).
915. D. Psaltis, D. Perrodin, K.R. Dienes, I. Mocioiu, *Phys. Rev. Lett.* **100**, 091101 (2008); *Erratum* **100**, 091101 (2008).
916. R. Punzi, F.S. Schuller, M.N.R. Wohlfarth, *J. Cosmol. Astropart. Phys.* **0712**, 013 (2008).
917. R. Punzi, F.S. Schuller, M.N.R. Wohlfarth, *J. High Energy Phys.* **0702**, 030 (2008).
918. R. Punzi, F.S. Schuller, M.N.R. Wohlfarth, *Phys. Lett. B* **670**, 161 (2008).
919. R. Punzi, F.S. Schuller, M.N.R. Wohlfarth, *Phys. Rev. D* **76**, 101501 (2007).
920. I. Quandt, H.-J. Schmidt, *Astron. Nachr.* **312**, 97 (1991).
921. V.U.M. Rao, D.R.K. Reddy, *Gen. Relat. Gravit.* **14**, 1017 (1982).
922. S. Räsänen, *J. Cosmol. Astropart. Phys.* **02**, 003 (2004).
923. B. Ratra, P.J.E. Peebles, *Phys. Rev. D* **37**, 3406 (1988).
924. A.K. Raychaudhuri, B. Bagchi, *Phys. Lett. B* **124**, 168 (1983).
925. A.K. Raychaudhuri, N. Bandyopadhyay, *Prog. Theor. Phys.* **59**, 414 (1978).
926. R. Rebolo *et al.*, *Mon. Not. R. Astron. Soc.* **353**, 747 (2004).
927. D.R.K. Reddy, *Astrophys. Sp. Sci.* **140**, 85 (1988).
928. D.R.K. Reddy, *J. Math. Phys.* **20**, 23 (1979).
929. D.R.K. Reddy, *J. Phys. A* **6**, 1867 (1973).
930. D.R.K. Reddy, *J. Phys. A* **10**, 185 (1977).
931. D.R.K. Reddy, M.B. Avadhanulu, R. Venkateswarlu, *Astrophys. Sp. Sci.* **134**, 201 (1987).
932. O.A. Reula, *Living Rev. Relativity* **1**, 3 (1998).
933. M. Reuter, *Phys. Rev. D* **49**, 6379 (1994).
934. A. Riazuelo, J.-P. Uzan, *Phys. Rev. D* **62**, 083506 (2000).
935. A. Riazuelo, J.-P. Uzan, *Phys. Rev. D* **66**, 023535 (2002).
936. A.G. Riess *et al.*, *Astron. J.* **116**, 1009 (1998).
937. A.G. Riess *et al.*, *Astron. J.* **118**, 2668 (1999).

938. A.G. Riess *et al.*, *Astrophys. J.* **536**, 62 (2000).
939. A.G. Riess *et al.*, *Astrophys. J.* **560**, 49 (2001).
940. A.G. Riess *et al.*, *Astrophys. J.* **607**, 665 (2004).
941. A.G. Riess *et al.*, *Astrophys. J.* **659**, 98 (2007).
942. S. Rippl, H. van Elst, R.K. Tavakol, D.R. Taylor, *Gen. Relat. Gravit.* **28**, 193 (1996).
943. D.C. Robinson, *Gen. Relat. Gravit.* **38**, 153 (2006).
944. C. Romero, A. Barros, *Astr. Sp. Sci.* **192**, 263 (1992).
945. C. Romero, A. Barros, *Gen. Relat. Gravit.* **23**, 491 (1993).
946. C. Romero, A. Barros, *Gen. Relat. Gravit.* **25**, 1305 (1993).
947. C. Romero, A. Barros, preprint DF–CCEN–UFPb No. 9 (1992).
948. C. Romero, A. Barros, *Phys. Lett. A* **173**, 243 (1993).
949. F. Rosati, *Nucl. Phys. Proc. Suppl.* **95**, 74 (2001).
950. M.L. Ruggiero, L. Iorio, *J. Cosmol. Astropart. Phys.* **0701**, 010 (2007).
951. T.V. Ruzmaikina, A.A. Ruzmaikin, *JETP* **30**, 372 (1970).
952. A. Saa *et al.*, *Int. J. Theor. Phys.* **40**, 2295 (2001).
953. A. Saa, *Phys. Rev. D* **53**, 7377 (1996).
954. E. Sagi, J.D. Bekenstein, *Phys. Rev. D* **77**, 024010 (2008).
955. V. Sahni, *Class. Quantum Grav.* **19**, 3435 (2002).
956. V. Sahni, A.A. Starobinsky, *Int. J. Mod. Phys. D* **9**, 373 (2000).
957. H. Saida, T. Harada, H. Maeda, *Class. Quantum Grav.* **24**, 4711 (2007).
958. T.D. Saini, S. Raychaudhuri, V. Sahni, A.A. Starobinsky, *Phys. Rev. Lett.* **85**, 1162 (2000).
959. N. Sakai, J.D. Barrow, *Class. Quantum Grav.* **18**, 4717 (2001).
960. A.D. Sakharov, *Sov. Phys. Dokl.* **12**, 1040 (1968).
961. A. Salam, J. Strathdee, *Phys. Rev. D* **2**, 2869 (1970).
962. M. Salgado, *Class. Quantum Grav.* **23**, 4719 (2006).
963. D.S. Salopek, *Phys. Rev. Lett.* **69**, 3602 (1992).
964. D.S. Salopek, J.R. Bond, J.M. Bardeen, *Phys. Rev. D* **40**, 1753 (1989).
965. M. Sami, A. Toporensky, P.V. Tretjakov, S. Tsujikawa, *Phys. Lett. B* **619**, 193 (2005).
966. R.H. Sanders, *Ann. Rev. Astron. Astrophys.* **2**, 1 (1990).
967. D.I. Santiago, D. Kalligas, R.V. Wagoner, *Phys. Rev. D* **56**, 7627 (1997).
968. D.I. Santiago, D. Kalligas, R.V. Wagoner, *Phys. Rev. D* **58**, 124005 (1999).
969. C. Santos, arXiv:gr-qc/0009054.
970. C. Santos, R. Gregory, *Ann. Phys. (NY)* **258**, 111 (1997).
971. D. Sapone, arXiv:1006.5694.
972. D. Sarkar, A. Amblard, D.E. Holz, A. Cooray, *Astrophys. J.* **678**, 1 (2008).
973. R. Saunders *et al.*, *Mon. Not. R. Astron. Soc.* **341**, 937 (2003).
974. P. Schuecker *et al.*, *Astron. Astrophys.* **402**, 53 (2003).
975. M.A. Scheel, S.L. Shapiro, S.A. Teukolsky, *Phys. Rev. D* **51**, 4208 (1995).
976. M.A. Scheel, S.L. Shapiro, S.A. Teukolsky, *Phys. Rev. D* **51**, 4236 (1995).
977. J. Scherk, H.J. Schwarz, *Nucl. Phys. B* **81**, 118 (1974).
978. R. Schimming, H.-J. Schmidt, *NTM-Schriftenr. Gesch. Naturw. (Leipzig)* **27**, 48 (1990).
979. S. Schindler, *Astrophys. Sp. Sci.* **28**, 419 (2004).
980. K. Schleich, D.M. Witt, arXiv:0908.4110.
981. K. Schleich, D.M. Witt, arXiv:0910.5194.
982. B.R. Schmidt *et al.*, *Astrophys. J.* **507**, 46 (1998).
983. B.P. Schmidt *et al.*, *Bull. Am. Astron. Soc.* **37**, 457 (2005).
984. H.-J. Schmidt, *Astr. Nachr.* **308**, 183 (1987).
985. H.-J. Schmidt, *Class. Quantum Grav.* **7**, 1023 (1990).
986. H.-J. Schmidt, *Int. J. Geom. Meth. Mod. Phys.* **4**, 209 (2007).
987. H.-J. Schmidt, *Lectures on Mathematical Cosmology*, arXiv:gr-qc/0407095.
988. H.-J. Schmidt, *Phys. Rev. D* **53**, 7906 (1996).
989. R.W. Schmidt, S.W., Allen, *Mont. Not. R. Astron. Soc.* **379**, 209 (2007).
990. R.W. Schmidt, S.W. Allen, A.C. Fabian, *Mont. Not. R. Astron. Soc.* **352**, 1413 (2004).
991. P. Schneider, J. Ehlers, E.E. Falco, *Gravitational Lenses* (Springer, Berlin, 1992).
992. D.N. Schramm, M.S. Turner, *Rev. Mod. Phys.* **70**, 303 (1998).

993. D.N. Schramm, R.V. Wagoner, *Ann. Rev. Nucl. Sci.* **27**, 37 (1977).
994. E. Schrödinger, *Space-Time Structure* (Cambridge University Press, Cambridge, 1960).
995. E. Schücking, *Z. Phys.* **148**, 72 (1957).
996. F.P. Schuller, M.N.R. Wohlfarth, *J. High Energy Phys.* **0602**, 059 (2006).
997. A.E. Schulz, M. White, *Phys. Rev. D* **64**, 043514 (2001).
998. B.F. Schutz, F. Ricci, "Gravitational Waves, Sources and Detectors", in *Gravitational Waves*, ed. by I. Ciufolini *et al.* (Institute of Physics, Bristol, 2001), arXiv:1005.4735.
999. D.W. Sciama, *Mont. Not. R. Astron. Soc.* **113**, 34 (1953).
1000. C. Sealfon, L. Verde, R. Jimenez, *Phys. Rev. D* **71**, 083004 (2005).
1001. M.D. Seifert, *Phys. Rev. D* **76**, 064002 (2007).
1002. A.A. Sen, R.T. Seshadri, *Int. J. Mod. Phys. D* **12**, 445 (2000).
1003. D.K. Sen, *Can. Math. Bull.* **3**, 255 (1960).
1004. D.K. Sen, *Z. Phys.* **149**, 311 (1957).
1005. D.K. Sen, K.A. Dunn, *J. Math. Phys.* **12**, 578 (1971).
1006. D.K. Sen, J.R. Vanstone, *J. Math. Phys.* **13**, 990 (1972).
1007. A. Serna, J.-M. Alimi, *Phys. Rev. D* **50**, 7304 (1996).
1008. A. Serna, J.-M. Alimi, A. Navarro, *Class. Quantum Grav.* **19**, 857 (2002).
1009. A. Serna, R., Dominguez-Tenreiro, G. Yepes, *Astrophys. J.* **391**, 433 (1992).
1010. C.-G. Shao, R.-G. Cai, B. Wang, R.-K. Su, *Phys. Lett. B* **633**, 164 (2006).
1011. M. Shibata, K. Nakao, T. Nakamura, *Phys. Rev. D* **50**, 7304 (1994).
1012. A. Shirata, T. Shiromizu, N. Yoshida, *Phys. Rev. D* **71**, 064030 (2005).
1013. A. Shirata, Y. Suto, C. Hikage, T. Shiromizu, N. Yoshida, *Phys. Rev. D* **76**, 044026 (2007).
1014. D. Sigg *et al.*, www.ligo.org/pdfpublic/P050036.pdf
1015. T. Singh, *J. Math. Phys.* **16**, 219 (1975).
1016. T. Singh, *J. Math. Phys.* **27**, 1074 (1986).
1017. T. Singh, *Phys. Rev. D* **34**, 646 (1986).
1018. T. Singh, L.N. Rai, *Indian J. Pure Appl. Math.* **10**, 1432 (1979).
1019. C. Skordis, D.F. Mota, P.G. Ferreira, C. Boehm, *Phys. Rev. Lett.* **96**, 011301 (2006).
1020. G.F. Smoot *et al.*, *Astrophys. J. Lett.* **396**, L1 (1992).
1021. Y. Sobouti, *Astron. Astrophys.* **464**, 921 (2007).
1022. L.M. Sokolowski, *Acta Phys. Polon. B* **35**, 587 (2004).
1023. L.M. Sokolowski, *Class. Quantum Grav.* **6**, 2045 (1989).
1024. L.M. Sokolowski, *Class. Quantum Grav.* **24**, 3391 (2007).
1025. L.M. Sokolowski, *Class. Quantum Grav.* **24**, 3713 (2007).
1026. H.H. Soleng, *Class. Quantum Grav.* **5**, 1489 (1988).
1027. P. Solín, *Partial Differential Equations and The Finite Element Method* (Wiley, New York, 2006).
1028. S. Sonego, V. Faraoni, *Class. Quantum Grav.* **10**, 1185 (1993).
1029. Y.-S. Song, W. Hu, I. Sawicki, *Phys. Rev. D* **75**, 044004 (2007).
1030. T.P. Sotiriou, *Class. Quantum Grav.* **23**, 1253 (2006).
1031. T.P. Sotiriou, *Gen. Relat. Gravit.* **38**, 1407 (2006).
1032. T.P. Sotiriou, E. Barausse, *Phys. Rev. D* **75**, 084007 (2007).
1033. T.P. Sotiriou, V. Faraoni, *Rev. Mod. Phys.* **82**, 451 (2010).
1034. T.P. Sotiriou, S. Liberati, *Ann. Phys. (NY)* **322**, 935 (2007).
1035. T.P. Sotiriou, S. Liberati, V. Faraoni, *Int. J. Mod. Phys. D* **17**, 393 (2008).
1036. M.E. Soussa, R.P. Woodard, *Gen. Relat. Gravit.* **36**, 855 (2004).
1037. C.C. Speake, T.J. Quinn, *Phys. Rev. Lett.* **61**, 1340 (1988).
1038. D.N. Spergel *et al.*, *Astrophys. J. (Suppl.)* **148**, 175 (2003).
1039. D.N. Spergel *et al.* [WMAP Collaboration], *Astrophys. J. (Suppl.)* **170**, 377 (2007).
1040. R. Stabell, S. Refsdal, *Mon. Not. R. Astron. Soc.* **132**, 379 (1966).
1041. H.F. Stabenau, B. Jain, *Phys. Rev. D* **74**, 084007 (2006).
1042. A.A. Starobinsky, *Gravit. Cosmol.* **6**, 157 (2000).
1043. A.A. Starobinsky, *JETP Lett.* **68**, 757 (1998).
1044. A.A. Starobinsky, *Phys. Lett. B* **91**, 99 (1980).
1045. A.A. Starobinsky, *Sov. Astron. Lett.* **7**, 36 (1981).

1046. A.A. Starobinsky, *Sov. Astron. Lett.* **9**, 302 (1983).
1047. A.A. Starobinsky, in *Proceedings of the 4th Seminar on Quantum Gravity*, ed. by M.A. Markov, V.P. Frolov (World Scientific, Singapore, 1987).
1048. A.A. Starobinsky, J. Yokoyama, in *Proceedings of the 4th Workshop on General Relativity and Gravitation*, ed. by K. Nakao *et al.*, (Yukawa Institute, Tokyo, 1995), p. 381.
1049. P.J. Steinhardt, F.S. Accetta, *Phys. Rev. Lett.* **64**, 2740 (1990).
1050. P.J. Steinhardt, L. Wang, I. Zlatev, *Phys. Rev. D* **59**, 123504 (1999).
1051. K.S. Stelle, *Gen. Relat. Gravit.* **9**, 343, (1978).
1052. K.S. Stelle, *Phys. Rev. D* **16**, 953 (1977).
1053. H. Stephani, D. Kramer, M. MacCallum, C. Hoenselaers, E. Herlt, *Exact Solutions of the Einstein Field Equations* (Cambridge University Press, Cambridge, 2003).
1054. W.R. Stoeger, R. Maartens, G.F.R. Ellis, *Astrophys. J.* **443**, 1 (1995).
1055. R. Stompor *et al.*, *Astrophys. J. Lett.* **561**, L7 (2001).
1056. M.A. Strauss *et al.*, *Astron. J.* **124**, 1810 (2002).
1057. A. Strominger, *Phys. Rev. D* **30**, 2257 (1984).
1058. W.-M. Suen, P.R. Anderson, *Phys. Rev. D* **35**, 2940 (1987).
1059. J. Sultana, C.C. Dyer, *Gen. Relat. Gravit.* **37**, 1347 (2005).
1060. M. Susperregi, arXiv:gr-qc/9712031.
1061. M. Susperregi, *Phys. Rev. D* **55**, 560 (1997).
1062. M. Susperregi, *Phys. Rev. D* **58**, 083512 (1998).
1063. M. Susperregi, *Phys. Lett. B* **440**, 257 (1998).
1064. L. Susskind, in *Universe or Multiverse?*, ed. by B.J. Carr (Cambridge University Press, Cambridge, 2007), p. 247, arXiv:hep-th/0302219.
1065. J.L. Synge, *Relativity: The General Theory* (Amsterdam: North Holland, 1955).
1066. D. Tatsumi, Y. Tsunesada, *Class. Quantum Grav.* **21**, S451 (2004).
1067. D.R. Taylor, R. Maartens, *Gen. Relat. Gravit.* **27**, 1309 (1995).
1068. T.R. Taylor, G. Veneziano, *Phys. Lett. B* **213**, 450 (1988).
1069. J.H. Taylor, J.M. Weinberg, *Astrophys. J.* **235**, 908 (1982).
1070. M. Tegmark *et al.*, *Phys. Rev. D* **69**, 103501 (2004).
1071. M. Tegmark, A.N. Taylor, A.F. Heavens, *Astrophys. J.* **4802**, 22 (2007).
1072. P. Teyssandier, P. Tourrenc, *J. Math. Phys.* **24**, 2793 (1983).
1073. K.S. Thorne, J.J. Dykla, *Astrophys. J.* **166**, 235 (1971).
1074. F.J. Tipler, *Phys. Lett. A* **64**, 81 (1977).
1075. M.E. Tobar, T. Suzuki, K. Kuroda, *Phys. Rev. D* **79**, 082002 (2009).
1076. R.C. Tolman, *Proc. Nat. Acad. Sci. U.S.A.* **20**, 169 (1934).
1077. J.L. Tonry *et al.*, arXiv:astro-ph/0105413.
1078. J.L. Tonry *et al.*, *Astrophys. J.* **594**, 1 (2003).
1079. A.V. Toporensky, *Int. J. Mod. Phys. D* **8**, 739 (1999).
1080. D.F. Torres, arXiv:gr-qc/9612048.
1081. D.F. Torres, *Phys. Lett. A* **225**, 13 (1997).
1082. D.F. Torres, *Phys. Lett. B* **359**, 249 (1995).
1083. D.F. Torres, *Phys. Rev. D* **66**, 043522 (2002).
1084. D.F. Torres, H. Vucetich, *Phys. Rev. D* **54**, 7373 (1996).
1085. P.K. Townsend, *J. High Energy Phys.* **0111**, 042 (2001).
1086. A.A. Tseytlin, *Int. J. Mod. Phys. A* **4**, 1257 (1989).
1087. A.A. Tseytlin, C. Vafa, *Nucl. Phys. B* **372**, 443 (1992).
1088. S. Tsujikawa, *Phys. Rev. D* **76**, 023514 (2007).
1089. S. Tsujikawa, *Phys. Rev. D* **77**, 023507 (2008).
1090. S. Tsujikawa, B. Bassett, *Phys. Lett. B* **536**, 9 (2002).
1091. S. Tsujikawa, K. Maeda, T. Torii, *Phys. Rev. D* **61**, 103501 (2000).
1092. S. Tsujikawa, K. Uddin, R. Tavakol, *Phys. Rev. D* **77**, 043007 (2008).
1093. M.S. Turner, A.G. Riess, *Astrophys. J.* **569**, 18 (2002).
1094. M.S. Turner, D.S. Salopek, E.W. Kolb, *Phys. Rev. D* **42**, 3925 (1990).
1095. M.S. Turner, E.W. Widrow, *Phys. Rev. D* **37**, 3428 (1988).

1096. J.A. Tyson, in Procedings of *Survey and Other Telescope Technologies and Discoveries*, Hawaii 2002, ed. by J.A. Tyson and S. Wolff (SPIE, 2002), p. 10.
1097. K. Uddin, J.E. Lidsey, R. Tavakol, *Class. Quantum Grav.* **24**, 3951 (2007).
1098. C. Ungarelli, P. Corasaniti, R.A. Mercer, A. Vecchio, *Class. Quantum Grav.* **22**, S955 (2005).
1099. W.G. Unruh, *Phys. Rev. Lett.* **46**, 1351 (1981).
1100. R. Utiyama, B.S. DeWitt, *J. Math. Phys.* **3**, 608 (1962).
1101. J.-P. Uzan, *Phys. Rev. D* **59**, 123510 (1999).
1102. H. van Dam, M.J.G. Veltman, *Nucl. Phys. B* **22**, 397 (1970).
1103. N. Van den Bergh, *Gen. Relat. Gravit.* **12**, 863 (1980).
1104. N. Van den Bergh, *Gen. Relat. Gravit.* **14**, 17 (1982).
1105. N. Van den Bergh, *Gen. Relat. Gravit.* **15**, 441 (1983).
1106. N. Van den Bergh, *Gen. Relat. Gravit.* **15**, 449 (1983).
1107. N. Van den Bergh, *Gen. Relat. Gravit.* **15**, 1043 (1983).
1108. N. Van den Bergh, *Gen. Relat. Gravit.* **16**, 2191 (1983).
1109. J.J. Van der Bij, M. Gleiser, *Phys. Lett. B* **194**, 482 (1987).
1110. J.J. Van der Bij, E. Radu, *Nucl. Phys. B* **585**, 637 (2000).
1111. G. Veneziano, *Phys. Lett. B* **265**, 287 (1991).
1112. R. Venkateswarlu, D.R.K. Reddy, *Astrophys. Sp. Sci.* **159**, 173 (1989).
1113. A. Viceré, *Phys. Rev. D* **66**, 062002 (2002).
1114. A. Vikhlinin, A. Kravtsov, W. Forman, C. Jones, M. Markevitch, S.S. Murray, L. Van Speybroeck, *Astrophys. J.* **640**, 691 (2006).
1115. A. Vikhlinin, M. Markevitch, S.S. Murray, C. Jones, W. Forman, L. Van Speybroeck, *Astrophys. J.* **628**, 655 (2005).
1116. A. Vikhlinin, private communication to S. Capozziello (2008).
1117. A. Vilenkin, *Phys. Lett. B* **117**, 25 (1982).
1118. A. Vilenkin, *Phys. Rev. D* **30**, 509 (1984).
1119. A. Vilenkin, *Phys. Rev. D* **32**, 2511 (1985).
1120. A. Vilenkin, *Phys. Rev. D* **33**, 3560 (1986).
1121. A. Vilenkin, *Phys. Rev. D* **39**, 1116 (1989).
1122. G. Vilkovisky, *Class. Quantum Grav.* **9**, 895 (1992).
1123. *VIRGO* homepage **http://www.virgo.infn.it/**
1124. M. Visser, arXiv:0901.4365.
1125. M. Visser, *Class. Quantum Grav.* **15**, 1767 (1998).
1126. M. Visser, *Class. Quantum Grav.* **21**, 2603 (2004).
1127. M. Visser, *Phys. Rev. D* **48**, 5697 (1993).
1128. V. Vitagliano, S. Liberati, V. Faraoni, *Class. Quantum Grav.* **26**, 215005 (2009).
1129. D.N. Vollick, *Phys. Rev. D* **68**, 063510 (2003).
1130. D.N. Vollick, *Phys. Rev. D* **76**, 124001 (2007).
1131. M.B. Voloshin, A.D. Dolgov, *Sov. J. Nucl. Phys.* **35**, 120 (1982).
1132. R.V. Wagoner, *Astrophys. J.* **179**, 343 (1973).
1133. R.V. Wagoner, *Phys. Rev. D* **1**, 3209 (1970).
1134. R.V. Wagoner, W.A. Fowler, F. Hoyle, *Astrophys. J.* **148**, 3 (1967).
1135. R.V. Wagoner, D. Kalligas, in *Relativistic Gravitation and Gravitational Radiation*, Proceedings of the Les Houches School of Physics 1995, ed. by J.-A. Marck, J.-P. Lasota (Cambridge University Press, Cambridge, 1997).
1136. J. Wainwright, G.F.R. Ellis, *Dynamical Systems in Cosmology* (Cambridge University Press, Cambridge, 1997).
1137. R.M. Wald, arXiv:gr-qc/9710068.
1138. R.M. Wald, *Class. Quantum Grav.* **16**, A177 (1999).
1139. R.M. Wald, *General Relativity* (Chicago University Press, Chicago, 1984).
1140. R.M. Wald, *Living. Rev. Relativity* **4**, 6 (2001).
1141. R.M. Wald, *Phys. Rev. D* **48**, R3427 (1993).
1142. R.M. Wald, *Quantum Field Theory in Curved Spacetime and Black Hole Thermodynamics* (Chicago University Press, Chicago, 1995).
1143. T.P. Walker *et al.*, *Astrophys. J.* **376**, 51 (1991).

1144. D. Wands, *Class. Quantum Grav.* **11**, 269 (1994).
1145. D. Wands, J.P. Mimoso, *Phys. Rev. D* **52**, 5612 (1995).
1146. P. Wang, *Phys. Rev. D* **72**, 024030 (2005).
1147. Y. Wang, *Astrophys. J.* **536**, 531 (2000).
1148. L.-M. Wang, R.R. Caldwell, J.P. Ostriker, P.J. Steinhardt, *Astrophys. J.* **530**, 17 (2000).
1149. Y. Wang, P. Mukherejee, *Astrophys. J.* **606**, 654 (2004).
1150. Y. Wang, M. Tegmark, *Phys. Rev. Lett.* **92**, 241302 (2004).
1151. H. Wei, S.N. Zhang, *Phys. Rev. D* **78**, 023011 (2008).
1152. E.J. Weinberg, *Phys. Rev. D* **40** (1989) 3950.
1153. S. Weinberg, *Gravitation and Cosmology* (Wiley, New York, 1972).
1154. S. Weinberg, *Rev. Mod. Phys.* **61**, 1 (1989).
1155. S. Weinberg, in *Critical Dialogues in Cosmology*, Proceedings, Princeton, NJ, 1996, ed. by N. Turok (World Scientific, Singapore, 1996), p. 195.
1156. C. Wetterich, *Nucl. Phys. B* **302**, 668 (1988).
1157. J.A. Wheeler, *Rev. Mod. Phys.* **29**, 463 (1957).
1158. J.A. Wheeler, in *Battelle Rencontres*, ed. by C. DeWitt, J.A. Wheeler (Benjamin, New York, 1968).
1159. M.J. White, C.S. Kochanek, *Astrophys. J.* **560**, 539 (2001).
1160. B. Whitt, *Phys. Lett. B* **145**, 176 (1984).
1161. B. Whitt, *Phys. Rev. D* **32**, 379 (1985).
1162. B. Willke *et al.*, *Class. Quantum Grav.* **23**, S207 (2006).
1163. D.L. Wiltshire, *New J. Phys.* **9**, 377 (2007).
1164. D.L. Wiltshire, *Phys. Rev. Lett.* **99**, 251101 (2007).
1165. G. Worthey, Ph.D. Thesis, University of California, Santa Cruz (1992).
1166. C.M. Will, *Living Rev. Relativity* **9**, 3 (2006).
1167. C.M. Will, *Theory and Experiment in Gravitational Physics*, 2nd edn. (Cambridge University Press, Cambridge, 1993).
1168. E. Winstanley, arXiv:gr-qc/0408046.
1169. G. Woan, *The Cambridge Handbook of Physics Formulas* (Cambridge University Press, Cambridge, 2000).
1170. S.S. Yazadjiev, arXiv:gr-qc/0305053.
1171. A.F. Zakharov, A.A. Nucita, F. De Paolis, G. Ingrosso, *Phys. Rev. D* **74**, 107101 (2006).
1172. P. Zhang, *Phys. Rev. D* **76**, 024007 (2007).
1173. P. Zhang, M. Liguori, R. Bean, S. Dodelson, *Phys. Rev. Lett.* **99**, 141302 (2007).
1174. O.B. Zavlavskii, *Class. Quantum Grav.* **19**, 3783 (2002).
1175. Ya.B. Zel'dovich, *Sov. Phys. JETP Lett.* **12**, 307 (1970).
1176. Ya.B. Zeldovich, A.A. Starobinsky, *Sov. Phys. JETP* **34**, 1159 (1972).
1177. H. Ziaeepour, arXiv:astro-ph/0301640.
1178. H. Ziaeepour, arXiv:astro-ph/0002400.
1179. W. Zimdahl, D.J. Schwarz, A.B. Balakin, D. Pavon, *Phys. Rev. D* **64**, 063501 (2001).
1180. I. Zlatev, L. Wang, P.J. Steinhardt, *Phys. Rev. Lett.* **82**, 896 (1999).

# Index